DATE DUE

Demco, Inc. 38-293

VOLUME FOUR HUNDRED AND FIFTY-TWO

METHODS IN ENZYMOLOGY

Autophagy in Mammalian Systems, Part B

METHODS IN ENZYMOLOGY

Editors-in-Chief

JOHN N. ABELSON AND MELVIN I. SIMON

Division of Biology
California Institute of Technology
Pasadena, California, USA

Founding Editors

SIDNEY P. COLOWICK AND NATHAN O. KAPLAN

VOLUME FOUR HUNDRED AND FIFTY-TWO

Methods in ENZYMOLOGY

Autophagy in Mammalian Systems, Part B

EDITED BY

DANIEL J. KLIONSKY
Life Sciences Institute
University of Michigan
Ann Arbor, Michigan, USA

ELSEVIER

AMSTERDAM • BOSTON • HEIDELBERG • LONDON
NEW YORK • OXFORD • PARIS • SAN DIEGO
SAN FRANCISCO • SINGAPORE • SYDNEY • TOKYO
Academic Press is an imprint of Elsevier

Academic Press is an imprint of Elsevier
525 B Street, Suite 1900, San Diego, California 92101-4495, USA
30Corporate Drive, Suite 400, Burlington,MA01803, USA
32 Jamestown Road, LondonNW17BY,UK

Copyright © 2009, Elsevier Inc. All Rights Reserved.

No part of this publication may be reproduced or transmitted in any form or by any means, electronic or mechanical, including photocopy, recording, or any information storage and retrieval system, without permission in writing from the Publisher.

The appearance of the code at the bottom of the first page of a chapter in this book indicates the Publisher's consent that copies of the chapter may be made for personal or internal use of specific clients. This consent is given on the condition, however, that the copier pay the stated per copy fee through the Copyright Clearance Center, Inc. (www.copyright.com), for copying beyond that permitted by Sections 107 or 108 of the U.S. Copyright Law. This consent does not extend to other kinds of copying, such as copying for general distribution, for advertising or promotional purposes, for creating new collective works, or for resale. Copy fees for pre-2008 chapters are as shown on the title pages. If no fee code appears on the title page, the copy fee is the same as for current chapters. 0076-6879/2009 $35.00

Permissions may be sought directly from Elsevier's Science & Technology Rights Department in Oxford, UK: phone: (+44) 1865 843830, fax: (+44) 1865 853333, E-mail: permissions@elsevier. com. You may also complete your request on-line via the Elsevier homepage (http://elsevier. com), by selecting "Support & Contact" then "Copyright and Permission" and then "Obtaining Permissions."

For information on all Elsevier Academic Press publications visit our Web site at elsevierdirect.com

ISBN-13: 978-0-12-374547-7

PRINTED IN THE UNITED STATES OF AMERICA
09 10 11   9 8 7 6 5 4 3 2 1

Working together to grow libraries in developing countries

www.elsevier.com | www.bookaid.org | www.sabre.org

ELSEVIER   BOOK AID International   Sabre Foundation

Contents

Contributors	xiii
Preface	xxi
Volumes in Series	xxiii

1. Monitoring Autophagy in Mammalian Cultured Cells through the Dynamics of LC3 — 1

Shunsuke Kimura, Naonobu Fujita, Takeshi Noda, and Tamotsu Yoshimori

1. Introduction	2
2. Estimation of Autophagy Induction by LC3 Puncta Formation	4
3. The tfLC3 Assay	7
4. Determination of LC3 Turnover by Western Blotting	9
5. Concluding Remarks	10
References	11

2. Methods for Monitoring Autophagy Using GFP-LC3 Transgenic Mice — 13

Noboru Mizushima

1. Introduction	14
2. Genetic Features of GFP-LC3 Mice	15
3. Mouse Maintenance	16
4. Genotyping of GFP-LC3 Mice	17
5. Sample Preparation	18
6. Fluoresence Microscopy	19
7. Precautions	20
References	21

3. Using Photoactivatable Proteins to Monitor Autophagosome Lifetime — 25

Dale W. Hailey and Jennifer Lippincott-Schwartz

1. Introduction	26
2. Photoactivatable Fluorescent Protein Labeling	27
3. Experimental Example: Setting up a Sample and Optimizing Photoactivation	32
4. Photobleaching	35
5. Carrying Out the Photochase Assay	36

6. Pulse-Labeling Induced Autophagosomes	38
7. Determining the Half-Life of Pulse-Labeled Autophagosomes	40
8. Controls	42
9. Conclusions	43
References	44

4. Assaying of Autophagic Protein Degradation — 47

Chantal Bauvy, Alfred J. Meijer, and Patrice Codogno

1. Introduction	48
2. Assaying the Degradation of Long-Lived Proteins in Cultured Cell Lines	48
3. Assay of the Degradation of Long-Lived Protein in Isolated Rat Hepatocytes	50
4. Other Methods for Measuring Autophagic Flux	57
5. Pitfalls	58
References	59

5. Sequestration Assays for Mammalian Autophagy — 63

Per O. Seglen, Anders Øverbye, and Frank Sætre

1. Introduction	64
2. Membrane-Impermeant Autophagy Probes: Introduction into Cytosol by Reversible Electropermeabilization (Electroporation) of the Plasma Membrane	67
3. Damage-Induced Cell Permeabilization	69
4. Electrodisruption: A Simple Method for the Separation of Sedimentable from Soluble Cell Components	70
5. Electroinjected Sugars as Autophagic Sequestration Probes	71
6. Cytosolic Proteins as Autophagic Sequestration Probes	74
7. Autophagic Fragment Generation: An Autophagic Cargo Assay Applicable to Whole Cells	77
8. Concluding Remarks	79
References	80

6. Assays to Assess Autophagy Induction and Fusion of Autophagic Vacuoles with a Degradative Compartment, Using Monodansylcadaverine (MDC) and DQ-BSA — 85

Cristina Lourdes Vázquez and María Isabel Colombo

1. Overview	86
2. Assessing Autophagy Induction by Fluorescence Microscopy and Fluorometry	86
3. Monitoring the Formation of an Autolysosome	92
4. Concluding Remarks	94
Acknowledgments	94
References	94

7. The GST-BHMT Assay and Related Assays for Autophagy — 97
Patrick B. Dennis and Carol A. Mercer

1. Introduction — 98
2. Measurement of Macroautophagy Using the GST-BHMT Assay — 100
3. Measurement of Autophagy Using a Linker-Specific Cleavage Site (LSCS) — 107
4. Measurement of Reticulophagy and Mitophagy Using LSCS-Based Reporters — 110
5. Concluding Remarks — 114
Acknowledgments — 116
References — 116

8. Monitoring Starvation-Induced Reactive Oxygen Species Formation — 119
Ruth Scherz-Shouval and Zvulun Elazar

1. Introduction — 120
2. Detection of ROS in Living Cells Using Fluorescent Dyes — 121
3. Interfering with ROS Formation: The Use of Antioxidants to Inhibit Autophagy — 126
4. Concluding Remarks — 128
Acknowledgments — 128
References — 129

9. Flow Cytometric Analysis of Autophagy in Living Mammalian Cells — 131
Elena Shvets and Zvulun Elazar

1. Introduction — 132
2. Qualitative Analysis of LC3 Levels in Cells — 133
3. Establishing Flow Cytometry and FACS Analysis to Quantify GFP-LC3 Levels During Amino Acid Deprivation — 134
4. Quantification of Decline in GFP-LC3 Using Flow Cytometry and FACS Analysis — 137
5. Using FACS to Quantify Autophagy in Response to Specific Treatments — 138
6. Concluding Remarks — 139
Acknowledgments — 140
References — 140

10. Monitoring Autophagy by Electron Microscopy in Mammalian Cells — 143
Päivi Ylä-Anttila, Helena Vihinen, Eija Jokitalo, and Eeva-Liisa Eskelinen

1. Introduction — 144
2. Methods — 146

3.	Results and Discussion	155
4.	Concluding Remarks	162
	Acknowledgments	163
	References	163

11. Monitoring Mammalian Target of Rapamycin (mTOR) Activity   165

Tsuneo Ikenoue, Sungki Hong, and Ken Inoki

1.	Introduction	166
2.	Methods	168
3.	Concluding Remarks	177
	Acknowledgment	178
	References	178

12. Monitoring Autophagic Degradation of p62/SQSTM1   181

Geir Bjørkøy, Trond Lamark, Serhiy Pankiv, Aud Øvervatn, Andreas Brech, and Terje Johansen

1.	Introduction	182
2.	Monitoring Autophagy-Mediated Degradation of Endogenous p62/SQSTM1	183
3.	Monitoring the Autophagic Degradation of p62/SQSTM1 by Live Cell Imaging	192
4.	Concluding Remarks	194
	Acknowledgments	195
	References	195

13. Cytosolic LC3 Ratio as a Quantitative Index of Macroautophagy   199

Motoni Kadowaki and Md. Razaul Karim

1.	Introduction	200
2.	Measurement of Proteolysis	201
3.	Measurement of the Cytosolic LC3 Ratio	203
4.	Concluding Remarks	211
	References	211

14. Method for Monitoring Pexophagy in Mammalian Cells   215

Junji Ezaki, Masaaki Komatsu, Sadaki Yokota, Takashi Ueno, and Eiki Kominami

1.	Introduction	216
2.	Experimental Models for the Study of Pexophagy in Mammals	216
3.	Monitoring Degradation of Excess Peroxisomes	217

4. Concluding Remarks	223
Acknowledgment	225
References	225

15. Mitophagy in Mammalian Cells: The Reticulocyte Model   227

Ji Zhang, Mondira Kundu, and Paul A. Ney

1. Introduction	228
2. Reticulocyte Production and Maturation	229
3. Methods	230
4. Mouse Models of Mitophagy	239
5. Conclusions	240
Acknowledgments	241
References	241

16. Assessing Mammalian Autophagy by WIPI-1/Atg18 Puncta Formation   247

Tassula Proikas-Cezanne and Simon G. Pfisterer

1. The Human WIPI Protein Family	248
2. WIPI-1 Puncta-Formation Assay	250
3. Indirect Immunofluorescence to Visualize Endogenous WIPI-1 by Confocal Microscopy	251
4. Direct Fluorescence of Transiently Expressed GFP-WIPI-1	255
5. Generation of Stable GFP-WIPI-1 Cell Lines	256
6. Live-Cell Imaging of GFP-WIPI-1	257
7. WIPI-1 Protein-Lipid Overlay Assay	258
8. Concluding Remarks	259
Acknowledgments	259
References	259

17. Correlative Light and Electron Microscopy   261

Minoo Razi and Sharon A. Tooze

1. Introduction	262
2. CLEM	263
3. Cell Culture	264
4. Light Microscopy	266
5. Electron Microscopy	268
6. Results, Interpretation, and Presentation	271
7. Tomography and CLEM	273
8. Cryoimmunogold EM and CLEM—the Future?	273
9. Conclusions	274
References	274

18. Semiconductor Nanocrystals in Autophagy Research: Methodology Improvement at Nanosized Scale   277

Oleksandr Seleverstov, James M. Phang, and Olga Zabirnyk

1. Introduction   278
2. Autophagy Probing in Living Cells Using Different-Sized Fluorescent NP   279
3. Application of Nanotechnology-Based Products for Other Methods in Autophagy Research. The Perspectives   285
4. Conclusion and Future Perspectives   290
Acknowledgments   292
References   293

19. Methods to Monitor Chaperone-Mediated Autophagy   297

Susmita Kaushik and Ana Maria Cuervo

1. Introduction   298
2. Experimental Models for the Study of CMA   300
3. Properties of CMA Substrates   301
4. Methods to Measure CMA   303
5. Measurement of Protein Degradation Rates   303
6. Measurement of Levels of Key CMA Components   308
7. Analysis of the Subcellular Location of CMA-Active Lysosomes   312
8. *In Vitro* Assay to Measure Translocation of CMA Substrates   315
9. Concluding Remarks   320
Acknowledgments   321
References   321

20. Methods to Monitor Autophagy of *Salmonella enterica* serovar Typhimurium   325

Cheryl L. Birmingham and John H. Brumell

1. Introduction   326
2. Characteristics of Autophagy of *S.* Typhimurium   330
3. *S.* Typhimurium Infection and Autophagy   334
4. Controls for Autophagy of Bacteria   337
5. Concluding Remarks   339
References   339

21. Monitoring Autophagy during *Mycobacterium tuberculosis* Infection   345

Marisa Ponpuak, Monica A. Delgado, Rasha A. Elmaoued, and Vojo Deretic

1. Introduction   346
2. Methods   347

Acknowledgments 359
References 359

22. *Streptococcus*-, *Shigella*-, and *Listeria*-Induced Autophagy 363

Michinaga Ogawa, Ichiro Nakagawa, Yuko Yoshikawa, Torsten Hain, Trinad Chakraborty, and Chihiro Sasakawa

1. Introduction 364
2. GAS Infection and Autophagy 365
3. *Shigella* Infection and Autophagy 371
4. *Listeria* Infection and Autophagy 376
5. Concluding Remarks 379
Acknowledgments 379
References 379

23. Kinetic Analysis of Autophagosome Formation and Turnover in Primary Mouse Macrophages 383

Michele S. Swanson, Brenda G. Byrne, and Jean-Francois Dubuisson

1. Introduction 384
2. Monitoring Flux Through the Autophagy Pathway 386
3. Assessing the Impact of Infection on the Autophagy Pathway 388
4. LC3 Localization in Primary Mouse Macrophages 390
5. Isolation of Bone Marrow-Derived Macrophages 393
6. Conclusion 400
References 400

24. Monitoring Macroautophagy by Major Histocompatibility Complex Class II Presentation of Targeted Antigens 403

Monique Gannagé and Christian Münz

1. Introduction 404
2. Antigen Presentation on MHC Class II Molecules After Macroautophagy 404
3. Substrates of Endogenous Antigen Processing for MHC Class II Presentation via Macroautophagy 405
4. Methods to Measure Antigen Processing for MHC Class II Presentation After Macroautophagy 407
5. Concluding Remarks 419
Acknowledgments 419
References 419

25. Detachment-Induced Autophagy In Three-Dimensional Epithelial Cell Cultures 423

Jayanta Debnath

1. Introduction 424

2.	Monitoring and Manipulating Autophagy In Mammalian Epithelial Cells	425
3.	Autophagy During Lumen Formation In 3D Epithelial Cultures	428
4.	Measuring Autophagy In Traditional Anoikis Assays	434
5.	Incubation of Cells With Integrin Function-Blocking Antibodies	437
6.	Conclusions	437
	Acknowledgments	438
	References	438

26. Methods for Inducing and Monitoring Liver Autophagy Relative to Aging and Antiaging Caloric Restriction in Rats — 441

Alessio Donati, Gabriella Cavallini, and Ettore Bergamini

1.	Overview	442
2.	Models of Caloric Restriction	442
3.	Methods to Induce and Monitor Liver Autophagy	444
4.	A Minimally Invasive Procedure for the Rapid Evaluation of the Induced Intensification of Autophagy	448
5.	Autophagic Protein Degradation in Isolated Liver Cells	449
6.	Concluding Remarks	453
	References	454

27. Physiological Autophagy in the Syrian Hamster Harderian Gland — 457

Ignacio Vega-Naredo and Ana Coto-Montes

1.	Introduction	458
2.	Methods to Detect Autophagy by Morphological Techniques	459
3.	Methods to Detect Autophagy by Biochemical and Molecular Biology Techniques	465
4.	Measurement of Cell Viability: Ratio of Cathepsins B:D	466
5.	Analysis of the Lysosomal Pathway	467
6.	Western Blot Analysis of Autophagy-Related Proteins	471
7.	Methods to Detect Chaperone-Mediated Autophagy	472
8.	Concluding Remarks	473
	Acknowledgments	474
	References	474

Author Index — *477*
Subject Index — *499*

Contributors

Chantal Bauvy
INSERM U756, Université Paris-Sud 11, Châtenay-Malabry, France

Ettore Bergamini
Centro di Ricerca Interdipartimentale di Biologia e Patologia dell'Invecchiamento, Università di Pisa, Pisa, Italy

Cheryl L. Birmingham
Department of Molecular Genetics and Institute of Medical Science, University of Toronto, Toronto, Ontario, Canada

Geir Bjørkøy
Biochemistry Department, Institute of Medical Biology, University of Tromsø, Norway

Andreas Brech
Department of Biochemistry, Institute for Cancer Research, Norwegian Radium Hospital, Oslo, Norway

John H. Brumell
Department of Molecular Genetics and Institute of Medical Science, University of Toronto, Toronto, Ontario, Canada, and Cell Biology Program, Hospital for Sick Children, Toronto, Ontario, Canada

Brenda G. Byrne
Department of Microbiology and Immunology, University of Michigan Medical School, Ann Arbor, Michigan, USA

Gabriella Cavallini
Centro di Ricerca Interdipartimentale di Biologia e Patologia dell'Invecchiamento, Università di Pisa, Pisa, Italy

Trinad Chakraborty
Institute of Medical Microbiology, Justus-Liebig University Giessen, Giessen, Germany

Patrice Codogno
INSERM U756, Université Paris-Sud 11, Châtenay-Malabry, France

María Isabel Colombo
Laboratorio de Biología Celular y Molecular, Instituto de Histología y Embriología (IHEM), Facultad de Ciencias Médicas, Universidad Nacional de Cuyo-CONICET, Mendoza, Argentina

Ana Coto-Montes
Departamento de Morfología y Biología Celular, Facultad de Medicina, Universidad de Oviedo, Oviedo, Spain

Ana Maria Cuervo
Department of Developmental and Molecular Biology, Marion Bessin Liver Research Center, Institute for Aging Research, Albert Einstein College of Medicine, Bronx, New York, USA

Jayanta Debnath
Department of Pathology, University of California, San Francisco, California, USA

Monica A. Delgado
Department of Molecular Genetics and Microbiology, University of New Mexico, Health Sciences Center, Albuquerque, New Mexico, USA

Patrick B. Dennis
University of Cincinnati, Genome Research Institute, Department of Cancer and Cell Biology, Cincinnati, Ohio, USA

Vojo Deretic
Department of Molecular Genetics and Microbiology, University of New Mexico, Health Sciences Center, Albuquerque, New Mexico, USA

Alessio Donati
Centro di Ricerca Interdipartimentale di Biologia e Patologia dell'Invecchiamento, Università di Pisa, Pisa, Italy

Jean-Francois Dubuisson
Department of Microbiology and Immunology, University of Michigan Medical School, Ann Arbor, Michigan, USA

Zvulun Elazar
Department of Biological Chemistry, The Weizmann Institute of Science, Rehovot, Israel

Rasha A. Elmaoued
Department of Molecular Genetics and Microbiology, University of New Mexico, Health Sciences Center, Albuquerque, New Mexico, USA

Eeva-Liisa Eskelinen
Department of Biological and Environmental Sciences, Division of Biochemistry, University of Helsinki, Helsinki, Finland

Junji Ezaki
Department of Biochemistry, Juntendo University School of Medicine, Hongo, Tokyo, Japan

Naonobu Fujita
Department of Cellular Regulation, Research Institute for Microbial Diseases, Osaka University, Osaka, Japan

Monique Gannagé
Viral Immunobiology, Institute of Experimental Immunology, University Hospital of Zürich, Zürich, Switzerland

Dale W. Hailey
Section on Organelle Biology, Cell Biology and Metabolism Branch, NICHD, National Institutes of Health, Bethesda, Maryland, USA

Torsten Hain
Institute of Medical Microbiology, Justus-Liebig University Giessen, Giessen, Germany

Sungki Hong
Life Sciences Institute, University of Michigan, Ann Arbor, Michigan, USA

Tsuneo Ikenoue
Life Sciences Institute, University of Michigan, Ann Arbor, Michigan, USA

Ken Inoki
Department of Molecular and Integrative Physiology, University of Michigan Medical School, Ann Arbor, Michigan, USA, and Life Sciences Institute, University of Michigan, Ann Arbor, Michigan, USA

Terje Johansen
Biochemistry Department, Institute of Medical Biology, University of Tromsø, Norway

Eija Jokitalo
Institute of Biotechnology, Electron Microscopy Unit, University of Helsinki, Helsinki, Finland

Motoni Kadowaki
Department of Applied Biological Chemistry, Faculty of Agriculture, Graduate School of Science and Technology, Niigata University, Niigata, Japan

Md. Razaul Karim
Department of Applied Biological Chemistry, Faculty of Agriculture, Graduate School of Science and Technology, Niigata University, Niigata, Japan

Susmita Kaushik
Department of Developmental and Molecular Biology, Marion Bessin Liver

Research Center, Institute for Aging Research, Albert Einstein College of Medicine, Bronx, New York, USA

Shunsuke Kimura
Department of Cellular Regulation, Research Institute for Microbial Diseases, Osaka University, Osaka, Japan

Masaaki Komatsu
Department of Biochemistry, Juntendo University School of Medicine, Hongo, Tokyo, Japan

Eiki Kominami
Department of Biochemistry, Juntendo University School of Medicine, Hongo, Tokyo, Japan

Mondira Kundu
Department of Pathology and Laboratory Medicine, Abramson Family Cancer Research Institute, University of Pennsylvania, Philadelphia, Pennsylvania, USA

Trond Lamark
Biochemistry Department, Institute of Medical Biology, University of Tromsø, Norway

Jennifer Lippincott-Schwartz
Section on Organelle Biology, Cell Biology and Metabolism Branch, NICHD, National Institutes of Health, Bethesda, Maryland, USA

Alfred J. Meijer
Department of Medical Biochemistry, Academic Medical Center, University of Amsterdam, Amsterdam, The Netherlands

Carol A. Mercer
University of Cincinnati, Genome Research Institute, Department of Cancer and Cell Biology, Cincinnati, Ohio, USA

Noboru Mizushima
Department of Physiology and Cell Biology, Tokyo Medical and Dental University, Tokyo, Japan

Christian Münz
Viral Immunobiology, Institute of Experimental Immunology, University Hospital of Zürich, Zürich, Switzerland

Ichiro Nakagawa
Division of Bacteriology, Department of Infectious Disease Control, International Research Center for Infectious Diseases, Institute of Medical Science, University of Tokyo, Tokyo, Japan

Paul A. Ney
Department of Biochemistry, St. Jude Children's Research Hospital, Memphis, Tennessee, USA

Takeshi Noda
Department of Cellular Regulation, Research Institute for Microbial Diseases, Osaka University, Osaka, Japan

Michinaga Ogawa
Division of Bacterial Infection, Department of Microbiology and Immunology, Institute of Medical Science, University of Tokyo, Tokyo, Japan

Anders Øverbye
Proteomics Section DNR, Department of Cell Biology, Institute for Cancer Research, The Norwegian Radium Hospital, Oslo, Norway

Aud Øvervatn
Biochemistry Department, Institute of Medical Biology, University of Tromsø, Norway

Serhiy Pankiv
Biochemistry Department, Institute of Medical Biology, University of Tromsø, Norway

Simon G. Pfisterer
Autophagy Laboratory, Department of Molecular Biology, Interfaculty Institute for Cell Biology, University of Tübingen, Tübingen, Germany

James M. Phang
Metabolism and Cancer Susceptibility Section, Laboratory of Comparative Carcinogenesis, Center for Cancer Research, National Cancer Institute, Frederick, Maryland, USA

Marisa Ponpuak
Department of Molecular Genetics and Microbiology, University of New Mexico, Health Sciences Center, Albuquerque, New Mexico, USA

Tassula Proikas-Cezanne
Autophagy Laboratory, Department of Molecular Biology, Interfaculty Institute for Cell Biology, University of Tübingen, Tübingen, Germany

Minoo Razi
London Research Institute, Cancer Research UK, London, UK

Frank Sætre
Proteomics Section DNR, Department of Cell Biology, Institute for Cancer Research, The Norwegian Radium Hospital, Oslo, Norway

Chihiro Sasakawa
Division of Bacterial Infection, Department of Microbiology and Immunology, Institute of Medical Science, University of Tokyo, Tokyo, Japan, and Department of Infectious Disease Control, International Research Center for Infectious Diseases, Institute of Medical Science, University of Tokyo, Tokyo, Japan, and CREST, Japan Science and Technology Agency, Kawaguchi, Japan

Ruth Scherz-Shouval
Department of Biological Chemistry, The Weizmann Institute of Science, Rehovot, Israel

Per O. Seglen
Proteomics Section DNR, Department of Cell Biology, Institute for Cancer Research, The Norwegian Radium Hospital, Oslo, Norway

Oleksandr Seleverstov
Department of Animal Science, College of Agriculture, University of Wyoming, Laramie, Wyoming, USA

Elena Shvets
Department of Biological Chemistry, The Weizmann Institute of Science, Rehovot, Israel

Michele S. Swanson
Department of Microbiology and Immunology, University of Michigan Medical School, Ann Arbor, Michigan, USA

Sharon A. Tooze
London Research Institute, Cancer Research UK, London, UK

Takashi Ueno
Department of Biochemistry, Juntendo University School of Medicine, Hongo, Tokyo, Japan

Cristina Lourdes Vázquez
Laboratorio de Biología Celular y Molecular, Instituto de Histología y Embriología (IHEM), Facultad de Ciencias Médicas, Universidad Nacional de Cuyo-CONICET, Mendoza, Argentina

Ignacio Vega-Naredo
Departamento de Morfología y Biología Celular, Facultad de Medicina, Universidad de Oviedo, Oviedo, Spain

Helena Vihinen
Institute of Biotechnology, Electron Microscopy Unit, University of Helsinki, Helsinki, Finland

Päivi Ylä-Anttila
Department of Biological and Environmental Sciences, Division of Biochemistry, University of Helsinki, Helsinki, Finland

Sadaki Yokota
Faculty of Parmaceutical Science, Nagasaki International University, Sasebo, Nagasaki, Japan

Yuko Yoshikawa
Division of Bacterial Infection, Department of Microbiology and Immunology, Institute of Medical Science, University of Tokyo, Tokyo, Japan

Tamotsu Yoshimori
Department of Cellular Regulation, Research Institute for Microbial Diseases, Osaka University, Osaka, Japan

Olga Zabirnyk
Metabolism and Cancer Susceptibility Section, Laboratory of Comparative Carcinogenesis, Center for Cancer Research, National Cancer Institute, Frederick, Maryland, USA

Ji Zhang
Department of Biochemistry, St. Jude Children's Research Hospital, Memphis, Tennessee, USA

Preface

Research into the topic of autophagy started in the late 1950s. At that time, and for the following several decades, there were few methods available for studying this process. The initial methodology relied primarily on electron microscopy, sometimes coupled with subcellular fractionation, and electron microscopy remains one of the principal methods of analysis. Additional techniques were eventually added, which included sequestration and protein degradation assays, and all of these are described in this volume of *Methods in Enzymology*. Overall, however, the methods for examining autophagy in mammalian cells have been relatively limited.

In the 1990s, the autophagy-related (*ATG*) genes were identified in various fungi, which in part opened a new era, allowing an understanding of the molecular mechanism of autophagy. In addition, the identification of homologues to the fungal genes in higher eukaryotes provided evidence for the role of autophagy in a growing number of processes, in pathophysiology and also in development.

Although there is tremendous conservation among the autophagy-related genes across species, the relative difficulty of using molecular genetic approaches in mammalian cell culture and, in some cases, the need to monitor autophagy in intact organisms or tissue samples, has limited the development of techniques for assessing autophagy in mammalian systems. By far, the most versatile methodology among the molecular approaches relies on detection of the Atg8 homologue microtubule-associated protein 1 light chain 3, or LC3. The primary reasons for the utility of this protein are that it is often upregulated following autophagy induction, it undergoes posttranslational modifications that can be used to monitor certain aspects of autophagy, and it is presently the only Atg protein that is reliably associated with the completed autophagosome membrane. The LC3 protein can be followed by western blot, fluorescence and immunoelectron microscopy. Accordingly, many of the chapters in this volume describe different ways to use LC3 for following autophagy, and one of the main benefits of these chapters is that they describe specific variations that are applicable to particular systems and/or questions.

Despite the overall robustness of LC3-dependent assays, the entrance of many researchers into this field has also led to the development of new techniques, as well as modifications of earlier approaches. Accordingly, the chapters in this volume describe the use of dyes including DQ-BSA that can be used to monitor amphisome fusion with the lysosome, alternative sequestration substrates such as GST-BHMT, the analysis of p62/SQSTM1 that links LC3 with ubiquitin, the use of WIPI-1/Atg18 as a phagophore marker, and methods to monitor selective peroxisome and mitochondrial degradation.

Finally, the chapters in this volume are concerned not just with methodology, but also provide the background that allows the reader to appreciate the importance of monitoring autophagy with regard to the particular questions being asked. This second volume concludes with several chapters that are focused on the analysis of autophagy in connection with microbial pathogenesis and the immune response, as well as autophagy in tissues and intact organisms; these chapters set the stage for the third volume that will be devoted to connections with disease and clinical applications. Indeed, there is growing interest in manipulating autophagy for therapeutic purposes. One hope is that this and the companion volumes of *Methods in Enzymology* will stimulate researchers to pursue ongoing and new lines of investigation into autophagy so that we may continue to understand, and ultimately manipulate to our advantage, this complex and ubiquitous process.

DANIEL J. KLIONSKY

METHODS IN ENZYMOLOGY

VOLUME I. Preparation and Assay of Enzymes
Edited by SIDNEY P. COLOWICK AND NATHAN O. KAPLAN

VOLUME II. Preparation and Assay of Enzymes
Edited by SIDNEY P. COLOWICK AND NATHAN O. KAPLAN

VOLUME III. Preparation and Assay of Substrates
Edited by SIDNEY P. COLOWICK AND NATHAN O. KAPLAN

VOLUME IV. Special Techniques for the Enzymologist
Edited by SIDNEY P. COLOWICK AND NATHAN O. KAPLAN

VOLUME V. Preparation and Assay of Enzymes
Edited by SIDNEY P. COLOWICK AND NATHAN O. KAPLAN

VOLUME VI. Preparation and Assay of Enzymes *(Continued)*
Preparation and Assay of Substrates
Special Techniques
Edited by SIDNEY P. COLOWICK AND NATHAN O. KAPLAN

VOLUME VII. Cumulative Subject Index
Edited by SIDNEY P. COLOWICK AND NATHAN O. KAPLAN

VOLUME VIII. Complex Carbohydrates
Edited by ELIZABETH F. NEUFELD AND VICTOR GINSBURG

VOLUME IX. Carbohydrate Metabolism
Edited by WILLIS A. WOOD

VOLUME X. Oxidation and Phosphorylation
Edited by RONALD W. ESTABROOK AND MAYNARD E. PULLMAN

VOLUME XI. Enzyme Structure
Edited by C. H. W. HIRS

VOLUME XII. Nucleic Acids (Parts A and B)
Edited by LAWRENCE GROSSMAN AND KIVIE MOLDAVE

VOLUME XIII. Citric Acid Cycle
Edited by J. M. LOWENSTEIN

VOLUME XIV. Lipids
Edited by J. M. LOWENSTEIN

VOLUME XV. Steroids and Terpenoids
Edited by RAYMOND B. CLAYTON

VOLUME XVI. Fast Reactions
Edited by KENNETH KUSTIN

VOLUME XVII. Metabolism of Amino Acids and Amines (Parts A and B)
Edited by HERBERT TABOR AND CELIA WHITE TABOR

VOLUME XVIII. Vitamins and Coenzymes (Parts A, B, and C)
Edited by DONALD B. MCCORMICK AND LEMUEL D. WRIGHT

VOLUME XIX. Proteolytic Enzymes
Edited by GERTRUDE E. PERLMANN AND LASZLO LORAND

VOLUME XX. Nucleic Acids and Protein Synthesis (Part C)
Edited by KIVIE MOLDAVE AND LAWRENCE GROSSMAN

VOLUME XXI. Nucleic Acids (Part D)
Edited by LAWRENCE GROSSMAN AND KIVIE MOLDAVE

VOLUME XXII. Enzyme Purification and Related Techniques
Edited by WILLIAM B. JAKOBY

VOLUME XXIII. Photosynthesis (Part A)
Edited by ANTHONY SAN PIETRO

VOLUME XXIV. Photosynthesis and Nitrogen Fixation (Part B)
Edited by ANTHONY SAN PIETRO

VOLUME XXV. Enzyme Structure (Part B)
Edited by C. H. W. HIRS AND SERGE N. TIMASHEFF

VOLUME XXVI. Enzyme Structure (Part C)
Edited by C. H. W. HIRS AND SERGE N. TIMASHEFF

VOLUME XXVII. Enzyme Structure (Part D)
Edited by C. H. W. HIRS AND SERGE N. TIMASHEFF

VOLUME XXVIII. Complex Carbohydrates (Part B)
Edited by VICTOR GINSBURG

VOLUME XXIX. Nucleic Acids and Protein Synthesis (Part E)
Edited by LAWRENCE GROSSMAN AND KIVIE MOLDAVE

VOLUME XXX. Nucleic Acids and Protein Synthesis (Part F)
Edited by KIVIE MOLDAVE AND LAWRENCE GROSSMAN

VOLUME XXXI. Biomembranes (Part A)
Edited by SIDNEY FLEISCHER AND LESTER PACKER

VOLUME XXXII. Biomembranes (Part B)
Edited by SIDNEY FLEISCHER AND LESTER PACKER

VOLUME XXXIII. Cumulative Subject Index Volumes I-XXX
Edited by MARTHA G. DENNIS AND EDWARD A. DENNIS

VOLUME XXXIV. Affinity Techniques (Enzyme Purification: Part B)
Edited by WILLIAM B. JAKOBY AND MEIR WILCHEK

VOLUME XXXV. Lipids (Part B)
Edited by JOHN M. LOWENSTEIN

VOLUME XXXVI. Hormone Action (Part A: Steroid Hormones)
Edited by BERT W. O'MALLEY AND JOEL G. HARDMAN

VOLUME XXXVII. Hormone Action (Part B: Peptide Hormones)
Edited by BERT W. O'MALLEY AND JOEL G. HARDMAN

VOLUME XXXVIII. Hormone Action (Part C: Cyclic Nucleotides)
Edited by JOEL G. HARDMAN AND BERT W. O'MALLEY

VOLUME XXXIX. Hormone Action (Part D: Isolated Cells, Tissues, and Organ Systems)
Edited by JOEL G. HARDMAN AND BERT W. O'MALLEY

VOLUME XL. Hormone Action (Part E: Nuclear Structure and Function)
Edited by BERT W. O'MALLEY AND JOEL G. HARDMAN

VOLUME XLI. Carbohydrate Metabolism (Part B)
Edited by W. A. WOOD

VOLUME XLII. Carbohydrate Metabolism (Part C)
Edited by W. A. WOOD

VOLUME XLIII. Antibiotics
Edited by JOHN H. HASH

VOLUME XLIV. Immobilized Enzymes
Edited by KLAUS MOSBACH

VOLUME XLV. Proteolytic Enzymes (Part B)
Edited by LASZLO LORAND

VOLUME XLVI. Affinity Labeling
Edited by WILLIAM B. JAKOBY AND MEIR WILCHEK

VOLUME XLVII. Enzyme Structure (Part E)
Edited by C. H. W. HIRS AND SERGE N. TIMASHEFF

VOLUME XLVIII. Enzyme Structure (Part F)
Edited by C. H. W. HIRS AND SERGE N. TIMASHEFF

VOLUME XLIX. Enzyme Structure (Part G)
Edited by C. H. W. HIRS AND SERGE N. TIMASHEFF

VOLUME L. Complex Carbohydrates (Part C)
Edited by VICTOR GINSBURG

VOLUME LI. Purine and Pyrimidine Nucleotide Metabolism
Edited by PATRICIA A. HOFFEE AND MARY ELLEN JONES

VOLUME LII. Biomembranes (Part C: Biological Oxidations)
Edited by SIDNEY FLEISCHER AND LESTER PACKER

VOLUME LIII. Biomembranes (Part D: Biological Oxidations)
Edited by SIDNEY FLEISCHER AND LESTER PACKER

VOLUME LIV. Biomembranes (Part E: Biological Oxidations)
Edited by SIDNEY FLEISCHER AND LESTER PACKER

VOLUME LV. Biomembranes (Part F: Bioenergetics)
Edited by SIDNEY FLEISCHER AND LESTER PACKER

VOLUME LVI. Biomembranes (Part G: Bioenergetics)
Edited by SIDNEY FLEISCHER AND LESTER PACKER

VOLUME LVII. Bioluminescence and Chemiluminescence
Edited by MARLENE A. DELUCA

VOLUME LVIII. Cell Culture
Edited by WILLIAM B. JAKOBY AND IRA PASTAN

VOLUME LIX. Nucleic Acids and Protein Synthesis (Part G)
Edited by KIVIE MOLDAVE AND LAWRENCE GROSSMAN

VOLUME LX. Nucleic Acids and Protein Synthesis (Part H)
Edited by KIVIE MOLDAVE AND LAWRENCE GROSSMAN

VOLUME 61. Enzyme Structure (Part H)
Edited by C. H. W. HIRS AND SERGE N. TIMASHEFF

VOLUME 62. Vitamins and Coenzymes (Part D)
Edited by DONALD B. MCCORMICK AND LEMUEL D. WRIGHT

VOLUME 63. Enzyme Kinetics and Mechanism (Part A: Initial Rate and Inhibitor Methods)
Edited by DANIEL L. PURICH

VOLUME 64. Enzyme Kinetics and Mechanism
(Part B: Isotopic Probes and Complex Enzyme Systems)
Edited by DANIEL L. PURICH

VOLUME 65. Nucleic Acids (Part I)
Edited by LAWRENCE GROSSMAN AND KIVIE MOLDAVE

VOLUME 66. Vitamins and Coenzymes (Part E)
Edited by DONALD B. MCCORMICK AND LEMUEL D. WRIGHT

VOLUME 67. Vitamins and Coenzymes (Part F)
Edited by DONALD B. MCCORMICK AND LEMUEL D. WRIGHT

VOLUME 68. Recombinant DNA
Edited by RAY WU

VOLUME 69. Photosynthesis and Nitrogen Fixation (Part C)
Edited by ANTHONY SAN PIETRO

VOLUME 70. Immunochemical Techniques (Part A)
Edited by HELEN VAN VUNAKIS AND JOHN J. LANGONE

VOLUME 71. Lipids (Part C)
Edited by JOHN M. LOWENSTEIN

VOLUME 72. Lipids (Part D)
Edited by JOHN M. LOWENSTEIN

VOLUME 73. Immunochemical Techniques (Part B)
Edited by JOHN J. LANGONE AND HELEN VAN VUNAKIS

VOLUME 74. Immunochemical Techniques (Part C)
Edited by JOHN J. LANGONE AND HELEN VAN VUNAKIS

VOLUME 75. Cumulative Subject Index Volumes XXXI, XXXII, XXXIV–LX
Edited by EDWARD A. DENNIS AND MARTHA G. DENNIS

VOLUME 76. Hemoglobins
Edited by ERALDO ANTONINI, LUIGI ROSSI-BERNARDI, AND EMILIA CHIANCONE

VOLUME 77. Detoxication and Drug Metabolism
Edited by WILLIAM B. JAKOBY

VOLUME 78. Interferons (Part A)
Edited by SIDNEY PESTKA

VOLUME 79. Interferons (Part B)
Edited by SIDNEY PESTKA

VOLUME 80. Proteolytic Enzymes (Part C)
Edited by LASZLO LORAND

VOLUME 81. Biomembranes (Part H: Visual Pigments and Purple Membranes, I)
Edited by LESTER PACKER

VOLUME 82. Structural and Contractile Proteins (Part A: Extracellular Matrix)
Edited by LEON W. CUNNINGHAM AND DIXIE W. FREDERIKSEN

VOLUME 83. Complex Carbohydrates (Part D)
Edited by VICTOR GINSBURG

VOLUME 84. Immunochemical Techniques (Part D: Selected Immunoassays)
Edited by JOHN J. LANGONE AND HELEN VAN VUNAKIS

VOLUME 85. Structural and Contractile Proteins (Part B: The Contractile Apparatus and the Cytoskeleton)
Edited by DIXIE W. FREDERIKSEN AND LEON W. CUNNINGHAM

VOLUME 86. Prostaglandins and Arachidonate Metabolites
Edited by WILLIAM E. M. LANDS AND WILLIAM L. SMITH

VOLUME 87. Enzyme Kinetics and Mechanism (Part C: Intermediates, Stereo-chemistry, and Rate Studies)
Edited by DANIEL L. PURICH

VOLUME 88. Biomembranes (Part I: Visual Pigments and Purple Membranes, II)
Edited by LESTER PACKER

VOLUME 89. Carbohydrate Metabolism (Part D)
Edited by WILLIS A. WOOD

VOLUME 90. Carbohydrate Metabolism (Part E)
Edited by WILLIS A. WOOD

VOLUME 91. Enzyme Structure (Part I)
Edited by C. H. W. HIRS AND SERGE N. TIMASHEFF

VOLUME 92. Immunochemical Techniques (Part E: Monoclonal Antibodies and General Immunoassay Methods)
Edited by JOHN J. LANGONE AND HELEN VAN VUNAKIS

VOLUME 93. Immunochemical Techniques (Part F: Conventional Antibodies, Fc Receptors, and Cytotoxicity)
Edited by JOHN J. LANGONE AND HELEN VAN VUNAKIS

VOLUME 94. Polyamines
Edited by HERBERT TABOR AND CELIA WHITE TABOR

VOLUME 95. Cumulative Subject Index Volumes 61–74, 76–80
Edited by EDWARD A. DENNIS AND MARTHA G. DENNIS

VOLUME 96. Biomembranes [Part J: Membrane Biogenesis: Assembly and Targeting (General Methods; Eukaryotes)]
Edited by SIDNEY FLEISCHER AND BECCA FLEISCHER

VOLUME 97. Biomembranes [Part K: Membrane Biogenesis: Assembly and Targeting (Prokaryotes, Mitochondria, and Chloroplasts)]
Edited by SIDNEY FLEISCHER AND BECCA FLEISCHER

VOLUME 98. Biomembranes (Part L: Membrane Biogenesis: Processing and Recycling)
Edited by SIDNEY FLEISCHER AND BECCA FLEISCHER

VOLUME 99. Hormone Action (Part F: Protein Kinases)
Edited by JACKIE D. CORBIN AND JOEL G. HARDMAN

VOLUME 100. Recombinant DNA (Part B)
Edited by RAY WU, LAWRENCE GROSSMAN, AND KIVIE MOLDAVE

VOLUME 101. Recombinant DNA (Part C)
Edited by RAY WU, LAWRENCE GROSSMAN, AND KIVIE MOLDAVE

VOLUME 102. Hormone Action (Part G: Calmodulin and Calcium-Binding Proteins)
Edited by ANTHONY R. MEANS AND BERT W. O'MALLEY

VOLUME 103. Hormone Action (Part H: Neuroendocrine Peptides)
Edited by P. MICHAEL CONN

VOLUME 104. Enzyme Purification and Related Techniques (Part C)
Edited by WILLIAM B. JAKOBY

VOLUME 105. Oxygen Radicals in Biological Systems
Edited by LESTER PACKER

VOLUME 106. Posttranslational Modifications (Part A)
Edited by FINN WOLD AND KIVIE MOLDAVE

VOLUME 107. Posttranslational Modifications (Part B)
Edited by FINN WOLD AND KIVIE MOLDAVE

VOLUME 108. Immunochemical Techniques (Part G: Separation and Characterization of Lymphoid Cells)
Edited by GIOVANNI DI SABATO, JOHN J. LANGONE, AND HELEN VAN VUNAKIS

VOLUME 109. Hormone Action (Part I: Peptide Hormones)
Edited by LUTZ BIRNBAUMER AND BERT W. O'MALLEY

VOLUME 110. Steroids and Isoprenoids (Part A)
Edited by JOHN H. LAW AND HANS C. RILLING

VOLUME 111. Steroids and Isoprenoids (Part B)
Edited by JOHN H. LAW AND HANS C. RILLING

VOLUME 112. Drug and Enzyme Targeting (Part A)
Edited by KENNETH J. WIDDER AND RALPH GREEN

VOLUME 113. Glutamate, Glutamine, Glutathione, and Related Compounds
Edited by ALTON MEISTER

VOLUME 114. Diffraction Methods for Biological Macromolecules (Part A)
Edited by HAROLD W. WYCKOFF, C. H. W. HIRS, AND SERGE N. TIMASHEFF

VOLUME 115. Diffraction Methods for Biological Macromolecules (Part B)
Edited by HAROLD W. WYCKOFF, C. H. W. HIRS, AND SERGE N. TIMASHEFF

VOLUME 116. Immunochemical Techniques
(Part H: Effectors and Mediators of Lymphoid Cell Functions)
Edited by GIOVANNI DI SABATO, JOHN J. LANGONE, AND HELEN VAN VUNAKIS

VOLUME 117. Enzyme Structure (Part J)
Edited by C. H. W. HIRS AND SERGE N. TIMASHEFF

VOLUME 118. Plant Molecular Biology
Edited by ARTHUR WEISSBACH AND HERBERT WEISSBACH

VOLUME 119. Interferons (Part C)
Edited by SIDNEY PESTKA

VOLUME 120. Cumulative Subject Index Volumes 81–94, 96–101

VOLUME 121. Immunochemical Techniques (Part I: Hybridoma Technology and Monoclonal Antibodies)
Edited by JOHN J. LANGONE AND HELEN VAN VUNAKIS

VOLUME 122. Vitamins and Coenzymes (Part G)
Edited by FRANK CHYTIL AND DONALD B. MCCORMICK

VOLUME 123. Vitamins and Coenzymes (Part H)
Edited by FRANK CHYTIL AND DONALD B. MCCORMICK

VOLUME 124. Hormone Action (Part J: Neuroendocrine Peptides)
Edited by P. MICHAEL CONN

VOLUME 125. Biomembranes (Part M: Transport in Bacteria, Mitochondria, and Chloroplasts: General Approaches and Transport Systems)
Edited by SIDNEY FLEISCHER AND BECCA FLEISCHER

VOLUME 126. Biomembranes (Part N: Transport in Bacteria, Mitochondria, and Chloroplasts: Protonmotive Force)
Edited by SIDNEY FLEISCHER AND BECCA FLEISCHER

VOLUME 127. Biomembranes (Part O: Protons and Water: Structure and Translocation)
Edited by LESTER PACKER

VOLUME 128. Plasma Lipoproteins (Part A: Preparation, Structure, and Molecular Biology)
Edited by JERE P. SEGREST AND JOHN J. ALBERS

VOLUME 129. Plasma Lipoproteins (Part B: Characterization, Cell Biology, and Metabolism)
Edited by JOHN J. ALBERS AND JERE P. SEGREST

VOLUME 130. Enzyme Structure (Part K)
Edited by C. H. W. HIRS AND SERGE N. TIMASHEFF

VOLUME 131. Enzyme Structure (Part L)
Edited by C. H. W. HIRS AND SERGE N. TIMASHEFF

VOLUME 132. Immunochemical Techniques (Part J: Phagocytosis and Cell-Mediated Cytotoxicity)
Edited by GIOVANNI DI SABATO AND JOHANNES EVERSE

VOLUME 133. Bioluminescence and Chemiluminescence (Part B)
Edited by MARLENE DELUCA AND WILLIAM D. MCELROY

VOLUME 134. Structural and Contractile Proteins (Part C: The Contractile Apparatus and the Cytoskeleton)
Edited by RICHARD B. VALLEE

VOLUME 135. Immobilized Enzymes and Cells (Part B)
Edited by KLAUS MOSBACH

VOLUME 136. Immobilized Enzymes and Cells (Part C)
Edited by KLAUS MOSBACH

VOLUME 137. Immobilized Enzymes and Cells (Part D)
Edited by KLAUS MOSBACH

VOLUME 138. Complex Carbohydrates (Part E)
Edited by VICTOR GINSBURG

VOLUME 139. Cellular Regulators (Part A: Calcium- and Calmodulin-Binding Proteins)
Edited by ANTHONY R. MEANS AND P. MICHAEL CONN

VOLUME 140. Cumulative Subject Index Volumes 102–119, 121–134

VOLUME 141. Cellular Regulators (Part B: Calcium and Lipids)
Edited by P. MICHAEL CONN AND ANTHONY R. MEANS

VOLUME 142. Metabolism of Aromatic Amino Acids and Amines
Edited by SEYMOUR KAUFMAN

VOLUME 143. Sulfur and Sulfur Amino Acids
Edited by WILLIAM B. JAKOBY AND OWEN GRIFFITH

VOLUME 144. Structural and Contractile Proteins (Part D: Extracellular Matrix)
Edited by LEON W. CUNNINGHAM

VOLUME 145. Structural and Contractile Proteins (Part E: Extracellular Matrix)
Edited by LEON W. CUNNINGHAM

VOLUME 146. Peptide Growth Factors (Part A)
Edited by DAVID BARNES AND DAVID A. SIRBASKU

VOLUME 147. Peptide Growth Factors (Part B)
Edited by DAVID BARNES AND DAVID A. SIRBASKU

VOLUME 148. Plant Cell Membranes
Edited by LESTER PACKER AND ROLAND DOUCE

VOLUME 149. Drug and Enzyme Targeting (Part B)
Edited by RALPH GREEN AND KENNETH J. WIDDER

VOLUME 150. Immunochemical Techniques (Part K: *In Vitro* Models of B and T Cell Functions and Lymphoid Cell Receptors)
Edited by GIOVANNI DI SABATO

VOLUME 151. Molecular Genetics of Mammalian Cells
Edited by MICHAEL M. GOTTESMAN

VOLUME 152. Guide to Molecular Cloning Techniques
Edited by SHELBY L. BERGER AND ALAN R. KIMMEL

VOLUME 153. Recombinant DNA (Part D)
Edited by RAY WU AND LAWRENCE GROSSMAN

VOLUME 154. Recombinant DNA (Part E)
Edited by RAY WU AND LAWRENCE GROSSMAN

VOLUME 155. Recombinant DNA (Part F)
Edited by RAY WU

VOLUME 156. Biomembranes (Part P: ATP-Driven Pumps and Related Transport: The Na, K-Pump)
Edited by SIDNEY FLEISCHER AND BECCA FLEISCHER

VOLUME 157. Biomembranes (Part Q: ATP-Driven Pumps and Related Transport: Calcium, Proton, and Potassium Pumps)
Edited by SIDNEY FLEISCHER AND BECCA FLEISCHER

VOLUME 158. Metalloproteins (Part A)
Edited by JAMES F. RIORDAN AND BERT L. VALLEE

VOLUME 159. Initiation and Termination of Cyclic Nucleotide Action
Edited by JACKIE D. CORBIN AND ROGER A. JOHNSON

VOLUME 160. Biomass (Part A: Cellulose and Hemicellulose)
Edited by WILLIS A. WOOD AND SCOTT T. KELLOGG

VOLUME 161. Biomass (Part B: Lignin, Pectin, and Chitin)
Edited by WILLIS A. WOOD AND SCOTT T. KELLOGG

VOLUME 162. Immunochemical Techniques (Part L: Chemotaxis and Inflammation)
Edited by GIOVANNI DI SABATO

VOLUME 163. Immunochemical Techniques (Part M: Chemotaxis and Inflammation)
Edited by GIOVANNI DI SABATO

VOLUME 164. Ribosomes
Edited by HARRY F. NOLLER, JR., AND KIVIE MOLDAVE

VOLUME 165. Microbial Toxins: Tools for Enzymology
Edited by SIDNEY HARSHMAN

VOLUME 166. Branched-Chain Amino Acids
Edited by ROBERT HARRIS AND JOHN R. SOKATCH

VOLUME 167. Cyanobacteria
Edited by LESTER PACKER AND ALEXANDER N. GLAZER

VOLUME 168. Hormone Action (Part K: Neuroendocrine Peptides)
Edited by P. MICHAEL CONN

VOLUME 169. Platelets: Receptors, Adhesion, Secretion (Part A)
Edited by JACEK HAWIGER

VOLUME 170. Nucleosomes
Edited by PAUL M. WASSARMAN AND ROGER D. KORNBERG

VOLUME 171. Biomembranes (Part R: Transport Theory: Cells and Model Membranes)
Edited by SIDNEY FLEISCHER AND BECCA FLEISCHER

VOLUME 172. Biomembranes (Part S: Transport: Membrane Isolation and Characterization)
Edited by SIDNEY FLEISCHER AND BECCA FLEISCHER

VOLUME 173. Biomembranes [Part T: Cellular and Subcellular Transport: Eukaryotic (Nonepithelial) Cells]
Edited by SIDNEY FLEISCHER AND BECCA FLEISCHER

VOLUME 174. Biomembranes [Part U: Cellular and Subcellular Transport: Eukaryotic (Nonepithelial) Cells]
Edited by SIDNEY FLEISCHER AND BECCA FLEISCHER

VOLUME 175. Cumulative Subject Index Volumes 135–139, 141–167

VOLUME 176. Nuclear Magnetic Resonance (Part A: Spectral Techniques and Dynamics)
Edited by NORMAN J. OPPENHEIMER AND THOMAS L. JAMES

VOLUME 177. Nuclear Magnetic Resonance (Part B: Structure and Mechanism)
Edited by NORMAN J. OPPENHEIMER AND THOMAS L. JAMES

VOLUME 178. Antibodies, Antigens, and Molecular Mimicry
Edited by JOHN J. LANGONE

VOLUME 179. Complex Carbohydrates (Part F)
Edited by VICTOR GINSBURG

VOLUME 180. RNA Processing (Part A: General Methods)
Edited by JAMES E. DAHLBERG AND JOHN N. ABELSON

VOLUME 181. RNA Processing (Part B: Specific Methods)
Edited by JAMES E. DAHLBERG AND JOHN N. ABELSON

VOLUME 182. Guide to Protein Purification
Edited by MURRAY P. DEUTSCHER

VOLUME 183. Molecular Evolution: Computer Analysis of Protein and Nucleic Acid Sequences
Edited by RUSSELL F. DOOLITTLE

VOLUME 184. Avidin-Biotin Technology
Edited by MEIR WILCHEK AND EDWARD A. BAYER

VOLUME 185. Gene Expression Technology
Edited by DAVID V. GOEDDEL

VOLUME 186. Oxygen Radicals in Biological Systems (Part B: Oxygen Radicals and Antioxidants)
Edited by LESTER PACKER AND ALEXANDER N. GLAZER

VOLUME 187. Arachidonate Related Lipid Mediators
Edited by ROBERT C. MURPHY AND FRANK A. FITZPATRICK

VOLUME 188. Hydrocarbons and Methylotrophy
Edited by MARY E. LIDSTROM

VOLUME 189. Retinoids (Part A: Molecular and Metabolic Aspects)
Edited by LESTER PACKER

VOLUME 190. Retinoids (Part B: Cell Differentiation and Clinical Applications)
Edited by LESTER PACKER

VOLUME 191. Biomembranes (Part V: Cellular and Subcellular Transport: Epithelial Cells)
Edited by SIDNEY FLEISCHER AND BECCA FLEISCHER

VOLUME 192. Biomembranes (Part W: Cellular and Subcellular Transport: Epithelial Cells)
Edited by SIDNEY FLEISCHER AND BECCA FLEISCHER

VOLUME 193. Mass Spectrometry
Edited by JAMES A. MCCLOSKEY

VOLUME 194. Guide to Yeast Genetics and Molecular Biology
Edited by CHRISTINE GUTHRIE AND GERALD R. FINK

VOLUME 195. Adenylyl Cyclase, G Proteins, and Guanylyl Cyclase
Edited by ROGER A. JOHNSON AND JACKIE D. CORBIN

VOLUME 196. Molecular Motors and the Cytoskeleton
Edited by RICHARD B. VALLEE

VOLUME 197. Phospholipases
Edited by EDWARD A. DENNIS

VOLUME 198. Peptide Growth Factors (Part C)
Edited by DAVID BARNES, J. P. MATHER, AND GORDON H. SATO

VOLUME 199. Cumulative Subject Index Volumes 168–174, 176–194

VOLUME 200. Protein Phosphorylation (Part A: Protein Kinases: Assays, Purification, Antibodies, Functional Analysis, Cloning, and Expression)
Edited by TONY HUNTER AND BARTHOLOMEW M. SEFTON

VOLUME 201. Protein Phosphorylation (Part B: Analysis of Protein Phosphorylation, Protein Kinase Inhibitors, and Protein Phosphatases)
Edited by TONY HUNTER AND BARTHOLOMEW M. SEFTON

VOLUME 202. Molecular Design and Modeling: Concepts and Applications (Part A: Proteins, Peptides, and Enzymes)
Edited by JOHN J. LANGONE

VOLUME 203. Molecular Design and Modeling: Concepts and Applications (Part B: Antibodies and Antigens, Nucleic Acids, Polysaccharides, and Drugs)
Edited by JOHN J. LANGONE

VOLUME 204. Bacterial Genetic Systems
Edited by JEFFREY H. MILLER

VOLUME 205. Metallobiochemistry (Part B: Metallothionein and Related Molecules)
Edited by JAMES F. RIORDAN AND BERT L. VALLEE

VOLUME 206. Cytochrome P450
Edited by MICHAEL R. WATERMAN AND ERIC F. JOHNSON

VOLUME 207. Ion Channels
Edited by BERNARDO RUDY AND LINDA E. IVERSON

VOLUME 208. Protein–DNA Interactions
Edited by ROBERT T. SAUER

VOLUME 209. Phospholipid Biosynthesis
Edited by EDWARD A. DENNIS AND DENNIS E. VANCE

VOLUME 210. Numerical Computer Methods
Edited by LUDWIG BRAND AND MICHAEL L. JOHNSON

VOLUME 211. DNA Structures (Part A: Synthesis and Physical Analysis of DNA)
Edited by DAVID M. J. LILLEY AND JAMES E. DAHLBERG

VOLUME 212. DNA Structures (Part B: Chemical and Electrophoretic Analysis of DNA)
Edited by DAVID M. J. LILLEY AND JAMES E. DAHLBERG

VOLUME 213. Carotenoids (Part A: Chemistry, Separation, Quantitation, and Antioxidation)
Edited by LESTER PACKER

VOLUME 214. Carotenoids (Part B: Metabolism, Genetics, and Biosynthesis)
Edited by LESTER PACKER

VOLUME 215. Platelets: Receptors, Adhesion, Secretion (Part B)
Edited by JACEK J. HAWIGER

VOLUME 216. Recombinant DNA (Part G)
Edited by RAY WU

VOLUME 217. Recombinant DNA (Part H)
Edited by RAY WU

VOLUME 218. Recombinant DNA (Part I)
Edited by RAY WU

VOLUME 219. Reconstitution of Intracellular Transport
Edited by JAMES E. ROTHMAN

VOLUME 220. Membrane Fusion Techniques (Part A)
Edited by NEJAT DÜZGÜNEŞ

VOLUME 221. Membrane Fusion Techniques (Part B)
Edited by NEJAT DÜZGÜNEŞ

VOLUME 222. Proteolytic Enzymes in Coagulation, Fibrinolysis, and Complement Activation (Part A: Mammalian Blood Coagulation Factors and Inhibitors)
Edited by LASZLO LORAND AND KENNETH G. MANN

VOLUME 223. Proteolytic Enzymes in Coagulation, Fibrinolysis, and Complement Activation (Part B: Complement Activation, Fibrinolysis, and Nonmammalian Blood Coagulation Factors)
Edited by LASZLO LORAND AND KENNETH G. MANN

VOLUME 224. Molecular Evolution: Producing the Biochemical Data
Edited by ELIZABETH ANNE ZIMMER, THOMAS J. WHITE, REBECCA L. CANN, AND ALLAN C. WILSON

VOLUME 225. Guide to Techniques in Mouse Development
Edited by PAUL M. WASSARMAN AND MELVIN L. DEPAMPHILIS

VOLUME 226. Metallobiochemistry (Part C: Spectroscopic and Physical Methods for Probing Metal Ion Environments in Metalloenzymes and Metalloproteins)
Edited by JAMES F. RIORDAN AND BERT L. VALLEE

VOLUME 227. Metallobiochemistry (Part D: Physical and Spectroscopic Methods for Probing Metal Ion Environments in Metalloproteins)
Edited by JAMES F. RIORDAN AND BERT L. VALLEE

VOLUME 228. Aqueous Two-Phase Systems
Edited by HARRY WALTER AND GÖTE JOHANSSON

VOLUME 229. Cumulative Subject Index Volumes 195–198, 200–227

VOLUME 230. Guide to Techniques in Glycobiology
Edited by WILLIAM J. LENNARZ AND GERALD W. HART

VOLUME 231. Hemoglobins (Part B: Biochemical and Analytical Methods)
Edited by JOHANNES EVERSE, KIM D. VANDEGRIFF, AND ROBERT M. WINSLOW

VOLUME 232. Hemoglobins (Part C: Biophysical Methods)
Edited by JOHANNES EVERSE, KIM D. VANDEGRIFF, AND ROBERT M. WINSLOW

VOLUME 233. Oxygen Radicals in Biological Systems (Part C)
Edited by LESTER PACKER

VOLUME 234. Oxygen Radicals in Biological Systems (Part D)
Edited by LESTER PACKER

VOLUME 235. Bacterial Pathogenesis (Part A: Identification and Regulation of Virulence Factors)
Edited by VIRGINIA L. CLARK AND PATRIK M. BAVOIL

VOLUME 236. Bacterial Pathogenesis (Part B: Integration of Pathogenic Bacteria with Host Cells)
Edited by VIRGINIA L. CLARK AND PATRIK M. BAVOIL

VOLUME 237. Heterotrimeric G Proteins
Edited by RAVI IYENGAR

VOLUME 238. Heterotrimeric G-Protein Effectors
Edited by RAVI IYENGAR

VOLUME 239. Nuclear Magnetic Resonance (Part C)
Edited by THOMAS L. JAMES AND NORMAN J. OPPENHEIMER

VOLUME 240. Numerical Computer Methods (Part B)
Edited by MICHAEL L. JOHNSON AND LUDWIG BRAND

VOLUME 241. Retroviral Proteases
Edited by LAWRENCE C. KUO AND JULES A. SHAFER

VOLUME 242. Neoglycoconjugates (Part A)
Edited by Y. C. LEE AND REIKO T. LEE

VOLUME 243. Inorganic Microbial Sulfur Metabolism
Edited by HARRY D. PECK, JR., AND JEAN LEGALL

VOLUME 244. Proteolytic Enzymes: Serine and Cysteine Peptidases
Edited by ALAN J. BARRETT

VOLUME 245. Extracellular Matrix Components
Edited by E. RUOSLAHTI AND E. ENGVALL

VOLUME 246. Biochemical Spectroscopy
Edited by KENNETH SAUER

VOLUME 247. Neoglycoconjugates (Part B: Biomedical Applications)
Edited by Y. C. LEE AND REIKO T. LEE

VOLUME 248. Proteolytic Enzymes: Aspartic and Metallo Peptidases
Edited by ALAN J. BARRETT

VOLUME 249. Enzyme Kinetics and Mechanism (Part D: Developments in Enzyme Dynamics)
Edited by DANIEL L. PURICH

VOLUME 250. Lipid Modifications of Proteins
Edited by PATRICK J. CASEY AND JANICE E. BUSS

VOLUME 251. Biothiols (Part A: Monothiols and Dithiols, Protein Thiols, and Thiyl Radicals)
Edited by LESTER PACKER

VOLUME 252. Biothiols (Part B: Glutathione and Thioredoxin; Thiols in Signal Transduction and Gene Regulation)
Edited by LESTER PACKER

VOLUME 253. Adhesion of Microbial Pathogens
Edited by RON J. DOYLE AND ITZHAK OFEK

VOLUME 254. Oncogene Techniques
Edited by PETER K. VOGT AND INDER M. VERMA

VOLUME 255. Small GTPases and Their Regulators (Part A: Ras Family)
Edited by W. E. BALCH, CHANNING J. DER, AND ALAN HALL

VOLUME 256. Small GTPases and Their Regulators (Part B: Rho Family)
Edited by W. E. BALCH, CHANNING J. DER, AND ALAN HALL

VOLUME 257. Small GTPases and Their Regulators (Part C: Proteins Involved in Transport)
Edited by W. E. BALCH, CHANNING J. DER, AND ALAN HALL

VOLUME 258. Redox-Active Amino Acids in Biology
Edited by JUDITH P. KLINMAN

VOLUME 259. Energetics of Biological Macromolecules
Edited by MICHAEL L. JOHNSON AND GARY K. ACKERS

VOLUME 260. Mitochondrial Biogenesis and Genetics (Part A)
Edited by GIUSEPPE M. ATTARDI AND ANNE CHOMYN

VOLUME 261. Nuclear Magnetic Resonance and Nucleic Acids
Edited by THOMAS L. JAMES

VOLUME 262. DNA Replication
Edited by JUDITH L. CAMPBELL

VOLUME 263. Plasma Lipoproteins (Part C: Quantitation)
Edited by WILLIAM A. BRADLEY, SANDRA H. GIANTURCO, AND JERE P. SEGREST

VOLUME 264. Mitochondrial Biogenesis and Genetics (Part B)
Edited by GIUSEPPE M. ATTARDI AND ANNE CHOMYN

VOLUME 265. Cumulative Subject Index Volumes 228, 230–262

VOLUME 266. Computer Methods for Macromolecular Sequence Analysis
Edited by RUSSELL F. DOOLITTLE

VOLUME 267. Combinatorial Chemistry
Edited by JOHN N. ABELSON

VOLUME 268. Nitric Oxide (Part A: Sources and Detection of NO; NO Synthase)
Edited by LESTER PACKER

VOLUME 269. Nitric Oxide (Part B: Physiological and Pathological Processes)
Edited by LESTER PACKER

VOLUME 270. High Resolution Separation and Analysis of Biological Macromolecules (Part A: Fundamentals)
Edited by BARRY L. KARGER AND WILLIAM S. HANCOCK

VOLUME 271. High Resolution Separation and Analysis of Biological Macromolecules (Part B: Applications)
Edited by BARRY L. KARGER AND WILLIAM S. HANCOCK

VOLUME 272. Cytochrome P450 (Part B)
Edited by ERIC F. JOHNSON AND MICHAEL R. WATERMAN

VOLUME 273. RNA Polymerase and Associated Factors (Part A)
Edited by SANKAR ADHYA

VOLUME 274. RNA Polymerase and Associated Factors (Part B)
Edited by SANKAR ADHYA

VOLUME 275. Viral Polymerases and Related Proteins
Edited by LAWRENCE C. KUO, DAVID B. OLSEN, AND STEVEN S. CARROLL

VOLUME 276. Macromolecular Crystallography (Part A)
Edited by CHARLES W. CARTER, JR., AND ROBERT M. SWEET

VOLUME 277. Macromolecular Crystallography (Part B)
Edited by CHARLES W. CARTER, JR., AND ROBERT M. SWEET

VOLUME 278. Fluorescence Spectroscopy
Edited by LUDWIG BRAND AND MICHAEL L. JOHNSON

VOLUME 279. Vitamins and Coenzymes (Part I)
Edited by DONALD B. MCCORMICK, JOHN W. SUTTIE, AND CONRAD WAGNER

VOLUME 280. Vitamins and Coenzymes (Part J)
Edited by DONALD B. MCCORMICK, JOHN W. SUTTIE, AND CONRAD WAGNER

VOLUME 281. Vitamins and Coenzymes (Part K)
Edited by DONALD B. MCCORMICK, JOHN W. SUTTIE, AND CONRAD WAGNER

VOLUME 282. Vitamins and Coenzymes (Part L)
Edited by DONALD B. MCCORMICK, JOHN W. SUTTIE, AND CONRAD WAGNER

VOLUME 283. Cell Cycle Control
Edited by WILLIAM G. DUNPHY

VOLUME 284. Lipases (Part A: Biotechnology)
Edited by BYRON RUBIN AND EDWARD A. DENNIS

VOLUME 285. Cumulative Subject Index Volumes 263, 264, 266–284, 286–289

VOLUME 286. Lipases (Part B: Enzyme Characterization and Utilization)
Edited by BYRON RUBIN AND EDWARD A. DENNIS

VOLUME 287. Chemokines
Edited by RICHARD HORUK

VOLUME 288. Chemokine Receptors
Edited by RICHARD HORUK

VOLUME 289. Solid Phase Peptide Synthesis
Edited by GREGG B. FIELDS

VOLUME 290. Molecular Chaperones
Edited by GEORGE H. LORIMER AND THOMAS BALDWIN

VOLUME 291. Caged Compounds
Edited by GERARD MARRIOTT

VOLUME 292. ABC Transporters: Biochemical, Cellular, and Molecular Aspects
Edited by SURESH V. AMBUDKAR AND MICHAEL M. GOTTESMAN

VOLUME 293. Ion Channels (Part B)
Edited by P. MICHAEL CONN

VOLUME 294. Ion Channels (Part C)
Edited by P. MICHAEL CONN

VOLUME 295. Energetics of Biological Macromolecules (Part B)
Edited by GARY K. ACKERS AND MICHAEL L. JOHNSON

VOLUME 296. Neurotransmitter Transporters
Edited by SUSAN G. AMARA

VOLUME 297. Photosynthesis: Molecular Biology of Energy Capture
Edited by LEE MCINTOSH

VOLUME 298. Molecular Motors and the Cytoskeleton (Part B)
Edited by RICHARD B. VALLEE

VOLUME 299. Oxidants and Antioxidants (Part A)
Edited by LESTER PACKER

VOLUME 300. Oxidants and Antioxidants (Part B)
Edited by LESTER PACKER

VOLUME 301. Nitric Oxide: Biological and Antioxidant Activities (Part C)
Edited by LESTER PACKER

VOLUME 302. Green Fluorescent Protein
Edited by P. MICHAEL CONN

VOLUME 303. cDNA Preparation and Display
Edited by SHERMAN M. WEISSMAN

VOLUME 304. Chromatin
Edited by PAUL M. WASSARMAN AND ALAN P. WOLFFE

VOLUME 305. Bioluminescence and Chemiluminescence (Part C)
Edited by THOMAS O. BALDWIN AND MIRIAM M. ZIEGLER

VOLUME 306. Expression of Recombinant Genes in Eukaryotic Systems
Edited by JOSEPH C. GLORIOSO AND MARTIN C. SCHMIDT

VOLUME 307. Confocal Microscopy
Edited by P. MICHAEL CONN

VOLUME 308. Enzyme Kinetics and Mechanism (Part E: Energetics of Enzyme Catalysis)
Edited by DANIEL L. PURICH AND VERN L. SCHRAMM

VOLUME 309. Amyloid, Prions, and Other Protein Aggregates
Edited by RONALD WETZEL

VOLUME 310. Biofilms
Edited by RON J. DOYLE

VOLUME 311. Sphingolipid Metabolism and Cell Signaling (Part A)
Edited by ALFRED H. MERRILL, JR., AND YUSUF A. HANNUN

VOLUME 312. Sphingolipid Metabolism and Cell Signaling (Part B)
Edited by ALFRED H. MERRILL, JR., AND YUSUF A. HANNUN

VOLUME 313. Antisense Technology (Part A: General Methods, Methods of Delivery, and RNA Studies)
Edited by M. IAN PHILLIPS

VOLUME 314. Antisense Technology (Part B: Applications)
Edited by M. IAN PHILLIPS

VOLUME 315. Vertebrate Phototransduction and the Visual Cycle (Part A)
Edited by KRZYSZTOF PALCZEWSKI

VOLUME 316. Vertebrate Phototransduction and the Visual Cycle (Part B)
Edited by KRZYSZTOF PALCZEWSKI

VOLUME 317. RNA–Ligand Interactions (Part A: Structural Biology Methods)
Edited by DANIEL W. CELANDER AND JOHN N. ABELSON

VOLUME 318. RNA–Ligand Interactions (Part B: Molecular Biology Methods)
Edited by DANIEL W. CELANDER AND JOHN N. ABELSON

VOLUME 319. Singlet Oxygen, UV-A, and Ozone
Edited by LESTER PACKER AND HELMUT SIES

VOLUME 320. Cumulative Subject Index Volumes 290–319

VOLUME 321. Numerical Computer Methods (Part C)
Edited by MICHAEL L. JOHNSON AND LUDWIG BRAND

VOLUME 322. Apoptosis
Edited by JOHN C. REED

VOLUME 323. Energetics of Biological Macromolecules (Part C)
Edited by MICHAEL L. JOHNSON AND GARY K. ACKERS

VOLUME 324. Branched-Chain Amino Acids (Part B)
Edited by ROBERT A. HARRIS AND JOHN R. SOKATCH

VOLUME 325. Regulators and Effectors of Small GTPases (Part D: Rho Family)
Edited by W. E. BALCH, CHANNING J. DER, AND ALAN HALL

VOLUME 326. Applications of Chimeric Genes and Hybrid Proteins (Part A: Gene Expression and Protein Purification)
Edited by JEREMY THORNER, SCOTT D. EMR, AND JOHN N. ABELSON

VOLUME 327. Applications of Chimeric Genes and Hybrid Proteins (Part B: Cell Biology and Physiology)
Edited by JEREMY THORNER, SCOTT D. EMR, AND JOHN N. ABELSON

VOLUME 328. Applications of Chimeric Genes and Hybrid Proteins (Part C: Protein–Protein Interactions and Genomics)
Edited by JEREMY THORNER, SCOTT D. EMR, AND JOHN N. ABELSON

VOLUME 329. Regulators and Effectors of Small GTPases (Part E: GTPases Involved in Vesicular Traffic)
Edited by W. E. BALCH, CHANNING J. DER, AND ALAN HALL

VOLUME 330. Hyperthermophilic Enzymes (Part A)
Edited by MICHAEL W. W. ADAMS AND ROBERT M. KELLY

VOLUME 331. Hyperthermophilic Enzymes (Part B)
Edited by MICHAEL W. W. ADAMS AND ROBERT M. KELLY

VOLUME 332. Regulators and Effectors of Small GTPases (Part F: Ras Family I)
Edited by W. E. BALCH, CHANNING J. DER, AND ALAN HALL

VOLUME 333. Regulators and Effectors of Small GTPases (Part G: Ras Family II)
Edited by W. E. BALCH, CHANNING J. DER, AND ALAN HALL

VOLUME 334. Hyperthermophilic Enzymes (Part C)
Edited by MICHAEL W. W. ADAMS AND ROBERT M. KELLY

VOLUME 335. Flavonoids and Other Polyphenols
Edited by LESTER PACKER

VOLUME 336. Microbial Growth in Biofilms (Part A: Developmental and Molecular Biological Aspects)
Edited by RON J. DOYLE

VOLUME 337. Microbial Growth in Biofilms (Part B: Special Environments and Physicochemical Aspects)
Edited by RON J. DOYLE

VOLUME 338. Nuclear Magnetic Resonance of Biological Macromolecules (Part A)
Edited by THOMAS L. JAMES, VOLKER DÖTSCH, AND ULI SCHMITZ

VOLUME 339. Nuclear Magnetic Resonance of Biological Macromolecules (Part B)
Edited by THOMAS L. JAMES, VOLKER DÖTSCH, AND ULI SCHMITZ

VOLUME 340. Drug–Nucleic Acid Interactions
Edited by JONATHAN B. CHAIRES AND MICHAEL J. WARING

VOLUME 341. Ribonucleases (Part A)
Edited by ALLEN W. NICHOLSON

VOLUME 342. Ribonucleases (Part B)
Edited by ALLEN W. NICHOLSON

VOLUME 343. G Protein Pathways (Part A: Receptors)
Edited by RAVI IYENGAR AND JOHN D. HILDEBRANDT

VOLUME 344. G Protein Pathways (Part B: G Proteins and Their Regulators)
Edited by RAVI IYENGAR AND JOHN D. HILDEBRANDT

VOLUME 345. G Protein Pathways (Part C: Effector Mechanisms)
Edited by RAVI IYENGAR AND JOHN D. HILDEBRANDT

VOLUME 346. Gene Therapy Methods
Edited by M. IAN PHILLIPS

VOLUME 347. Protein Sensors and Reactive Oxygen Species (Part A: Selenoproteins and Thioredoxin)
Edited by HELMUT SIES AND LESTER PACKER

VOLUME 348. Protein Sensors and Reactive Oxygen Species (Part B: Thiol Enzymes and Proteins)
Edited by HELMUT SIES AND LESTER PACKER

VOLUME 349. Superoxide Dismutase
Edited by LESTER PACKER

VOLUME 350. Guide to Yeast Genetics and Molecular and Cell Biology (Part B)
Edited by CHRISTINE GUTHRIE AND GERALD R. FINK

VOLUME 351. Guide to Yeast Genetics and Molecular and Cell Biology (Part C)
Edited by CHRISTINE GUTHRIE AND GERALD R. FINK

VOLUME 352. Redox Cell Biology and Genetics (Part A)
Edited by CHANDAN K. SEN AND LESTER PACKER

VOLUME 353. Redox Cell Biology and Genetics (Part B)
Edited by CHANDAN K. SEN AND LESTER PACKER

VOLUME 354. Enzyme Kinetics and Mechanisms (Part F: Detection and Characterization of Enzyme Reaction Intermediates)
Edited by DANIEL L. PURICH

VOLUME 355. Cumulative Subject Index Volumes 321–354

VOLUME 356. Laser Capture Microscopy and Microdissection
Edited by P. MICHAEL CONN

VOLUME 357. Cytochrome P450, Part C
Edited by ERIC F. JOHNSON AND MICHAEL R. WATERMAN

VOLUME 358. Bacterial Pathogenesis (Part C: Identification, Regulation, and Function of Virulence Factors)
Edited by VIRGINIA L. CLARK AND PATRIK M. BAVOIL

VOLUME 359. Nitric Oxide (Part D)
Edited by ENRIQUE CADENAS AND LESTER PACKER

VOLUME 360. Biophotonics (Part A)
Edited by GERARD MARRIOTT AND IAN PARKER

VOLUME 361. Biophotonics (Part B)
Edited by GERARD MARRIOTT AND IAN PARKER

VOLUME 362. Recognition of Carbohydrates in Biological Systems (Part A)
Edited by YUAN C. LEE AND REIKO T. LEE

VOLUME 363. Recognition of Carbohydrates in Biological Systems (Part B)
Edited by YUAN C. LEE AND REIKO T. LEE

VOLUME 364. Nuclear Receptors
Edited by DAVID W. RUSSELL AND DAVID J. MANGELSDORF

VOLUME 365. Differentiation of Embryonic Stem Cells
Edited by PAUL M. WASSAUMAN AND GORDON M. KELLER

VOLUME 366. Protein Phosphatases
Edited by SUSANNE KLUMPP AND JOSEF KRIEGLSTEIN

VOLUME 367. Liposomes (Part A)
Edited by NEJAT DÜZGÜNEŞ

VOLUME 368. Macromolecular Crystallography (Part C)
Edited by CHARLES W. CARTER, JR., AND ROBERT M. SWEET

VOLUME 369. Combinational Chemistry (Part B)
Edited by GUILLERMO A. MORALES AND BARRY A. BUNIN

VOLUME 370. RNA Polymerases and Associated Factors (Part C)
Edited by SANKAR L. ADHYA AND SUSAN GARGES

VOLUME 371. RNA Polymerases and Associated Factors (Part D)
Edited by SANKAR L. ADHYA AND SUSAN GARGES

VOLUME 372. Liposomes (Part B)
Edited by NEJAT DÜZGÜNEŞ

VOLUME 373. Liposomes (Part C)
Edited by NEJAT DÜZGÜNEŞ

VOLUME 374. Macromolecular Crystallography (Part D)
Edited by CHARLES W. CARTER, JR., AND ROBERT W. SWEET

VOLUME 375. Chromatin and Chromatin Remodeling Enzymes (Part A)
Edited by C. DAVID ALLIS AND CARL WU

VOLUME 376. Chromatin and Chromatin Remodeling Enzymes (Part B)
Edited by C. DAVID ALLIS AND CARL WU

VOLUME 377. Chromatin and Chromatin Remodeling Enzymes (Part C)
Edited by C. DAVID ALLIS AND CARL WU

VOLUME 378. Quinones and Quinone Enzymes (Part A)
Edited by HELMUT SIES AND LESTER PACKER

VOLUME 379. Energetics of Biological Macromolecules (Part D)
Edited by JO M. HOLT, MICHAEL L. JOHNSON, AND GARY K. ACKERS

VOLUME 380. Energetics of Biological Macromolecules (Part E)
Edited by JO M. HOLT, MICHAEL L. JOHNSON, AND GARY K. ACKERS

VOLUME 381. Oxygen Sensing
Edited by CHANDAN K. SEN AND GREGG L. SEMENZA

VOLUME 382. Quinones and Quinone Enzymes (Part B)
Edited by HELMUT SIES AND LESTER PACKER

VOLUME 383. Numerical Computer Methods (Part D)
Edited by LUDWIG BRAND AND MICHAEL L. JOHNSON

VOLUME 384. Numerical Computer Methods (Part E)
Edited by LUDWIG BRAND AND MICHAEL L. JOHNSON

VOLUME 385. Imaging in Biological Research (Part A)
Edited by P. MICHAEL CONN

VOLUME 386. Imaging in Biological Research (Part B)
Edited by P. MICHAEL CONN

VOLUME 387. Liposomes (Part D)
Edited by NEJAT DÜZGÜNEŞ

VOLUME 388. Protein Engineering
Edited by DAN E. ROBERTSON AND JOSEPH P. NOEL

VOLUME 389. Regulators of G-Protein Signaling (Part A)
Edited by DAVID P. SIDEROVSKI

VOLUME 390. Regulators of G-Protein Signaling (Part B)
Edited by DAVID P. SIDEROVSKI

VOLUME 391. Liposomes (Part E)
Edited by NEJAT DÜZGÜNEŞ

VOLUME 392. RNA Interference
Edited by ENGELKE ROSSI

VOLUME 393. Circadian Rhythms
Edited by MICHAEL W. YOUNG

VOLUME 394. Nuclear Magnetic Resonance of Biological Macromolecules (Part C)
Edited by THOMAS L. JAMES

VOLUME 395. Producing the Biochemical Data (Part B)
Edited by ELIZABETH A. ZIMMER AND ERIC H. ROALSON

VOLUME 396. Nitric Oxide (Part E)
Edited by LESTER PACKER AND ENRIQUE CADENAS

VOLUME 397. Environmental Microbiology
Edited by JARED R. LEADBETTER

VOLUME 398. Ubiquitin and Protein Degradation (Part A)
Edited by RAYMOND J. DESHAIES

VOLUME 399. Ubiquitin and Protein Degradation (Part B)
Edited by RAYMOND J. DESHAIES

VOLUME 400. Phase II Conjugation Enzymes and Transport Systems
Edited by HELMUT SIES AND LESTER PACKER

VOLUME 401. Glutathione Transferases and Gamma Glutamyl Transpeptidases
Edited by HELMUT SIES AND LESTER PACKER

VOLUME 402. Biological Mass Spectrometry
Edited by A. L. BURLINGAME

VOLUME 403. GTPases Regulating Membrane Targeting and Fusion
Edited by WILLIAM E. BALCH, CHANNING J. DER, AND ALAN HALL

VOLUME 404. GTPases Regulating Membrane Dynamics
Edited by WILLIAM E. BALCH, CHANNING J. DER, AND ALAN HALL

VOLUME 405. Mass Spectrometry: Modified Proteins and Glycoconjugates
Edited by A. L. BURLINGAME

VOLUME 406. Regulators and Effectors of Small GTPases: Rho Family
Edited by WILLIAM E. BALCH, CHANNING J. DER, AND ALAN HALL

VOLUME 407. Regulators and Effectors of Small GTPases: Ras Family
Edited by WILLIAM E. BALCH, CHANNING J. DER, AND ALAN HALL

VOLUME 408. DNA Repair (Part A)
Edited by JUDITH L. CAMPBELL AND PAUL MODRICH

VOLUME 409. DNA Repair (Part B)
Edited by JUDITH L. CAMPBELL AND PAUL MODRICH

VOLUME 410. DNA Microarrays (Part A: Array Platforms and Web-Bench Protocols)
Edited by ALAN KIMMEL AND BRIAN OLIVER

VOLUME 411. DNA Microarrays (Part B: Databases and Statistics)
Edited by ALAN KIMMEL AND BRIAN OLIVER

VOLUME 412. Amyloid, Prions, and Other Protein Aggregates (Part B)
Edited by INDU KHETERPAL AND RONALD WETZEL

VOLUME 413. Amyloid, Prions, and Other Protein Aggregates (Part C)
Edited by INDU KHETERPAL AND RONALD WETZEL

VOLUME 414. Measuring Biological Responses with Automated Microscopy
Edited by JAMES INGLESE

VOLUME 415. Glycobiology
Edited by MINORU FUKUDA

VOLUME 416. Glycomics
Edited by MINORU FUKUDA

VOLUME 417. Functional Glycomics
Edited by MINORU FUKUDA

VOLUME 418. Embryonic Stem Cells
Edited by IRINA KLIMANSKAYA AND ROBERT LANZA

Volume 419. Adult Stem Cells
Edited by Irina Klimanskaya and Robert Lanza

Volume 420. Stem Cell Tools and Other Experimental Protocols
Edited by Irina Klimanskaya and Robert Lanza

Volume 421. Advanced Bacterial Genetics: Use of Transposons and Phage for Genomic Engineering
Edited by Kelly T. Hughes

Volume 422. Two-Component Signaling Systems, Part A
Edited by Melvin I. Simon, Brian R. Crane, and Alexandrine Crane

Volume 423. Two-Component Signaling Systems, Part B
Edited by Melvin I. Simon, Brian R. Crane, and Alexandrine Crane

Volume 424. RNA Editing
Edited by Jonatha M. Gott

Volume 425. RNA Modification
Edited by Jonatha M. Gott

Volume 426. Integrins
Edited by David Cheresh

Volume 427. MicroRNA Methods
Edited by John J. Rossi

Volume 428. Osmosensing and Osmosignaling
Edited by Helmut Sies and Dieter Haussinger

Volume 429. Translation Initiation: Extract Systems and Molecular Genetics
Edited by Jon Lorsch

Volume 430. Translation Initiation: Reconstituted Systems and Biophysical Methods
Edited by Jon Lorsch

Volume 431. Translation Initiation: Cell Biology, High-Throughput and Chemical-Based Approaches
Edited by Jon Lorsch

Volume 432. Lipidomics and Bioactive Lipids: Mass-Spectrometry–Based Lipid Analysis
Edited by H. Alex Brown

Volume 433. Lipidomics and Bioactive Lipids: Specialized Analytical Methods and Lipids in Disease
Edited by H. Alex Brown

Volume 434. Lipidomics and Bioactive Lipids: Lipids and Cell Signaling
Edited by H. Alex Brown

Volume 435. Oxygen Biology and Hypoxia
Edited by Helmut Sies and Bernhard Brüne

VOLUME 436. Globins and Other Nitric Oxide-Reactive Protiens (Part A)
Edited by ROBERT K. POOLE

VOLUME 437. Globins and Other Nitric Oxide-Reactive Protiens (Part B)
Edited by ROBERT K. POOLE

VOLUME 438. Small GTPases in Disease (Part A)
Edited by WILLIAM E. BALCH, CHANNING J. DER, AND ALAN HALL

VOLUME 439. Small GTPases in Disease (Part B)
Edited by WILLIAM E. BALCH, CHANNING J. DER, AND ALAN HALL

VOLUME 440. Nitric Oxide, Part F Oxidative and Nitrosative Stress in Redox Regulation of Cell Signaling
Edited by ENRIQUE CADENAS AND LESTER PACKER

VOLUME 441. Nitric Oxide, Part G Oxidative and Nitrosative Stress in Redox Regulation of Cell Signaling
Edited by ENRIQUE CADENAS AND LESTER PACKER

VOLUME 442. Programmed Cell Death, General Principles for Studying Cell Death (Part A)
Edited by ROYA KHOSRAVI-FAR, ZAHRA ZAKERI, RICHARD A. LOCKSHIN, AND MAURO PIACENTINI

VOLUME 443. Angiogenesis: *In Vitro* Systems
Edited by DAVID A. CHERESH

VOLUME 444. Angiogenesis: *In Vivo* Systems (Part A)
Edited by DAVID A. CHERESH

VOLUME 445. Angiogenesis: *In Vivo* Systems (Part B)
Edited by DAVID A. CHERESH

VOLUME 446. Programmed Cell Death, The Biology and Therapeutic Implications of Cell Death (Part B)
Edited by ROYA KHOSRAVI-FAR, ZAHRA ZAKERI, RICHARD A. LOCKSHIN, AND MAURO PIACENTINI

VOLUME 447. RNA Turnover in Prokaryotes, Archae and Organelles
Edited by LYNNE E. MAQUAT AND CECILIA M. ARRAIANO

VOLUME 448. RNA Turnover in Eukaryotes: Nucleases, Pathways and Anaylsis of mRNA Decay
Edited by LYNNE E. MAQUAT AND MEGERDITCH KILEDJIAN

VOLUME 449. RNA Turnover in Eukaryotes: Analysis of Specialized and Quality Control RNA Decay Pathways
Edited by LYNNE E. MAQUAT AND MEGERDITCH KILEDJIAN

VOLUME 450. Fluorescence Spectroscopy
Edited by LUDWING BRAND AND MICHAEL JOHNSON

VOLUME 451. Autophagy: Lower Eukaryotes and Non-Mammalian Systems (Part A)
Edited by DANIEL J. KLIONSKY

VOLUME 452. Autophagy in Mammalian Systems (Part B)
Edited by DANIEL J. KLIONSKY

CHAPTER ONE

Monitoring Autophagy in Mammalian Cultured Cells through the Dynamics of LC3

Shunsuke Kimura,* Naonobu Fujita,* Takeshi Noda,* *and* Tamotsu Yoshimori*

Contents

1. Introduction	2
2. Estimation of Autophagy Induction by LC3 Puncta Formation	4
2.1. Immunofluorescence	4
2.2. Examination of autophagy induction by observing GFP-LC3	6
2.3. Considerations for immunofluorescence and GFP-LC3	6
3. The tfLC3 Assay	7
3.1. The procedure	7
3.2. Interpretation of the results	7
4. Determination of LC3 Turnover by Western Blotting	9
4.1. The procedure	9
4.2. Interpretation of the results	10
5. Concluding Remarks	10
References	11

Abstract

In this chapter, we introduce several methods that rely on the analysis of LC3, a versatile marker protein of autophagic structures in mammalian cultured cells. The appearance of LC3-positive puncta is indicative of the induction of autophagy, and it is observed either by immunofluorescence or by GFP-based microscopy. The maturation process by which autophagosomes are converted into autolysosomes can be monitored by the GFP and RFP tandemly tagged LC3 (tfLC3) method. Lysosomal turnover of LC3 is a good index of the proceeding of autophagy and can be assessed by Western blotting. These methods will provide a relatively easy assessment of autophagy, and the details of the procedure will be described along with possible pitfalls.

* Department of Cellular Regulation, Research Institute for Microbial Diseases, Osaka University, Osaka, Japan

1. INTRODUCTION

In the history of autophagy studies, electron microscopy observations of the appearance of double membrane–enclosed structures, autophagosomes, in the cytoplasm had been mostly the sole morphological assessment of the autophagic process in mammalian cells for a long period. Recently, identification of microtubule-associated protein 1 light chain 3 (LC3) as the homolog of yeast Atg8, an autophagy-related protein, opened up a new era in the study of mammalian autophagy because of its usefulness as an autophagy marker (Kabeya *et al.*, 2000). In this chapter, we introduce several methods to monitor autophagy in mammalian cultured cells using LC3 as a marker.

LC3 is specifically localized to autophagic structures, including the autophagosome and its precursor structure, the isolation membrane (also called the phagophore), and its derivative, the autolysosome; although there is also a cytosolic fraction. Under conditions in which autophagy is not induced, such as nutrient-rich conditions, LC3-positive puncta scarcely exist. Therefore, the appearance of multiple LC3-positive puncta suggests the induction of autophagy (see also subsequent sections). One of the best methods to monitor LC3-positive puncta formation is immunofluorescence using an anti-LC3 antibody. The standard staining procedure is described in section 2.1. Another convenient and widely used way to assess autophagy induction is through green fluorescent protein (GFP)–fused LC3, which is described in section 2.2. General considerations to these approaches are described in section 2.3.

There is, however, a pitfall in the interpretation of the increase in the number of LC3 puncta. The autophagosome is a dynamic and transient structure, and it eventually matures into an autolysosome. LC3 is associated with the outer face of the outer membrane and inner face of the inner membrane of the double-membrane autophagosome. When the autophagosome matures into an autolysosome, the outer membrane LC3 is liberated into the cytosol, whereas the inner membrane LC3 is trapped within the autolysosome and finally degraded by lysosomal proteases. If the maturation process does not normally proceed for some specific reason, the degradation of the inner membrane–localized LC3 does not take place, and the number of LC3-positive puncta also increases as a result of the accumulation of autophagosomes, which form during cell propagation. Therefore once the increase of LC3 puncta is observed, the maturation process must be further examined to determine whether or not autophagy is induced, and whether or not it goes to completion. Another point needed of critical consideration is that if the experimental condition yields incomplete autophagosomes or intermediate structures, it is still possible that some LC3-positive puncta will be generated (Hara et al., 2008). The tfLC3 (tandem fluorescent tagged LC3) method introduced in section 3 is one of the methods able to discriminate between these possibilities (Kimura, 2007). This assay is based on the nature of the fluorescent signal of GFP, which is highly susceptible to the lysosomal acidic/proteolytic environment, whereas that of red fluorescent protein (RFP) is not. LC3 tandemly tagged with GFP and RFP shows both GFP and RFP signals before fusion with the lysosome, but once the maturation to an autolysosome occurs, it exhibits only the RFP signal. Therefore, the appearance of the puncta exhibiting only an RFP signal indicates the normal autophagic maturation process.

Western blotting of LC3 is also a very informative approach to assess these problems (Kabeya et al., 2000). Atg8/LC3 is a ubiquitin-like protein, and it undergoes posttranslational modification (Ichimura et al., 2000; Kirisako et al., 2000). The carboxyl-terminal flanking region of nascent LC3 (proLC3) is cleaved off by the Atg4 protease to become LC3-I, which then has an exposed carboxyl-terminal glycine. At the glycine residue, LC3-I is modified with phosphatidylethanolamine, and this is referred to as the LC3-II form; this step involves a ubiquitination-like reaction mediated by Atg7 (E1-like activating enzyme), Atg3 (E2-like conjugating enzyme), and the Atg16L complex (E3-like ligase enzyme). These two forms of LC3 can be discriminated by their difference in mobility during SDS-PAGE. The LC3-I form is soluble and exists in the cytosol, while LC3-II is bound to the membrane. There is a general tendency that the amount of LC3-II increases when autophagy is induced, reflecting that the lipidation reaction is enhanced. Thus, the amount of LC3-II provides a good index of autophagy induction, but the rate of the increase depends on the cell type.

As mentioned previously, the fate of individual LC3-II molecules depends on their localization. LC3-II that is located on the outer face of the autophagosome is delipidated through a second cleavage by the Atg4 protease and is converted back into the LC3-I form in the cytosol; this population of LC3 is presumably reused for another round of conjugation to phosphatidylethanolamine. In contrast, the LC3-II that localizes to the inner autophagosome is degraded by lysosomal proteases after the fusion of the autophagosome with the lysosome as part of the normal maturation process. Therefore, treatment of cells that are undergoing autophagy with the protease inhibitors E64d and pepstatin A blocks the decrease of the LC3-II form, even if the maturation process otherwise proceeds normally (Tanida, 2005). This type of analysis can be used to determine whether autophagic maturation occurs normally (i.e., whether autophagy goes to completion). That is, if there is no increase in the amount of LC3-II when cells are treated with lysosomal protease inhibitors, autophagic maturation is not occurring normally (Mizushima, 2007; Tanida, 2008).

2. ESTIMATION OF AUTOPHAGY INDUCTION BY LC3 PUNCTA FORMATION

2.1. Immunofluorescence

1. Sterilize the cover slips (Matsunami glass; 12-mm round, No. 1S thickness) using a dry-heat sterilizer. Then coat the cover slips with collagen by immersing in 0.1 mg/ml collagen (Nitta gelatin, Cell matrix Type I-C) for 2 h at room temperature and subsequently washing with phosphate-buffered saline (PBS) (10 mM Na$_2$HPO$_4$, 1.8 mM KH$_2$PO$_4$, 137 mM NaCl, 2.7 mM KCl, pH 7.4).
2. The day before the experiment, plate the appropriate number of cells in 6-well dishes containing Dulbecco's Modified Eagle Medium (DMEM) (Sigma, D6546) containing 10% fetal bovine serum (FBS) heat-inactivated by incubation at 56 °C for 45 min on the preceding cover slips placed on the bottom of a culture plate to become 60%–80% confluent at the time of the experiment. Incubate the cells at 37 °C in a CO$_2$ incubator for 18–24 h. We usually use adherent cell lines like mouse embryonic fibroblast (MEF), NIH3T3, or A549 cells for this experiment.
3. Discard the medium and wash the cells once with PBS prewarmed at 37 °C. For starvation treatment, culture the cells in Earle's Balanced Salt Solutions (EBSS) (Sigma, E2888) at 37 °C for 1–4 h depending on the cell line used (e.g., 1 h for MEF and NIH3T3, and 4 h for A549 cells). For nutrient-rich conditions, culture the cells in DMEM with 10% heat-inactivated FBS for 1–4 h. For the negative control of LC3 puncta formation, add wortmannin (Wako, 532-81051) at the start of starvation to a final concentration of 100 nM from a 100 μM stock solution in dimethylsulfoxide (DMSO). The stock can be stored at −20 °C for 3 months.
4. After the incubations in rich or starvation conditions, rinse the cells with PBS at room temperature. Fix the cells in 3% paraformaldehyde in PBS, pH 7.4, for 15 min at room temperature. Aliquots of 3% paraformaldehyde solution can be stored at −30 °C. Do not use 3% paraformaldehyde that has been stored at 4 °C for over 1 month. Wash the cells twice with PBS.
5. To quench the fluorescence signal from free aldehyde groups in paraformaldehyde, incubate the cells with 50 mM NH$_4$Cl in PBS for 10 min at room temperature. NH$_4$Cl should be dissolved just prior to use. Wash the cells twice with 0.1% gelatin (Wako, 077-03155) in PBS.
6. Permeabilize the cells by incubation with 50 μg/mL digitonin (Wako, 043-21371) in 0.1% gelatin-PBS, pH 7.2, for 10 min. Wash the samples

twice with 0.1% gelatin-PBS. Block the cells by incubation with 0.1% gelatin-PBS for 30 min at room temperature.
7. Incubate the cells on cover slips with 30 µl of polyclonal anti-LC3 antibody (Medical and Biological Laboratories, PM036) diluted 1:1000 in 0.1% gelatin in PBS for 1 h at room temperature. Wash the samples 3 times with 0.1% gelatin-PBS. Anti-LC3 antibody is now commercially available from many companies, and to our knowledge, rabbit polyclonal antibody PM036 (MBL) yields a good signal in human, mouse, and rat cell lines.
8. Incubate the cells on cover slips with 30 µl of Alexa488-conjugated anti-rabbit antibody (Invitrogen, A-11008), or another equivalent, diluted 1:1000 in 0.1% gelatin-PBS for 40 min at room temperature. Wash the samples three times with 0.1% gelatin-PBS.
9. Mount on glass slides with 5 µl of slow-fade gold mounting medium (Molecular Probes, S36936). Observe the LC3 signals either by epifluorescence microscopy or by confocal laser scanning fluorescence microscopy equipped with x60 and/or x100 magnification.
10. Count the number of LC3 puncta per cell (Fig. 1.1). Using computer software such as ImageJ and WatershedCounting3D will aid in yielding reproducible counting (Gniadek and Warren, 2007). Note that it is generally better to count the number of puncta per cell rather than the number of cells with puncta. If doing the latter, it is important to set some threshold for the number of puncta that constitutes an autophagically active cell. Along these lines, the wortmannin-treated cells provide a control for the background level (i.e., nonautophagic) of puncta formation.

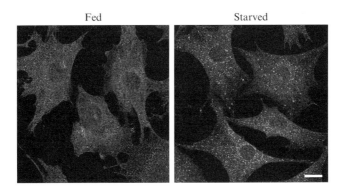

Figure 1.1 Immunocytochemistry of endogenous LC3. MEF cells were cultured in DMEM containing 10% FCS (rich medium, left panel) or in EBSS (starvation medium, right panel) for 1 h. Cells were fixed and the subjected to immunocytochemical analysis using anti-LC3 antibody according to the procedures described in the text.

2.2. Examination of autophagy induction by observing GFP-LC3

LC3 fused with GFP labels the autophagic structures similar to endogenous LC3 (Kabeya et al., 2000). Exogenous expression of GFP-LC3 is therefore a useful alternative for observing autophagy induction. Using a cell line stably expressing GFP-LC3 is preferred (see next section). The following is a protocol for transient transfection of GFP-LC3.

1. The day before the transfection, plate the appropriate number of cells in 6-cm dishes containing DMEM with 10% heat-inactivated FBS to become 60%–80% confluent at the time of transfection. Incubate the cells at 37 °C in a CO_2 incubator for 18–24 h.
2. Transfect the cell with GFP-LC3 plasmid using a lipofection reagent such as lipofectAMINE 2000 (Invitrogen, 11668–019) according to the manufacturer's protocol for as short a time as possible, preferably no more than 4 h.
3. Replate the cells onto 6-well dishes containing DMEM with 10% heat-inactivated FBS on collagen-coated cover slips (as described previously) placed on the bottom of culture plates to become 50%–60% confluent at the time of the experiment.
4. Follow steps 3 and 4 in section 2.1.
5. Wash the cells with PBS 3 times.
6. Follow steps 9 and 10 in section 2.1.

2.3. Considerations for immunofluorescence and GFP-LC3

First of all, it should be noted that autophagy can be observed generally in most cells, and the difference in each cell is quite limited. This concept stands apart from other phenomena such as apoptosis. If you see the LC3 puncta only in a part of a cell population, for example, in less than half of the cells, you should be cautious about the integrity of the experimental system. For example, it is reported that highly abundant GFP-LC3 is prone to be contained in aggregates, which sometimes resemble an autophagosome signal and can lead to incorrect interpretation of the results (Kuma, 2007). Observing the samples with another filter set will sometimes discriminate the true signal because these aggregates are prone to emit a broad range of fluorescence. It is advisable to use a cell line that stably expresses a modest level of GFP-LC3. However, it is important to include proper negative controls to ensure that the signal is derived from the autophagosome, especially if you use transient transfection of the GFP-LC3 expression plasmid. Using a conjugation-incompetent mutant form of GFP-LC3 whose carboxyl terminus is changed from glycine to alanine will be adequate as a negative control (Tanida et al., 2008). Treatment of the cell culture with wortmannin, an inhibitor for autophagosome formation, is also desirable. One of the best controls is using an Atg5 or Atg7 knockout cell line, which is known to be defective in LC3 puncta formation (Komatsu *et al.*, 2005; Mizushima *et al.*, 2001). Successful knockdown of these genes is not generally very easy, however, because a very small amount of Atg5 is sufficient for the LC3 lipidation reaction (Hosokawa *et al.*, 2007). Another caution is that the lipofection-mediated transfection procedure itself is prone to induce autophagy. You may need to shorten the period during which the cells are exposed to the lipofection reagent as much as possible. Finally, when the cell culture becomes confluent, autophagy is induced without a specific stimulation; therefore, the cell culture should be maintained as subconfluent.

3. The tfLC3 Assay

If the cells are stably expressing tfLC3, they show red-only puncta even in nutrient-rich conditions in the absence of starvation-induced autophagy. This may be due to the basal autophagy that occurs constitutively during cell propagation. Therefore, tfLC3 should be expressed after the condition of interest has taken effect to test the appearance of the red-only signal. Accordingly, transient transfection of the tfLC3 plasmid is needed. If you need to transfect some other gene, it is important to do the transfection simultaneously with the tfLC3 plasmid.

3.1. The procedure

The procedure is the same as section 2.1 (Fig. 1.2).

3.2. Interpretation of the results

We use this assay for HeLa cells, and other cell types should be tested to determine whether this assay is compatible. The degree of colocalization of mRFP and GFP signals can be quantified using the colocalization finder plug-in of ImageJ software (http://rsb.info.nih.gov/ij/) (Figs. 1.2B and 1.2C).

If there is a significant population of red-only signal, then the autophagosomes are normally matured into autolysosomes (where the GFP will be relatively unstable). If most puncta exhibit both red and green signals, autophagy is impaired at some step. In the latter case, the first possibility is that abnormal autophagosomes or autophagosome intermediates are accumulated. A second possibility is that the fusion between autophagosomes and endosomes/lysosomes is impaired. These two possibilities can be

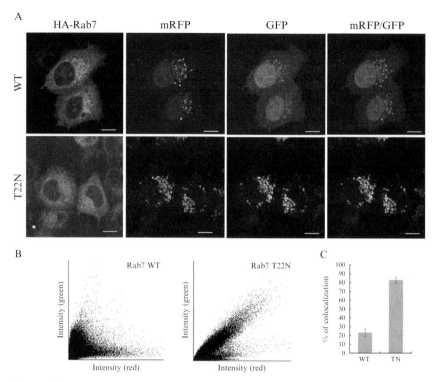

Figure 1.2 The tfLC3 assay. (A) HeLa cells were co-transfected with plasmids expressing mRFP-GFP-LC3 and either wild-type or the T22N mutant of HA-Rab7. Twenty-four h after transfection, the cells were subjected to starvation for 2 h, fixed, and subjected to immunocytochemistry using anti-HA antibody. Antimouse antibody conjugated with Alexa405 was used as the secondary antibody. Bar indicates 10 μm. (B) Each correlation plot is derived from a field of view shown in panel A. (C) Colocalization efficiency of mRFP with GFP signals of tfLC3 puncta was measured using ImageJ software, and shown as the percentage of the total number of mRFP puncta. The value indicates average and standard deviation from at least 5 images. (See Color Insert.)

distinguished further by other approaches such as electron microscopy (see the chapter by Ylä-Anttila *et al.*, in this volume). A third possibility is that the autophagosomes normally fuse with endosomes/lysosomes, but the nature of the lysosomes is altered, such as loss of acidification and/or lysosomal proteases, and the organelle is therefore unable to decrease the GFP-derived green signal. This possibility can be tested by staining the lysosome. For example, the lysosome can be labeled with the endocytosed Cascade Blue Dextran (MW 10,000) (Molecular Probes, D1976) added to the medium the day before the experiment (Kimura, 2007). This dye stains lysosomes because it is the final destination of endocytosis. Note, the acidotropic dye LysoTracker Red may be inappropriate to observe the lysosome under conditions in which lysosomal function may be altered.

4. DETERMINATION OF LC3 TURNOVER BY WESTERN BLOTTING

4.1. The procedure

1. We routinely use NIH3T3, HEK293, MEF, PC12 and A549 cells for this experiment. One day before the experiment, plate the appropriate number of cells in 6-cm dishes containing DMEM with 10% heat-inactivated FBS to become 60%–80% confluent at the time of the experiment. Incubate the cells at 37 °C in a CO_2 incubator for 18–24 h.
2. Discard the medium and wash the cells once with PBS prewarmed at 37 °C. For the nutrient-starvation treatment, add EBSS prewarmed at 37 °C and incubate for 1–4 h depending on the cell lines used. For the nutrient condition, culture the cells in DMEM with 10% heat-inactivated FBS for 1–4 h. For monitoring autophagy-dependent degradation of LC3-II that correlates with autophagic flux, add E64d (Peptide Institute, 4321-v) and pepstatin A (Peptide Institute; 4397-v) to final concentrations of 10 μg/mL from 10 mg/ml stocks in DMSO. These protease inhibitors should be prepared just prior to use, but in any event, do not use a stock solution that has been stored for over 2 weeks at −20 °C.
3. Discard the medium and rinse the cells once with ice-cold PBS. Place the plate on ice, and add 1 ml of ice-cold PBS. Scrape the cells using a cell scraper and collect the cell suspension into a microcentrifuge tube on ice. Collect the cells by centrifuging at $2000 \times g$ for 1 min at 4 °C. Discard the supernatant fraction.
4. Suspend the cells with 100 mL of ice-cold lysis buffer (20 mM Tris-HCl, pH 7.6, 150 mM NaCl, 2% Triton X-100 with phenylmethanesulfonyl fluoride (PMSF; Wako, 1548–5) added to a final concentration of 2 mM from a 100 mM stock in 2-propanol and $1 \times$ Complete Protease Inhibitor Cocktail (Roche, 1697498) added just prior to use; note that PMSF is highly unstable in H_2O) by pipetting. Place the tube on ice for 5 min. Centrifuge the tube at $16,000 \times g$ for 10 min at 4 °C and collect the supernatant fraction to a new microcentrifuge tube.
5. Measure the protein concentration by the Bradford method or an equivalent protein assay system. Add ⅕ volume of 6x Laemmli sample buffer (10% SDS, 1 M DTT, 0.5 M Tris-HCl, pH 6.8, 0.006% bromophenol blue, 30% glycerol) and boil the tubes for 5 min.
6. Run the standard Laemmli SDS-PAGE using 13% minisize gels. We usually load samples equivalent to 10 μg of protein for 1 lane.
7. For electrotransfer, we use a wet-type transfer system (Mini-Trans Blot; Biorad) with the buffer system of 24 mM Tris, 190 mM glycine, 20% methanol, under the condition of 150 mA constant current for 1 h to

PVDF membrane (Amersham Hybond-P; RPN303F). Staining the membrane with 0.1% (w/v) Ponceau S (Sigma; P7170) in 5% acetic acid for 5 min with constant agitation is advisable to verify the electrotransfer efficiency.

8. Incubate the membrane with 1% (w/v) skim milk in TBS/T (25 mM Tris-HCl, pH 7.4, 137 mM NaCl, 2.7 mM KCl, 0.08% Tween 20) at room temperature for 1 h with constant agitation.
9. Incubate the membrane with polyclonal anti-LC3 antibody (MBL, PM036) diluted 1:2000 in 1% (w/v) skim milk in TBS/T with constant agitation. Monoclonal antibody 5F10 from Nanotools antikörpertechik GmbH & Co., KG-Teningen, Germany, is also good for Western blotting. Incubation may be carried out either at room temperature for 1 h or at 4 °C overnight. Wash the membrane 3 times with TBS/T for 10 min at room temperature.
10. Incubate the membrane with HRP-conjugated antirabbit IgG (Jackson Laboratories, 711-036-152) diluted 1:10,000 in 1% (w/v) skim milk in TBS/T at room temperature for 1 h. Wash the membrane 3 times with TBS/T for 10 min at room temperature.
11. Using an HRP-based Western blotting detection system such as ECL (GE Healthcare, RPN2106), develop the signal according to the manufacturer's instructions.
12. Using the standard X-ray film development procedure or an equivalent apparatus such as a CCD camera (e.g., the LAS-3000s, Fuji Film), detect the luminescent signals on the membrane according to the manufacturer's instructions.

4.2. Interpretation of the results

Some antibodies against LC3 react preferentially with the LC3-II form rather than the LC3-I form. There may be some difference in the steric structure of LC3 that is preserved even after Western blotting that accounts for these differences (Kabeya *et al.*, 2004). Therefore, the signal ratio between LC3-I and LC3-II provides little information. If there is some increase in the amount of LC3-II following protease inhibitor treatment, the occurrence of autophagic turnover can be concluded (Fig. 1.3).

5. Concluding Remarks

The assays introduced in this section are relatively easy and therefore have been widely accepted. However, as many aspects of the nature of LC3 have become gradually apparent, several specific pitfalls that may bring about incorrect interpretation of the results may be encountered, as

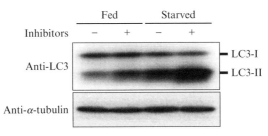

Figure 1.3 Western blotting of LC3 with protease inhibitor treatment. MEF cells were cultured in fed or starved conditions for 4 h with or without protease inhibitors (E64d and pepstatin). Ten μg lysates of MEF cells were resolved by SDS-PAGE and subjected to Western blotting with LC3 antibody, as described in text, and anti-tubulin antibody as a loading control.

described previously. Researchers should undertake several approaches in combination, including these assays, before finally concluding whether autophagy is affected under their experimental conditions.

REFERENCES

Gniadek, T. J., and Warren, G. (2007). WatershedCounting3D: A new method for segmenting and counting punctate structures from confocal image data. *Traffic* **8,** 339–346.

Hara, T., Takamura, A., Kishi, C., Iemura, S., Natsume, T., Guan, J. L., and Mizushima, N. (2008). FIP200, a ULK-interacting protein, is required for autophagosome formation in mammalian cells. *J. Cell Biol.* **181,** 497–510.

Hosokawa, N., Hara, Y., and Mizushima, N. (2007). Generation of cell lines with tetracycline-regulated autophagy and a role for autophagy in controlling cell size. *FEBS Lett.* **581,** 2623–2629.

Ichimura, Y., Kirisako, T., Takao, T., Satomi, Y., Shimonishi, Y., Ishihara, N., Mizushima, N., Tanida, I., Kominami, E., Ohsumi, M., Noda, T., and Ohsumi, Y. (2000). A ubiquitin-like system mediates protein lipidation. *Nature* **408,** 488–492.

Kabeya, Y., Mizushima, N., Ueno, T., Yamamoto, A., Kirisako, T., Noda, T., Kominami, E., Ohsumi, Y., and Yoshimori, T. (2000). LC3, a mammalian homologue of yeast Apg8p, is localized in autophagosome membranes after processing. *EMBO J.* **19,** 5720–5728.

Kabeya, Y., Mizushima, N., Yamamoto, A., Oshitani-Okamoto, S., Ohsumi, Y., and Yoshimori, T. (2004). LC3, GABARAP and GATE16 localize to autophagosomal membrane depending on form-II formation. *J. Cell Sci.* **117,** 2805–2812.

Kimura, S. N., and Yoshimori, T. T. (2007). Dissection of the autophagosome maturation process by a novel reporter protein, tandem fluorescent-tagged LC3. *Autophagy* **3,** 452–460.

Kirisako, T., Ichimura, Y., Okada, H., Kabeya, Y., Mizushima, N., Yoshimori, T., Ohsumi, M., Takao, T., Noda, T., and Ohsumi, Y. (2000). The reversible modification regulates the membrane-binding state of Apg8/Aut7 essential for autophagy and the cytoplasm to vacuole targeting pathway. *J. Cell Biol.* **151,** 263–276.

Komatsu, M., Waguri, S., Ueno, T., Iwata, J., Murata, S., Tanida, I., Ezaki, J., Mizushima, N., Ohsumi, Y., Uchiyama, Y., Kominami, E., Tanaka, K., and Chiba, T.

(2005). Impairment of starvation-induced and constitutive autophagy in Atg7-deficient mice. *J. Cell Biol.* **169,** 425–434.

Kuma, M., and Mizushima, N. (2007). LC3, an autophagosome marker, can be incorporated into protein aggregates independent of autophagy: Caution in the interpretation of LC3 localization. *Autophagy* **3,** 323–328.

Mizushima, N. (2004a). Methods for monitoring autophagy. *Int. J. Biochem. Cell Biol.* **36,** 2491–2502.

Mizushima, N., Yamamoto, A., Hatano, M., Kobayashi, Y., Kabeya, Y., Suzuki, K., Tokuhisa, T., Ohsumi, Y., and Yoshimori, T. (2001). Dissection of autophagosome formation using Apg5-deficient mouse embryonic stem cells. *J. Cell Biol.* **152,** 657–668.

Mizushima, N. Y., Matsui, A., Yoshimori, M., and Ohsumi, Y. (2004b). *In vivo* analysis of autophagy in response to nutrient starvation using transgenic mice expressing a fluorescent autophagosome marker. *Mol. Biol. Cell* **15,** 1101–1111.

Mizushima, N. Y. T. (2007). How to interpret LC3 immunoblotting. *Autophagy* **3,** 542–545.

Tanida, I., Yamaji, T., Ueno, T., Ishiura, S., Kominami, E., and Hanada, K. (2008). Consideration about negative controls for LC3 and expression vectors for four colored fluorescent protein-LC3 negative controls. *Autophagy* **4,** 131–134.

Tanida, I. Minematsu-Ikeguchi., Ueno, N., and Kominami, E. (2005). Lysosomal turnover, but not a cellular level, of endogenous LC3 is a marker for autophagy. *Autophagy* **1,** 84–91.

Tanida, I., and Kominami, E. (2008). LC3 and Autophagy. *Methods Mol. Biol.* **445,** 77–88.

CHAPTER TWO

METHODS FOR MONITORING AUTOPHAGY USING GFP-LC3 TRANSGENIC MICE

Noboru Mizushima*

Contents

1. Introduction	14
2. Genetic Features of GFP-LC3 Mice	15
3. Mouse Maintenance	16
4. Genotyping of GFP-LC3 Mice	17
4.1. GFP-LC3#53 mice	17
4.2. Homo/hemi determination by polymerase chain reaction	17
4.3. Detection of fluorescent signal	18
5. Sample Preparation	18
5.1. Tissue fixation	18
5.2. Cryosection preparation	19
6. Fluorescence Microscopy	19
6.1. GFP observation	19
6.2. Dot counting and quantification	20
7. Precautions	20
7.1. GFP-LC3 puncta may represent protein aggregates/inclusion bodies	20
7.2. Distinguishing GFP-LC3 dot signals from autofluorescent signals	20
7.3. Induction of autophagy versus blockage of degradation	21
References	21

Abstract

Several methods are now available for monitoring autophagy. Although biological methods are useful for cultured cells and homogenous tissues, these methods are not suitable for determining the autophagic activity of each cell type in heterogenous tissues. Furthermore, intracellular localization of autophagosomes often provides valuable information. Thus, morphological assays are still important in many studies. Although electron microscopy has been the gold standard, recent studies of the molecular mechanism of

---

* Department of Physiology and Cell Biology, Tokyo Medical and Dental University, Tokyo, Japan

autophagy have led to the development of several marker proteins for autophagosomes, the most widely used of which is LC3, a mammalian homolog of Atg8. These marker proteins allow identification of autophagic structures by fluorescence microscopy. This method has been applied to whole animals by generating green fluorescent protein (GFP)–LC3 transgenic mice. This chapter describes the background and practicality of, and possible precautions in the application of, this method using the GFP-LC3 transgenic mouse model.

1. Introduction

Since the discovery of autophagic vacuoles approximately 50 years ago, these structures have been mainly observed and characterized by electron microscopy. However, recent genetic studies performed in yeast have identified many autophagy-related (Atg) proteins (Klionsky et al., 2003; Suzuki and Ohsumi, 2007). More than half of these Atg proteins are required for autophagosome formation and have counterparts in mammals. Most of them are specifically localized to an intermediate structure called the isolation membrane or phagophore; these include Atg12-Atg5 (Mizushima et al., 2001), Atg16L (Itakura et al., 2008), ULK1/2 (Hara et al., 2008), Atg14 (our unpublished data), and FIP200 (Hara et al., 2008). In contrast, LC3 family proteins (MAP1-LC3, GATE-16, and GABARAP), which are mammalian counterparts of yeast Atg8, are present on both the isolation membrane and as complete autophagosomes, occurring in phosphatidylethanolamine (PE)–conjugated forms (Kabeya et al., 2000; Kabeya et al., 2004). LC3 appears to be degraded and/or dissociated after autophagosome fusion with lysosomes, leaving less LC3 on the autolysosome membrane. When green fluorescent protein (GFP)–fused LC3 (GFP-LC3) is expressed in cultured cells, punctate signals are observed simply by fluorescence microscopy, which represent isolation membranes and autophagosomes. If these structures are sufficiently large, they can be observed as ring-shaped structures. Another merit of this method is that we can monitor the kinetics of autophagosome generation, movement, and disappearance in living cells, which can hardly be performed by electron microscopy (Jahreiss et al., 2008; Kimura et al., 2007; Köchl et al., 2006).

The GFP-LC3 method is not only simple but also generally highly specific. Specificity for GFP-LC3 puncta was previously determined by several studies. PE conjugation of LC3 requires the Atg12-Atg5 conjugate (Mizushima et al., 2001). Accordingly, LC3 punctate structures are not generated in Atg5-deficient cells (Kuma et al., 2004; Kuma et al., 2007; Mizushima et al., 2001). PE conjugation also requires the C-terminal glycine residue in the LC3 family proteins. If this residue is replaced with alanine in LC3, GATE-16, or GABARAP, puncta formation is impaired, confirming that the puncta represent biologically relevant structures (Kabeya *et al.*, 2004; Tanida *et al.*, 2008). However, as will be discussed subsequently, several important precautions have been pointed out regarding this GFP-LC3 method and should be kept in mind whenever it is used.

This simple fluorescence method was previously applied to various organisms including yeast (Suzuki *et al.*, 2001), *Caenorhabditis elegans* (Melendez *et al.*, 2003), *Dictyostelium discoideum* (Otto *et al.*, 2003), *Drosophila melanogaster* (Rusten *et al.*, 2004; Scott *et al.*, 2004), *Arabidopsis thaliana* (Yoshimoto *et al.*, 2004), and mice (Mizushima *et al.*, 2004). This chapter focuses on the GFP-LC3 transgenic mouse and presents the application and pitfalls of the fluorescence method.

2. Genetic Features of GFP-LC3 Mice

A transgenic vector contains an enhanced GFP (EGFP)–LC3 cassette inserted between the CAG promoter (cytomegalovirus immediate-early (CMVie) enhancer and chicken β-actin promoter) (Niwa *et al.*, 1991) and the SV40 late polyadenylation signal. In this construct, EGFP is fused to the N terminus of rat LC3B (U05784) so as not to affect C-terminal PE conjugation. The 3.4-kbp CAG-EGFP-LC3-SV40 polyA fragment was microinjected into C57BL/6N Crj × BDF1 fertilized oocytes. The initial screen gave 8 transgenic lines. One of the transgenic lines, GFP-LC3#53, was selected because GFP-LC3 was ubiquitously expressed (Mizushima *et al.*, 2004).

As judged from immunoblot analysis, the level of GFP-LC3 expression is comparable to that of endogenous LC3 expression in the brain, whereas GFP-LC3 is overexpressed in other tissues. In particular, the levels of GFP-LC3 expression are much higher than those of endogenous LC3 in the heart, liver, pancreas, and skeletal muscle. However, a cell culture experiment showed that such levels of GFP-LC3 overexpression do not affect endogenous autophagy activity (Mizushima *et al.*, 2004).

As the GFP-LC3 fragment is randomly integrated into the mouse genome, there might be a concern that some important genes are disrupted by the transgene insertion. To address this concern, the insertion site was identified by the genomic walking technique (Kuma and Mizushima, 2008). The GFP-LC3 transgene is inserted at a very distal portion of chromosome 2 (2H4), which is 106 bp upstream of a putative gene named "similar to transcriptional adaptor 3-like (LOC665264)" (Fig. 2.1A). This gene sequence was derived by automated computational analysis using the gene prediction method (GNOMON) and is predicted to be a pseudogene. Indeed, no transcript corresponding to this gene has yet been reported. We also failed to detect mRNA derived from this putative gene (our unpublished data). Taken together, it is very likely that LOC665264 is indeed a pseudogene.

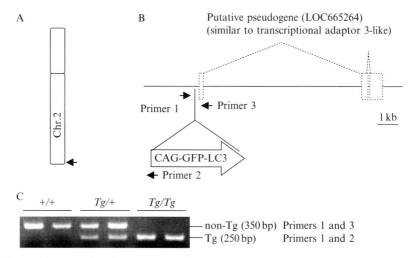

Figure 2.1 PCR-based genotyping of GFP-LC3 mice. (A) Chromosomal mapping of the GFP-LC3 transgene. (B) Positions of PCR primers (1, 2, and 3) used for genotyping. (C) Zygosity determination of GFP-LC3 transgenic mice by genomic PCR. This figure was modified from data previously published in Kuma *et al.* (2008).

Because there are no other genes near this region, this GFP-LC3 transgene unlikely affects the function of other genes.

Because we can now detect autophagosomes in most tissues using this mouse model by fluorescence microscopy analysis of cryosections, this method has been applied in many studies. For example, autophagy is upregulated in various tissues, except in the brain following food withdrawal (Fig. 2.2) and birth (Kuma *et al.*, 2004; Marino *et al.*, 2007; Mizushima *et al.*, 2004; Qu *et al.*, 2003), and in early embryos shortly after fertilization (Tsukamoto *et al.*, 2008). Induction of autophagy is also detected in disease models such as acute pancreatitis (Hashimoto *et al.*, 2008).

3. Mouse Maintenance

In initial experiments, it is recommended that GFP-LC3 mice be maintained as heterozygous, because it is very important to compare transgenic mice with wild-type (nonfluorescent) siblings to distinguish true GFP-LC3 signals from background autofluorescent signals (see subsequent sections).

After finishing the initial experiments or if littermate control is not necessary, GFP-LC3 mice can be maintained as homozygous. Because the transgenic locus has already been identified, we can easily distinguish homozygous mice from heterozygous mice by simple polymerase chain reaction (PCR) analysis (described subsequently). As the GFP-LC3 transgene in this line does not affect other genes, homozygous mice (*tg/tg*) are healthy, fertile,

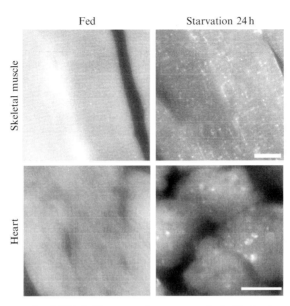

Figure 2.2 Example of GFP-LC3 transgenic mouse analysis. Skeletal muscle (extensor digitorum longus) and heart samples were prepared from GFP-LC3 transgenic mice before (*left*) or after 24-h starvation (*right*). The heart images were modified from data previously published in Mizushima *et al.* (2004). Bar: 10 μm.

and do not show any abnormal phenotype. Therefore, we can maintain this line as homozygous, which can reduce the number of animals needed to maintain this line and the time required for genotyping. Moreover, maintaining them as homozygous facilitates crossing with other animal models, and this will greatly increase the chance of obtaining offspring with a combination of the GFP-LC3 alleles and other alleles of interest.

4. Genotyping of GFP-LC3 Mice

4.1. GFP-LC3#53 mice

GFP-LC3#53 mice are now distributed through the RIKEN Bio-Resource Center in Japan (http://www.brc.riken.jp/lab/animal/en/dist.shtml). The registration number is #BRC00806. They have been backcrossed to C57BL/6 mice.

4.2. Homo/hemi determination by polymerase chain reaction

Transgenic and nontransgenic wild-type alleles can be distinguished by simple PCR (Fig. 2.1B and C). Primers 1 and 2 amplify a 250-bp fragment of the transgenic allele and primers 1 and 3 a 350-bp fragment of the nontransgenic allele. The PCR reaction can be conducted in 1 tube containing all 3 primers. The PCR steps are as follows:

1. Cut 0.2–0.5 cm of the tail and place in a 1.5-mL tube.
2. Incubate the tails in 100–400 μL of tail digestion solution* at 55 °C for 8 h (or overnight). Mix occasionally.
3. Boil for 5 min.
4. Centrifuge at 15,000 rpm for 10 min.
5. Take 1 μL of the supernatant fraction for running a PCR reaction. Mix 1 μL of the tail sample, 0.2 μL of primer 1** 0.2 μL of primer 2, 0.2 μL of primer 3, 2 μL of PCR buffer, 1.6 μL of dNTP mix (2.5 mM each), 0.2 μL of rTaq (Takara Bio Inc., Japan), and 14.6 μL of distilled water.
6. PCR reaction. Step 1: 94 °C for 4 min; step 2: 94 °C for 0.5 min; step 3: 60 °C for 0.5 min; step 4: 72 °C for 1 min, 30 cycles to step 2; step 5: 72 °C for 7 min; and step 6: 4 °C.

*Tail digestion solution: 10 mM Tris-HCl, pH 8.4, 50 mM KCl, 2.5 mM MgCl$_2$, 0.45% NP-40, and 0.45% Tween-20. Autoclave and store at room temperature. Add {1/100} vol of 20 mg/ml proteinase K (final concentration, 0.2 mg/ml) prior to use. The proteinase K solution should be made fresh to avoid self-degradation.

**Primers

Primer 1: 5'-ATAACTTGCTGGCCTTTCCACT-3'
Primer 2: 5'-CGGGCCATTTACCGTAAGTTAT-3'
Primer 3: 5'-GCAGCTCATTGCTGTTCCTCAA -3'

4.3. Detection of fluorescent signal

GFP-LC3 transgenic neonates can be distinguished from wild-type neonates using conventional fluorescence microscopy. If this method needs to be performed in mouse cages in animal facilities, a portable GFP macroscope (Model GFsP-5 from Biological Laboratory Equipment, Maintenance and Service [http://www.bls-ltd.com/]) is convenient. These direct observation methods are highly efficient for neonates but can also be applied to adult mice by checking the GFP fluorescent signal of mouse palms. In the case of our GFP-LC3 mice, a portable UV illuminator does not work.

5. SAMPLE PREPARATION

5.1. Tissue fixation

1. Perfuse mice transcardially with about 3 times the volume (body weight) of 4% paraformaldehyde (PFA) dissolved in 0.1 M Na-phosphate buffer,

pH 7.4. A peristaltic pump (e.g., ATTO AC-2110) can be used in this procedure. Dipping tissues in PFA may be sufficient, but quick fixation is important to prevent artificial induction of autophagy during sample preparation.
2. After perfusion, remove tissues and further fix them in the same fixative for an additional 4 h or overnight (depending on the antibodies used for double staining).
3. Immerse the fixed tissues in 15% sucrose/phosphate buffered saline (PBS) for at least 4 h, then in 30% sucrose/PBS for at least an additional 4 h (or overnight).
4. Embed the tissue samples using OCT compound (Tissue-Tek) and store at $-70\,°C$.

5.2. Cryosection preparation

1. Prepare 5-7-μm-thick tissue sections using a cryostat (e.g., LEICA CM3050S).
2. Air-dry the sections at room temperature for 30 min. The sections can be stored at $-70\,°C$ (or $-20\,°C$) until use.
3. Wash the well-dried cryosections in PBS and mount on glass slides. If necessary, antibody staining (either direct or indirect) can be performed before mounting.

6. FLUORESCENCE MICROSCOPY

6.1. GFP observation

1. Select a 60x or 100x oil-immersion objective (OIO) lens and place a small drop of immersion oil on the objective lens. We usually use a fluorescence microscope (Olympus IX81) equipped with a 60x oil-immersion objective lens (Plan Apo, 1.40 NA) and a cooled charge-coupled device camera (Hamamatsu Photonics, ORCA-ER [1360×1024] or Roper, CoolSNAP HQ2).
2. Place a glass slide on the microscope stage and focus on the cells by transmitted light imaging (usually using differential interference contrast).
3. Select an appropriate dichroic filter set (fluorescein isothiocyanate [FITC] or GFP).
4. Capture images. Fluorescent exposure should be as short as possible to prevent quenching. It is also recommended to take images using unrelated filter sets such as RFP or Cy5 as controls to check the specificity of the green signals.

6.2. Dot counting and quantification

The number of GFP-LC3 puncta can be manually counted; however, using computer software facilitates the analysis and improves objectivity. Uneven cytosolic background signals sometimes make conventional thresholding difficult. For example, weak dot signals cannot be extracted, and brighter dots tend to stick to one another. To better extract each dot signal, the "Top Hat" algorithm of MetaMorph Series Version 6 or later (Molecular Device) is useful. Small dot peaks can be extracted from the surrounding relatively lower background signals irrespective of absolute signal intensity. Using this method, more than 100 GFP-LC3 dots are usually detected in starved mouse embryonic fibroblasts (Kuma et al., 2007).

7. Precautions

7.1. GFP-LC3 puncta may represent protein aggregates/inclusion bodies

GFP-LC3 can be incorporated into protein aggregates independently of autophagy (Kuma et al., 2007). Thus, LC3 localization should be carefully interpreted in cells having protein aggregates or inclusion bodies, such as cells expressing aggregate-prone proteins and those defective in autophagy machinery (Hara et al., 2006; Komatsu et al., 2006; Komatsu et al., 2005). Also, even GFP-LC3 overexpression can generate aggregation (Kuma et al., 2007). In these cells, the occurrence of autophagy should be examined using additional methods such as electron microscopy.

7.2. Distinguishing GFP-LC3 dot signals from autofluorescent signals

Cells sometimes possess autofluorescent punctate structures that are detectable using a GFP filter set. The best example is lipofuscin showing a broad range of fluorescent signals, which are often detected in neurons, macrophages and heart muscles of elderly animals. Therefore, it is quite important to distinguish true GFP-LC3 dot signals from these autofluorescent signals. Two basic procedures are recommended to prevent the formation of these artifacts. First, as described previously, it is essential to compare GFP-LC3 transgenic samples with nontransgenic control samples. Second, true GFP-LC3 signals should be detected specifically by the GFP or FITC filter set but not by others. If similar signals are detected even when fluorescence filter sets for rhodamine, Cy5, or UV are used, these signals are likely nonspecific autofluorescent ones. We usually use the U-MGFPHQ unit (Olympus) for GFP observation and the U-MWIG2 (Olympus) unit to check for autofluorescence.

7.3. Induction of autophagy versus blockage of degradation

GFP-LC3 puncta indicate the presence of autophagosomes but do not provide information on their kinetics. For example, accumulation of larger numbers of GFP-LC3 puncta indicates either enhancement of autophagosome formation (induction) or a decrease in autophagosome turnover. If autophagosome-lysosome fusion is blocked, a number of GFP-LC3 puncta should be detected. In contrast, very rapid fusion of autophagosomes with lysosomes may result in a smaller number of GFP-LC3 dots, which would underestimate autophagic activity. Therefore, in some cases, detection of steady-state levels is not sufficient, and analysis of autophagy flux is required instead. Autophagy flux can be measured by inhibiting lysosome degradation activity using a lysosome protease inhibitor (Tanida *et al.*, 2005). This method can be applied to whole animals using chloroquine to inhibit lysosomal degradation (Iwai-Kanai *et al.*, 2008).

REFERENCES

Hara, T., Nakamura, K., Matsui, M., Yamamoto, A., Nakahara, Y., Suzuki-Migishima, R., Yokoyama, M., Mishima, K., Saito, I., Okano, H., and Mizushima, N. (2006). Suppression of basal autophagy in neural cells causes neurodegenerative disease in mice. *Nature* **441,** 885–889.

Hara, T., Takamura, A., Kishi, C., Iemura, S., Natsume, T., Guan, J. L., and Mizushima, N. (2008). FIP200, a ULK-interacting protein, is required for autophagosome formation in mammalian cells. *J. Cell Biol.* **181,** 497–510.

Hashimoto, D., Ohmuraya, M., Hirota, M., Yamamoto, A., Suyama, K., Ida, S., Okumura, Y., Takahashi, E., Kido, H., Araki, K., Baba, H., Mizushima, N., *et al.* (2008). Involvement of autophagy in trypsinogen activation within the pancreatic acinar cells. *J. Cell Biol.* **181,** 1065–1072.

Itakura, E., Kishi, C., Inoue, K., and Mizushima, N. (2008). Beclin I forms two distinct phosphatidylinositol 3-kinase complexes with mammalian Atg14 and UVRAG *Mol. Biol. Cell* in press.

Iwai-Kanai, E., Yuan, H., Huang, C., Sayen, M. R., Perry-Garza, C. N., Kim, L., and Gottlieb, R. A. (2008). A method to measure cardiac autophagic flux *in vivo*. *Autophagy* **4,** 322–329.

Jahreiss, L., Menzies, F. M., and Rubinsztein, D. C. (2008). The itinerary of autophagosomes: from peripheral formation to kiss-and-run fusion with lysosomes. *Traffic* **9,** 574–587.

Kabeya, Y., Mizushima, N., Ueno, T., Yamamoto, A., Kirisako, T., Noda, T., Kominami, E., Ohsumi, Y., and Yoshimori, T. (2000). LC3, a mammalian homologue of yeast Apg8p, is localized in autophagosome membranes after processing. *EMBO J.* **19,** 5720–5728.

Kabeya, Y., Mizushima, N., Yamamoto, A., Oshitani-Okamoto, S., Ohsumi, Y., and Yoshimori, T. (2004). LC3, GABARAP and GATE16 localize to autophagosomal membrane depending on form-II formation. *J. Cell Sci.* **117,** 2805–2812.

Kimura, S., Noda, T., and Yoshimori, T. (2007). Dissection of the autophagosome maturation process by a novel reporter protein, tandem fluorescent-tagged LC3. *Autophagy* **3,** 452–460.

Klionsky, D. J., Cregg, J. M., Dunn, W. A. Jr., Emr, S. D., Sakai, Y., Sandoval, I. V., Sibirny, A., Subramani, S., Thumm, M., Veenhuis, M., and Ohsumi, Y. (2003). A unified nomenclature for yeast autophagy-related genes. *Dev. Cell* **5**, 539–545.

Köchl, R., Hu, X. W., Chan, E. Y. W., and Tooze, S. A. (2006). Microtubules facilitate autophagosome formation and fusion of autophagosomes with endosomes. *Traffic* **7**, 129–145.

Komatsu, M., Waguri, S., Chiba, T., Murata, S., Iwata, J. I., Tanida, I., Ueno, T., Koike, M., Uchiyama, Y., Kominami, E., and Tanaka, K. (2006). Loss of autophagy in the central nervous system causes neurodegeneration in mice. *Nature* **441**, 880–884.

Komatsu, M., Waguri, S., Ueno, T., Iwata, J., Murata, S., Tanida, I., Ezaki, J., Mizushima, N., Ohsumi, Y., Uchiyama, Y., Kominami, E., Tanaka, K., and Chiba, T. (2005). Impairment of starvation-induced and constitutive autophagy in Atg7-deficient mice. *J. Cell. Biol.* **169**, 425–434.

Kuma, A., Hatano, M., Matsui, M., Yamamoto, A., Nakaya, H., Yoshimori, T., Ohsumi, Y., Tokuhisa, T., and Mizushima, N. (2004). The role of autophagy during the early neonatal starvation period. *Nature* **432**, 1032–1036.

Kuma, A., Matsui, M., and Mizushima, N. (2007). LC3, an autophagosome marker, can be incorporated into protein aggregates independent of autophagy: Caution in the interpretation of LC3 localization. *Autophagy* **3**, 323–328.

Kuma, A., and Mizushima, N. (2008). Chromosomal mapping of the GFP-LC3 transgene in GFP-LC3 mice. *Autophagy* **4**, 61–62.

Marino, G., Salvador-Montoliu, N., Fueyo, A., Knecht, E., Mizushima, N., and Lopez-Otin, C. (2007). Tissue-specific autophagy alterations and increased tumorigenesis in mice deficient in ATG4C/autophagin-3. *J. Biol. Chem.* **282**, 18573–18583.

Melendez, A., Tallóczy, Z., Seaman, M., Eskelinen, E.-L., Hall, D. H., and Levine, B. (2003). Autophagy genes are essential for dauer development and life-span extension in. *C. elegans. Science.* **301**, 1387–1391.

Mizushima, N., Kuma, A., Kobayashi, Y., Yamamoto, A., Matsubae, M., Takao, T., Natsume, T., Ohsumi, Y., and Yoshimori, T. (2003). Mouse Apg16L, a novel WD-repeat protein, targets to the autophagic isolation membrane with the Apg12-Apg5 conjugate. *J. Cell. Sci.* **116**, 1679–1688.

Mizushima, N., Yamamoto, A., Hatano, M., Kobayashi, Y., Kabeya, Y., Suzuki, K., Tokuhisa, T., Ohsumi, Y., and Yoshimori, T. (2001). Dissection of autophagosome formation using Apg5-deficient mouse embryonic stem cells. *J. Cell. Biol.* **152**, 657–667.

Mizushima, N., Yamamoto, A., Matsui, M., Yoshimori, T., and Ohsumi, Y. (2004). In vivo analysis of autophagy in response to nutrient starvation using transgenic mice expressing a fluorescent autophagosome marker. *Mol. Biol. Cell* **15**, 1101–1111.

Niwa, H., Yamamura, K., and Miyazaki, J. (1991). Efficient selection for high-expression transfectants with a novel eukaryotic vector. *Gene* **108**, 193–199.

Otto, G. P., Wu, M. Y., Kazgan, N., Anderson, O. R., and Kessin, R. H. (2003). Macroautophagy is required for multicellular development of the social amoeba *Dictyostelium discoideum. J. Biol. Chem.* **278**, 17636–17645.

Qu, X., Yu, J., Bhagat, G., Furuya, N., Hibshoosh, H., Troxel, A., Rosen, J., Eskelinen, E.-L., Mizushima, N., Ohsumi, Y., Cattoretti, G., and Levine, B. (2003). Promotion of tumorigenesis by heterozygous disruption of the *beclin 1* autophagy gene. *J. Clin. Invest.* **112**, 1809–1820.

Rusten, T. E., Lindmo, K., Juhasz, G., Sass, M., Seglen, P. O., Brech, A., and Stenmark, H. (2004). Programmed autophagy in the *Drosophila* fat body is induced by ecdysone through regulation of the PI3K pathway. *Dev. Cell* **7**, 179–192.

Scott, R. C., Schuldiner, O., and Neufeld, T. P. (2004). Role and regulation of starvation-induced autophagy in the *Drosophila* fat body. *Dev. Cell* **7**, 167–178.

Suzuki, K., Kirisako, T., Kamada, Y., Mizushima, N., Noda, T., and Ohsumi, Y. (2001). The pre-autophagosomal structure organized by concerted functions of *APG* genes is essential for autophagosome formation. *EMBO J.* **20,** 5971–5981.

Suzuki, K., and Ohsumi, Y. (2007). Molecular machinery of autophagosome formation in yeast, *Saccharomyces cerevisiae*. *FEBS. Lett.* **581,** 2156–2161.

Tanida, I., Minematsu-Ikeguchi, N., Ueno, T., and Kominami, E. (2005). Lysosomal turnover, but not a cellular level, of endogenous LC3 is a marker for autophagy. *Autophagy* **1,** 84–91.

Tanida, I., Yamaji, T., Ueno, T., Ishiura, S., Kominami, E., and Hanada, K. (2008). Consideration about negative controls for LC3 and expression vectors for four colored fluorescent protein-LC3 negative controls. *Autophagy* **4,** 131–134.

Tsukamoto, S., Kuma, A., Murakami, M., Kishi, C., Yamamoto, A., and Mizushima, N. (2008). Autophagy is essential for preimplantation development of mouse embryos. *Science* **321,** 117–120.

Yoshimoto, K., Hanaoka, H., Sato, S., Kato, T., Tabata, S., Noda, T., and Ohsumi, Y. (2004). Processing of ATG8s, ubiquitin-like proteins, and their deconjugation by ATG4s are essential for plant autophagy. *Plant. Cell* **16,** 2967–2983.

CHAPTER THREE

Using Photoactivatable Proteins to Monitor Autophagosome Lifetime

Dale W. Hailey* *and* Jennifer Lippincott-Schwartz*

Contents

1. Introduction	26
2. Photoactivatable Fluorescent Protein Labeling	27
2.1. Choosing the biological system	28
2.2. Choosing an appropriate photoactivatable protein	29
2.3. Spectral considerations	30
2.4. Photoconversion considerations	31
3. Experimental Example: Setting up a Sample and Optimizing Photoactivation	32
3.1. Setting up chambers	33
3.2. Optimizing photoactivation	34
4. Photobleaching	35
4.1. Experimental example: Assessing photobleaching	36
5. Carrying Out the Photochase Assay	36
5.1. Inducing autophagy	36
5.2. Experimental example: Generating starvation-induced autophagosomes	38
6. Pulse-Labeling Induced Autophagosomes	38
7. Determining the Half-Life of Pulse-Labeled Autophagosomes	40
8. Controls	42
9. Conclusions	43
References	44

Abstract

Many conditions are now known to cause autophagosome proliferation in cells and organisms including amino acid and serum starvation, ER and oxidative stress, and pathogen infection. Autophagosome proliferation is also observed in disease states and developmental programs. The widespread use of GFP-Atg8 fusion molecules has provided a simple way to visualize the proliferation

---

* Section on Organelle Biology, Cell Biology and Metabolism Branch, NICHD, National Institutes of Health, Bethesda, Maryland, USA

of autophagosomes in cells. However, GFP-Atg8 markers do not reveal the underlying cause of autophagosome proliferation. Two processes regulate the number of autophagosomes present in cells: (1) formation of the structures and (2) their turnover through fusion with lysosomes. Here we describe the use of photoactivatable proteins to decouple the processes of autophagosome formation from autophagosome turnover. Photoactivatable proteins fused to Atg8 homologs make it possible to pulse-label existing populations of autophagosomes in living cells. The fate of those pulse-labeled autophagosomes can then be monitored to determine autophagosome lifetime. This assay is applicable to both engineered tissue culture models and transgenic organisms expressing photoactivatable proteins fused to Atg8 homologs.

1. Introduction

Since Atg8 was first characterized in *Saccharomyces cerevisiae* in the mid-1990s, its related homologs in higher eukaryotes (collectively referred to as Atg8 herein) have been used to directly visualize autophagic vesicles in cells and tissues, replacing monodansylcadaverine staining as a standard autophagosome marker. During autophagosome formation, the C terminus of Atg8 is processed and covalently bound to phosphotidylethanolamine in membranes (Ichimura *et al.*, 2000). The resulting shift of Atg8 from the cytosol to autophagosomal membranes provides a convenient, easily assayed readout for cellular conditions that proliferate autophagosomes. However, because Atg8 and its homologs persist on autophagic structures from their inception through fusion with lysosomes (Tanida *et al.*, 2005), Atg8-positive structures can accumulate either because of increased rates of autophagosome formation or decreased rates in their degradation (Klionsky *et al.*, 2008a). Many conditions cause Atg8-positive structures to proliferate, including amino acid starvation, serum deprivation, ER stress, proteasome inhibition, calcium dysregulation, mitochondrial damage, and some viral and protozoan infections (Ding and Yin, 2008; Kirkegaard *et al.*, 2004; Kuma *et al.*, 2004; Lum *et al.*, 2005). Autophagosomes also accumulate during developmental programs and the progression of diseases including Alzheimer's and Huntington's (Fimia *et al.*, 2007; Levine and Kroemer, 2008). With such diverse conditions reported to induce Atg8-positive structures, a straightforward method is needed to evaluate the role these structures play in cell homeostasis—in particular, a reliable means to determine whether autophagosome proliferation indicates increased formation or alternatively a failure of the structures to degrade. The assay presented here decouples autophagic induction from degradation to reveal turnover rates of Atg8-positive autophagosomes in a straightforward manner.

Catabolism of autophagic substrates was initially assessed by radiolabel pulse-chase experiments. These experiments provided a direct readout for autophagic activity. A number of approaches have since been developed to analyze the fate of autophagic substrates (Tasdemir et al., 2008). Autophagy is most commonly described as a nonselective degradation process that turns over small fractions of diverse protein populations by bulk capture. To confidently assess autophagic turnover of a substrate population, protein populations often must be tracked over long periods of time. Assessing the effects of drugs over such time periods increases the risk of confounding off-target effects. Additionally, pulse-labeling of autophagic substrates is often laborious and difficult to apply in a high-throughput fashion. We present an alternative method to assess the lifetime of autophagosomes by using a photoactivatable fluorescent protein-based assay. This technique requires substitution of GFP in well-characterized GFP-Atg8 fusion proteins with either photoactivatable GFP (PAGFP) or other genetically encoded photoactivatable fluorescent proteins to pulse label a population of autophagosomes and quantify their lifetime.

2. Photoactivatable Fluorescent Protein Labeling

Fluorescent protein (FP) labeling is a ubiquitous cell biological tool that enables researchers to monitor a protein of interest in living cells or organisms. Fluorescent protein fusions must be shown to phenocopy the activity of the untagged endogenous protein of interest. Once characterized, they are invaluable for their ease of use and diverse utility. Fluorescent protein fusions minimize artifacts of sample preparation and enable researchers to monitor dynamic processes in living cells. Autophagy is particularly well suited to studies using fluorescent protein fusions because core autophagy components are recruited from the cytosol to autophagosome membranes upon induction (Xie and Klionsky, 2007). The now-common use of fluorescent protein Atg8 fusions to monitor autophagy is further described in this volume (see the chapter by Kimura et al.).

Conventional fluorescent proteins behave like fluorescent dyes. They absorb a defined range of wavelengths of light (the excitation spectra) and emit lower energy, longer wavelengths (the emission spectra). In the case of GFP, maximum excitation occurs around 490 nm and maximum emission occurs around 525 nm. For most commonly used fluorescent proteins, excitation and emission spectra are a consistent property of the fluorescent protein. The spectra change little in response to environmental factors such as pH, temperature, proximal redox potential, and so on. The chromophore of GFP (the portion of the protein that absorbs light energy and reemits light) maintains a stable structure under most conditions, and the consequent stable spectral behavior of GFP is in fact critical for its use in many quantitative imaging applications (Zimmer, 2002).

Recently, a number of fluorescent proteins have been characterized that can undergo alterations in chromophore structure. These structural alterations are induced by exposure to high-energy light and result in irreversible changes in the spectral properties of the chromophores. So-called photoactivatable proteins behave much like caged compounds that can be light-activated (Lippincott-Schwartz and Patterson, 2008; Lukyanov et al., 2005). Once photoactivated, the spectral properties of a targeted fluorescent protein population are distinct from the unactivated population. Therefore, a set of proteins can be highlighted. With appropriate imaging configurations, this highlighted population can then be uniquely followed. Proteins outside of the activated region as well as proteins folded and/or translated after the photoactivation will be spectrally distinct and can be distinguished from activated proteins ((Lippincott-Schwartz and Patterson, 2008).

By substituting the photoactivatable protein PAGFP for GFP in GFP-Atg8 fusions, autophagosomes can be pulse-labeled. A brief pulse of 400nm light can be used to highlight a population of autophagosomes present at a given time. Fluorescence from this population is lost as the pulse-labeled autophagosomes fuse with lysosomes. Following lysosomal fusion, decreased pH and proteolysis abolish fluorescence of activated PAGFP-Atg8. An autophagosomal population can therefore be monitored in live cells at the microscope to quantify how quickly the pulse-labeled autophagosome population disappears. By knowing how long Atg8-positive structures persist within cells, the role autophagosomes play in active catabolism of substrates can be deduced. Using this approach, it is relatively simple to determine whether a condition or drug treatment accumulates autophagic structures via inducing autophagosome formation or alternatively blocking autophagosome degradation. This assay can be used in diverse research models ranging from tissue culture cells to whole organisms.

Here we present an example of this photo-chase assay using a *Rattus norvegicus* tissue culture cell line referred to below as NRK144. This NRK (normal rat kidney) line stably expresses PAGFP-LC3 (the photoactivatable protein PAGFP fused to microtubule-associated protein 1, light chain 3, the best-characterized mammalian Atg8 homolog). (Kabeya et al., 2000). We discuss considerations in choosing a photoactivatable protein and present an example of how the photo-chase assay is carried out.

2.1. Choosing the biological system

In this chapter we use a stable tissue culture line as a simple tool to evaluate the behavior of autophagosomes. We note that based on studies of the GFP-LC3-expressing mouse line and many other reports, autophagy is a tissue-specific, context-dependent process (Mizushima et al., 2004). Many questions will need to be addressed in more complicated systems. However, initial studies to explore regulatory controls, mechanistic aspects of autophagy, and drug effects may be more tractable in tissue culture models. There are important caveats to using these systems. Overexpression of Atg8 homologs has been reported to induce formation of Atg8-positive structures that may generate aggregates of Atg8 or induce formation of autophagosomes (Kuma *et al.*, 2007). Anecdotally, cells transiently expressing Atg8 homologs also have nonhomogenous responses, possibly because of the range of Atg8 levels or variable stress responses to transfection reagents. To minimize these issues, we have used cell lines that stably express photoactivatable protein fusions to Atg8 homologs.

Once an expression system is established, growth conditions can be adjusted to minimize basal autophagy. To assess the lifetime of autophagosomes formed in response to a condition of interest, there should be very few Atg8-positive structures present in untreated cells. Serum concentration, media pH, and confluency all affect the basal level of autophagosome formation. We find that increasing serum concentrations from 10% to 12% and decreasing cell confluency decrease basal autophagosome formation, consistent with serum factors suppressing autophagy (Furuta *et al.*, 2004). Additionally, autophagosome formation is a known response to a range of pathogen infections. If there is any question about whether cells are free of mycoplasm or other intracellular pathogens, cells should be cleaned with a broad range antibiotic such as BM Cyclin (Roche, Cat. No. 10 799 050 001). Establishing conditions that minimize the number of autophagosomes present in the cells under basal conditions ensures that induced autophagic structures represent autophagosomes formed in response to the induction being studied.

2.2. Choosing an appropriate photoactivatable protein

In the following example we use PAGFP as the photoactivatable genetically encoded marker. PAGFP was the first published photoactivatable protein—a variant of wild-type GFP (Patterson and Lippincott-Schwartz, 2002). Since PAGFP was reported in 2002, the catalog of photoactivatable proteins has grown. To choose the best available photoactivatable protein, we advise surveying the current literature. The properties of many of these proteins are now being optimized for biological applications. At the time of this publication, a review by Patterson (2008) provided a convenient, comprehensive overview of the current catalog of photoactivatable proteins. Although some of the new photoactivatable proteins show much promise, there are two notes of caution.

First, most photoactivatable proteins have not to date been used extensively. Atg8 fusions must function in both cytosolic and membrane-bound forms (Kabeya *et al.*, 2000). When properly intercalated into autophagosome membranes, these fusion proteins are locally concentrated and are sterically constrained. This environment may promote oligomerization, and it is especially important to confirm that novel Atg8 fusion proteins are fully functional. New fluorescent protein Atg8 fusions should be evaluated empirically by careful comparison with either endogenous Atg8 or the well-characterized GFP fusions to Atg8. Of course, it must be kept in mind that the C terminus of Atg8 is proteolytically processed by Atg4 for subsequent lipidation; fluorescent protein fusions therefore must be at the N terminus.

Second, some spectral properties of photoactivatable proteins are undesirable for the photo-chase assay. The assay uses the fact that fluorescence from the activated population is readily quenched as lysosomes fuse with pulse-labeled autophagosomes. The effective disappearance of pulse-labeled autophagosomes enables one to quantify the lifetime of autophagosomes. Accordingly, photoactivatable proteins that fluoresce in low pH (lysosomal) environments are not appropriate; lysosomal fusion will not abolish fluorescent signal from these proteins. (Note that a second fluorescent protein-based technique to quantify autophagosome turnover uses different pH sensitivities and is previously reported (Kimura *et al.*, 2007) and described in this volume (see the chapter by Kimura *et al.*).

With these caveats in mind, choosing the optimal photoactivatable protein involves three considerations: Are the spectra appropriate for the particular application? Are the photoactivation properties optimal? Is the photoactivated fluorescent protein sufficiently photostable to allow repetitive imaging during the monitoring of fluorescence?

2.3. Spectral considerations

Photoactivatable proteins currently fall into two categories: those that turn on and those that undergo a spectral shift (Patterson, 2008). Proteins that turn on are effectively invisible prior to activation. Visualizing these proteins requires using one excitation and emission set. Consequently, a large range of available spectra can be used to simultaneously image a second fluorophore. However, photoactivatable proteins that turn on are difficult to detect prior to activation. Using these proteins, it can be challenging to identify transfected cells or cells of interest in a mosaic tissue. It is also difficult to evaluate the state of the cells prior to activation (i.e., whether autophagosomes have been formed).

A second class of photoactivatable proteins are those that undergo a profound spectral shift upon photoactivation. Most described proteins in this category undergo a green to red (GFP- to RFP-like) shift. Because these proteins are fluorescent prior to activation, cells expressing these proteins can be readily identified. This can expedite locating cells expressing a transfected marker or cells in an organism expressing a transgene. However, because they fluoresce in two spectral ranges, they may limit applications that require monitoring additional tagged proteins of interest or other fluors.

Selecting an appropriate photoactivatable protein also requires considering autofluorescence within cells. We observe that starvation conditions increase autofluorescence in many tissue cultures lines. Ideally, spectra of photoactivatable proteins should have minimal overlap with intracellular autofluorescence. This is particularly important because autofluorescence frequently appears in lysosomes. To evaluate the contribution of autofluorescence, nonactivated cells should be imaged with identical imaging parameters under identical culture conditions as those that will be used to track photoactivated autophagosomes. Although we have not observed this in the NRK144 line, one should also address whether the photoactivation induces autofluorescence by photoactivating cells that do not express the photoactivatable protein, and subsequently monitoring these cells. Ideally no signal should be visible in either case. If significant autofluorescence is detected, narrowing the band pass of the emitted signal may help restrict its contribution.

As indicated previously, in the example described in this chapter, we use photoactivatable GFP (PAGFP) fused to LC3. PAGFP is derived from GFP. It differs by a few amino acids that specifically affect the chromophore (Patterson and Lippincott-Schwartz, 2002); structurally, PAGFP-LC3 is nearly identical to the commonly used GFP-LC3. Whereas unactivated PAGFP can be visualized with filters customized to its blue-shifted excitation spectra, for practical purposes it behaves like a switchable protein. It is turned on by light in the 400-nm range, and following activation it can be monitored with commonly used GFP or FITC imaging configurations. Similar to GFP, its fluorescence is rapidly destroyed at low pH. Its fluorescence is therefore lost once pulse-labeled autophagosomes fuse with lysosomes.

2.4. Photoconversion considerations

2.4.1. Calibrating laser targeting

To highlight a population of photoactivatable fluorescent proteins, the proteins are exposed to a brief pulse of light. For most photoactivatable proteins thus far reported, exposure to 400-nm range light efficiently activates the proteins. Relatively inexpensive low-wavelength 405-nm diode lasers are now available on many microscopy systems. These lasers are ideal for photoactivation experiments. On scanning confocal systems, light from a 405-nm diode laser is typically directed along the imaging light path; it is therefore not usually necessary to calibrate targeting of the photoactivation laser. A number of nonscanning high-resolution microscopy systems (i.e., spinning disk, line-scan, and deconvolution systems) are now also configured for photoactivation. On these systems, the light path of the photoactivation laser is separate from the imaging light path. Therefore, laser targeting must be calibrated. To calibrate targeting, a region is defined and exposed to 405 nm of light and an image is immediately captured.

The defined region is then compared to the actual bleached region in the captured image. Most systems have software adjustments that allow the user to easily offset the laser targeting if the position of the bleached region does not match the defined region. Note that these offsets are specific for the wavelength. For this calibration we recommend using a fixed sample stained with a ubiquitous probe such as FITC-αActin.

2.4.2. Optimizing photoactivation

Low-wavelength light can photo-oxidize cellular components and generate reactive molecules, particularly in tissue culture cells that we typically culture in relatively high oxygen environments (Baier et al., 2006). Ideally, the 405-nm light pulse used to photoactivate fluorescent protein fusions should be minimized as much as possible without excessively compromising the signal, to avoid phototoxicity. For optimal photoactivation, one would like as little signal as possible from the fluorescent protein prior to photoactivation, followed by a robust amount of signal following photoactivation. The difference between these points is the fold activation, and the ideal photoactivation produces the best fold activation for the least light exposure.

Because the amount of energy needed to effectively activate a population of photoactivatable proteins depends on specific biological considerations and hardware configurations, we suggest empirically optimizing photoactivation conditions. To optimize photoconversion, cells expressing the photoactivatable fluorescent protein are repetitively targeted with a low level of photoactivation light. Each iteration will result in an increase in signal until the process of photobleaching overtakes the process of photoactivation. (400 nm of light will photobleach the majority of fluorescent proteins.) Plotting the total fluorescence of the photoactivated protein against the iteration number will generate a plot with a maximum that indicates the maximum fluorescence and the associated optimal number of iterations for photoactivation (Fig. 3.1). This optimization procedure is easily carried out with photoactivatable protein fusions to Atg8 homologs. Prior to autophagic induction, these proteins are essentially cytosolic. The entire cell can therefore be targeted by the photoactivation light, and subsequently total cell fluorescence can be monitored.

3. Experimental Example: Setting up a Sample and Optimizing Photoactivation

In the example presented here, we begin by seeding NRK144 cells into a chamber with a cover-glass bottom. Following are step-by-step instructions to set up the chamber and optimize photoactivation. This

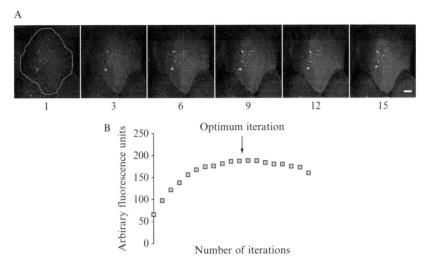

Figure 3.1 Example of an experiment to optimize photoactivation parameters. (A) A live cell stably expressing PAGFP-LC3 was repetitively exposed to light pulses from a 405-nm diode laser. The white dashed line indicates the targeted region. An image (using parameters for the GFP signal) was captured prior to each exposure. Numbers below the images indicate the iteration number. (B) Mean fluorescent intensity was quantified within the targeted region for each image to track the increase in fluorescence as a function of the number of exposures to the 405-nm pulse. The highest fluorescence occurred at the 11th repetition. Loss in sequential frames indicates that photoactivation is nearly complete and photobleaching is then the dominant effect. Therefore, for the given laser intensity, 11 cycles of 405-nm exposure will produce the maximum fold activation of PAGFP-LC3. Note that this may need to be decreased if phototoxicity is observed. Scale bar: 20 μm.

example assumes use of an inverted microscope and imaging system capable of targeting low-wavelength laser light to a user-defined region of interest.

3.1. Setting up chambers

1. Grow a T25 flask of NRK144 cells to 90% confluency in DMEM/10% FBS.
2. Transfer 350 μL of 37 °C DMEM/10%FBS to each well of an 8-well Nunc LabTek chamber (Nalge Nunc International, 155411).
3. Remove the medium from the T25 flask and rinse twice with 37 °C phosphate-buffered saline (PBS).
4. Remove PBS and add 2 mL of 37 °C 0.05% Trypsin/0.53 mM EDTA (CellGro, 25-051-Cl); wait for cells to lift (<2 min).
5. Transfer 1.0 mL of trypsinized cells to a sterile 1.5 mL microcentrifuge tube to simplify handling in step 6.

6. Transfer 10 μL of trypsinized cells from the microcentrifuge tube to each well in the LabTek chamber. Move the pipette back and forth while dispensing the cells into the DMEM to ensure even cell distribution.
7. Return the chamber to a 37 °C/5% CO_2 incubator and culture 16 h.

3.2. Optimizing photoactivation

1. Warm CO_2-independent medium (Gibco, 18045-088) to 37 °C.
2. Remove the LabTek chamber with cells from the incubator and place the chamber on the surface of a T25 flask filled with 37 °C water to maintain temperature while out of the incubator.
3. Replace the DMEM with the preheated CO_2-independent medium.
4. Transfer the chamber to a preheated 37 °C microscope stage.
5. Locate a field of cells using transmitted light.
6. Take a transmitted light image to identify the location of the cells and an image with GFP parameters to assess the preactivation PAGFP-LC3 signal. This should be close to background (*black*). *Note*: PAGFP does have some emission in the GFP range prior to photoactivation; adjust the intensity of the excitation light and the gain on the system to minimize this signal.
7. Draw a bleach box around the field of cells to define a bleach region.
8. Set up bleach parameters. Typically this involves setting the percent power, duration of exposure, and number of iterations for the targeted laser. Both of these settings should be significantly less than saturation levels (see subsequent steps).
9. Set up a time lapse with the following sequence:
   a. Capture an image using a configuration appropriate for the GFP signal.
   b. Photoactivate by targeting 405 nm of light to the user-defined bleach box.
   c. Repeat for 20 cycles.
10. Run this series. Initially, fluorescence should increase with each iteration.
11. Identify the number of iterations where optimal fluorescence occurs. This roughly indicates the number of iterations required at the selected power level, scan zoom, and so on, for optimal photoactivation.

*Note*1: If no increase in fluorescence is observed, either the laser power is excessive and bleaching following photoactivation is already occurring in the first iteration, or there is insufficient laser power for photoactivation. Modulate the bleach settings (i.e., laser power, duration of exposure, or number of iterations) and repeat the experiment.

*Note*2: Although low-wavelength lasers are typically used to photoactivate, photoactivation can also be achieved using noncoherent light sources such as mercury and mercury-xenon lamps. Low-wavelength filters such as those for DAPI will typically work. However, on most microscope systems it is difficult to regulate the sample exposure using nonlaser sources. This may affect reproducibility and the ability to control phototoxicity.

Note3: If phototoxicity is observed, we find that defining the targeted region such that the cell nucleus is not exposed reduces potential phototoxic effects of photoactivation. This also expedites photobleaching cytosolic activated PAGFP-LC3 (discussed subsequently).

4. Photobleaching

Once photoactivation parameters have been optimized, it is important to establish imaging parameters that minimize photobleaching of the photoactivated signal. Because the autophagosome photo-chase assay assumes that disappearance of signal from pulse-labeled autophagosomes indicates fusion of pulse-labeled autophagosomes with lysosomes and subsequent degradation, it is important to know that loss of signal is not due to photobleaching during image acquisition. Photobleaching of fluorophores depends on properties of the fluorophores as well as specific imaging configurations. Again, we advise using an empirical assay to assess the effect of photobleaching. Following photoactivation with the parameters identified previously, a field should be repetitively imaged using identical image capture settings as those that will be used in the actual experiment. To determine how many iterations to use, consider the appropriate time sampling and experimental duration to confidently assess turnover rates. If autophagosome loss is predicted to occur over the course of 2 h, 10-min time points are appropriate to generate a curve defined by a total of 12 points from the 12 captured images. Given this scenario, photobleaching resulting from the collection of 12 images then needs to be assessed.

Photobleaching can be evaluated by setting up a time series to repetitively capture signal from the photoactivated field. For the purpose of assessing photobleaching, the delay between sequential frames can be reduced from 10 min to 5 s. This dramatically reduces the experimental duration to roughly 1 min. Following capture of this time series, the fluorescence of the photoactivated field can be assessed. This photobleaching assay is carried out prior to inducing autophagy and is completed in a narrow time frame. Loss of photoactivated signal due to degradation of photoactivated proteins is therefore negligible. Fluorescence of the field at the start and end of this time series should consequently be nearly unchanged. If this is not the case, the imaging parameters should be changed. A number of parameters can be optimized to minimize photobleaching during image acquisition. Minimizing photobleaching is easiest to achieve by maximizing the amount of light emitted from the sample that reaches the detector. Objectives and filters with efficient light throughput dramatically help. Long-pass filters may be useful, provided they do not pass signal from autofluorescence or other fluorophores present in the sample. Also, decreased noise in detectors will improve signal to noise. We note that the photo-chase assay only requires resolution sufficient to identify PAGFP-Atg8-positive structures. Therefore, increasing the pinhole on scanning confocal microscopes or using a lower magnification lens may be desirable because better resolution generally comes at the expense of signal. Finally, we note that whereas time sampling should be sufficient to average out fluctuations, excessive sampling will unnecessarily bleach the sample. We therefore advise minimizing the number of images captured during the assay.

4.1. Experimental example: Assessing photobleaching

1. Using the photoactivation parameters established previously, photoactivate a field of cells.
2. Set up a time-lapse series with the total number of images based on the anticipated time duration for autophagosome turnover. We suggest starting with 15 data points.
3. Run the time series below for 20 cycles:
   a. Capture an image using a configuration appropriate for the GFP signal.
   b. Pause for 5 s.
   c. Repeat the sequence.
4. Following image capture, measure the total fluorescence by drawing a region of interest around the activated cells in the first field. Determine whether the fluorescence in sequential frames decreases.
5. If little or no bleaching is measured, proceed. If the sample exhibits appreciable bleaching, modify the GFP image capture parameters as discussed previously.

5. Carrying Out the Photochase Assay

5.1. Inducing autophagy

Autophagy is commonly induced by replacing growth the medium with buffered solutions that lack critical nutrients. Media formulations that induce autophagy include Hanks Buffered Saline, Earle's Balanced Salt Solution, serum-free media, and DPBS (Tasdemir et al., 2008). Incubation of cells in these minimal media leads to robust formation of starvation-induced autophagosomes. Punctate structures that label with PAGFP-LC3 are typically evident within 1 h following medium replacement with these formulations. Stresses other than starvation are also known to produce autophagosomes. Some of these stresses are induced by broadly affecting cell homeostasis (i.e., H_2O_2 incubation); others are induced by targeting specific molecular processes (i.e., kinase inhibitors). The current catalog of small molecule inducers of autophagy is relatively short. However, ongoing autophagy drug screens are likely to uncover many more pharmacological modulators of the process (Zhang et al., 2007).

When evaluating the role small molecules play in autophagosome proliferation, it is important to bear in mind that the autophagy pathway responds to a broad range of cellular stresses, and commonly used solvents such as DMSO can subtly affect cell homeostasis. To minimize off-target effects of drugs or solvents, the duration of treatments should not be excessive, and solvent controls should always be run in parallel in adjacent wells seeded from the same cell stock. Ideally solvent controls should show no induction of autophagy relative to cells in optimal growth conditions.

Treating cells either with media formulations or small molecules can be easily performed using commercially available microscope chambers with cover-glass bottoms. We have described seeding 8-well LabTek chambers. Low-volume microscope chambers such as these are convenient because they minimize required reagents. Having multiple chambers on a single slide is also convenient for drug titrations and parallel controls, as all chambers are exposed to the same seeding and handling conditions. The chambers have removable plastic lids that maintain sterility during transfer and are accessible for media exchange and drug addition at the microscope.

The condition and density of cells is an important factor in photo-chase experiments. Because cell density can affect the autophagic response to stresses, cell density should be controlled to ensure it is not excessively high when the experiment is carried out. For the NRK144 line, we find that basal autophagy is minimized when the cells are roughly 90% confluent. For autophagic inductions known to be rapid, growth medium can be replaced with either incomplete medium, or medium containing a compound of interest on the day of the experiment with cells at 90% confluency. For inductions that require prolonged treatment, treatments should be started at a lower cell density to prevent overgrowth during the treatment. Similarly, if expression of a transgene is being tested, the transgene should be introduced at a cell density that accommodates cell growth while the gene is expressed.

Note that preliminary testing of drugs, media formulations, and culture conditions can be done using GFP-Atg8 expression systems and standard GFP imaging on conventional microscopes. There is very little structural difference between PAGFP and GFP (Patterson and Lippincott-Schwartz, 2002); in our experience with NRK-derived cell lines, PAGFP-LC3 and GFP-LC3 fusions are functionally identical.

5.2. Experimental example: Generating starvation-induced autophagosomes

1. Warm DPBS to 37 °C.
2. Remove the seeded LabTek chamber from the incubator (set up as previously) and place it on the surface of a T25 flask filled with 37 °C water to maintain temperature.
3. Wash cells twice with DPBS using 500 μL per wash.
4. Remove and replace with 350 μL of DPBS.
5. Transfer the chamber to a 37 °C incubator for 2 h.
6. Transfer the chamber to the stage of an inverted microscope maintained at 37 °C.
7. Locate a field of cells with transmitted light.

6. PULSE-LABELING INDUCED AUTOPHAGOSOMES

Once autophagy is robustly induced, autophagosomes can be pulse-labeled by photoactivating PAGFP-Atg8 in the cells. We have here outlined a simple protocol to optimize photoactivation parameters. These parameters are now used to activate PAGFP-Atg8-positive autophagosomes. Fig. 3.2 presents a schematic outline of the steps described previously to pulse-label a population of autophagosomes. Note that when 405 nm of light is targeted to cells of interest, both membrane-bound PAGFP-Atg8 and cytosolic PAGFP-Atg8 are activated. The cytosolic PAGFP-Atg8 pool can be incorporated into autophagosomes that form later. To pulse label only autophagosomes, it may therefore be necessary to bleach the fluorescence from the cytosolic PAGFP-Atg8 pool following the photoactivation. Activated PAGFP, like GFP, can be photobleached by 490 nm of light. Cytosolic PAGFP-Atg8 rapidly transits the volume of the cell and can be selectively bleached by repetitively targeting a small region of the cytosol.

To deplete activated cytosolic PAGFP-LC3 in the NRK144 line, we bleach a 5- × 5-micron box in the cytosol with 488 nm of light every 10 s for 2 m. Note this will also bleach labeled autophagosomes in the bleach region; keeping the bleach box small minimizes the number of autophagosomes affected. The bleaching parameters depend on laser power, cell shape, and so on, and are best tested empirically. Following the photobleaching step, the signal in the bleached region of interest should be roughly equivalent to the signal in this region of interest prior to photoactivation. Because autophagosomes recruit cytosolic PAGFP-LC3 to their membranes and thereby enrich signal on the membrane relative to the cytosol, one should confirm that photobleaching parameters are sufficient to rule out subdetectable levels of cytosolic activated PAGFP-LC3-labeling autophagosomes formed after pulse-labeling. To do this, noninduced cells can be photoactivated and subsequently photobleached using identical parameters.

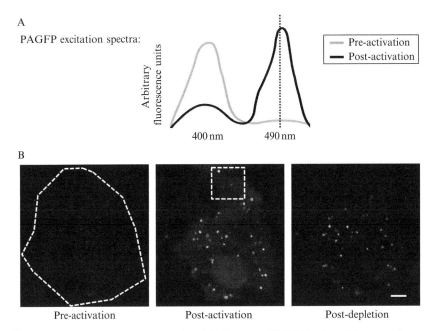

Figure 3.2 PAGFP spectra and pulse-labeling of a PAGFP-LC3 positive autophagosome population. (A) Prior to photoactivation, the excitation spectrum of PAGFP is centered near 400 nm. Following photoactivation, the excitation spectrum shifts such that its peak absorbance is like that of GFP. Therefore, before photoactivation, signal captured from PAGFP using GFP imaging parameters is minimal; the fluorophore does not efficiently absorb light in the 490-nm range. Following photoactivation, the fluorophore efficiently absorbs 490-nm-range light and consequently emits a signal. (B) An induced cell was first photoactivated by targeting the outlined region with 405-nm light (*left panel*). The cytosolic activated signal was then bleached by repetitively targeting a small region of the cytosol (*middle panel; dashed box*) with 490-nm light. This depleted signal from the cytosolic activated PAGFP-LC3. At this stage an existing population of autphagosomes was pulse-labeled (*right panel*). Scale bar: 25 μm.

Subsequently, autophagy can be induced. If photobleaching is sufficient, no labeled autophagosomes should be detected during the course of the incubation. For more details about FLIP (fluorescence loss in photobleaching) approaches to deplete signal, see the FLIP section of Snapp *et al.* (2003).

Following selective bleaching of the activated cytosolic PAGFP-Atg8 pool, a distinct set of autophagosomes is pulse-labeled (Fig. 3.3). Autophagosomes that form after this time can incorporate either newly synthesized and unactivated PAGFP-LC3, or cytosolic activated and bleached PAGFP-LC3. In either case, these molecules do not fluorescently label new autophagosomes. Therefore, the fate of the labeled autophagosomes can now be uniquely tracked by monitoring the disappearance of this population as a function of time.

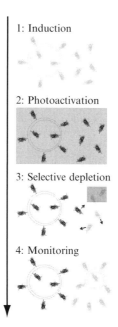

Figure 3.3 Schematic representation of the steps needed to pulse-label an existing autophagosome population using photoactivation. Cells are first induced (1); induction leads to recruitment of PAGFP-Atg8 molecules (in gray) from the cytosol onto autophagosome membranes. Cells are next photoactivated with 405-nm light (2; targeted region indicated by gray box). This step activates both membrane bound and cytosolic PAGFP-Atg8 (activated PAGFP-Atg8 is shown in black). Subsequently, signal from activated cytosolic PAGFP-Atg8 is photobleached by repetitively targeting a small region of the cytosol with 490-nm light (3; targeted region indicated by gray box). A population of autophagosomes is now pulse-labeled (4). New autophagosome formation will incorporate either activated and bleached PAGFP-Atg8 or nonactivated PAGFP-Atg8; therefore, newly formed autophagosomes will not be detected.

7. Determining the Half-Life of Pulse-Labeled Autophagosomes

Determining the lifetime of pulse-labeled autophagosomes following pulse labeling simply involves imaging the pulse-labeled structures at the microscope. Previously we discussed a simple protocol to assess potential photobleaching during image acquisition. Because autophagosomes move dynamically (Jahreiss *et al.*, 2008), it is unrealistic to expect to track individual autophagosomes over their lifetime. Instead, capturing images at distinct time points and quantifying the number of pulse-labeled autophagosomes present at each point provides a reasonable statistical picture of the rate of turnover of the pulse-labeled population. Because photobleaching is an issue (see previous sections), we do not advise attempting to capture three-dimensional stacks to track the total number of autophagosomes in a cell at a given time. Instead, we suggest taking a single plane with a pinhole set to capture 2 microns of information. For the NRK144 line, this provides sufficient resolution to easily resolve autophagosomes and a broad enough sampled volume to generate clear reproducible data. Fig. 3.4 shows a panel of time points taken following pulse-labeling of an autophagosome population.

The ease and confidence with which pulse-labeled autophagosomes can be counted depends on cell type; for example, this affects autophagosome size and number. We have chosen to work with NRK-derived cells in part because these cells generate easily resolvable and quantifiable autophagosomes. For large-scale analyses, when possible we advise establishing a

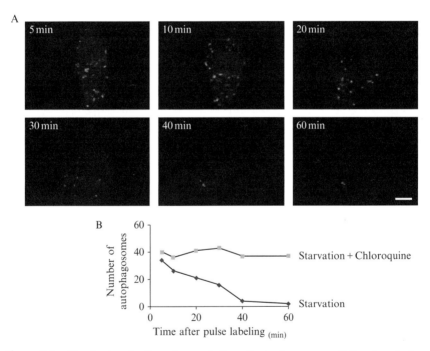

Figure 3.4 Monitoring the fate of a population of autophagosomes. Following photoactivation and selective depletion, a pulse-labeled population of autophagosomes was imaged every 5 min for 1 h (A) Single-plane images were captured using parameters previously shown to result in no appreciable photobleaching. (B) PAGFP-LC3-positive autophagosomes were counted in sequential images. The number of autophagosomes observed in each image was plotted as a function of elapsed time (black line). The autophagosome population exhibited a half-life of approximately 25 min. Autophagosomes in a cell treated with chloroquine did not degrade in this time frame. Following chloroquine treatment, an autophagosome population was pulse-labeled and tracked using identical photoactivation, bleaching, and imaging conditions (gray line). Scale bar: 25 μm.

system that generates easily quantifiable signals. Reasonably discrete autophagosomal signals may also expedite automating quantification. A number of software algorithms can be applied to count autophagosomes (e.g., Metamorph TopHat, Improvision Volocity segmentation). While these analyses remove bias, they should always be evaluated against counting by eye to confirm that the algorithms used do not either misidentify or discount significant numbers of autophagosomes.

8. Controls

To demonstrate that loss of signal from pulse-labeled autophagosomes is in fact due to fusion of the structures with lysosomes, the photo-chase assay can be repeated in the presence of lysosomal inhibitors. Many lysosomal inhibitors (e.g., bafilomycin A_1, chloroquine, ammonium chloride) act by increasing intralysosomal pH and consequently inactivating pH-dependent lysosomal proteases (Tasdemir et al., 2008). Although the mechanism is unclear at this time, increase in intralysosomal pH is reported to perturb fusion of lysosomes with autophagosomes (Klionsky et al., 2008b; Yamamoto et al., 1998). Consistent with this, we observe that disappearance of starvation-induced autophagosomes in the NRK144 line is completely blocked by lysosomal inhibition with chloroquine. Persistent PAGFP-LC3-labeled autophagosomes do not appear to be lysosomes that fail to quench the PAGFP-LC3 signal, as almost all of these structures do not label with the lysosomal marker LysoTracker Red DND-99. (Note that in our experience the LysoTracker label functions despite pH elevation by chloroquine; however, a non pH-sensitive marker such as LAMP-1-RFP could also be used.) This control is important to demonstrate that activated PAGFP-Atg8 is not released from autophagosomal membranes in a manner independent of their conversion to autolysosomes. Although there is debate about whether Atg4 activity might remove LC3 from autophagosomal membranes prior to lysosomal fusion (Tanida et al., 2004), notably at least the activated signal on the inner membrane will persist irrespective of Atg4 activity.

To further confirm that pulse-labeled autophagosomes disappear via lysosomal fusion, one can also track individual autophagosomal structures to visualize their fusion with lysosomes. Lysosomes can be easily labeled using lysotropic dyes such as LysoTracker Red DND-99; these dyes are trapped in low pH environments due to their inability to cross lysosomal membranes once they are protonated. If pulse-labeled autophagosomes are degraded or quenched once in the lysosomal environment, very little if any overlap of the autophagosome marker with the lysosome marker should be observed. The activated PAGFP-LC3 signal in NRK144 cells persists on autophagosomes through fusion with lysosomes. Upon lysosomal fusion, the signal is rapidly lost.

A final point: the preceding controls address whether the disappearance of the PAGFP-Atg8 signal on autphagosomes is the result of fusion with lysosomes. The photo-chase assay described here indicates the fate of autophagosomes up to the point when lysosomal fusion occurs. However, we note that this assay does not address the future fate of the structures once they convert to autolysosomes and lose the Atg8 signal.

9. Conclusions

The photo-chase experiment we describe here is a straightforward approach to study how autophagosomes in cells or tissues contribute to cell homeostasis. Simply observing increased numbers of autophagosomes does not indicate that the autophagy pathway is catabolically active. This fact has been a challenge in the field since early autophagic studies in the 1960s (De Duve and Wattiaux, 1966). For autophagic substrates to be degraded, they must be exposed to lysosomal hydrolases. The photo-chase assay described here addresses whether fusion with lysosomes occurs by quantifying how quickly fluorescently pulse-labeled autophagosomes disappear in response to lysosomal fusion events. Unlike assays that quantify the catabolic activity of autophagosomes by substrate analyses, the photo-chase assay evaluates the fate of the autophagosomal structures themselves. Because of this, the assay is a sensitive readout for autophagic catabolism. Additionally, because it is carried out in live cells expressing a transgenic marker, it does not require excessive cell handling or large amounts of reagents needed for standard pulse-labeling.

The assay presented here uses photoactivatable GFP (PAGFP). Since the initial description of PAGFP, many new photoactivatable proteins have been reported, and the photo-chase assay will likely benefit from these new proteins. These proteins already address phototoxicity and detection issues by more efficiently activating and exhibiting a higher fold signal increase. Additionally, some of these proteins have both pre- and postactivation fluorescent signals; in the future, the photo-chase assay may be adapted to simultaneously track turnover of a pulse-labeled population while also tracking formation of new autophagosomes (labeled with the unactivated fluorescent protein).

Autophagy is a very promising field for biomedical applications. Its so-called housekeeping function is a promising target to address diseases caused by general accumulations of deleterious proteins or organelles, and its role in cell death is a potential cancer therapy target (Hoyer-Hansen and Jäättelä, 2008). Many screens for effectors of autophagy are either recently completed or underway. Approaches to determine how these newly identified effectors function are needed. The photo-chase assay is one approach to rapidly address this challenge. Using the photo-chase assay and complimentary approaches described in this volume and elsewhere (Tasdemir et al., 2008), we can delineate what autophagosome proliferation functionally means.

REFERENCES

Baier, J., et al. (2006). Singlet oxygen generation by UVA light exposure of endogenous photosensitizers. Biophys. J. **91**, 1452–1459.

de Duve, C., and Wattiaux, R. (1966). Functions of lysosomes. Annu. Rev. Physiol. **28**, 435–492.

Ding, W. X., and Yin, X. M. (2008). Sorting, recognition and activation of the misfolded protein degradation pathways through macroautophagy and the proteasome. Autophagy **4**, 141–50.

Fimia, G. M., et al. (2007). Ambra1 regulates autophagy and development of the nervous system. Nature **447**, 1121–1125.

Furuta, S., et al. (2004). Ras is involved in the negative control of autophagy through the class I PI3-kinase. Oncogene **23**, 3898–3904.

Høyer-Hansen, M., and Jäättelä, M. (2008). Autophagy: An emerging target for cancer therapy. Autophagy **4**, 574–580.

Ichimura, Y., et al. (2000). A ubiquitin-like system mediates protein lipidation. Nature **408**, 488–492.

Jahreiss, L., et al. (2008). The itinerary of autophagosomes: From peripheral formation to kiss-and-run fusion with lysosomes. Traffic **9**, 574–587.

Kabeya, Y., et al. (2000). LC3, a mammalian homologue of yeast Apg8p, is localized in autophagosome membranes after processing. EMBO J. **19**, 5720–5728.

Kimura, S., et al. (2007). Dissection of the autophagosome maturation process by a novel reporter protein, tandem fluorescent-tagged LC3. Autophagy **3**, 452–460.

Kirkegaard, K., et al. (2004). Cellular autophagy: surrender, avoidance and subversion by microorganisms. Nat. Rev. Microbiol. **2**, 301–314.

Klionsky, D. J., et al. (2008a). Guidelines for the use and interpretation of assays for monitoring autophagy in higher eukaryotes. Autophagy **4**, 151–175.

Klionsky, D. J., et al. (2008b). Does bafilomycin A_1 block the fusion of autophagosomes with lysosomes? Autophagy **4**, 849—850.

Kuma, A., et al. (2004). The role of autophagy during the early neonatal starvation period. Nature **432**, 1032–1036.

Kuma, A., et al. (2007). LC3, an autophagosome marker, can be incorporated into protein aggregates independent of autophagy: Caution in the interpretation of LC3 localization. Autophagy **3**, 323–328.

Levine, B., and Kroemer, G. (2008). Autophagy in the pathogenesis of disease. Cell **132**, 27–42.

Lippincott-Schwartz, J., and Patterson, G. H. (2008). Fluorescent proteins for photoactivation experiments. Methods Cell. Biol. **85**, 45–61.

Lukyanov, K. A., et al. (2005). Innovation: Photoactivatable fluorescent proteins. Nat. Rev. Mol. Cell Biol. **6**, 885–891.

Lum, J. J., et al. (2005). Autophagy in metazoans: Cell survival in the land of plenty. Nat. Rev. Mol. Cell Biol. **6**, 439–448.

Mizushima, N., et al. (2004). *In vivo* analysis of autophagy in response to nutrient starvation using transgenic mice expressing a fluorescent autophagosome marker. *Mol. Biol. Cell* **15,** 1101–1111.

Patterson, G. H. (2008). Photoactivation and imaging of photoactivatable fluorescent proteins. *Curr. Protoc. Cell. Biol.* Chap. 21, unit 21.6.

Patterson, G. H., and Lippincott-Schwartz, J. (2002). A photoactivatable GFP for selective photolabeling of proteins and cells. *Science* **297,** 1873–1877.

Snapp, E. L., et al. (2003). Measuring protein mobility by photobleaching GFP chimeras in living cells. *Curr. Protoc. Cell Biol.* **21,** Chap. 21, unit 21.1.

Tanida, I., et al. (2005). Lysosomal turnover, but not a cellular level, of endogenous LC3 is a marker for autophagy. *Autophagy* **1,** 84–91.

Tanida, I., et al. (2004).). HsAtg4B/HsApg4B/autophagin-1 cleaves the carboxyl termini of three human Atg8 homologues and delipidates microtubule-associated protein light chain 3- and $GABA_A$ receptor-associated protein-phospholipid conjugates. *J. Biol. Chem.* **279,** 36268–36276.

Tasdemir, E., et al. (2008). Methods for assessing autophagy and autophagic cell death. *Methods Mol. Biol.* **445,** 29–76.

Xie, Z., and Klionsky, D. J. (2007). Autophagosome formation: Core machinery and adaptations. *Nat. Cell Biol.* **9,** 1102–1109.

Yamamoto, A., et al. (1998). Bafilomycin A_1 prevents maturation of autophagic vacuoles by inhibiting fusion between autophagosomes and lysosomes in rat hepatoma cell line, H-4-II-E cells. *Cell Struct. Funct.* **23,** 33–42.

Zhang, L., et al. (2007). Small molecule regulators of autophagy identified by an image-based high-throughput screen. *Proc. Natl. Acad. Sci. USA* **104,** 19023–19028.

Zimmer, M. (2002). Green fluorescent protein (GFP): Applications, structure, and related photophysical behavior. *Chem. Rev.* **102,** 759–781.

CHAPTER FOUR

Assaying of Autophagic Protein Degradation

Chantal Bauvy,[*] Alfred J. Meijer,[†] *and* Patrice Codogno[*]

Contents

1. Introduction	48
2. Assaying the Degradation of Long-Lived Proteins in Cultured Cell Lines	48
2.1. Materials and reagents	48
2.2. Methods	49
2.3. Notes	50
3. Assay of the Degradation of Long-Lived Protein in Isolated Rat Hepatocytes	50
3.1. Materials and reagents	51
3.2. Methods	52
3.3. Notes	56
4. Other Methods for Measuring Autophagic Flux	57
5. Pitfalls	58
References	59

Abstract

Macroautophagy is a three-step process: (1) autophagosomes form and mature, (2) the autophagosomes fuse with lysosomes, and (3) the autophagic cargo is degraded in the lysosomes. It is this lysosomal degradation of the autophagic cargo that constitutes the autophagic flux. As in the case of metabolic pathways, the steady-state concentration of the intermediary autophagic structures alone is insufficient for investigating the flux. Assaying the degradation of long-lived proteins as described in this chapter is one of the methods that can be used to measure autophagic flux.

---

[*] INSERM U756, Université Paris-Sud 11, Châtenay-Malabry, France
[†] Alfred J. Meijer, Department of Medical Biochemistry, Academic Medical Center, University of Amsterdam, Amsterdam, The Netherlands

1. Introduction

Macroautophagy (hereafter called autophagy) is a mechanism conserved among eukaryotic cells that starts with the formation of a multimembrane-bound vacuole, known as an autophagosome, which ultimately fuses with the lysosomal compartment to degrade the sequestered material (Klionsky, 2007; Levine and Klionsky, 2004; Mizushima et al., 2008; Ohsumi, 2001). Autophagy has two nonexclusive functions: (1) to sequester and then degrade cytoplasmic structures; (2) to produce amino acids (and other small molecular compounds) within the lysosomes to sustain metabolic processes outside these organelles (Mizushima, 2005). The rate at which intracellular material is transported via the autophagic pathway to the lysosomes, where it is degraded, is defined as the autophagic flux. The sequestration function is sometimes the most important aspect, for example for segregating harmful or redundant structures from the cytoplasm (Komatsu et al., 2007; Rubinsztein, 2006). The second function of autophagy becomes important when there is a shortage of extracellular nutrients, for instance during starvation-induced autophagy: under such conditions, autophagy provides the amino acids needed to produce ATP and to maintain metabolic processes (Kuma et al., 2004; Lum et al., 2005). Obviously, a dysfunction of autophagy causes greater damage if the ATP level and metabolism are already compromised.

The autophagic flux cannot be determined by simply measuring the number of autophagosomes. This is analogous to the situation for metabolic pathways; that is, just knowing the amounts of a particular intermediate will not tell you whether the pathway is proceeding to completion. Autophagy is responsible for the degradation of long-lived proteins, and so measuring their rate of degradation is one of the methods frequently used to monitor the flux through the autophagic pathway in cell cultures and freshly isolated cells.

In this chapter, we describe methods used to assay protein degradation in cultured cell lines and freshly isolated rat hepatocytes, and we briefly discuss other methods that can be used to monitor the autophagic flux.

2. Assaying the Degradation of Long-Lived Proteins in Cultured Cell Lines

2.1. Materials and reagents

2.1.1. Medium used for cell culture and treatment

1. Dulbecco's Modified Eagle's Medium (DMEM, 4.5 g/l glucose) (Invitrogen, Fisher Biosciences) supplemented with 10% fetal bovine serum (Invitrogen, Fisher Biosciences) and 1% penicillin-streptomycin (Invitrogen, Fisher Biosciences).

2. 3-Methyladenine (3-MA) is routinely used at a final concentration of 10 mM. It is a good idea to prepare a more concentrated stock solution (e.g., 50 mM) in water, culture medium, or balanced salt solution. It may be necessary to heat the solution under a warm-water faucet to dissolve such a high concentration of 3-MA. The prepared solution can be stored frozen, but the 3-MA may precipitate on thawing.

2.1.2. Solutions used for the degradation of long-lived proteins

1. Phosphate-buffered saline (PBS): 137 mM NaCl, 2.7 mM KCl, 4.3 mM Na$_2$HPO$_4$, 1.4 mM KH$_2$PO$_4$, pH 7.3.
2. Hanks's balanced salt solution without sodium bicarbonate (HBSS) or Earle balanced salt solution (EBSS) (Invitrogen, Fisher Biosciences).
3. L-[U-^{14}C] valine (266 mCi/mmol, Amersham Biosciences).
4. Trichloroacetic acid (TCA).
5. 0.2 M NaOH.

2.2. Methods

The protocol described subsequently was originally validated for use in HT-29 human colon cancer cells, and then in various different cancer cells, in particular in MCF7 human breast cancer cells (Lavieu *et al.*, 2006; Scarlatti *et al.*, 2004), and can be optimized depending on the cell system used. Cells are seeded in 6-well plates at 10^6 cells/plate and used near confluence.

1. Intracellular proteins are labeled for 18 h at 37 °C with 0.2 μCi/ml of L-[U-^{14}C] valine (spec. activity 266 mCi/mmol) in complete medium.
2. Any unincorporated radioactivity is eliminated by rinsing the cells 3 times with PBS.
3. The cells are then incubated with fresh complete medium, and 10 mM valine for 1 h (see Note A) to degrade short-lived proteins (in some cell lines this period can be extended to 24 h). Throughout the chase period, 3-MA can be added to inhibit *de novo* formation of autophagic vacuoles (Seglen and Gordon, 1982; also see Note B).
4. Next, the medium is removed by aspiration and replaced; the cells are then incubated in HBSS (or EBSS) (plus 0.1% of bovine serum albumin [see Note C] and 10 mM valine) to stimulate autophagy, or with the appropriate fresh complete medium supplemented with 10 mM valine, and incubated for a further 4 h (longer incubations in the chase medium are also possible, and in some cases it may be desirable to take multiple time points).
5. The medium is then precipitated overnight after adding TCA at a final concentration of 10%.

6. After centrifuging the culture medium for 10 min at 470×g at 4 °C, the acid-soluble radioactivity is measured by liquid scintillation counting.
7. The cells are washed twice with cold 10% TCA (w/v), plus 10 mM valine, to make sure that no radioactivity has remained adsorbed to the denatured proteins. The cell pellet is then dissolved at 37 °C in 0.2 M NaOH for 2 h. Radioactivity is then measured by liquid scintillation counting. The rate of degradation of long-lived proteins is calculated from the ratio of the acid-soluble radioactivity in the medium to that in the acid-precipitable cell fraction.

2.3. Notes

A. Amino acids are physiological inhibitors of autophagy (van Sluijters et al., 2000). The choice of the amino acid used during the pulse-chase is important because some amino acids, such as leucine, are potent inhibitors of autophagy. Valine is frequently used because it does not interfere with autophagy in most cell types. The inhibitory cocktail of amino acids is given in Section 3.1.2.

B. 3-MA blocks autophagy by inhibiting class-III phosphatidylinositol 3-kinase (Petiot et al., 2000). However, it should be kept in mind that 3-MA is a phosphatidylinositol 3-kinase inhibitor (Blommaart et al., 1997), which interferes with other intracellular trafficking pathways dependent on the phosphatidylinositol 3-kinases (Punnonen et al., 1994). 3-MA also affects some other intracellular events (Tolkovsky et al., 2002).

C. HBSS is used when the cells are incubated in a humidified chamber at 37 °C in the absence of CO_2, otherwise EBSS is used. The reason is that EBSS, but not HBSS, contains bicarbonate, and the pH of the medium is governed by the Henderson-Hasselbalch equation: $pH = pK_a + \log[HCO_3^-]/[CO_2] = 6.1 + \log[HCO_3^-]/[CO_2]$.

3. Assay of the Degradation of Long-Lived Protein in Isolated Rat Hepatocytes

Because autophagy in liver is rapid, preparations of isolated hepatocytes have been very popular in the study of autophagy in the past. Especially for animals (e.g., mice) harboring transgenic or null mutations in the genes coding for Atg proteins or signal-transduction pathway components controlling autophagy, the study of autophagic proteolysis in hepatocytes is very convenient (e.g., Ueno et al., 2008; also see the chapter by Seglen et al. in this volume).

Autophagic proteolysis in hepatocytes in culture can be studied as described in section 2. It must be stressed, however, that hepatocytes rapidly dedifferentiate in culture and lose many liver-specific functions. Primary hepatocytes (i.e., freshly isolated hepatocytes) do possess most, but not all (Klionsky *et al.*, 2008), of the *in vivo* functions of the liver: intermediary metabolism and the regulation of autophagy in this preparation closely approaches the *in vivo* situation. It must be pointed out that hepatocyte proteolysis, measured as described subsequently, largely represents the breakdown of long-lived protein.

3.1. Materials and reagents

3.1.1. Solutions used during the preparation of hepatocytes

1. Salt solution, 20x concentrated. Prepare a mixture containing 140.2 g NaCl, 7.16 g KCl, 3.26 g KH_2PO_4, and 5.92 g $MgSO_4 \cdot 7H_2O$ in 1 liter of water. This solution can be stored at 4 °C.
2. Na^+-bicarbonate, 20x concentrated (480 mM): 40.3 g/l of water. Store at 4 °C.
3. 1.3 M $CaCl_2 \cdot 2H_2O$: 19.1 g per 100 ml of water. Store at 4 °C.
4. 1 M Na^+-HEPES (4-(2-hydroxyethyl)-1-piperazineethanesulfonic acid): Dissolve 23.8 g of HEPES in approximately 80 ml of water. Adjust the pH of this solution to 7.4 with 5 M of NaOH, and bring the final volume to 100 ml by adding water. Store at 4 °C.
5. Krebs–Henseleit bicarbonate buffer plus Na^+-HEPES, without Ca^{2+} (120 mM NaCl, 4.8 mM KCl, 1.2 mM KH_2PO_4, 1.2 mM $MgSO_4$, 24 mM $NaHCO_3$, and 10 mM Na^+-HEPES pH 7.4) (see Note 1), prepared daily as follows: To 890 ml H_2O add: 50 ml 20x concentrated salt solution, 50 ml 20x concentrated Na^+-bicarbonate and 10 ml 1 M Na^+-HEPES, pH 7.4. This solution must be gassed for about 10 min with 95% (v/v) O_2 plus 5% (v/v) CO_2 at 37 °C to reach a final pH of 7.4.
6. Collagenase (Sigma type VIII).
7. Trypan blue (0.25% w/v) (in 0.9% (w/v) NaCl + 10 mM Na^+-HEPES, pH 7.4).

3.1.2. Incubation of hepatocytes

1. Krebs–Henseleit bicarbonate buffer (120 mM NaCl, 4.8 mM KCl, 1.2 mM KH_2PO_4, 1.2 mM $MgSO_4$, 24 mM $NaHCO_3$, 1.3 mM $CaCl_2$), fortified with 10 mM Na^+-HEPES, pH 7.4 (see Note 2).

   It is convenient to prepare a stock solution that is 2x concentrated, as follows: To 389 ml H_2O add, 50 ml of 20x concentrated salt solution (see section 3.1.1.), 50 ml of 20x concentrated Na^+-bicarbonate, 10 ml of 1 M Na^+-HEPES, pH 7.4, and 1 ml of 1.3 M $CaCl_2$. Using this concentrated stock solution makes it possible to add other components to the incubation medium that have to be dissolved in water

(e.g., glucose). If the more concentrated Krebs-Henseleit-Na$^+$-HEPES solution is not used, remember that any other components added to the incubation medium must be dissolved in this solution, and not in water.
2. Glucose, 20 mM; prepare a stock solution of 1 M glucose in water. Store frozen.
3. Cycloheximide, 20 μM; prepare a stock solution of 1 mM in water. Store frozen.
4. A mixture of amino acids from which valine is omitted (in μM: Asparagine: 60, isoleucine: 100, leucine: 250, lysine: 300, methionine: 40, phenylalanine: 50, proline: 100, threonine: 180, tryptophan: 70, alanine: 400, aspartate: 30, glutamate: 100, glutamine: 350, glycine: 300, cysteine: 60, histidine: 60, serine: 200, tyrosine: 75, ornithine: 100). In this mixture (1xPMAA minus valine, physiological mixture of amino acids), the concentration of each of the amino acids is equal to that found in the portal vein of a 24 h–fasted rat (Meijer et al., 1985) (see Note 3). It is convenient to use a stock solution of this mixture of amino acids that is 20 times more concentrated (i.e., the concentration of each of the amino acids in this solution is then 20x the concentration found in the portal vein of a fasted rat). When the amino acid mixture is being prepared, it is important to realize that some amino acids are acidic (e.g., glutamic acid); consequently, the pH of the final solution needs to be adjusted to 7.4 by adding 1 M NaOH. This stock solution can be stored frozen for several weeks.
5. 3-Methyladenine, 10 mM (See section 2.1.1).
6. HClO$_4$ 14% (w/v).
7. A mixture of 2 M KOH and 0.3 M MOPS (3-(N-morpholino)propanesulfonic acid). Warning: Be careful when preparing a concentrated solution of KOH (or of NaOH) because the solution becomes hot during preparation. Cool on ice.

3.2. Methods

3.2.1. Preparation of hepatocytes

Hepatocytes are prepared from 18–24h starved rats (or mice) (see Note 4), according to established procedures (Groen et al., 1982; Seglen, 1976). The procedure of Groen et al. (1982) is followed here, but other methods are equally good (Berry, 2000; Seglen, 1976).

1. After cannulating the portal vein, the liver is perfused at 37 °C with Ca^{2+}-free Krebs-Henseleit bicarbonate + 10 mM Na$^+$-HEPES (50 ml) (see section 3.1.1.) to remove blood, followed by a 10-min linear perfusion at a flow of 40 ml/min in the reverse direction through the vena cava to disrupt the desmosomal junctions.

2. The liver is then circularly perfused for 10–15 min in the same direction with 100 ml of the same medium, to which 0.1 ml of 1.3 M $CaCl_2$ and 20 mg of collagenase have been added. The medium is continuously saturated with 95% O_2 and 5% CO_2 by bubbling this gas mixture through the solution.
3. The liver is then removed and, after proper digestion, virtually falls apart, so that the cells can easily be dispersed on a Petri dish by applying light pressure or cutting the liver with scissors.
4. The cells are transferred to a plastic Erlenmeyer flask (250 ml), and shaken for 2–3 min under continuous gassing in the same perfusion medium.
5. The cell suspension is then filtered through a nylon filter to remove cell debris (pore size, 120 μm), before washing the cells three times with ice-cold Krebs-Henseleit-bicarbonate (plus 1.3 mM Ca^{2+}) + 10 mM Na^+-HEPES, pH 7.4, to remove collagenase. Approximately 90% of the hepatocytes isolated in this way exclude trypan blue.
6. The cells may be stored before use for approximately 2 h on ice at a concentration of 20–30 mg dry mass/ml (see Note 5). The dry mass of the cells is determined by weighing 1 ml of cell suspension in a pre-weighed glass vial after heating to complete dryness in an oven at 90 °C. As a blank control, 1 ml of Krebs-Henseleit-bicarbonate + 10 mM Na^+-HEPES (without cells) is subjected to the same procedure. To obtain a quick estimate of the dry mass, 0.25 ml of cell suspension can be spun for 1 s in a microcentrifuge in a previously weighed tube. After discarding the extracellular fluid, the increase in weight is roughly equal to the dry mass of cells per milliliter of cell suspension. This is because 1 g of dry mass of cells equals 3.8 g of wet weight.

3.2.2. Incubating the hepatocytes

The basal medium consists of Krebs-Henseleit bicarbonate medium (Ca^{2+} included), to which 10 mM Na^+-HEPES, pH 7.4, and 10–20 mM glucose are also added (see Note 6). To prevent the reincorporation of valine (and other amino acids) formed during proteolysis into protein, cycloheximide (20 μM) is also added to the medium. In hepatocytes, low concentrations of cycloheximide do not affect autophagy, at least not for the first 2 h (Kovács and Seglen, 1981; Meijer et al., 1993) (see Note 7). Other authors prefer to add cycloheximide (which acts almost instantaneously) for a short time only, and to measure the rate of proteolysis during that short period (Kanazawa et al., 2004). A higher concentration of cycloheximide must be avoided, because it could inhibit mitochondrial electron transport (Klionsky et al., 2008).

Incubation can conveniently be carried out in closed plastic scintillation vials (or in stoppered 25-ml Nalgene Erlenmeyer flasks) in a shaking water bath at 37 °C with a frequency of at least 70 rpm to ensure proper oxygenation and mixing (see Note 8). Hepatocytes are used at a final concentration of approximately 5 mg of dry mass/ml, in a final incubation volume of 2–4 ml.

Typical flasks may contain: 1.50 ml of 2x concentrated Krebs-Henseleit-bicarbonate-Na^+-HEPES, 0.08 ml of 1 M glucose, 0.08 ml of 1 mM cycloheximide, 1.34 ml of water, and 1 ml of cell suspension. Or in the presence of amino acids: 1.50 ml of 2x concentrated Krebs-Henseleit-bicarbonate-Na^+-HEPES, 0.08 ml of 1 M glucose, 0.08 ml of 1 mM cycloheximide, 0.80 ml of 20xPMAA (-valine), 0.54 ml of water, and 1 ml of cell suspension. The cell suspension must not be added to the flasks until all the other components are already present, and the medium must be in equilibrium with 95% O_2 + 5% CO_2 (the gas mixture is humidified by allowing it first to pass through a gas-washing bottle filled with water).

1. The flasks are gassed with 95% O_2 + 5% CO_2, the caps closed, and transferred to the water bath to be incubated for at least 10 min before adding the cells. Once the gas equilibrium has been established, the cap is opened and the cells are added (be sure to mix the cells well before pipetting them) while gassing the flasks; the flask is then closed again.
2. The cells are incubated for the appropriate time (e.g., 0, 30, 60, 90, or 120 min). It is advisable to use a separate flask for each time point. The reactions are terminated by removal of, for example, 1 ml of the suspension, which is added to 0.3 ml of 14% $HClO_4$ in a microcentrifuge tube (on ice) (see Note 9).
3. Allow the cells to denature completely for at least 15 min before centrifuging them for 1 min in a microcentrifuge.
4. Then transfer 1 ml of the supernatant fraction into an empty microcentrifuge tube (on ice), and adjust the pH of the sample to approximately pH 7 by adding 2 M KOH + 0.3 M MOPS. The MOPS present in this mixture buffers the solution at around pH 7, which makes it possible to carry out a titration with KOH. The KOH must be added to the tube slowly while mixing on a vortex (hold the tube firmly) to avoid local alkalinization of the sample, which could impair the stability of some metabolites (e.g., ATP). During neutralization, a crystalline precipitate of $KClO_4$ will form that can easily be removed later by centrifuging.
5. The neutralized samples can then be used to analyze amino acids or other metabolites. They can also be stored frozen safely.

Valine (and other amino acids) can be analyzed by HPLC, after pre-column derivatization with *o*-phthaldialdehyde, using a Supelco(sil) 40-μm reversed-phase C18 column (4.6 × 150 mm) with a Supelco 40-μm reversed-phase C18 guard column (4.6 × 50 mm) (Wu et al., 1996; Wu and Knabe, 1994). Another method uses precolumn derivatization with dansyl chloride (Niioka et al., 1998). Quantification involves comparison with amino acid standards. The HPLC method is much more sensitive than conventional amino acid analysis.

Another way to measure proteolysis is to label long-lived proteins by administering an intravenous injection of 50 μCi [^{14}C]valine to the rats 24 h before preparing the hepatocytes (Seglen et al., 1979). Proteolysis in these labeled hepatocytes can be determined from the appearance of acid-soluble radioactivity, as described in section 2.2.

The inhibition of total proteolysis produced by the amino acid mixture (final concentration, 4xPMAA) or by 3-methyladenine (10 mM) in the hepatocytes, as measured by the appearance of valine, is 60%–70%. The 3-methyladenine-sensitive (or amino acid–sensitive) fraction corresponds to the autophagic proteolysis.

A short overview of the various solutions required for the preparation and incubation of the hepatocytes can be found in Table 4.1.

Table 4.1 Solutions required for the preparation and incubation of rat hepatocytes

Solution	Use
Preparation of hepatocytes	
Salt solution (20x concentrated)	Component of perfusion medium
NaHCO$_3$ (20x concentrated)	Component of perfusion medium
Na$^+$-HEPES, pH 7.4 (100x concentrated)	Component of perfusion medium
CaCl$_2$ (1,000x concentrated)	Component of perfusion medium, to be added after Ca^{2+}-free perfusion
Collagenase	Component of perfusion medium, to be added together with Ca^{2+}
Trypan blue	Hepatocyte quality test
Incubation of hepatocytes	
Krebs-Henseleit bicarbonate plus Na-HEPES (2x concentrated)	Component of incubation medium
Glucose 1 M (50x concentrated)	Component of incubation medium; ATP-producing substrate
Cycloheximide 1 mM (50x concentrated)	Component of incubation medium; inhibitor of protein synthesis
3-Methyladenine 50 mM (5x concentrated)	Component of incubation medium (optional); autophagy inhibitor
PMAA (20x concentrated)	Component of incubation medium (optional); amino acids inhibit autophagy
HClO$_4$	Incubation stop
Analysis	
KOH 2 M + 0.3 M MOPS	Neutralization of samples

3.2.3. Perfusion of hepatocytes

Another way to measure proteolysis is to incubate freshly isolated hepatocytes in a perifusion chamber in a flow-through system, resembling the isolated perfused liver (Häussinger *et al.*, 1999). A full description of the system has been provided elsewhere (Groen *et al.*, 1982; van der Meer and Tager, 1976). The advantage of this method is that autophagic proteolysis (reflected by the appearance of valine in the effluent perifusate) can be studied under true steady-state conditions (Leverve *et al.*, 1987). Samples of the cell suspension can also be taken and used to analyze the intracellular content of metabolites, if necessary. The disadvantage of the method is that it requires special equipment, and that each hepatocyte preparation can be used to investigate only a few experimental conditions.

3.3. Notes

1. Adding 10 mM Na$^+$-HEPES helps to dampen changes in pH. Such a change can occur when the CO_2 concentration in the gas phase drops e.g., when the liver is removed and dispersed after collagenase perfusion, or later while the hepatocytes are being incubated, when the incubation flasks are opened for cell sampling. In these situations the CO_2 concentration drops rapidly, and the pH will rise immediately.
2. It is essential not to mix the $CaCl_2$ with the concentrated bicarbonate solution before adding the water; this is because a precipitate of $CaCO_3$ would be formed, which is difficult to dissolve, even after the final volume has been adjusted to 1 liter. In the original recipe (Krebs and Henseleit, 1932), the $CaCl_2$ concentration used was 2.5 mM. However, this concentration was combined with 4% (w/v) serum albumin: under these conditions approximately half of the Ca^{2+} would be bound to the serum albumin, so that the free, unbound, Ca^{2+} concentration is actually 1.3 mM.
3. In this amino acid mixture, arginine has been replaced by ornithine, because in hepatocytes arginine is immediately converted to ornithine and urea. This does not mean that arginine is entirely absent: hepatocytes are able to maintain a low steady-state concentration of arginine via the urea cycle. However, other cell types have no urea cycle, and so arginine must be included in the mix of amino acids.

    It is also possible to eliminate isoleucine instead of valine, because isoleucine, similar to valine, does not affect autophagy. This may be preferred if the separation of valine by the HPLC method of amino acid analysis is not perfect. Branched-chain amino acids undergo very limited catabolism in rat hepatocytes, so the ratio in which they are produced by proteolysis is the same as that at which they are present in liver proteins (Leverve *et al.*, 1987; Ward and Mortimore, 1978).

4. Glycogen stores are depleted in hepatocytes obtained from fasted animals. This can be an advantage, because when hepatocytes from fed animals are incubated, the glycogen stores may become depleted during the incubation, depending on the operating conditions. This results in a drastic change in cell metabolism, which may go unnoticed unless one is aware of this problem.
5. Some people prefer to wash and store the cells at room temperature. However, considerable metabolism occurs at room temperature. This means that adequate oxygenation may become crucial if local anoxia develops in the cell sediment. Another important point to note is that at room temperature the amount of amino acid (e.g., valine) formed as a result of proteolysis increases with the storage time, and so several zero-time controls have to be carried out throughout the experiment.
6. Glucose is added as energy-donating substrate. Its concentration is a matter of choice (e.g., 10–20 mM; the concentration of glucose in the portal vein of small animals is higher than the 5 mM found in blood in man). Because glucokinase has a high K_m for glucose, the rate of glucose consumption is approximately proportional to the concentration of glucose added. It is important to note that the endogenous fatty acids present in hepatocytes can also be used for ATP production.
7. Synthesis of Atg proteins, which is governed by the FoxO3 transcription factor (Mammucari *et al.*, 2007), probably occurs after more prolonged incubation of hepatocytes.
8. In our hands, glass vials (unless collagen coated) are not suitable, because the cells do not remain viable; presumably the cells are mechanically damaged by shear forces under these conditions.
9. Although the acidified samples do not need to be kept on ice for amino acid analysis, it is advisable to keep them on ice because it may be necessary to measure other metabolites that may be acid labile and/or temperature sensitive. One such example is ATP, the measurement of which provides useful information about the potential toxicity of compounds being tested for their effect on autophagy. ATP is extremely unstable in an acid environment at room temperature.

4. Other Methods for Measuring Autophagic Flux

Other methods have been developed for measuring autophagic flux (Klionsky *et al.*, 2008). Some of these methods are based on the turnover of specific autophagy cargoes. The protein LC3-II, which is involved in the formation of autophagosomes, remains associated with the inner membrane of the resulting autophagosome. This LC3-II fraction is then transported into the lysosomal compartment, where it is subsequently degraded. The turnover of LC3-II can be assayed by immunoblotting in the presence and absence of lysosomal inhibitors (inhibitors of cathepsins or of lysosomal acidification) (Sarkar et al., 2005; Tanida et al., 2005; also see the chapter by Kimura et al., in this volume). The lysosomal turnover of protein p62, which interacts with LC3-II, can also be used to monitor the autophagic flux (Mizushima and Yoshimori, 2007). Recently the use of an mRFP-GFP tandem fluorescent-tagged LC3 has been proposed for visualizing the autophagic flux (Kimura et al., 2007). The method is based on the differential quenching of the fluorescence emitted by the red and green probes in the lysosomal environment. This method is discussed in detail elsewhere (Kimura et al., 2007). Briefly, autophagosomes appear yellow and lysosomes red, because of the disappearance of the GFP-LC3 signal at an acidic pH. When the autophagic flux is interrupted, the red fluorescence is no longer detectable, and only yellow dots are visible. Recently, Elazar and colleagues (Shvets et al., 2008) introduced a fluorescence-activated cell sorter (FACS)–based method to quantify the turnover of GFP-LC3 as a measure of the autophagic activity in mammalian cells (see the chapter by Shvets and Elazar in this volume). In this method, the decrease of the fluorescence reflects the lysosomal delivery of LC3 to the lysosomal compartment. FACS analysis presents several advantages. For example, this method is quantitative and enables one to perform large-scale screens for unknown modulators of autophagy.

Another interesting method for estimating autophagic flux is based on sequestration of the autophagic cargo (e.g., cytosolic markers such as lactate dehydrogenase; combined with an inhibitor of lysosomal function to prevent the intralysosomal degradation of lactate dehydrogenase) or electroinjected, metabolically inert [^{3}H]raffinose (see the chapter by Seglen et al., in this volume). The assay measures the rate of transfer of the marker from the soluble (cytosol) to the particulate (sedimentable) cell fraction that contains the autophagic compartments. The method was originally developed by Seglen and coworkers for hepatocytes but can, in principle, also be applied to other cell types (see Klionsky et al., 2007, and references therein).

The advantages and limitations of all these methods, including proteolysis, have been discussed in detail in a recent review of the guidelines to monitor autophagy (Klionsky et al., 2008). Readers can consult this review for more information about the various aspects of autophagic flux.

5. Pitfalls

A potential pitfall not discussed in the review by Klionsky et al. (2008) is that ammonia is produced in cell cultures as a result of the hydrolysis of glutamine (a major component of culture media) via glutaminase and by amino acid oxidation. With the exception of hepatocytes, cells cannot synthesize urea, and so this ammonia may accumulate to reach concentrations that increase with culture time and cell density. Because ammonia is an acidotropic agent, the lysosomes may swell, the lysosomal pH increase, and intralysosomal degradation may be inhibited, depending on the concentration of ammonia that is reached. Under such conditions, intermediate autophagic structures, such as LC3-II, may also accumulate, depending on the relative inhibitory effects of amino acids on the autophagic sequestration step and of ammonia on hydrolysis in the lysosomes. This can lead to the erroneous conclusion that autophagy has increased, when in fact flux through the autophagic pathway has actually stopped at the lysosome level. The same may be true of hepatocytes incubated under conditions with insufficient ornithine present for the urea cycle to operate properly or incubated in media that do not contain bicarbonate, one of the substrates required for urea synthesis. It is important to point out that even under anoxic conditions, some ammonia will still be formed via the glutaminase reaction, which does not require oxygen.

REFERENCES

Berry, M. N. (2000). Isolated hepatocytes: Forty years on. *In* "The hepatocyte review." (M. N. Berry and A. M. Edwards, eds.), pp. 1–19. Kluwer Academic publishers, Dordrecht, The Netherlands.

Blommaart, E. F., Krause, U., Schellens, J. P., Vreeling-Sindelarova, H., and Meijer, A. J. (1997). The phosphatidylinositol 3-kinase inhibitors wortmannin and LY294002 inhibit autophagy in isolated rat hepatocytes. *Eur. J. Biochem.* **243,** 240–246.

Groen, A. K., Sips, H. J., Vervoorn, R. C., and Tager, J. M. (1982). Intracellular compartmentation and control of alanine metabolism in rat liver parenchymal cells. *Eur. J. Biochem.* **122,** 87–93.

Häussinger, D., Schliess, F., Dombrowski, F., and vom Dahl, S. (1999). Involvement of p38MAPK in the regulation of proteolysis by liver cell hydration. *Gastroenterology* **116,** 921–935.

Kanazawa, T., Taneike, I., Akaishi, R., Yoshizawa, F., Furuya, N., Fujimura, S., and Kadowaki, M. (2004). Amino acids and insulin control autophagic proteolysis through different signaling pathways in relation to mTOR in isolated rat hepatocytes. *J. Biol. Chem.* **279,** 8452–8459.

Kimura, S., Noda, T., and Yoshimori, T. (2007). Dissection of the autophagosome maturation process by a novel reporter protein, tandem fluorescent-tagged LC3. *Autophagy* **3,** 452–460.

Klionsky, D. J. (2007). Autophagy: from phenomenology to molecular understanding in less than a decade. *Nat. Rev. Mol. Cell Biol.* **8,** 931–937.

Klionsky, D. J., Abeliovich, H., Agostinis, P., Agrawal, D. K., Aliev, G., Askew, D. S., Baba, M., Baehrecke, E. H., Bahr, B. A., Ballabio, A., *et al.* (2008). Guidelines for the use and interpretation of assays for monitoring autophagy in higher eukaryotes. *Autophagy* **4,** 151–175.

Klionsky, D. J., Cuervo, A. M., and Seglen, P. O. (2007). Methods for monitoring autophagy from yeast to human. *Autophagy* **3,** 181–206.

Komatsu, M., Ueno, T., Waguri, S., Uchiyama, Y., Kominami, E., and Tanaka, K. (2007). Constitutive autophagy: vital role in clearance of unfavorable proteins in neurons. *Cell Death Differ.* **14,** 887–894.

Kovács, A. L., and Seglen, P. O. (1981). Inhibition of hepatocytic protein degradation by methylaminopurines and inhibitors of protein synthesis. *Biochim. Biophys. Acta* **676,** 213–220.

Krebs, H. A., and Henseleit, K. (1932). Untersuchungen über die Harnstoffbildung im Tierkörper: Hoppe-Seyler's. *Z. Physiol. Chem.* **210,** 33–66.

Kuma, A., Hatano, M., Matsui, M., Yamamoto, A., Nakaya, H., Yoshimori, T., Ohsumi, Y., Tokuhisa, T., and Mizushima, N. (2004). The role of autophagy during the early neonatal starvation period. *Nature* **432,** 1032–1036.

Lavieu, G., Scarlatti, F., Sala, G., Carpentier, S., Levade, T., Ghidoni, R., Botti, J., and Codogno, P. (2006). Regulation of autophagy by sphingosine kinase 1 and its role in cell survival during nutrient starvation. *J. Biol. Chem.* **281,** 8518–8527.

Leverve, X. M., Caro, L. H., Plomp, P. J., and Meijer, A. J. (1987). Control of proteolysis in perifused rat hepatocytes. *FEBS Lett.* **219,** 455–458.

Levine, B., and Klionsky, D. J. (2004). Development by self-digestion: molecular mechanisms and biological functions of autophagy. *Dev. Cell* **6,** 463–477.

Lum, J. J., DeBerardinis, R. J., and Thompson, C. B. (2005). Autophagy in metazoans: cell survival in the land of plenty. *Nat. Rev. Mol. Cell Biol.* **6,** 439–448.

Mammucari, C., Milan, G., Romanello, V., Masiero, E., Rudolf, R., Del Piccolo, P., Burden, S. J., Di Lisi, R., Sandri, C., Zhao, J., *et al.* (2007). FoxO3 controls autophagy in skeletal muscle in vivo. *Cell Metab.* **6,** 458–471.

Meijer, A. J., Gustafson, L. A., Luiken, J. J., Blommaart, P. J., Caro, L. H., Van Woerkom, G. M., Spronk, C., and Boon, L. (1993). Cell swelling and the sensitivity of autophagic proteolysis to inhibition by amino acids in isolated rat hepatocytes. *Eur. J. Biochem.* **215,** 449–454.

Meijer, A. J., Lof, C., Ramos, I. C., and Verhoeven, A. J. (1985). Control of ureogenesis. *Eur. J. Biochem.* **148,** 189–196.

Mizushima, N. (2005). The pleiotropic role of autophagy: from protein metabolism to bactericide. *Cell Death Differ.* **12**(suppl. 2), 1535–1541.

Mizushima, N., Levine, B., Cuervo, A. M., and Klionsky, D. J. (2008). Autophagy fights disease through cellular self-digestion. *Nature* **451,** 1069–1075.

Mizushima, N., and Yoshimori, T. (2007). How to interpret LC3 immunoblotting. *Autophagy* **3,** 542–545.

Niioka, S., Goto, M., Ishibashi, T., and Kadowaki, M. (1998). Identification of autolysosomes directly associated with proteolysis on the density gradients in isolated rat hepatocytes. *J. Biochem.* **124,** 1086–1093.

Ohsumi, Y. (2001). Molecular dissection of autophagy: two ubiquitin-like systems. *Nat. Rev. Mol. Cell Biol.* **2,** 211–216.

Petiot, A., Ogier-Denis, E., Blommaart, E. F., Meijer, A. J., and Codogno, P. (2000). Distinct classes of phosphatidylinositol 3′-kinases are involved in signaling pathways that control macroautophagy in HT-29 cells. *J. Biol. Chem.* **275,** 992–998.

Punnonen, E. L., Marjomaki, V. S., and Reunanen, H. (1994). 3-Methyladenine inhibits transport from late endosomes to lysosomes in cultured rat and mouse fibroblasts. *Eur. J. Cell Biol.* **65,** 14–25.

Rubinsztein, D. C. (2006). The roles of intracellular protein-degradation pathways in neurodegeneration. *Nature* **443,** 780–786.

Sarkar, S., Floto, R. A., Berger, Z., Imarisio, S., Cordenier, A., Pasco, M., Cook, L. J., and Rubinsztein, D. C. (2005). Lithium induces autophagy by inhibiting inositol monophosphatase. *J. Cell Biol.* **170,** 1101–1111.

Scarlatti, F., Bauvy, C., Ventruti, A., Sala, G., Cluzeaud, F., Vandewalle, A., Ghidoni, R., and Codogno, P. (2004). Ceramide-mediated macroautophagy involves inhibition of protein kinase B and up-regulation of Beclin 1. *J. Biol. Chem.* **279,** 18384–18391.

Seglen, P. O. (1976). Preparation of isolated rat liver cells. *Methods Cell Biol.* **13,** 29–83.

Seglen, P. O., and Gordon, P. B. (1982). 3-Methyladenine: specific inhibitor of autophagic/lysosomal protein degradation in isolated rat hepatocytes. *Proc. Natl. Acad. Sci. USA* **79,** 1889–1892.

Seglen, P. O., Grinde, B., and Solheim, A. E. (1979). Inhibition of the lysosomal pathway of protein degradation in isolated rat hepatocytes by ammonia, methylamine, chloroquine and leupeptin. *Eur. J. Biochem.* **95,** 215–225.

Shvets, E., Fass, E., and Elazar, Z. (2008). Utilizing flow cytometry to monitor autophagy in living mammalian cells. *Autophagy* **4,** 621–628.

Tanida, I., Minematsu-Ikeguchi, N., Ueno, T., and Kominami, E. (2005). Lysosomal turnover, but not a cellular level, of endogenous LC3 is a marker for autophagy. *Autophagy* **1,** 84–91.

Tolkovsky, A. M., Xue, L., Fletcher, G. C., and Borutaite, V. (2002). Mitochondrial disappearance from cells: a clue to the role of autophagy in programmed cell death and disease? *Biochimie.* **84,** 233–240.

Ueno, T., Sato, W., Horie, Y., Komatsu, M., Tanida, I., Yoshida, M., Ohshima, S., Mak, T. W., Watanabe, S., and Kominami, E. (2008). Loss of Pten, a tumor suppressor, causes the strong inhibition of autophagy without affecting LC3 lipidation. *Autophagy* **4,** 692–700.

van der Meer, R., and Tager, J. M. (1976). A simple method for the perfusion of isolated liver cells. *FEBS Lett.* **67,** 36–40.

van Sluijters, D. A., Dubbelhuis, P. F., Blommaart, E. F., and Meijer, A. J. (2000). Amino-acid-dependent signal transduction. *Biochem. J.* **351,** 545–550.

Ward, W. F., and Mortimore, G. E. (1978). Compartmentation of intracellular amino acids in rat liver. Evidence for an intralysosomal pool derived from protein degradation. *J. Biol. Chem.* **253,** 3581–3587.

Wu, G., Bazer, F. W., Tuo, W., and Flynn, S. P. (1996). Unusual abundance of arginine and ornithine in porcine allantoic fluid. *Biol. Reprod.* **54,** 1261–1265.

Wu, G., and Knabe, D. A. (1994). Free and protein-bound amino acids in sow's colostrum and milk. *J. Nutr.* **124,** 415–424.

CHAPTER FIVE

Sequestration Assays for Mammalian Autophagy

Per O. Seglen,* Anders Øverbye,* *and* Frank Sætre*

Contents

1. Introduction	64
2. Membrane-Impermeant Autophagy Probes: Introduction into Cytosol by Reversible Electropermeabilization (Electroporation) of the Plasma Membrane	67
2.1. Preparation and incubation of rat hepatocytes	67
2.2. Electropermeabilization equipment	68
2.3. Electropermeabilization procedure	69
3. Damage-Induced Cell Permeabilization	69
4. Electrodisruption: A Simple Method for the Separation of Sedimentable from Soluble Cell Components	70
4.1. Electrodisruption procedure	70
5. Electroinjected Sugars as Autophagic Sequestration Probes	71
5.1. Raffinose sequestration	72
5.2. Sucrose sequestration: Selective digitonin extraction of autophagocytosed sucrose	72
5.3. Use of sucrose and invertase to study autophagic-endocytic interactions	73
5.4. Lactose sequestration: Addressing individual steps in the autophagic-lysosomal pathway	73
5.5. Autophagic lactolysis: An assay of autophagic flux	74
6. Cytosolic Proteins as Autophagic Sequestration Probes	74
6.1. Intralysosomal accumulation of autophagocytosed lactic dehydrogenase (LDH)	75
6.2. LDH sequestration assay	75
6.3. Prelysosomal accumulation of LDH	77

* Proteomics Section DNR, Department of Cell Biology, Institute for Cancer Research, The Norwegian Radium Hospital, Oslo, Norway

7. Autophagic Fragment Generation: An Autophagic Cargo Assay
   Applicable to Whole Cells   77
   7.1. Measurement of p10(BHMT) formation in hepatocytes
        by immunoblotting   78
8. Concluding Remarks   79
References   80

Abstract

Macroautophagic activity is most directly and precisely measured by a cargo sequestration assay. Long-lived, cytosolic proteins that are degraded exclusively by the autophagic-lysosomal pathway, such as lactate dehydrogenase (LDH) are suitable as endogenous sequestration probes. Autophagic sequestration is measured as transfer of the protein from the soluble (cytosolic) to the sedimentable (organelle-containing) cell fraction, using leupeptin or other proteinase inhibitors to block inactivation and degradation of the protein inside autophagic vacuoles. A convenient separation method is electrodisruption of the cells, followed by sedimentation of the organelle fraction through a Nycodenz density cushion.

 A promising variant of the cargo assay is to use a protein probe that is processed by the autophagic-lysosomal pathway so as to generate an intravacuolar fragment. Because there is no cytosolic background, subcellular fractionation is unnecessary, allowing the use of the autophagic fragment assay to measure autophagic activity in whole cells. In hepatocytes, a small fragment, p10(BHMT), made by autophagic processing of the enzyme betaine:homocysteine methyltransferase, thus accumulates in an autophagy-dependent manner in the presence of leupeptin.

 Autophagic sequestration can also be measured by using exogenous cargo probes, such as radiolabeled di- and trisaccharides, which can be loaded into the cytosol of hepatocytes by reversible electrodisruption or mechanical stress. Raffinose is the preferable probe for measurement of autophagic activity, whereas sucrose (which can be hydrolyzed in amphisomes and lysosomes by added endocytosed invertase) and lactose (which is hydrolyzed in lysosomes by the endogenous β-galactosidase) are useful for dissection of the various steps in the autophagic-lysosomal pathway and for studying autophagic-endocytic interactions. Furthermore, the intralysosomal hydrolysis of autophagocytosed lactose can be measured in whole cells (as formation of the hydrolysis product, galactose), thus providing a background-free assay (autophagic lactolysis) of the overall autophagic-lysosomal pathway.

1. INTRODUCTION

 For a quantitative measurement of macroautophagic activity, the most relevant functional parameter is the fraction of cytoplasm autophagically sequestered per unit time. Cytoplasmic volume fractions are difficult to measure directly except by relatively cumbersome ultrastructural-morphometric methods (Kovács et al., 1981, 1982; Seglen and Reith, 1976), but in principle any cargo marker (i.e., an individual cytoplasmic component sequestered in bulk along with the remainder of the cytoplasm) can serve as an indicator of macroautophagic activity. The indicator should, however, not be subject to sequestration by any selective autophagic process such as pexophagy, mitophagy, reticulophagy, aggrephagy, and so on (Klionsky et al., 2007a). This requirement would tend to exclude most—probably all—organelle markers, in particular those associated with the autophagic machinery itself (Tanida et al., 2005).

Endogenous autophagy probes should thus preferably be sought among the cytosolic elements of the cell. Although a few cytosolic proteins are selectively sequestered by autophagy (Xiao, 2007; Yu et al., 2006), the majority are not (Fengsrud et al., 2000), and a handful of cytosolic enzymes, chosen on the basis of their widely different overall half-lives, are found to be autophagically sequestered at identical rates (Kopitz et al., 1990). The choice of cytosolic macroautophagy markers would, therefore, be legion and adaptable to any researcher's interests and capabilities. In the authors' laboratory, the ubiquitous cytosolic enzyme lactate dehydrogenase is routinely used, quantified by an enzymatic activity assay (Kopitz et al., 1990). Other researchers have used different proteins, such as fatty acid synthase (Egner et al., 1993) or neomycin phosphotransferase (Nimmerjahn et al., 2003).

An advantage of using a cytosolic autophagy marker is that, apart from the fraction present in autophagic vacuoles, the marker is soluble and nonsedimentable. It is, therefore, not necessary to perform any extensive subcellular fractionation to separate sequestered from nonsequestered marker; a mere high-speed centrifugation of a cellular homogenate, to separate the sedimentable from the nonsedimentable marker fraction, will do. In the authors' laboratory, an even simpler method is used: the plasma membranes of the cells (hepatocytes) are broken up by a single high-voltage electrodisruption in a nonionic medium, and the organelle- and cytoskeleton-containing cell corpses are separated from the soluble cell sap by low-speed centrifugation (Gordon and Seglen, 1982; Kopitz et al., 1990; Seglen and Gordon, 1984). A criterion for a good autophagy marker is that the sedimentable background (not due to autophagy) should be as low as possible, and it may, therefore, be useful to play around with the ionic composition, pH, and so on to reduce the nonspecific adsorption of the marker to sedimentable structures as much as possible (Kopitz et al., 1990).

An autophagic marker protein should ideally be targeted minimally by a degradation mechanism other than the autophagic-lysosomal pathway and should, therefore, be a long-lived protein (Seglen and Gordon, 1982). Its lysosomal degradation must of course be blocked to enable autophagic activity to be measured as the accumulation of sequestered marker. With lactate dehydrogenase (LDH) and several other cytosolic marker enzymes, a single proteinase inhibitor, leupeptin, prevents the disappearance of enzyme activity completely, presumably by blocking intralysosomal enzyme inactivation/degradation (Kopitz et al., 1990). Other markers may require other proteinase inhibitors or inhibitor combinations, in which case it must be ensured that the inhibitors do not affect the autophagic sequestration step (as has been checked in the case of leupeptin). In principle, more general inhibitors of lysosome function, such as neutralizing agents (ammonium chloride; various amines) (Seglen, 1975, 1983; Seglen and Gordon, 1980) or proton pump inhibitors like bafilomycin A (Mousavi et al., 2001) may be used, but these are likely to perturb cell function in a more nonspecific manner and may affect autophagic sequestration indirectly. Finally, vacuole fusion inhibitors such as vinblastine can be used to prevent the degradation of autophagy markers by suppressing their transfer from autophagosomes to amphisomes and lysosomes (Kovács et al., 1982; Strømhaug et al., 1998). Vinblastine may thus provide direct information about the very initial sequestration step, and although it suppresses the sequestration activity significantly (Gordon et al., 1992), it should be useful for the comparison of relative sequestration rates.

Proteolytic fragments specifically generated by the autophagic-lysosomal pathway (Furuya et al., 2001; Mercer et al., 2008; Øverbye, 2007) are attractive marker candidates, allowing, in principle, autophagy to be measured in whole cells. At the moment, autophagic fragment generation has been demonstrated only in the case of the abundant liver- and kidney-specific enzyme, betaine:homocysteine methyl transferase (BHMT), limiting the use of native fragments to cells from these tissues. It is, however, very likely that cytosolic proteins with a more ubiquitous tissue distribution may also be processed by the autophagic-lysosomal pathway, allowing for a more general applicability of this assay principle. Tissue specificity can, furthermore, be circumvented by using vector-mediated transfection of a BHMT fusion protein (Mercer et al., 2008), as Dennis and Mercer describe in this volume. It should be noted that proteolytic fragments can be expected to be further degraded in autophagic-lysosomal vacuoles, necessitating the use of cell-permeable proteinase inhibitors capable of blocking the processing enzyme(s) involved in fragment degradation without affecting the enzyme(s) involved in fragment generation.

In addition to endogenous, cytosolic autophagy markers, exogenous markers may be introduced into the cytosol by various reversible cell permeabilization methods, allowing, for example, the use of radioactive autophagy assays. Because the autophagy probe must be membrane impermeant to be sequestered exclusively by autophagy, the major problem is to get it across the plasma membrane without excessive damage to the cell. The authors' experience is limited to the use of reversible electropermeabilization of rat hepatocytes, which allows certain small molecules (<1000 Da) to diffuse into the cells, such as radioactive di- and trisaccharides

(MW 340–500) or eosin (a red stain, MW 648, suitable for visual inspection of permeabilization efficiency), but not NAD (MW 663) or ATP (MW 507), and trypan blue (MW 960) only partially (Gordon and Seglen, 1982; Høyvik *et al.*, 1986; Seglen *et al.*, 1986b). Other permeabilization or microinjection methods may allow the loading of larger molecules, such as proteins (Felgner *et al.*, 1987; Okada and Rechsteiner, 1982; Schaible *et al.*, 1999; Stacey and Allfrey, 1977; Walev *et al.*, 2001), but the reversibility of the permeabilization (in a quantitative sense) is not clear. Once the plasma membrane has become resealed, autophagic activity can be measured as the transfer of the internalized marker from a soluble to a sedimentable cell fraction, just as with endogenous autophagy markers. A general review of autophagic sequestration assays was recently published (Klionsky *et al.*, 2007b); in the present chapter, a more detailed description and discussion of the methods used in the authors' laboratory will be given.

2. Membrane-Impermeant Autophagy Probes: Introduction into Cytosol by Reversible Electropermeabilization (Electroporation) of the Plasma Membrane

Optimal conditions for reversible electropermeabilization (Zimmermann *et al.*, 1976) have been worked out for rat hepatocytes in suspension, as previously described in detail (Gordon and Seglen, 1982, 1986; Seglen and Gordon, 1984; Seglen *et al.*, 1986a). Because these are exceptionally big (mostly polyploid) cells, the permeabilization settings may have to be individually adapted to fit other cell types. For cultured cells, grown on a solid surface, other types of permeabilization equipment will have to be used.

2.1. Preparation and incubation of rat hepatocytes

A. Suspensions of isolated rat hepatocytes are prepared according to the standard two-step collagenase perfusion method, described in detail elsewhere (Seglen, 1976, 1993, 1994, 1998). In brief, we use the following protocol:

i. The liver of an approximately 280-g male Wistar rat is perfused at 37 °C (using a water-jacketed glass coil for heating) through a 1.5-mm portal vein cannula (nylon tubing), allowing efflux through the open-cut caval vein, while the liver is excised and placed in a glass dish with a support net and a conical outlet. The liver is perfused at 50 ml/min, first for 5 min one-way with a Ca^{2+}-free perfusion buffer (8.3 g/l NaCl, 0.5 g/l KCl, 2.4 g/l HEPES, 0.24 g/l NaOH, pH 7.4, at 37 °C) to disrupt cell-to-cell junctions, then for 5–10 min with a recirculating collagenase buffer (0.5 g/l collagenase, Sigma type IV, 0.7 g/l $CaCl_2 \cdot 2$ H_2O, 4.0 g/l NaCl, 0.5 g/l KCl, 24 g/l HEPES, 2.64 g/l NaOH, pH 7.6, at 37 °C) to disrupt cell-to-matrix attachment.

   ii. Finally, the liver is transferred to a square 10 × 10 cm Sterilin plastic dish containing 80 ml of ice-cold suspension buffer (1.8 g/l pyruvic acid, 4.0 g/l NaCl, 0.4 g/l KCl, 0.15 g/l KH_2PO_4, 0.1 g/l Na_2SO_4, 0.41 g/l $MgCl_2 \cdot 6H_2O$, 0.18 g/l $CaCl_2 \cdot 2$ H_2O, 7.2 g/l HEPES, 6.9 g/l TES, 6.5 g/l Tricine, 2.1 g/l NaOH, pH 7.6, at 37 °C), and the cells are gently raked out of the connective tissue with a stainless-steel comb while holding the organ with forceps.

   iii. After filtration through 250-μm nylon mesh, the cell suspension is incubated for 30 min at 37 °C in a wide (20 cm) glass Petri dish with gentle rocking. The cells are then cooled to 0 °C, carefully filtered through 250 and 100 μm nylon mesh, washed three times in ice-cold buffer (5 min at 200 rpm in 100-ml beakers), and finally resuspended in suspension buffer at approximately 100 mg cells (wet mass) per milliliter.

   B. The cells are incubated at 37 °C either as 0.4-ml aliquots (300 μl of cell suspension *plus* 100 μl of 0.9% NaCl or various additives in isotonic solution) in rapidly shaking 15-ml glass centrifuge tubes (215 rpm rotatory shaking suspension experiments, or as 200 μl of cell suspension *plus* 1.8 ml of 0.9% NaCl or isotonic additives in stationary 6-cm suspension dishes (Sarstedt, Germany) for short- or long-term monolayer experiments. Tube incubation allows the rapid processing of large sample numbers, but the incubation is limited to ~2.5 h due to the mechanical stress of shaking. Dish incubation can proceed for longer times (~6 h under nonsterile conditions; ~3 d in a sterile growth/maintenance medium), and lower drug concentrations can be used, but cell harvesting is more cumbersome, allowing fewer samples to be processed.

2.2. Electropermeabilization equipment

For electropermeabilization and electrodisruption (see subsequently), the authors' laboratory uses homemade equipment. The main unit is a Perspex block in which two flat parallel 1 × 5 cm stainless-steel electrode plates are spaced 1 cm apart, forming a chamber with a 1-cm^2 bottom and 5-cm height, allowing the filling of ~4.5 ml of cell suspension. The electrodes are connected through a load/unload switch to a capacitor, selected from an assembly of capacitors ranging from 0.2 to 1.2 microfarad (μF) capacitance, charged at the desired voltage (routinely 2 kV) from a power supply. At the turn of the load/unload switch, a permeabilizing pulse is delivered; for recharging of the capacitor, a wait period of ~5 s is allowed.

   In principle, any commercial electroporator equipment can be used for electropermeabilization. If the space between the electrodes is less than 1 cm, the voltage should be proportionately reduced. Because different commercial models have different pulse shapes and other electrical characteristics, optimal permeabilization conditions will have to be worked out experimentally.

2.3. Electropermeabilization procedure

a. In our routine procedure for hepatocyte permeabilization, 1 ml of cell suspension (100 mg of cellular wet wt/ml), which defines an effective electrode area of 1 cm^2, is permeabilized at 0 °C with 5 pulses at 0.6 μF and 2 kV, which gives a pulse duration of approximately 50 μs (exponential decay).
b. Radioactive sugar is then added (e.g., at 1 μCi/ml), and the suspension is left at 0 °C for 1 h for loading of the cells with radioactivity (no shaking needed at 0 °C).
c. The cells are then quickly warmed to 37 °C, and incubated (with gentle shaking to allow oxygenation; e.g., in a flat dish tilted on a rocking platform) for 20 min to achieve complete plasma membrane resealing.
d. The cells are then cooled to 0 °C again, sedimented by gentle centrifugation ($\sim$50g per min, e.g., 5 min at 200 rpm) and washed 3 times with 4 ml of wash buffer (per liter: 8.3 g NaCl, 0.5 g KCl, 0.18 g CaCl$_2\cdot$2H$_2$O, 2.4 g HEPES, 5.5 ml 1 M NaOH, pH 7.4, at 37 °C) to remove all extracellular radioactivity.
e. Finally, the cells are resuspended in the original volume (1 ml) of suspension buffer.

3. Damage-Induced Cell Permeabilization

The principle of electropermeabilization is (according to the present authors, anyway) to disturb the arrangement of the charged molecules in the plasma membrane to such an extent that small, readily reversible lesions are generated, at which the plasma membrane bulges out as hyperpermeable blebs (Gordon *et al.*, 1985b). Such blebs are also formed mechanically when hepatocytic cell contacts are broken during cell isolation (collagenase perfusion), and can in fact be utilized for sugar loading if the freshly prepared cells are incubated immediately at 0 °C (Gordon *et al.*, 1985b) and resealed after 60 min as described earlier. Although the loading efficiency is lower than with electropermeabilization, raising the specific radioactivity of the sugar probe to the desired level can compensate for this effect. Damage-induced sugar loading thus requires no special equipment, and represents the simplest method imaginable for reversible permeabilization of primary hepatocytes.

Cells freshly trypsinized from culture plates would presumably also carry small, reversible lesions and stay permeable at 0 °C. Although this possibility appears not to have been examined or utilized, treatment of fibroblasts with various proteinases, including collagenase and trypsin, has been shown to induce some partially reversible permeabilization (Lemons *et al.*, 1988). Other mechanical permeabilization methods, such as rolling glass beads (Fennell *et al.*, 1991) or scraping cell monolayers off the plate (Sanjuan *et al.*, 2007), probably work on the same principle (i.e., generation of reversible plasma membrane lesions).

4. Electrodisruption: A Simple Method for the Separation of Sedimentable from Soluble Cell Components

In a cargo assay for autophagy, based on the transfer of a soluble (cytosolic) probe to sedimentable (autophagic) organelles, it is advantageous to make the separation between the soluble and the sedimentable cell fraction as simple as possible. The method of electrodisruption, devised in the authors' laboratory (Gordon and Seglen, 1982), is a quick and practical technique that allows the facile processing of large numbers of cell samples. Although the current method requires the cells to be in suspension, it should be possible to develop an electrical apparatus suitable for the disruption of monolayer cells.

The basic principle of electrodisruption is to subject the cells to a single, strong electric pulse in a nonconductive (nonionic) aqueous medium such as, for example, unbuffered, isotonic sucrose. Because the only charges exposed to the fluid medium (apart from the few H^+ and OH^- ions from dissociated water) reside on the cell surface, the current is probably directed specifically to the cells, resulting in a uniform and extensive plasma membrane disruption. Unlike the electropermeabilization described previously, the electrodisruption is irreversible, and all cytosol leaks out (Gordon and Seglen, 1982). Intracellular structure, however, is remarkably unaffected by the electrical discharge, and all organelles, including the autophagic vacuoles (autophagosomes, amphisomes, and lysosomes) appear to be well preserved (Kopitz *et al.*, 1990). Any autophagically sequestered elements, be they electroinjected sugars or endogenous proteins, will thus stay inside the autophagic vacuoles, whereas their nonsequestered cytosolic counterparts will leak out of the electrodisrupted cells (cell corpses). By centrifuging the cell corpses through a density cushion, a complete separation between sedimentable and soluble autophagy probe is achieved.

4.1. Electrodisruption procedure

a. In our routine procedure for electrodisruption, the sample of incubated hepatocytes (0.4 ml containing 30 mg of cells from tubes, or 2 ml

containing 20 mg of cells from dishes) is rapidly cooled by the addition of 4 ml of ice-cold unbuffered sucrose and maintained for a few min at 0 °C.
b. The cells are then centrifuged at 1600 rpm for 5 min and the supernatant fraction aspirated.
c. The cell pellet is washed once with 4 ml of ice-cold unbuffered 10% sucrose and finally resuspended in 0.5 ml of ice-cold unbuffered 10% sucrose.
d. The entire sample is placed in the $1 \times 1 \times 5$-cm electrode chamber described earlier and given a single 2 kV/1.2 μF pulse, which will disrupt >99% of cells (Gordon and Seglen, 1982).
e. The disruptate is then diluted with an equal volume (0.5 ml) of double-strength buffer of a composition depending on the analytical purpose. For enzymatic assays, a phosphate-buffered sucrose (100 mM sodium phosphate, 2 mM dithiothreitol, 2 mM EDTA, and 1.75% sucrose, pH 7.5) is used; for some enzymes (not LDH), 0.01% Tween 20 and 100 μg/ml bovine serum albumin (BSA) are also included to minimize nonspecific adsorption to cell corpses (Kopitz *et al.*, 1990).
f. An aliquot (routinely, 0.4 ml) of the diluted disruptate is taken out for measurement of the total cellular amount of probe (enzyme, radioactivity, immunoblotting); the rest (0.6 ml) is placed on top of a 4-ml Nycodenz density cushion (8% Nycodenz, 2.2% sucrose, 50 mM sodium phosphate, 1 mM dithiothreitol and 1 mM EDTA, pH 7.5, and including as appropriate Tween 20 and BSA), and centrifuged at 4 °C for 30 min at 3750 rpm.
g. The cell corpse pellet is resuspended in 0.5–1.0 ml of buffer (e.g., a phosphate-buffered sucrose: 50 mM sodium phosphate, 1 mM dithiothreitol, 1 mM EDTA, 5.9% sucrose, pH 7.5), matching the composition of the total (diluted) disruptate. The cell corpse pellet contains essentially all of the organelles and autophagic compartments. The radioactivity, enzyme activity or marker protein amount of the pellet fraction is calculated as a percentage of the total to determine the efficiency of sequestration.

5. Electroinjected Sugars as Autophagic Sequestration Probes

The advantages of di- and trisaccharides as autophagy probes are several: many sugar species are available in radiolabeled form; they are small enough to enter cells through the stretched plasma membrane blebs formed by electropermeabilization, yet large enough to be unable to permeate intact biological membranes, thus remaining in the cytosol once the plasma membrane has resealed. In mammalian cells, sugars such as sucrose and raffinose are, conveniently, metabolically inert (Nyberg and Dingle, 1970), being neither synthesized nor metabolized (this will not be the case in yeast cells, where sugar probes, therefore, cannot be used). The disaccharide lactose is specifically cleaved in mammalian lysosomes by the lysosomal β-galactosidase, making this sugar useful as a probe of fusion between lysosomes and prelysosomal vacuoles, or as a probe of lysosomal function (see section 5.4).

5.1. Raffinose sequestration

The sugar found to be most useful as an autophagy probe is raffinose (Seglen *et al.*, 1986b). This trisaccharide was originally studied in the author's laboratory as [^{3}H]raffinose, obtained from New England Nuclear Co., but neither the company nor its tritiated raffinose product is any longer in existence. However, (galactose-6-^{3}H)raffinose is available from American Radiolabelled Chemicals, St. Louis, MO, USA, and should work just as well.

In contrast to the disaccharides (see subsequent sections), electroinjected raffinose is not detectably sequestered by any nonautophagic organelles (Seglen *et al.*, 1986b), hence no subcellular fractionation other than a separation between the nonsedimentable (cytosolic) and the sedimentable cell fraction is required, most easily achieved by electrodisruption and sedimentation of the resulting cell corpses (see section 4.1). The time-dependent transfer of raffinose to the sedimentable fraction that occurs in resealed hepatocytes incubated at 37 °C is completely suppressible by the autophagy inhibitor (Seglen and Gordon, 1982) 3-methyladenine (3MA) (Seglen *et al.*, 1986b).

5.2. Sucrose sequestration: Selective digitonin extraction of autophagocytosed sucrose

The sugar first used as an autophagy probe, the disaccharide sucrose (Gordon and Seglen, 1982), is available as uniformly labeled [^{14}C]sucrose from a number of manufacturers. Unfortunately, in addition to its specific sequestration by autophagy, sucrose is also taken up by mitochondria, by a poorly understood mechanism (Tolleshaug and Seglen, 1985; Tolleshaug *et al.*, 1984). To distinguish between mitochondrial sucrose and autophagically sequestered sucrose (in autophagosomes, amphisomes, and lysosomes) in sedimented cell corpses, a selective extraction step is required. The cholesterol-poor membranes of mitochondria are relatively resistant to the detergent digitonin (Zuurendonk *et al.*, 1979); this detergent will, therefore, selectively extract sucrose from the autophagic vacuoles at low concentrations (0.2–0.5 mg/ml) without affecting mitochondrial sucrose (Gordon *et al.*, 1985a).

a. In our routine procedure, a cell corpse pellet (from ~30 mg [^{14}C] sucrose-loaded hepatocytes) is suspended in 1 ml of 10% sucrose containing 0.3 mg/ml digitonin, 2 mM HEPES and 1 mM EDTA, pH 7.4.
b. The suspended pellet is gently agitated for 10 min at 0 °C.
c. The corpses are then resedimented (at ~90,000g × min) and the supernatant fraction, containing the autophagocytosed radioactivity, is aspirated and measured.

5.3. Use of sucrose and invertase to study autophagic-endocytic interactions

Unlike mammalian cells, yeasts have an enzyme, invertase, that can cleave sucrose. Added yeast invertase can be taken up endocytically by sucrose-loaded mammalian cells and can thus be used to probe the interrelationship between autophagy and endocytosis. By using this strategy, it has been shown that all hepatocellular lysosomes can receive simultaneous input from autophagy (sucrose) and endocytosis (invertase) (Høyvik et al., 1987). Furthermore, autophagocytosed (electroinjected) sucrose accumulated in amphisomes can be completely degraded by endocytosed invertase (Gordon et al., 1992). Nevertheless, intralysosomal sucrose is fully accessible to endocytosed invertase even in the presence of 3MA, indicating the existence of both autophagy-dependent and autophagy-independent endocytic pathways to the lysosome (Gordon et al., 1992).

5.4. Lactose sequestration: Addressing individual steps in the autophagic-lysosomal pathway

Lactose, available as (glucose-1-^{14}C)-labeled lactose from a number of suppliers, has the same drawback as sucrose: it is taken up by mitochondria as well as by autophagy and has to be extracted from cell corpses by digitonin to serve as a specific autophagy probe. Furthermore, lactose is metabolically inert only as long as it stays in the cytosol: upon entry into lysosomes, it is rapidly degraded by the lysosomal β-galactosidase (Høyvik et al., 1986). This opens up some interesting experimental possibilities. First, it makes autophagocytosed lactose a specific marker of prelysosomal autophagic vacuoles (autophagosomes and amphisomes) (Høyvik et al., 1986). Second, any perturbation of autophagic-lysosomal flux will cause an extra accumulation of lactose prior to the block point, enabling the dissection of individual steps in the autophagic pathway (Høyvik et al., 1991) and elucidating their requirements, such as for energy (Plomp et al., 1989). Finally, by allowing lactose-electroloaded cells to endocytose added β-galactosidase, prelysosomal encounters between autophagy and endocytosis can be studied (Gordon and Seglen, 1988). This strategy was used to discover and characterize the amphisome (Berg *et al.*, 1998; Gordon and Seglen, 1988; Høyvik *et al.*, 1991; Strømhaug and Seglen, 1993).

5.5. Autophagic lactolysis: An assay of autophagic flux

Electroinjected lactose can be used to directly measure the final step in the autophagic-lysosomal pathway (i.e., intralysosomal degradation) and can thus provide an assay for the overall autophagic flux. This assay is more clean cut than measurements of endogenous protein degradation, which require the use of an autophagy inhibitor such as 3MA to distinguish autophagic degradation from the high background of nonautophagic (e.g., proteasomal) degradation (Klionsky *et al.*, 2007b), which compromises accuracy. As autophagocytosed lactose reaches the lysosome, the lysosomal β-galactosidase cleaves off the radiolabeled glucose moiety, which can be separated from the remaining lactose (e.g., chromatographically), as detailed elsewhere (Høyvik *et al.*, 1991). Briefly, we use the following protocol:

a. [^{14}C]lactose-electroloaded hepatocytes that have been incubated as 0.4-ml aliquots at 37 °C are harvested by the addition of 0.1 ml of ice-cold 10% trichloroacetic acid (TCA) and kept on ice for 30 min.
b. Following centrifugation for 30 min at 5000 rpm, the supernatant fraction is filtered through a 0.45-μm Millipore filter and neutralized with NaOH.
c. Fifty μl of the neutralized TCA extract is applied onto a 5-μm Supelcosil LC-NH2 column (25 × 4.6 mm) and eluted with 75% acetonitrile at 1.0 ml/min.
d. The early fractions containing [^{14}C]glucose and the late fractions containing [^{14}C]lactose are pooled separately, and the glucose radioactivity expressed as a percentage of the total (glucose + lactose) radioactivity.

The accumulation of lactose-derived [^{14}C]glucose proceeds at a reasonably constant rate in isolated rat hepatocytes incubated under autophagic (amino acid deficient) conditions and is completely inhibitable by 3MA (Høyvik *et al.*, 1991).

6. Cytosolic Proteins as Autophagic Sequestration Probes

Although autophagy over the ages has become less and less selective because of the discovery of an increasing number of organelle- and protein-specific sequestration processes (Klionsky *et al.*, 2007a), basic macroautophagy still exhibits an impressive nonspecificity toward normal cytosolic proteins (Kopitz *et al.*, 1990). Many of these may be potential autophagy probes, but long-lived proteins, with little or no processing along nonautophagic pathways, would be preferable. Because the authors' laboratory measures autophagic sequestration as the transfer of cargo from the soluble (cytosolic) to the sedimentable cell fraction, we have chosen a cytosolic protein with low sedimentable (structure-associated) background: lactic dehydrogenase (LDH).

6.1. Intralysosomal accumulation of autophagocytosed lactic dehydrogenase (LDH)

To stop the intravacuolar inactivation and degradation of autophagocytosed LDH, the cysteine proteinase inhibitor leupeptin (0.3 mM) has been found to be effective. Leupeptin arrests all LDH inactivation/degradation almost immediately, causing the active enzyme to accumulate in autophagic vacuoles at a constant rate that, under steady-state conditions, reflects the rate of autophagic sequestration. An absolute requirement for this use of leupeptin is that the chosen protein probe is not degraded by any leupeptin-resistant proteinase, in which case other inhibitors must be applied. The sequestered LDH retains full enzymatic activity, allowing a standard coupled enzyme assay (based on the UV absorbance of NADH) to be used for quantification (Kopitz et al., 1990). The time of leupeptin addition defines the start of the measurement period, and the net amount of active LDH sequestered in the sedimentable fraction (e.g., cell corpses) during this period (usually 2 h; zero time background subtracted) is calculated as a percentage of the total cellular LDH activity (measured in whole-cell samples). Typical autophagic sequestration rates measured in amino acid–depleted hepatocytes by this method range from 2.5%/h to 4.0%/h (Fig. 5.1B). Activity measurements are preferable to immunoblotting, allowing much better quantification as well as a more precise time point of measurement. Fortunately, LDH is not sequestered by any nonautophagic process, and the measured autophagic sequestration is completely inhibited by 3MA (Kopitz et al., 1990). 3MA may in fact reduce the background of sedimentable LDH activity (measured in the absence of leupeptin) by some 30%, probably corresponding to the steady-state level of autophagocytosed enzyme that is active during transit through the prelysosomal autophagic vacuolar compartment.

6.2. LDH sequestration assay

The measurement of LDH sequestration is performed essentially as described in Kopitz et al. (1990).

a. Hepatocytes (30 mg wet wt. in the case of tube experiments, see section 2.1) that have been incubated for the desired period of time at

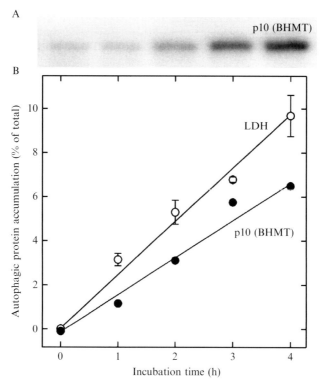

Figure 5.1 Accumulation of p10(BHMT) immunoreactivity in leupeptin-treated rat hepatocytes. Isolated rat hepatocytes were incubated at 37 °C in a saline buffer in the presence of leupeptin (0.3 mM) for the length of time indicated. After lysis of the cells in 1% SDS, the p10(BHMT) fragment was detected by immunoblotting (A) and quantified (in a different experiment) by image analysis (B, filled circles). LDH activity (B, open circles) was measured by an enzymatic assay (Kopitz *et al.*, 1990) as described earlier. Modified from Øverbye *et al.* (2008).

37 °C (in 300 µl of suspension buffer *plus* 100 µl of additives in 0.9% NaCl, including 0.3 mM leupeptin) are rapidly cooled by the addition of 4 ml of ice-cold 10% sucrose.

b. The cells are washed once with 4 ml of ice-cold 10% sucrose and finally resuspended in 0.5 ml of unbuffered 10% sucrose.

c. Following electrodisruption, the disruptate is diluted with an equal volume (0.5 ml) of a phosphate-buffered saline (100 mM sodium phosphate, 2 mM dithiothreitol, 2 mM EDTA and 1.75% sucrose, pH 7.5) and split into one 0.4-ml aliquot used to determine total cellular LDH activity and one 0.6-ml aliquot used to isolate cytosol-free cell corpses (section 4.1).

d. After isolation of the corpses, they are resuspended in 1 ml of a phosphate-buffered sucrose (50 mM sodium phosphate, 1 mM dithiothreitol,

1 mM EDTA and 5.9% sucrose, pH 7.5). A 17-μl sample is mixed with 7 μl of 14 mM pyruvate (with 0.3% Triton X-405) and 70 μl of 0.17 mM NADH (with 0.3% Triton X-405, 5.6 mM EDTA and 56 mM Trizma base, pH 7.5) in an AutoAnalyzer (Technicon SRA 2000). The 0.4-ml aliquot of total (50% diluted) disruptate is further diluted 21x with 50 mM phosphate-buffered sucrose before processing (of a 17-μl sample) in the AutoAnalyzer. LDH activity (Bergmeyer and Berndt, 1974) is measured by the decline in NADH absorbance at 340 nm during a 15-s period, and the absorbance change recorded for the cell corpse pellet is divided by the absorbance change recorded for the total disruptate and multiplied by 100/(0.6 × 21) for conversion to percentage of total.

6.3. Prelysosomal accumulation of LDH

As in the case of autophagocytosed lactose, any block in the autophagic-lysosomal pathway past autophagosome formation will cause the accumulation of autophagocytosed LDH. For example, the microtubule inhibitor vinblastine blocks the translocation-dependent fusion of autophagosomes with lysosomes or endosomes and induces a linear accumulation of sedimentable hepatocellular LDH that directly reflects the rate of autophagic sequestration (Kopitz et al., 1990). Although the measured rate is lower than with leupeptin, because of some inhibition of the sequestration step by vinblastine, the measurement is independent of subsequent lysosomal degradation and no proteinase inhibitor is required. Such an autophagosomal sequestration assay can be used to directly compare the rates of autophagic sequestration of different proteins (Kopitz et al., 1990), and allows, for example, the detection of any selective autophagy.

7. Autophagic Fragment Generation: An Autophagic Cargo Assay Applicable to Whole Cells

BHMT is an abundant liver-specific enzyme, present in hepatocytes as a large number of proteolytically truncated variants, several of which are selectively associated with autophagic membranes (Øverbye, 2007; Øverbye et al., 2007; Ueno et al., 1999). Some of the BHMT forms are cytosolic (Øverbye, 2007), whereas others are generated inside autophagic vacuoles, probably in lysosomes ((Furuya et al., 2001; Øverbye, 2007). Because the intravacuolar generation of BHMT fragments is strictly autophagy dependent, such fragments can be used as probes in cargo assays of autophagic sequestration. Furthermore, because the fragments are absent from the cytosol, they can be assayed without any need to separate vacuoles from cytosol (i.e., in whole cells).

In the authors' laboratory, we have identified a small (~10 kDa), native N-terminal BHMT fragment generated in an autophagy-dependent manner in isolated rat hepatocytes (Øverbye, 2007). This fragment, p10 (BHMT), probably corresponds to the N terminus removed from Ueno et al.'s (1999) large (32 kDa) fragment, and to the GST-conjugated fragment appearing in cells transfected with Mercer et al.'s (2008) GST-BHMT fusion protein. The p10(BHMT) appears to be generated by a leupeptin-resistant vacuolar serine proteinase (Øverbye et al., 2008) and is further processed by a leupeptin-sensitive cysteine proteinase; this fragment will, therefore, accumulate in hepatocytes treated with high concentrations of leupeptin (Øverbye, 2007). Because BHMT is prone to be fragmented by proteolysis *in vitro* (Ueno et al., 1999), a high concentration of the serine proteinase inhibitor 4-(2-aminoethyl)benzenesulfonyl fluoride (AEBSF) is required to suppress secondary p10 formation in hepatocyte extracts used for analysis.

7.1. Measurement of p10(BHMT) formation in hepatocytes by immunoblotting

a. Rat hepatocytes (~20 mg cells in 2 ml of suspension buffer) are incubated at 37 °C in stationary 6-cm suspension Petri dishes (section 2.1.1) for the desired period of time (up to 6 h) in the presence of leupeptin (0.3 mM).

b. At the end of incubation the cells are rapidly cooled to 0 °C, 2 ml of ice-cold wash buffer (2.1.3) is added, and the cell suspension is centrifuged at 1600 rpm for 5 min.

c. The supernatant fraction is discarded and the cell pellet is dissolved in 0.7 ml of lysis buffer with proteinase and phosphatase inhibitors (1% sodium dodecyl sulfate, 3 mM AEBSF, 0.3 mM leupeptin, 10 μM pepstatin A, 15 μM E-64, 50 μM bestatin, 0.8 μM aprotinin, 50 nM okadaic acid, 1 mM sodium orthovanadate, 2 mM sodium fluoride, 7.5% sucrose, 5 mM EDTA, 5 mM EGTA, 10 mM sodium pyrophosphate, 20 mM Tris, pH 7.2), incubated for 1 h at 0 °C, and heated for 5 min at 95 °C.

d. Five μl of lysate (~6 μg protein/μl) is applied to a 7-cm, 15% polyacrylamide gel and subjected to standard SDS-PAGE and immunoblotting as previously described (Samari et al., 2005). We use the N-terminal polyclonal BHMT antibody α-p44-10R (Ueno et al., 1999), a kind gift from Drs. T. Ueno and E. Kominami (Tokyo, Japan), at a dilution of 1:100,000 for p44(BHMT) and 1:8000 for p10(BHMT). For quantification, the integrated p10(BHMT) band intensity is expressed as percentage of the band intensity (blotted with a 10x more diluted lysate) of the full-length p44(BHMT), which is the probable precursor of p10 (Øverbye et al., 2008).

p10(BHMT) accumulation following leupeptin addition can be detected in whole hepatocytes by immunoblotting, with a very low background, and proceeds at a fairly constant rate for 4 h (Fig. 5.1). Subcellular fractionation and inhibitor studies have demonstrated that the accumulation is strictly autophagy dependent and occurs in the rate of p10(BHMT) accumulation (~2%/h) is of a similar order as the rate of accumulation of LDH activity, used in our standard autophagy assay (Kopitz et al., 1990), described previously, and most likely reflects the rate of autophagic sequestration (being the rate-limiting step). Autophagic fragment assays (Mercer et al., 2008; Øverbye, 2007) thus seem to represent a very useful type of cargo assay for the measurement of actual autophagic activity, as suggested by Ueno et al. (1999).

8. Concluding Remarks

Present-day autophagy research uses a large number of sophisticated methods capable of detecting components or changes in the autophagic machinery of the cell (Klionsky et al., 2008), but very few of these actually measure short-term autophagic activity in an unambiguous and quantitative manner. The cargo-based autophagy assays discussed in the present chapter should offer a sufficient number of options to be adaptable to any experimental system. For substratum-attached cultured cells, which are unsuitable for electropermeabilization, the probe of choice would be LDH or any other long-lived, endogenous cytosolic protein degraded in a leupeptin-sensitive manner by the autophagic-lysosomal pathway. Our laboratory routinely measures LDH by a simple enzymatic activity assay, which provides more precise quantification than immunoblotting. To separate sedimentable (autophagocytosed) from soluble (cytosolic) LDH, electrodisruption has been found to be very effective as applied to hepatocytes, but other cell disruption methods, including homogenization and a subsequent single-step density cushion centrifugation, may work satisfactorily with any cell type.

The p10(BHMT) fragment assay is a convenient immunoblotting-based alternative for hepatocytes, and presumably for other cell types (perhaps kidney cells) that express and process BHMT in a similar manner. For the majority of cells, which do not express endogenous BHMT, transfection with GST-BHMT seems to be a generally applicable method (Mercer et al., 2008).

Autophagy encompasses a number of different processes that are regulated at many levels (Klionsky et al., 2007a), including both long-term control of autophagic capacity through differential gene expression and short-term regulation of phagophore activity or substrate susceptibility (Meijer and Codogno, 2006; Suzuki and Ohsumi, 2007; Yorimitsu and Klionsky, 2005). For studies of short-term regulation, direct measurements of actual autophagic activity are indispensable. Hopefully, the present review may stimulate renewed interest in cargo-based assays of autophagic sequestration.

REFERENCES

Berg, T. O., Fengsrud, M., Strømhaug, P. E., Berg, T., and Seglen, P. O. (1998). Isolation and characterization of rat liver amphisomes: Evidence for fusion of autophagosomes with both early and late endosomes. *J. Biol. Chem.* **273,** 21883–21892.

Bergmeyer, H. U., and Berndt, E. (1974). *Methoden der Enzymatischen Analyse.* Verlag Chemie, Weinheim.

Egner, R., Thumm, M., Straub, M., Simeon, A., Schuller, H. J., and Wolf, D. H. (1993). Tracing intracellular proteolytic pathways: Proteolysis of fatty acid synthase and other cytoplasmic proteins in the yeast Saccharomyces cerevisiae. *J. Biol. Chem.* **268,** 27269–27276.

Felgner, P. L., Gadek, T. R., Holm, M., Roman, R., Chan, H. W., Wenz, M., Northrop, J. P., Ringold, G. M., and Danielsen, M. (1987). Lipofection: A highly efficient, lipid-mediated DNA-transfection procedure. *Proc. Natl. Acad. Sci. USA* **84,** 7413–7417.

Fengsrud, M., Raiborg, C., Berg, T. O., Strømhaug, P. E., Ueno, T., Erichsen, E. S., and Seglen, P. O. (2000). Autophagosome-associated variant isoforms of cytosolic enzymes. *Biochem. J.* **352,** 773–781.

Fennell, D. F., Whatley, R. E., McIntyre, T. M., Prescott, S. M., and Zimmerman, G. A. (1991). Endothelial cells reestablish functional integrity after reversible permeabilization. *Arterioscler. Thromb.* **11,** 97–106.

Furuya, N., Kanazawa, T., Fujimura, S., Ueno, T., Kominami, E., and Kadowaki, M. (2001). Leupeptin-induced appearance of partial fragment of betaine homocysteine methyltransferase during autophagic maturation in rat hepatocytes. *J. Biochem. (Tokyo)* **129,** 313–320.

Gordon, P. B., Høyvik, H., and Seglen, P. O. (1992). Prelysosomal and lysosomal connections between autophagy and endocytosis. *Biochem. J.* **283,** 361–369.

Gordon, P. B., and Seglen, P. O. (1988). Prelysosomal convergence of autophagic and endocytic pathways. *Biochem. Biophys. Res. Commun.* **151,** 40–47.

Gordon, P. B., and Seglen, P. O. (1986). Use of electrical methods in the study of hepatocytic autophagy. *Biomed. Biochim. Acta* **45,** 1635–1645.

Gordon, P. B., and Seglen, P. O. (1982). Autophagic sequestration of [^{14}C]sucrose, introduced into isolated rat hepatocytes by electropermeabilization. *Exp. Cell Res.* **142,** 1–14.

Gordon, P. B., Tolleshaug, H., and Seglen, P. O. (1985a). Use of digitonin extraction to distinguish between autophagic-lysosomal sequestration and mitochondrial uptake of [^{14}C]sucrose in hepatocytes. *Biochem. J.* **232,** 773–780.

Gordon, P. B., Tolleshaug, H., and Seglen, P. O. (1985b). Autophagic sequestration of [^{14}C]sucrose introduced into isolated rat hepatocytes by electrical and non-electrical methods. *Exp. Cell Res.* **160,** 449–458.

Høyvik, H., Gordon, P. B., Berg, T. O., Strømhaug, P. E., and Seglen, P. O. (1991). Inhibition of autophagic-lysosomal delivery and autophagic lactolysis by asparagine. *J. Cell Biol.* **113,** 1305–1312.

Høyvik, H., Gordon, P. B., and Seglen, P. O. (1986). Use of a hydrolysable probe, [^{14}C] lactose, to distinguish between pre-lysosomal and lysosomal steps in the autophagic pathway. *Exp. Cell Res.* **166,** 1–14.

Høyvik, H., Gordon, P. B., and Seglen, P. O. (1987). Convergence of autophagic and endocytic pathways at the level of the lysosome. *Biochem. Soc. Transact.* **15,** 964–965.

Klionsky, D. J., Abeliovich, H., Agostinis, P., Agrawal, D. K., Aliev, G., Askew, D. S., Baba, M., Baehrecke, E. H., Bahr, B. A., Ballabio, A., et al. (2008). Guidelines for the use and interpretation of assays for monitoring autophagy in higher eukaryotes. *Autophagy* **4,** 1–25.

Klionsky, D. J., Cuervo, A. M., Dunn, W. A., Jr., Levine, B., van der Klei, I. J., and Seglen, P. O. (2007a). How shall I eat thee? *Autophagy* **3,** 413–416.

Klionsky, D. J., Cuervo, A. M., and Seglen, P. O. (2007b). Methods for monitoring autophagy. *Autophagy* **3,** 181–206.

Kopitz, J., Kisen, G. Ø., Gordon, P. B., Bohley, P., and Seglen, P. O. (1990). Non-selective autophagy of cytosolic enzymes in isolated rat hepatocytes. *J. Cell Biol.* **111,** 941–953.

Kovács, A. L., Grinde, B., and Seglen, P. O. (1981). Inhibition of autophagic vacuole formation and protein degradation by amino acids in isolated hepatocytes. *Exp. Cell Res.* **133,** 431–436.

Kovács, A. L., Reith, A., and Seglen, P. O. (1982). Accumulation of autophagosomes after inhibition of hepatocytic protein degradation by vinblastine, leupeptin or a lysosomotropic amine. *Exp. Cell Res.* **137,** 191–201.

Lemons, R., Forster, S., and Thoene, J. (1988). Protein microinjection by protease permeabilization of fibroblasts. *Anal. Biochem.* **172,** 219–227.

Meijer, A. J., and Codogno, P. (2006). Signalling and autophagy regulation in health, aging and disease. *Mol. Aspects Med.* **27,** 411–425.

Mercer, C. A., Kaliappan, A., and Dennis, P. B. (2008). Macroautophagy-dependent, intralysosomal cleavage of a betaine homocysteine methyltransferase fusion protein requires stable multimerization. *Autophagy* **4,** 185–194.

Mousavi, S. A., Kjeken, R., Berg, T. O., Seglen, P. O., Berg, T., and Brech, A. (2001). Effects of inhibitors of the vacuolar proton pump on hepatic heterophagy and autophagy. *Biochim. Biophys. Acta* **1510,** 243–257.

Nimmerjahn, F., Milosevic, S., Behrends, U., Jaffee, E. M., Pardoll, D. M., Bornkamm, G. W., and Mautner, J. (2003). Major histocompatibility complex class II-restricted presentation of a cytosolic antigen by autophagy. *Eur. J. Immunol.* **33,** 1250–1259.

Nyberg, E., and Dingle, J. T. (1970). Endocytosis of sucrose and other sugars by cells in culture. *Exp. Cell Res.* **63,** 43–52.

Okada, C. Y., and Rechsteiner, M. (1982). Introduction of macromolecules into cultured mammalian cells by osmotic lysis of pinocytic vesicles. *Cell* **29,** 33–41.

Plomp, P. J. A. M., Gordon, P. B., Meijer, A. J., Høyvik, H., and Seglen, P. O. (1989). Energy dependence of different steps in the autophagic-lysosomal pathway. *J. Biol. Chem.* **264,** 6699–6704.

Samari, H. R., Møller, M. T. N., Holden, L., Asmyhr, T., and Seglen, P. O. (2005). Stimulation of hepatocytic AMP-activated protein kinase by okadaic acid and other autophagy-suppressive toxins. *Biochem. J.* **386,** 237–244.

Sanjuan, M. A., Dillon, C. P., Tait, S. W., Moshiach, S., Dorsey, F., Connell, S., Komatsu, M., Tanaka, K., Cleveland, J. L., Withoff, S., and Green, D. R. (2007). Toll-like receptor signalling in macrophages links the autophagy pathway to phagocytosis. *Nature* **450,** 1253–1257.

Schaible, U. E., Schlesinger, P. H., Steinberg, T. H., Mangel, W. F., Kobayashi, T., and Russell, D. G. (1999). Parasitophorous vacuoles of Leishmania mexicana acquire macromolecules from the host cell cytosol via two independent routes. *J. Cell Sci.* **112,** 681–693.

Seglen, P. O. (1994). In "Isolation of hepatocytes." *Handbook of cell biology,* (J. E. Celis, ed.), pp. 96–102. Academic Press, San Diego.

Seglen, P. O. (1983). Inhibitors of lysosomal function. *Methods Enzymol.* **96,** 737–764.

Seglen, P. O. (1998). In "Isolation of hepatocytes." *Cell biology: a laboratory handbook,* (J. E. Celis, ed.), pp. 119–124. Academic Press, San Diego.

Seglen, P. O. (1993). Isolation of hepatocytes by collagenase perfusion. *Methods Toxicol.* **1A,** 231–243.

Seglen, P. O. (1975). Protein degradation in isolated rat hepatocytes is inhibited by ammonia. *Biochem. Biophys. Res. Commun.* **66,** 44–52.

Seglen, P. O. (1976). Preparation of isolated rat liver cells. *Methods Cell Biol.* **13,** 29–83.

Seglen, P. O., and Gordon, P. B. (1984). Amino acid control of autophagic sequestration and protein degradation in isolated rat hepatocytes. *J. Cell Biol.* **99,** 435–444.

Seglen, P. O., and Gordon, P. B. (1980). Effects of lysosomotropic monoamines, diamines, amino alcohols, and other amino compounds on protein degradation and protein synthesis in isolated rat hepatocytes. *Mol. Pharmacol.* **18,** 468–475.

Seglen, P. O., and Gordon, P. B. (1982). 3-Methyladenine: A specific inhibitor of autophagic/lysosomal protein degradation in isolated rat hepatocytes. *Proc. Natl. Acad. Sci. USA* **79,** 1889–1892.

Seglen, P. O., Gordon, P. B., and Høyvik, H. (1986a). Radiolabelled sugars as probes of hepatocytic autophagy. *Biomed. Biochim. Acta* **45,** 1647–1656.

Seglen, P. O., Gordon, P. B., Tolleshaug, H., and Høyvik, H. (1986b). Use of [^{3}H]raffinose as a specific probe of autophagic sequestration. *Exp. Cell Res.* **162,** 273–277.

Seglen, P. O., and Reith, A. (1976). Ammonia inhibition of protein degradation in isolated rat hepatocytes: Quantitative ultrastructural alterations in the lysosomal system. *Exp. Cell Res.* **100,** 276–280.

Stacey, D. W., and Allfrey, V. G. (1977). Evidence for the autophagy of microinjected proteins in HeLa cells. *J. Cell Biol.* **75,** 807–817.

Strømhaug, P. E., Berg, T. O., Fengsrud, M., and Seglen, P. O. (1998). Purification and characterization of autophagosomes from rat hepatocytes. *Biochem. J.* **335,** 217–224.

Strømhaug, P. E., and Seglen, P. O. (1993). Evidence for acidity of prelysosomal autophagic/endocytic vacuoles (amphisomes). *Biochem. J.* **291,** 115–121.

Suzuki, K., and Ohsumi, Y. (2007). Molecular machinery of autophagosome formation in yeast, *Saccharomyces cerevisiae. FEBS Lett.* **581,** 2156–2161.

Tanida, I., Minematsu-Ikeguchi, N., and Kominami, E. (2005). Lysosomal turnover, but not a cellular level, of endogenous LC3 is a marker for autophagy. *Autophagy* **1,** 84–91.

Tolleshaug, H., Gordon, P. B., Solheim, A. E., and Seglen, P. O. (1984). Trapping of electro-injected [^{14}C]sucrose by hepatocyte mitochondria: A mechanism for cellular autofiltration? *Biochem. Biophys. Res. Commun.* **119,** 955–961.

Tolleshaug, H., and Seglen, P. O. (1985). Autophagic-lysosomal and mitochondrial sequestration of [^{14}C]sucrose: Density gradient distribution of sequestered radioactivity. *Eur. J. Biochem.* **153,** 223–229.

Ueno, T., Ishidoh, K., Mineki, R., Tanida, I., Murayama, K., Kadowaki, M., and Kominami, E. (1999). Autolysosomal membrane-associated betaine homocysteine methyltransferase: Limited degradation fragment of a sequestered cytosolic enzyme monitoring autophagy. *J. Biol. Chem.* **274,** 15222–15229.

Walev, I., Bhakdi, S. C., Hofmann, F., Djonder, N., Valeva, A., Aktories, K., and Bhakdi, S. (2001). Delivery of proteins into living cells by reversible membrane permeabilization with streptolysin-O. *Proc. Natl. Acad. Sci. USA* **98,** 3185–3190.

Xiao, G. (2007). Autophagy and NF-κB: Fight for fate. *Cytokine Growth Factor Rev.* **18,** 233–243.
Yorimitsu, T., and Klionsky, D. J. (2005). Autophagy: Molecular machinery for self-eating. *Cell Death. Differ.* **12,** 1542–1552.
Yu, L., Wan, F., Dutta, S., Welsh, S., Liu, Z., Freundt, E., Baehrecke, E. H., and Lenardo, M. (2006). Autophagic programmed cell death by selective catalase degradation. *Proc. Natl. Acad. Sci. USA* **103,** 4952–4957.
Zimmermann, U., Riemann, F., and Pilwat, G. (1976). Enzyme loading of electrically homogeneous human red blood cell ghosts prepared by dielectric breakdown. *Biochim. Biophys. Acta* **436,** 460–474.
Zuurendonk, P. F., Tischler, M. E., Akerboom, T. P., Van Der, M. R., Williamson, J. R., and Tager, J. M. (1979). Rapid separation of particulate and soluble fractions from isolated cell preparations (digitonin and cell cavitation procedures). *Methods Enzymol* **56,** 207–223.
Øverbye, A. (2007). Proteomic studies of hepatocytic autophagosomes Ph.D. thesis, University of Oslo. pp. 1–90.
Øverbye, A., Fengsrud, M., and Seglen, P. O. (2007). Proteomic analysis of membrane-associated proteins from rat liver autophagosomes. *Autophagy* **3,** 300–322.
Øverbye, A., Sætre, F., Hagen, L. K., and Seglen, P. O. (2008). Hepatocellular variants of betaine:homocysteine methyltransferase: Autophagy-dependent generation of a small fragment, p10(BHMT), suitable for measuring autophagy in whole cells : Manuscript submitted.

CHAPTER SIX

Assays to Assess Autophagy Induction and Fusion of Autophagic Vacuoles with a Degradative Compartment, Using Monodansylcadaverine (MDC) and DQ-BSA

Cristina Lourdes Vázquez* *and* María Isabel Colombo*

Contents

1. Overview	86
2. Assessing Autophagy Induction by Fluorescence Microscopy and Fluorometry	86
2.1. Labeling autophagic vacuoles with monodansylcadaverine (MDC) in cultured cells	86
2.2. Assessing autophagy induction by fluorometry in cells labeled with MDC	90
3. Monitoring the Formation of an Autolysosome	92
3.1. Labeling lysosomes with DQ-BSA	92
3.2. Monitoring fusion of an autophagosome/amphisome with a lysosomal compartment by real-time imaging	93
4. Concluding Remarks	94
Acknowledgments	94
References	94

Abstract

In this chapter we describe the use of monodasylcadaverine (MDC) and DQ-BSA, two practical and convenient tools to study the autophagic pathway. MDC is a lysosomotropic compound useful for the identification of autophagic vesicles by fluorescence microscopy and, in addition, to assess autophagy induction via the accumulation of MDC-labeled vacuoles. However, the increase of

* Laboratorio de Biología Celular y Molecular, Instituto de Histología y Embriología (IHEM), Facultad de Ciencias Médicas, Universidad Nacional de Cuyo–CONICET, Mendoza, Argentina

Methods in Enzymology, Volume 452
ISSN 0076-6879, DOI: 10.1016/S0076-6879(08)03606-9

autophagosomes does not necessarily reflect autophagosome maturation and degradation of the sequestered materials, thus the use of DQ-BSA in conjunction with and autophagic marker is an appropriate technique to monitor the formation of the autolysosome, the degradative compartment. Therefore, here we discuss the advantages of the utilization of these two methods to characterize the autophagy pathway.

1. Overview

Several methods have been developed to study the autophagy pathway and the steps involved in the maturation of autophagosomes to autolysosomes, an acidic hydrolase rich organelle in which the sequestered material is degraded. Monodasylcadaverine (MDC) and DQ-BSA are useful tools for identification and visualization of the autophagic process by fluorescence microscopy. The lysosomotropic compound MDC was initially characterized as a selective marker of autolysosomes (Biederbick et al., 1995). However, more recent studies indicate that MDC also works as a marker of earlier autophagic compartments, and the accumulation of MDC-positive vesicles responds to autophagy induction when this process is stimulated, both in cultured cells and in animals (Munafó and Colombo, 2001; Iwai-Kanai et al., 2008). On the other hand, the use of the bovine serum albumin derivative (DQ-BSA) conjugated to a self-quenched fluorophore, in conjunction with an autophagic marker, is a helpful bona fide method for detecting the formation of an active degradative compartment, the autolysosome, allowing the visualization of the completion of the autophagic pathway. Here, we detail both methods and attempt to clarify some controversial points.

2. Assessing Autophagy Induction by Fluorescence Microscopy and Fluorometry

2.1. Labeling autophagic vacuoles with monodansylcadaverine (MDC) in cultured cells

Monodansylcadaverine (MDC) has autofluorescent properties with an excitation wavelength at 365 nm, due to a dansyl group conjugated to cadaverine, a diamine-pentane (Fig. 6.1). MDC accumulates in lipid-rich membranous compartments and gets trapped in adjacent membranes. Elsässer and collaborators initiated the studies with MDC and established that it accumulates as a selective fluorescent marker for autophagic vacuoles under *in vivo* conditions by interacting with membrane lipids that are highly concentrated in the autophagic compartments. In our laboratory, we

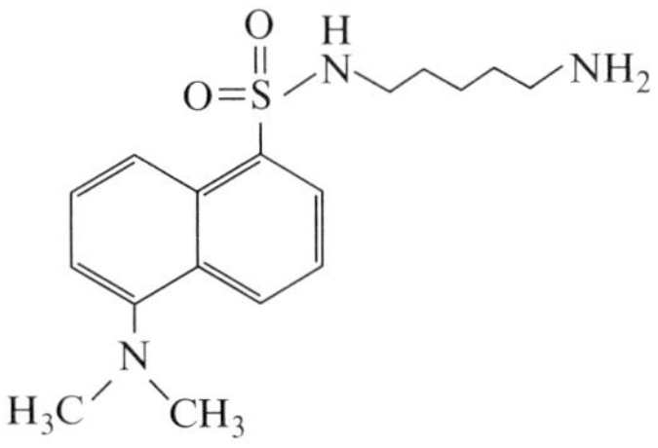

Figure 6.1 Formula of monodansylcadaverine (MDC). A dansyl group conjugated to a diamine-pentane is responsible for the autofluorescence.

characterized the labeling of cells with MDC and the incorporation of this compound by fluorometry (see subsequent sections) in cells subjected to different autophagy modulators (Munafó and Colombo, 2001).

MDC can be used to label autophagic vacuoles in several cell types (Biederbick et al., 1995). These investigators have indicated that dose-response experiments with PaTu 8902 cells show that the optimal concentration for in vivo labeling is 0.05 to 0.1 mM, whereas cells detached and disintegrated when the MDC concentration exceeded 0.1 mM. When MDC is incorporated into cells, the accumulation of this fluorescent reagent is observed in spherical compartments at the perinuclear region, in spots distributed throughout the cytoplasm, or both, depending on the cell type used. Biederbick et al. also demonstrated by ultrastructural analysis of fractions obtained on sucrose density gradients that the labeled compartments correspond to autophagic vacuoles in different stages of development. The MDC-labeled autophagic vacuoles do not contain fluid-phase markers and are spatially separated from endosomal compartments, but they do include lysosomal enzymes.

Our group demonstrated that in cells subjected to a physiological or pharmacological stimulus of autophagy, such as amino acid deprivation or rapamycin treatment, the size and number of MDC-fluorescent vesicles markedly increases (Munafó and Colombo, 2001). In contrast, when cells are previously treated with well-known inhibitors of autophagy, such as 3-methyladenine (3-MA) and wortmannin (WM), the MDC incorporation is almost completely inhibited, indicating that MDC-labeled compartments respond to the dynamic changes in autophagy that occur in response to several modulators. In addition, a population of the MDC-labeled vesicles colocalizes with LC3, a specific autophagosome marker. When CHO cells overexpressing GFP-LC3 are subjected to autophagy induction and subsequently labeled with MDC, a good colocalization of LC3 and MDC is observed, correlating with autophagosome formation in response to the autophagy stimulus (Munafó and Colombo, 2001). It has been suggested that MDC is accumulated only in autophagic vacuoles that have already fused with lysosomes, an acidic compartment. However, vinblastine, a microtubule despolymerizing agent that leads to the accumulation of autophagic vacuoles by blocking fusion with lysosomes alters the distribution and size of MDC-marked autophagosomes with an evident increase in the MDC labeling (Munafó and Colombo, 2001).

Because MDC is a basic compound and acts as a lysosomotropic agent, being concentrated into acidic compartments by an ion-trapping mechanism, a neutral derivative known as monodansylpentane (MDH) was synthesized (Niemann et al., 2001). Because of the properties of this new compound, MDH staining of autophagic vacuoles is independent of the acidic pH, and thus of an ion-trapping mechanism, but shows the same preferences for autophagic membrane lipids as MDC (Niemann et al., 2001). Furthermore, the authors indicate that under *in vivo* conditions MDH is a more stringent marker for autophagic vacuoles than MDC.

We and other colleagues have demonstrated that the GTPase Rab7 decorates autophagic vesicles and is required for the normal progression of the autophagy pathway (Gutierrez et al., 2004; Jager et al., 2004; Kimura et al., 2007). In cells overexpressing wild-type GFP-Rab7 and GFP-Rab7Q67L (a Rab7 constitutively active mutant), MDC associates with the Rab7-decorated vesicles incubated in either full-nutrient or starvation conditions. On the other hand, cells transfected with the dominant negative mutant (Rab7T22N) present a diffused localization in the cytosol in cells incubated under control conditions; but when autophagy is induced, the Rab7T22N protein localizes on the MDC-labeled vesicles. It is important to mention that the expression of the Rab7T22N mutant hampers fusion between autophagosomes and lysosomes (Gutierrez et al., 2004; Fader et al., 2008), thus these MDC-labeled vesicles cannot account for autolysosomes. We also tested MDH in transfected cells with wild-type Rab7 and the Rab7-negative mutant. Our results show that MDH colocalizes with both wild-type Rab7 and T22N-decorated vesicles in nutrient-deprived cells. Therefore, similar to MDC, MDH can be used as a good marker of autophagic vesicles. A significant problem with this compound is that it is not commercially available and thus needs to be synthesized (Niemann et al., 2001).

In a recent publication, Roberta Gottlieb and collaborators (Iwai-Kanai et al., 2008) investigated the use of MDC and the lysosomotropic drug chloroquine to measure autophagic flux in transgenic mice expressing mCherry-LC3 (a full description of this method is also presented in this volume). Interestingly, the results show mCherry-LC3-positive dots colocalizing with MDC in myocardium from hearts of mice injected with both rapamycin and chloroquine. Thus, chloroquine administration, which inhibits lysosomal activity by neutralizing its pH, does not prevent the accumulation of MDC, implying that MDC accumulates independent of the acidic pH. Furthermore, chloroquine hampers the starvation-induced colocalization of the lysosomal marker LAMP-2 and GFP-LC3 as well as the colocalization of mitochondria with lysosomes, indicating that the formation of the autolysosome is blocked in the presence of chloroquine (Boya et al., 2005). Taken together, these data indicate that MDC-labeled vesicles are not exclusively autolysosomes.

The individual steps of *in vivo* MDC staining when autophagy is induced by amino acid deprivation in Chinese hamster ovary (CHO) cells are described subsequently:

1. Plate CHO cells at 80% of confluence in 2 ml of culture media (α-MEM containing 10% fetal bovine serum (FBS), 2 mM glutamine, 100 U/ml penicillin, and 100 g/ml streptomycin) on a 6-well tissue-culture dish containing a cover glass for each well.
2. Incubate the cells 24 h at 37 °C in an atmosphere of 95% air and 5% CO_2.
3. Wash the cells 3 times with PBS to remove the media and add 2 ml of Earle's balanced salts solution (EBSS, starvation media; Sigma) to induce autophagy, or complete tissue culture medium for control conditions. Incubate the cells 2 h at 37 °C.
4. Subsequently, control and starved cells are incubated with 0.05 μM MDC in PBS at 37 °C for 10 min.
5. After incubation, the cells are washed 4 times with PBS and immediately analyzed by fluorescence microscopy (see Fig. 6.2). The microscope should be equipped with the adequate filters (MDC has an autofluorescence at 365 and 525 nm wavelength, for excitation and emission, respectively).

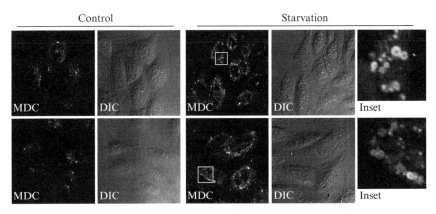

Figure 6.2 MDC-labeled vesicles are induced by starvation. CHO cells were incubated in α-MEM medium or in EBSS medium (starvation) at 37 °C for 2 h. Following this incubation period, both starved and control cells were incubated with 0.05 mM MDC for 10 min at 37 °C and then washed 4 times with PBS, pH 7.4. Cells were immediately analyzed by confocal microscopy. *Insets*: MDC-labeled vesicles with a clear ring-like structure, which is consistent with the preferential labeling of the double-membrane lipid-rich autophagic compartment.

Note: Even though MDC labeling can be visualized both in *in vivo* conditions and after fixation (Biederbick *et al.*, 1995), it is important to take into account that the staining intensity can be markedly decreased under certain fixation conditions. Although we recommend incubating the cells with 3% paraformaldehyde for 6 min, still the fluorescence intensity is very dim. Thus, it is preferable to visualize MDC in nonfixed cells.

2.2. Assessing autophagy induction by fluorometry in cells labeled with MDC

Another useful technique to determine autophagy induction is measuring intracellular MDC by fluorometry (Fig. 6.3).

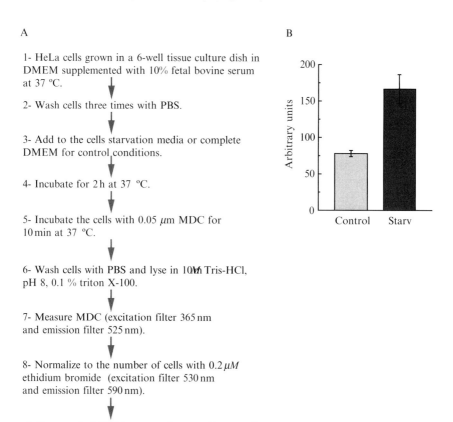

Figure 6.3 *Panel A*: Steps to incorporate and measure intracellular MDC by fluorescence photometry in HeLa cells. *Panel B*: Intracellular MDC was measured by fluorescence photometry in cells incubated in control or starvation media. The MDC uptake markedly increased when autophagy was induced by amino acid deprivation.

1. Cells are plated in 2 ml of culture medium (DMEM or α-MEM depending on the cell type used) on a 6-well tissue-culture dish to 80% of confluence.
2. Wash the cells 3 times with PBS to remove the media and add 2 ml of Earle's balanced salts solution (EBSS, starvation media; Sigma) to induce autophagy, or complete culture medium for control conditions. Incubate the cells 2 h at 37 °C.
3. Subsequently, the cells are incubated with 0.05 μM MDC in PBS for 10 min at 37 °C.
4. After incubation, the cells are washed 4 times with PBS pH 7.4, and then lysed in 10 mM Tris-HCl, pH 8, containing 0.1% Triton X-100.
5. Measure intracellular MDC by fluorescence photometry (excitation wavelength 365 nm, emission filter 525 nm) in a Packard Fluorocount microplate reader. To normalize for the amount of cells present in each well, a solution of ethidium bromide is added to a final concentration of 0.2 mM and the DNA fluorescence is measured with the adequate filters, 530 nm and 590 nm wavelength, for excitation and emission, respectively. The MDC incorporation is expressed as specific activity (arbitrary units).

In addition to the morphological analysis of the autophagy pathway by fluorescence microscopy, the biochemical assay described previously makes it possible to quantitatively monitor autophagy induction in a simple manner (Munafó and Colombo, 2001).

Special comments and clarifications: As indicated previously, there is evidence that MDC can interact with membrane lipids to function as a solvent polarity probe (Niemann*et al.*, 2005), and because one of the features of autophagic vacuoles is their high content of lipids, MDC has been widely used as an autophagic marker. However, one exhaustive study indicates that MDC labels only acidic compartments working just as a lysosomotropic compound (Bampton *et al.*, 2005). In addition, MDC dots can be detected in Atg5$^{-/-}$ mouse embryonic stem cells (Mizushima, 2004), indicating that MDC does not stain only autophagic compartments. Nevertheless, when the acidic pH is hampered by applying well-known inhibitors, the MDC fluorescence is reduced in some cell types (Niemann *et al.*, 2005; Bampton *et al.*, 2005); in others it is almost not affected (Iwai-Kanai *et al.*, 2008). Thus, the ion-trapping mechanism is not sufficient to explain the MDC labeling because only a certain fraction of MDC accumulation seems to depend on the acidic pH. In contrast, when the uptake of acridine orange is assessed under the same conditions (i.e., blocking the acidic pH) no labeling at all is observed, which is expected with a restrictive lysosomotropic compound. Furthermore, in paraformaldehyde-fixed cells, in which no proton gradients are maintained, the labeling of vesicles with MDC still occurs. Thus, MDC labeling in fixed cells, although weaker, is likely more specific as indicated in the work of Karla Kirkegaard (Jackson *et al.*, 2005). These differences suggest an additional mechanism for the MDC staining besides ion trapping (see also subsequent sections).

Furthermore, some studies suggest that MDC only marks autophagic compartments after fusion to acidic endo/lysosomes (Bampton *et al.*, 2005). However, we would like to point out that in the presence of vinblastine, which hampers fusion with lysosomes, a marked amount of enlarged MDC-labeled vesicles accumulate. Furthermore, in conditions that stimulate autophagy, several of the MDC-labeled vesicles appear as a ring-like structure (see Fig. 6.2; Munafó and Colombo, 2001), indicating that the drug is preferentially intercalating in the membranes more than just labeling a diffuse acidic compartment. These discrepancies, perhaps related to the cell type, have caused some concerns about the usefulness of MDC. Nevertheless, as indicated earlier, in the recent work by Gottlieb and collaborators (Iwai-Kanai *et al.*, 2008) in animals, MDC labeled LC3-marked vesicles even in the presence of the lysosomotropic agent chloroquine, which neutralizes the lysosomal pH. Although, it is likely that MDC can label autolysosomes, the observations described clearly suggest that MDC is able to label autophagic structures in spite of the disruption of their acidic pH.

3. Monitoring the Formation of an Autolysosome

3.1. Labeling lysosomes with DQ-BSA

It is well known that the autophagic and endocytic pathways join together to generate a vesicular compartment known as the amphisome (Klionsky, 2007). Subsequently, this compartment fuses with a lysosome, in which the trapped materials are degraded by lysosomal enzymes. Although amphisomes lack lysosomal proteases such as acid phosphatase and cathepsin, they are acidic (Stromhaug *et al.*, 1993). Thus, the use of an acidic pH marker such as LysoTracker does not allow the distinction between amphisomes and autolysosomes. Moreover, even the detection of lysosomal enzymes does not necessarily mean that this compartment is functional (i.e., degradative). To solve this problem, a red BODIPY dye conjugated to bovine serum albumin (DQ-BSA, Molecular Probes) can be used. This BSA derivative is so heavily labeled that the fluorophore is self-quenched. Proteolysis of this compound results in dequenching and release of brightly fluorescent fragments. Thus, the use of DQ-BSA is useful for the visualization of intracellular proteolytic activity (see Fig. 6.4).

To analyze the convergence between an autophagic compartment with a functional lysosome, the following protocol may be followed:

1. Cells are incubated for 1 h at 37 °C with DQ-BSA (10 μg/ml in complete tissue culture medium).

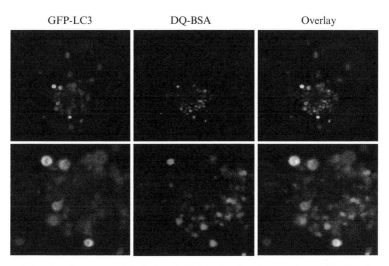

Figure 6.4 Autolysosomes labeled with DQ-BSA. Stably transfected K562 cells overexpressing pEGFP-LC3 were incubated with DQ-BSA (10μg/ml in RPMI + 10% fetal bovine serum) for 12 h at 37 °C to label the lysosomal compartment. Cells were then washed twice with PBS and incubated for 2 h at 37 °C in starvation media. Cells were mounted on coverslips and immediately analyzed by confocal microscopy. *Lower panels*: higher magnification of the upper panels (images from Fader and Colombo). (See Color Insert.)

2. Cells are then washed twice with PBS to remove excess probe.
3. The cells are then incubated in starvation medium (EBSS, starvation media; Sigma) to induce autophagy or under different experimental conditions. Incubation for 2–3 h is usually sufficient to allow for detection of DQ-BSA fluorescence.
4. Autophagosomes can be visualized by detecting the autophagic protein LC3 by indirect immunofluorescence, or alternatively by using transfected cells overexpressing GFP-LC3 (see Fig. 6.4).

An important recommendation is that when using inhibitors (e.g., wortmannin) or overexpression of proteins (e.g., Rab7 dominant negative mutants), the internalization or the transport of DQ-BSA to the lysosomes may be hampered. In this case the lysosomal compartment can be labeled with DQ-BSA overnight, before treating the cells with the drugs or prior to the transfection with the mutant proteins.

3.2. Monitoring fusion of an autophagosome/amphisome with a lysosomal compartment by real-time imaging

The degradation of the chromogenic self-quenched BSA can be also used to image the convergence of the autophagy and lysosomal pathways in real time. Cells overexpressing GFP-LC3 can be labeled with DQ-BSA as indicated previously. The fusion between an autophagosome/amphisome (e.g., labeled in green) with a compartment containing the highly red fluorescent products generated by hydrolytic cleavage (i.e., the lysosome) can be monitored by confocal microscopy.

For this purpose coverslips with the labeled cells are placed in a temperature-controlled stage and analyzed by time-lapse confocal microscopy. A total of 30 slides can be acquired every 5 s with the proper green and red filter sets using a laser-scanning confocal unit attached to an upright fluorescence microscope.

4. Concluding Remarks

Here we have described basic protocols to assay autophagy induction by fluorescence microscopy and fluorometry and a useful assay to monitor the last step of the pathway, the fusion with a degradative lysosomal compartment. Labeling and incorporation of MDC is an easy and fast method that reflects starvation-induced autophagic activity, in both cultured cells and animals. On the other hand, detecting the colocalization of fluorescent-labeled fragments, generated by the lysosomal degradation of DQ-BSA in conjunction with an autophagic protein, is the most reliable method for monitoring fusion between an autophagic vacuole with an active proteolytic compartment. It is important to take into account that in many publications the presence of a lysosomal enzyme has been taken as an indication of fusion with a lysosome. However, it is essential to consider that the detection of an enzyme by immunofluorescence does not necessarily mean that the compartment is degradative, as the enzyme might be present but in an inactive form or the compartment might be proteolytically nonfunctional.

ACKNOWLEDGMENTS

Our work in this area is funded by grants from Agencia Nacional de Promoción Científica y Tecnológica (PICT2004 # 20711 and PICT 2005 # 38420), CONICET (PIP # 5943), and SECTyP (Universidad Nacional de Cuyo).

REFERENCES

Bampton, E. T., Goemans, C. G., Niranja, D., Mizushima, N., and Tolkovsky, A. M. (2005). The dynamics of autophagy visualized in live cells: From autophagosome formation to fusion with endo/lysosomes. *Autophagy* **1**(1), 23–36.

Biederbick, A., Kern, H. F., and Elsasser, H. P. (1995). Monodansylcadaverine (MDC) is a specific *in vivo* marker for autophagic vacuoles. *Eur. J. Cell Biol.* **66**(1), 3–14.

Boya, P., González-Polo, R. A., Casares, N., Perfettini, J. L., Dessen, P., Larochette, N., Métivier, D., Meley, D., Souquere, S., Yoshimori, T., Pierron, G., Codogno, P., et al. (2005). Inhibition of macroautophagy triggers apoptosis. *Mol. Cell Biol.* **25**(3), 1025–1040.

Fader, C. M., Sanchez, D., Furlan, M., and Colombo, M. I. (2008). Induction of autophagy promotes fusion of multivesicular bodies with autophagic vacuoles in k562 cells. *Traffic* **9**(2), 230–250.

Gutierrez, M. G., Munafó, D. B., Beron, W., and Colombo, M. I. (2004). Rab7 is required for the normal progression of the autophagic pathway in mammalian cells. *J. Cell Sci.* **117**(13), 2687–2697.

Iwai-Kanai, E., Yuan, H., Huang, C., Sayen, M. R., Perry-Garza, C. N., Kim, L., and Gottlieb, R. A. (2008). A method to measure cardiac autophagic flux *in vivo*. *Autophagy* **4**(3), 322–329.

Jackson, W. T., Gidding, T. H., Jr., Taylor, M. P., Mulinyawe, S., Rabinovitch, M., Kopito, R. R., and Kirkegaard, K. (2005). Subversion of cellular autophagosomal machinery by RNA viruses. *PLoS Biol.* **3**(5), e156.

Jager, S., Bucci, C., Tanida, I., Ueno, T., Kominami, E., Saftig, P., and Eskelinen, E.-L. (2004). Role for Rab7 in maturation of late autophagic vacuoles. *J. Cell Sci.* **117**(20), 4837–4848.

Kimura, S., Noda, T., and Yoshimori, T. (2007). Dissection of the autophagosome maturation process by a novel reporter protein, tandem fluorescent-tagged LC3. *Autophagy* **3**(5), 452–460.

Klionsky, D. J. (2007). Autophagy: From phenomenology to molecular understanding in less than a decade. *Nat. Rev. Mol. Cell Biol.* **8**(11), 931–937.

Mizushima, N. (2004). Methods for monitoring autophagy. *Int. J. Biochem. Cell Biol.* **36**(12), 2491–2502.

Munafó, D. B., and Colombo, M. I. (2001). A novel assay to study autophagy: regulation of autophagosome vacuole size by amino acid deprivation. *J. Cell Sci.* **114**(20), 3619–3629.

Niemann, A., Baltes, J., and Elsässer, H. P. (2001). Fluorescence properties and staining behavior of monodansylpentane, a structural homologue of the lysosomotropic agent monodansylcadaverine. *J. Histochem. Cytochem.* **49**(2), 177–185.

Niemann, A., Takatsuki, A., and Elsasser, H. P. (2005). The lysosomotropic agent monodansylcadaverine also acts as a solvent polarity probe. *J. Histochem. Cytochem.* **48**(2), 251–258.

Stromhaug, P. E., and Seglen, P. O. (1993). Evidence for acidity of prelysosomal autophagic/endocytic vacuoles (amphisomes). *Biochem. J.* **291**(1), 115–121.

CHAPTER SEVEN

The GST-BHMT Assay and Related Assays for Autophagy

Patrick B. Dennis* *and* Carol A. Mercer*

Contents

1. Introduction	98
2. Measurement of Macroautophagy Using the GST-BHMT Assay	100
2.1. Basic cell culture conditions, transfection, and extraction	100
2.2. Purification and detection of GST-BHMT and proteolytic fragments	101
2.3. Selection of protease inhibitors for the GST-BHMT assay	102
2.4. GST-BHMT assay response to depletion of individual amino acids	103
2.5. The effect of time and reporter expression levels on GST-BHMT fragment accumulation	106
3. Measurement of Autophagy Using a Linker-Specific Cleavage Site (LSCS)	107
4. Measurement of Reticulophagy and Mitophagy Using LSCS-Based Reporters	110
5. Concluding Remarks	114
Acknowledgments	116
References	116

Abstract

The endpoint of the autophagic process is the breakdown of delivered cytoplasmic cargo in lysosomes. Therefore, assays based on degradation of cargo are of particular interest in that they can measure regulation of the entire autophagic process, including changes in cargo delivery and breakdown in the lytic compartment. Betaine homocysteine methyltransferase (BHMT) is one of many cytosolic proteins found in the mammalian autophagosome, and delivery of BHMT to the lysosome results in its proteolysis to discrete fragments under certain conditions. Making use of these observations, the GST-BHMT assay was developed as an endpoint, cargo-based autophagy assay. Using this assay as a

* University of Cincinnati, Genome Research Institute, Department of Cancer and Cell Biology, Cincinnati, Ohio, USA

starting point, additional cargo-based assays have been developed with the potential to measure autophagic degradation of specific subcellular compartments. Here we describe the development and validation of these assays.

1. Introduction

Macroautophagy is a complex process that requires the coordinated efforts of many gene products for the sequestration of cytoplasmic cargo and delivery of that cargo to the lysosome (Klionsky, 2005). Because the endpoint of macroautophagy is the destruction of sequestered cargo in the lytic compartment, assays that measure both sequestration and degradation of cargo are particularly attractive. A number of assays have been developed that are able to measure macroautophagy in both yeast and higher eukaryotes. Some, such as the Atg8/LC3 assay, measure changes in levels of an autophagosome membrane associated protein to assess the rate of autophagosome formation and destruction (Kabeya et al., 2000; Mizushima and Yoshimori, 2007; see also the chapter by Kimura et al., in this volume). Others, such as the lactate dehydrogenase (LDH) assay, measure the lysosomal sequestration of cytosolic proteins delivered through the macroautophagic pathway (Kopitz et al., 1990; see also the chapter by Seglen et al., in this volume). Related to the latter approach, another metabolic enzyme, betaine homocysteine methyltransferase (BHMT), is enriched in autolysosomes isolated from rat liver cells (Furuya et al., 2001; Overbye et al., 2007; Ueno et al., 1999). In addition, rat liver BHMT is cleaved at a specific site in response to amino acid starvation in a macroautophagy-dependent manner (Ueno et al., 1999). This suggests that BHMT proteolysis can be used as a marker for macroautophagic flux. However, because endogenous BHMT is expressed primarily in liver and kidney cells, it has limited potential as a macroautophagy marker in other cell types.

To test whether specific BHMT proteolysis can be used to measure macroautophagy in cell lines, we created a construct in which glutathione-S-transferase is fused to the amino terminus of human BHMT (GST-BHMT), and placed upstream of green fluorescent protein (GFP) under the translational control of an internal ribosomal entry sequence (IRES) (Fig. 7.1A). When this construct is transfected into HEK293 cells, a single GST-fused proteolytic fragment accumulates at an increased rate during nutrient starvation, and linear accumulation of the GST-BHMT proteolytic fragment is observed to occur up to 8 h after the switch to starvation media (Mercer et al., 2008). Additionally, the proteolytic fragment is found only in the light membrane fraction following differential centrifugation and cofractionates with autophagosome and lysosome markers on a sucrose gradient. Because of the size of the GST-fused fragment, proteolysis of

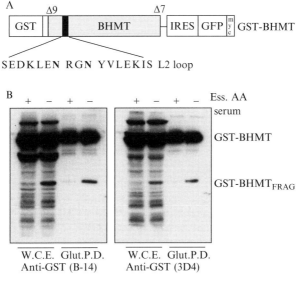

Figure 7.1 Detection of GST-BHMT fragmentation. (A) Schematic of the GST-BHMT macroautophagy reporter construct. The GST-BHMT reporter is missing the first nine and last seven amino acids (Δ9 and Δ7) of human BHMT. The position of the BHMT L2 loop is shown in black with the primary structure indicated below. The macroautophagy-dependent cleavage sites following asparagines in the L2 loop are indicated in bold. The internal ribosomal entry site (IRES) upstream of the myc-tagged green fluorescent protein (GFP-myc) is also indicated. (B) HEK293 cells were transiently transfected with the GST-BHMT reporter and then cultured for 6 h in full medium (+) or starvation medium (−) in the presence of leupeptin (11 μM) and E64d (6 μM) before extraction. GST-BHMT, and its accumulated proteolytic fragments were analyzed from 100 μg of total protein from the whole cell extracts (W.C.E.) blotted directly, or enriched by purification on glutathione agarose beads (Glut. P.D.) before blotting. Visualization of the blots was achieved using either the B-14 (*left panel*) or 3D4 (*right panel*) monoclonal antibodies against the glutathione-S-transferase (GST) tag. The positions of full-length GST-BHMT and the proteolytic fragment (GST-BHMT$_{FRAG}$) are indicated.

the GST-BHMT reporter likely occurs in the L2 loop of BHMT (Fig. 7.1A), consistent with the predicted cleavage site reported previously in rat BHMT (Ueno *et al.*, 1999). Subsequent studies have found that the asparaginyl endopeptidase legumain is responsible for lysosomal cleavage of GST-BHMT at two asparagines residues located in the L2 loop (Fig. 7.1A and manuscript in preparation). Increased accumulation of the proteolytic fragment is also observed in human tumor cells from different lineages as well as in nontransformed cell lines from rat and mouse sources. Importantly, starvation-induced fragment accumulation is potently blocked by pharmacological inhibitors of autophagy, as well as by depletion of the macroautophagy proteins Beclin 1 and ULK1 (Mercer *et al.*, 2008).

Together, the results indicate that site-specific proteolysis of GST-BHMT can be used as a marker for the induction of macroautophagy at its endpoint (i.e., delivery of cytoplasmic cargo to the lytic compartment) in numerous cell lines.

This chapter expands on the optimal conditions and reagents necessary to use GST-BHMT as an assay for macroautophagy. On the basis of the results obtained during the development of this assay, similar endpoint assays have been generated for macroautophagy of the endoplasmic reticulum (reticulophagy) and the mitochondria (mitophagy), using a strategy that involves compartment specific targeting of a reporter containing a linker-specific cleavage site (LSCS).

2. MEASUREMENT OF MACROAUTOPHAGY USING THE GST-BHMT ASSAY

2.1. Basic cell culture conditions, transfection, and extraction

The data demonstrating use and validation of the GST-BHMT assay have been reported previously (Mercer *et al.*, 2008). The basic protocol for assay use is recounted in this section.

1. HEK293 cells are first seeded at 50–65 × 10^4 cells per 6-cm culture dish (Corning) in DMEM containing 5 mM glutamine, 25 mM glucose, and 10% Nu-Serum I (BD) and cultured at 37 °C and 5% CO_2.
2. Approximately 42 h later, the cells are transfected with 1–2 μg of the GST-BHMT construct balanced with empty plasmid (PRK5) to a total of 5 μg of DNA. For HEK293 cells, the calcium phosphate transfection method is used (Chen and Okayama, 1988), resulting in approximately 60% transfection efficiency with plasmid. By this method, HEK293 cells can be transfected again on the second day for greater than 90% transfection efficiency with little toxicity. For cells that do not transfect well with calcium phosphate, a polyamine-based transfection reagent can be used, with the same DNA concentrations, per manufacturer's instructions.
3. Cells are incubated for 6 h with the transfection reagent at 37 °C and 3% CO_2 before being washed into fresh DMEM/10% Nu-Serum I.
4. Approximately 40 h later, transfected cells are treated by culturing them for 6 h in either full medium (DMEM with 5 mM glutamine, 25 mM glucose and 10% Nu-Serum I) or in starvation medium before extraction. The starvation medium is DMEM with 5 mM glutamine, other nonessential amino acids, and 25 mM glucose, but without the essential amino acids (Arg, Cys, His, Ile, Leu, Lys, Met, Phe, Thr, Trp, Tyr, and Val) and Nu-Serum I. The presence of glutamine in the starvation medium is necessary for a full starvation response of the GST-BHMT

assay (see subsequent sections). The protease inhibitors E64d (6 μM) and leupeptin (11 μM) are added to the medium at the beginning of the treatment to initiate accumulation of the lysosome-derived GST–BHMT proteolytic fragment, as described previously (Mercer et al., 2008).

5. After the 6-h treatment in the presence of the protease inhibitors, the cells are washed once with ice-cold PBS and extracted by Dounce homogenization (10 strokes) in 0.5 ml of extraction buffer (50 mM Tris-HCl, pH 8, 120 mM NaCl, 5 mM NaPPi, 10 mM NaF, 30 mM paranitrophenyl phosphate [PNPP], 1 mM benzamidine, 0.1% NP-40, 1% Triton X-100 and 0.2 mM PMSF). The phosphatase inhibitors, NaPPi, NaF and PNPP may be omitted if desired.

2.2. Purification and detection of GST-BHMT and proteolytic fragments

To measure the accumulation of the GST fused proteolytic fragment of BHMT during the treatment time, an anti-GST antibody is used to probe Western blots of either whole cell extracts or GST-fused proteins after purification with glutathione agarose (see subsequent sections). The decision to probe blotted whole cell extracts directly or use an affinity enrichment step depends on the transfection efficiency, the amount of reporter transfected, and the antibodies used for detection. Affinity purification in conjunction with a commercially available polyclonal antibody against GST (Z5, Santa Cruz) has been used successfully to measure fragment accumulation (Mercer et al., 2008). However, variations in antibody quality have been found in some lot preparations that affect the ability to cleanly detect the proteolytic fragment, even after enrichment on glutathione agarose beads (unpublished data). In an effort to make the detection of GST-BHMT fragment accumulation more standardized, various monoclonal anti-GST antibodies were tested. From these tests, two anti-GST monoclonal antibodies, B-14 and 3D4 (Santa Cruz), stood out as effective probes for GST-BHMT fragment detection (Fig. 7.1B).

1. The Western blot membranes are incubated with blocking buffer (Tris-buffered saline (TBS: 20 mM Tris-HCl, pH 7.5, 171 mM NaCl) with 0.05% Tween 20 and 1% BSA [fraction V, Fisher Scientific]). Membrane blocking is carried out at room temperature for 30 min with constant rocking.
2. Primary antibodies, diluted at a 1:1000 titer in blocking buffer, are added to the blot and incubation is carried out at 4 °C, overnight, with constant rocking. After the probing step, the blots are washed 3 times at room temperature, for a total of 30 min, in wash buffer (TBS with 0.5% Tween 20).
3. The blots are then probed for 1 h at room temperature with a biotin-conjugated secondary antibody (Zymed, 81-6540), diluted at a 1:4000

titer in blocking buffer. After the probing step, the blots are washed 3 times at room temperature, for a total of 30 min, in wash buffer.
4. The blots are incubated for 1 h at room temperature with horseradish peroxidase-conjugated streptavidin (Zymed, 43-8323) diluted at a titer of 1:10,000 in blocking buffer. After washing the blot as in steps 2 and 3, detection is achieved using an enhanced chemiluminescent (ECL) method.

Both B-14 and 3D4 GST monoclonal antibodies are able to detect starvation-induced fragment accumulation in blotted whole cell extracts, although significant background is observed with each antibody (Fig. 7.1B). To reduce the background, GST-BHMT from the same amount of extract is enriched with glutathione agarose.

1. Glutathione agarose beads (Sigma, G4510) are equilibrated in extraction buffer.
2. The extract (100 μg or more of whole cell extract diluted to 500 μl with extraction buffer) is incubated with 20 μl of a 1:1 glutathione agarose slurry for 2 h at 4 °C. This amount of glutathione agarose should be saturating.
3. The beads are then washed once with 1 ml of high-salt wash buffer (50 mM Tris-HCl, pH 8, 500 mM NaCl, 5 mM NaPPi, 10 mM NaF, 1 mM benzamidine, 30 mM β-glycerolphosphate, 0.1% NP-40, and 0.2 mM PMSF). Resuspended beads are pelleted in a bench-top microcentrifuge set at 2000 rpm for 1 min.
4. The beads are washed twice with 1 ml of the same wash buffer without NaCl, drained with a Hamilton syringe, and suspended in 15 μl of the wash buffer without NaCl. Five μl of concentrated SDS-PAGE sample buffer (500 mM Tris-HCl, pH 6.8, 250 mM dithiothreitol, 5% SDS, 25% glycerol, and 0.2% bromophenol blue) are added to the bead slurry, and the sample is heated to 95 °C in a heat block for 3 min.
5. The entire sample, including the beads, is loaded onto a 10% polyacrylamide gel, where the bound proteins are resolved by SDS-PAGE before analysis by Western blotting. The presence of beads in the sample well does not affect resolution of the proteins during electrophoresis.

Adsorption onto glutathione agarose beads is an effective method to reduce background observed on Western blots of crude extracts with both monoclonal antibodies and may be necessary if the transfection efficiency of a given cell line is low (Fig. 7.1B).

2.3. Selection of protease inhibitors for the GST-BHMT assay

On the basis of previous studies describing the conditions necessary to partially inhibit BHMT proteolysis, the protease inhibitors E64d and leupeptin are added together during the treatment period to ensure linear fragment accumulation (Furuya et al., 2001; Mercer et al., 2008; Ueno et al., 1999). However, injection of mice or treatment of hepatocytes with leupeptin alone is sufficient to observe BHMT fragment accumulation in liver cells (Furuya et al., 2001; Ueno et al., 1999). To determine the GST-BHMT assay requirement for protease inhibitors, E64d (6 μM) and leupeptin (11 μM) were individually tested for their ability to increase the GST-BHMT fragment. Either E64d or leupeptin treatment of cells alone is sufficient to increase GST-BHMT fragment accumulation, although leupeptin is more potent that E64d (Fig. 7.2A). Addition of E64d with leupeptin does not enhance fragment accumulation above that of leupeptin alone, indicating that proteolytic inhibition by leupeptin is sufficient for inducing fragment accumulation, consistent with what is observed in rat liver cells (Furuya et al., 2001; Ueno et al., 1999). Other commonly used protease inhibitors were also tested for their ability to either increase GST-BHMT fragment accumulation or inhibit fragment accumulation induced by serum and amino acid starvation in leupeptin treated cells. Treatment of transfected cells with pepstatin A (15 μM) and ebelactone B (20 μM), inhibitors of aspartate and serine proteases, respectively, does not promote fragment accumulation (Fig. 7.2B) and does not inhibit leupeptin-induced fragment accumulation in starved cells (Fig. 7.2C). However, treatment of cells with MG-132 (20 μM) increases GST-BHMT fragment accumulation, albeit to a lesser degree than leupeptin (Fig. 7.2A). As with E64d, addition of MG-132 with leupeptin does not change the level of GST-BHMT fragment accumulation during amino acid and serum starvation (unpublished data). Although MG-132 is commonly used as a proteasome inhibitor, it is also a known inhibitor of lysosomal proteases (Lee and Goldberg, 1998). Its similarity with leupeptin in inducing fragment accumulation is not surprising, as both molecules belong to the same class of tripeptide aldehyde inhibitors. Together, the protease inhibitor data are consistent with the finding that legumain is the protease responsible for cleavage of GST-BHMT in its L2 loop, as legumain activity is insensitive to leupeptin, E64d, pepstatin A and ebelactone B inhibition (Chen et al., 1997).

2.4. GST-BHMT assay response to depletion of individual amino acids

Serum starvation alone is not sufficient to increase GST-BHMT fragmentation, whereas depletion of the essential amino acids, even in the presence of serum, results in full fragment induction (Mercer et al., 2008). The individual removal of each essential amino acid from DMEM medium is not sufficient to increase GST-BHMT fragmentation to levels seen with the loss of all the essential amino acids (Figs. 7.3A and 7.3B). However, depletion of only leucine increases GST-BHMT fragment accumulation to ~50% of that observed with full essential amino acid depletion. Individual loss of the

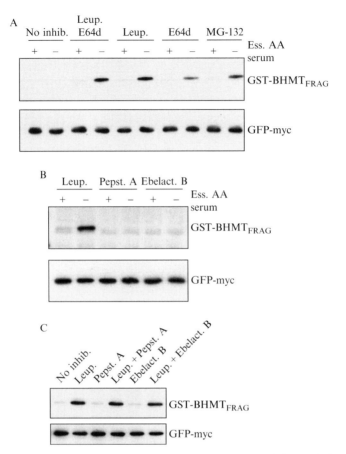

Figure 7.2 Sensitivity of the GST-BHMT assay to protease inhibitors. (A) HEK293 cells, transiently transfected with the GST-BHMT reporter, were maintained for 6 h in full medium (+) or starvation medium (−) and accumulation of the GST-BHMT fragment was tested under the following conditions: no protease inhibitors, with leupeptin (11 μM) and/or E64d (6 μM), or with MG-132 (20 μM). (B) Cells, transfected as in A, were maintained for 6 h in full medium or starvation medium in the presence of either leupeptin (11 μM), pepstatin A (15 μM), or ebelactone B (20 μM). (C) Cells, transfected as in A, were maintained for 6 h in starvation medium in the absence or presence of the indicated protease inhibitors at the same concentrations as in B. The accumulated proteolytic fragment of GST-BHMT is shown (GST-BHMT$_{FRAG}$) together with GFP-myc levels for normalization. Antibodies used for Western blotting are as follows: Z5 (Santa Cruz) to detect GST, and 9B11 (Cell Signaling) for the myc tag.

other branched-chain amino acids demonstrates little potential to increase fragmentation of the reporter (Fig. 7.3A), while depletion of all three branched chain amino acids results in a near full response (Fig. 7.3A). Removal of other essential amino acids, specifically arginine, lysine, methionine, and tryptophan, elicits a modest response from the GST–BHMT assay

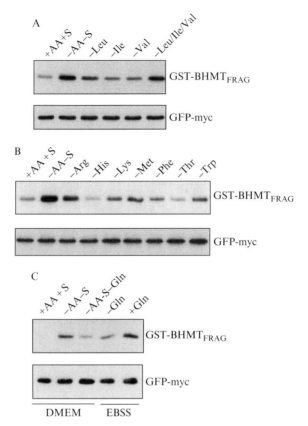

Figure 7.3 The effect of individual amino acid depletion on macroautophagy of GST-BHMT. (A and B) HEK293 cells were transiently transfected with the GST-BHMT reporter and cultured for 6 h in full medium (+AA+S) or starvation medium lacking the essential amino acids and serum (−AA−S) in the presence of leupeptin and E64d. In addition, the cells were cultured in serum-free starvation medium missing a specific essential amino acid as indicated above each figure. (C) Cells were transfected and treated for 6 h as follows: in complete (+AA+S) or starvation (−AA−S) medium as in A and B; in DMEM starvation medium without glutamine (−AA−S−Gln); in Earle's balanced salt solution (−Gln), or in EBSS with glutamine (+Gln). (EBSS contains 0.265 g/L $CaCl_2$, 0.098 g/L $MgSO_4$, 0.4 g/L KCl, 2.2 g/L $NaHCO_3$, 6.8 g/L NaCl, 0.122 g/L NaH_2PO_4, 1.0 g/L D-Glucose.) Levels of accumulated GST-BHMT proteolytic fragment were determined after enrichment on glutathione agarose followed by Western blotting and probing with Z5 antibody. Levels of GFP-myc are shown for normalization as in Fig. 7.2.

but not to the level observed with branched-chain amino acid depletion or starvation of all essential amino acids (Fig. 7.3B). Taken together, the results indicate that increased GST-BHMT fragmentation in the absence of essential amino acids is a composite response to the depletion of a handful of dominant essential amino acids. However, starvation from the branched-chain amino acids leads to a large fraction of the total response (Fig. 7.3A).

The starvation medium used to induce macroautophagy and increase GST-BHMT fragment accumulation lacks essential amino acids but contains glutamine and other non-essential amino acids. The accumulation of the GST-BHMT fragment is reduced in cells starved of essential amino acids and glutamine, when compared to fragmentation observed with depletion of only the essential amino acids (Fig. 7.3C). If other media components are removed (e.g., nonessential amino acids and vitamins) by starving the cells in Earl's balanced salt solution (EBSS), control levels of fragment accumulation are observed only if EBSS is supplemented with glutamine (Fig. 7.3C). This finding demonstrates that, though the essential amino acid levels are important for regulating the induction of GST-BHMT fragmentation, glutamine must be present in the starvation medium for full fragmentation to occur. This is consistent with the idea that glutamine can be used as an energy source in cultured cells (Neermann and Wagner, 1996; Reitzer et al., 1979), and autophagy is a strong consumer of intracellular energy in terms of the conjugation of autophagosome associating proteins and acidification of the lysosome (Marshansky and Futai, 2008; Ohsumi and Mizushima, 2004; Schellens and Meijer, 1991). Finally, it should be noted that if EBSS is used as starvation medium, the presence of sodium bicarbonate (26 mM) is essential to observe GST-BHMT fragment accumulation (unpublished data).

2.5. The effect of time and reporter expression levels on GST-BHMT fragment accumulation

The basis of the GST-BHMT assay is differential proteolytic fragment accumulation when cells are shifted from nutrient-rich to nutrient-poor conditions. Therefore, it is important to determine the effect that expression levels of GST-BHMT and treatment times may have on the differential accumulation of the proteolytic fragment. To determine whether the difference in fragment accumulation between fed and starved cells is influenced by expression levels of GST-BHMT, HEK293 cells, transfected with increasing amounts of GST-BHMT cDNA, were tested under the usual assay conditions. The results show that levels of accumulated fragment during starvation correlate well with expression of full length GST-BHMT and GFP-myc (Fig. 7.4A). Also, increasing expression of the GST-BHMT reporter has no effect on differential fragment accumulation observed when comparing fed and starved cells (Fig. 7.4A). Because levels of proteolytic fragment accumulation depend on the expression of full-length reporter, it is important to match GST-BHMT expression in control and test cultures when transient transfection assays are used. Stable expression of the GST-BHMT reporter in HEK293 cells has been achieved and greatly simplifies normalization of the assay results (Mercer et al., 2008).

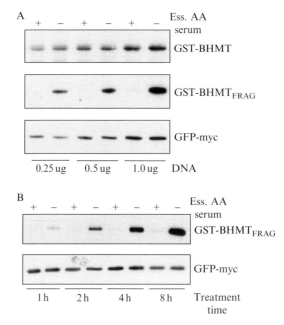

Figure 7.4 The effect of expression and treatment time on starvation-induced GST-BHMT proteolysis. (A) HEK293 cells were transiently transfected with increasing amounts of the GST-BHMT reporter construct as indicated. The cells were then cultured in full medium (+) or starvation medium (−) for 6 h in the presence of leupeptin and E64d, as in Fig. 7.1. (B) Cells transfected as in A were cultured for the times indicated in full medium or starvation medium in the presence of leupeptin and E64d. Levels of full length GST-BHMT are shown in A, and levels of accumulated GST-BHMT proteolytic fragment and expressed GFP-myc were analyzed as in Fig. 7.2. Monoclonal B-14 antibody was used for detection of GST. GFP-myc was probed with a 9B11 antibody.

A time-course analysis of GST-BHMT fragment accumulation demonstrates that the rate of accumulation in the absence of serum and essential amino acids appears steady up to 8 h (Fig. 7.4B) (Mercer et al., 2008). Importantly, increased fragment accumulation in starved HEK293 cells is apparent as early as 1 h into the treatment time course (Fig. 7.4B). This indicates that the GST-BHMT assay can measure the induction of macroautophagy at early and late time points.

3. Measurement of Autophagy Using a Linker-Specific Cleavage Site (LSCS)

During the development of the GST-BHMT assay, an early version of the GST-BHMT reporter demonstrated two fragments during amino acid and serum starvation, both of which accumulated with similar kinetics

(Mercer et al., 2008). The smaller fragment was approximately the size of endogenous GST and was specifically present in the light membrane fraction (LMF), suggesting a potential for autolysosomal generation. Given the size of the fragment and the likelihood that proteolysis occurred in the linker region between GST and BHMT, the cleavage site was termed linker-specific cleavage site (LSCS). As with the larger GST-BHMT fragment, accumulation of the smaller, LSCS fragment can be blocked by pharmacological inhibitors of autophagy, particularly the lysosomal inhibitor chloroquine (Mercer et al., 2008). The lysosomal stability and accumulation of GST-LSCS is similar to what has been reported earlier for GFP, which others have used successfully in the measurement of macroautophagy (Hosokawa et al., 2006; Shintani and Klionsky, 2004). Importantly, in contrast to cleavage in the L2 loop of BHMT, cleavage at the LSCS does not require the ability of the reporter to form stable oligomers (Mercer et al., 2008). This suggests that GST with the LSCS proteolytic site can be transferred to other proteins for the measurement of macroautophagy.

A $GST_{LSCS}GFP$ fusion protein was created to test whether the GST and LSCS could be used to detect macroautophagy of fused proteins (Fig. 7.5A). HEK293 cells were transfected with the $GST_{LSCS}GFP$ reporter and cultured in either full medium or starvation medium as with the GST-BHMT assay. Increased accumulation of released GST is observed when the cells are starved of serum and essential amino acids (Fig. 7.5B), consistent with what is observed when the GST and LSCS are fused with BHMT (Mercer et al., 2008). In agreement with others, release and accumulation of GFP during starvation is also observed (Hosokawa et al., 2006; Shintani and Klionsky, 2004). As with GST-BHMT L2 loop cleavage, legumain was identified as the lysosomal protease important for cleavage at the LSCS (manuscript in preparation). Unlike the BHMT L2 loop, the LSCS contains two aspartates but no asparagines (Fig. 7.5A); however, studies show that legumain can also cleave after aspartates (Li et al., 2003). Mutational analyses indicate that the aspartate closest to GFP is the primary site of cleavage, releasing GST and GFP under starvation conditions (Fig. 7.5C). Significantly less cleavage is observed at the aspartate nearer to GST (Fig. 7.5C and unpublished data). Mutation of the primary cleavage site to asparagine does not increase levels of free GST or GFP accumulation (Fig. 7.5C). In addition to the increased accumulation of GST and GFP during starvation, depletion of ULK1 or treatment with chloroquine reduces free GST accumulation (Fig. 7.5D), indicating that $GST_{LSCS}GFP$ is delivered to the lysosome by macroautophagy prior to cleavage at the LSCS. The results indicate that $GST_{LSCS}GFP$ can be used as a cytoplasmic macroautophagy reporter and support the idea that GST-LSCS can be fused to other proteins to assess their autophagic delivery and degradation in a biochemical endpoint-based assay.

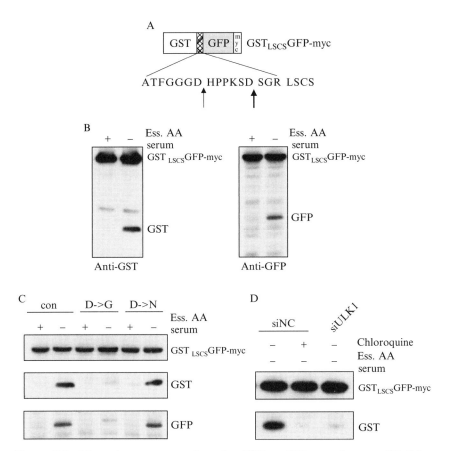

Figure 7.5 Measuring macroautophagy by GST$_{LSCS}$GFP-myc cleavage. (A) Schematic showing GST and GFP-myc linked with a linker-specific cleavage site (LSCS, hatched region) to form the GST$_{LSCS}$GFP reporter. The primary (*thick arrow*) and secondary (*thin arrow*) cleavage sites following aspartates in the linker region are shown. (B) HEK293 cells were transfected with the GST$_{LSCS}$GFP-myc construct and cultured for 6 h in either full medium (+) or starvation medium (−) in the presence of leupeptin and E64d as in Fig. 7.1. After extraction, released GST enriched on glutathione agarose was observed from extracts by Western blotting with anti-GST antibody (B14, Santa Cruz) (*left panel*). Released GFP was probed from Western-blotted crude extract (50 μg of total protein) using anti-GFP antibody (Santa Cruz, sc-8334) (*right panel*). (C) The parental GST$_{LSCS}$GFP-myc reporter (con) and GST$_{LSCS}$GFP constructs with Asp to Gly and Asp to Asn mutations of the LSCS aspartate closest to GFP were transfected into cells that were treated as in B. Expression of the full-length reporter (*upper panel*), released GST (*middle panel*), and released GFP (*lower panel*) were determined as in B. (D) GST$_{LSCS}$GFP-myc was cotransfected with nonsilencing siRNA (siNC) or siRNA against ULK1 and incubated for 6 h in starvation medium, as in B. One set of cells was treated with 100 μM chloroquine during the 6-h starvation period as indicated.

4. MEASUREMENT OF RETICULOPHAGY AND MITOPHAGY USING LSCS-BASED REPORTERS

GST$_{LSCS}$GFP demonstrates diffuse cytosolic localization by fluorescence microscopy (unpublished observation), making it ideal for measuring macroautophagy of the cytoplasm as well as providing a suitable background for targeted localization. Therefore, we reasoned that specific targeting of the GST$_{LSCS}$GFP to a subcellular location could be used to create an endpoint autophagy assay for that compartment. This concept is not without precedence in the measurement of macroautophagy. For instance, flux analysis of the autophagosome can be measured using the lytic release and accumulation of free GFP from a GFP-LC3 reporter (Hosokawa et al., 2006; Shintani and Klionsky, 2004). In addition, a yeast ribophagy assay uses a similar approach in which GFP is released from a GFP-fused ribosomal protein when the ribosomal subunit is delivered to the vacuole (Kraft et al., 2008). As with the GFP-Atg8/LC3 and ribophagy assays (Hosokawa et al., 2006; Kraft et al., 2008; Shintani and Klionsky, 2004), autophagy of a GST and LSCS-tagged subcellular compartment will be determined at the endpoint by release of GST. Organelle targeting sequences from the endoplasmic reticulum (ER) protein cytochrome b5 (cb5) and the listerial actin A (ActA) protein have been used by others to target expression of fusion proteins to the endoplasmic reticulum and the outer membrane of the mitochondria, respectively (Zhu et al., 1996; Zong et al., 2003). These sequences are of particular interest because of their potential application for the study of selective autophagy of the ER and mitochondria, termed *reticulophagy* and *mitophagy*, respectively.

To create an ER-specific, endpoint autophagy assay for reticulophagy, 35 amino acids of cb5 were fused to the carboxy terminus of the GST$_{LSCS}$GFP to create the GST$_{LSCS}$GFP-cb5 construct (Fig. 7.6A). To ensure that the targeted GST$_{LSCS}$GFP-cb5 reporter localizes at the ER, cells are transiently transfected with the reporter and then stained with ER-Tracker Red (Invitrogen) to fluorescently label the ER. Optimal concentration of ER-Tracker Red and incubation time are determined experimentally for each cell line. ER-Tracker Red (0.5 μM) in Hank's Buffered Salt Solution with Ca^{2+} and Mg^{2+} (HBSS$^{Ca2+Mg2+}$) is added to either HEK293 or T98G (glioblastoma) cells 24 h after transfection and incubated for 25 min at 37 °C, after which the live, costained cells are analyzed by confocal microscopy. The GST$_{LSCS}$GFP-cb5 reporter localizes to perinuclear structures consistent with proper targeting to the ER (Fig. 7.6B). Overlaying the images from the analysis of GST$_{LSCS}$GFP-cb5 with those of ER-Tracker Red demonstrates strong overlap of the signals, indicating that the GST$_{LSCS}$GFP-cb5 reporter is specifically localized to the ER in both cell types (Fig. 7.6B).

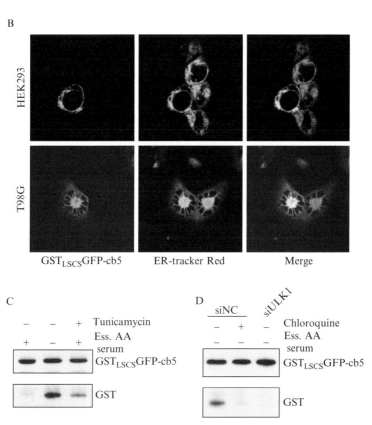

Figure 7.6 Measuring reticulophagy by GST$_{LSCS}$GFP-cb5 cleavage. (A) Schematic showing the GST$_{LSCS}$GFP reporter with the carboxy-terminal cb5 sequence and LSCS (*hatched region*). The primary structure of the cb5 sequence is shown. (B) HEK293 and T98G cells were transfected with the GST$_{LSCS}$GFP-cb5 construct (*green*) and cultured in full medium. ER-Tracker Red in HBSS$^{Ca2+Mg2+}$ was added to the transfected cells, to a final concentration of 0.5 μM, followed by incubation for 25 min at 37 °C before the live cells were analyzed by confocal microscopy. (C) The GST$_{LSCS}$GFP-cb5 construct was transfected into HEK293 cells and cultured for 13 h in either full medium (+) or starvation medium (−) in the presence of leupeptin (11 μM). One set of cells was cultured in full medium with 5 μg/ml tunicamycin for the 13-h treatment time. Full-length GST$_{LSCS}$GFP-cb5 and released GST were enriched on glutathione agarose and expression was analyzed by Western blotting, using the B-14 antibody to detect GST. (D) GST$_{LSCS}$GFP-cb5 was cotransfected with nonsilencing siRNA (siNC) or siRNA against ULK1 and treated as in C, except the starvation time was reduced to 6 h. One set of cells was treated with 100 μM chloroquine during the 6-h starvation period as indicated. Levels of full-length GST$_{LSCS}$GFP-cb5 and released GST were analyzed as in C. (See Color Insert.)

HEK293 cells transiently transfected with the $GST_{LSCS}GFP$-cb5 reporter are starved for 13 h from serum and essential amino acids or are maintained in full medium with tunicamycin (5 μg/ml) to induce ER stress and reticulophagy (Ogata et al., 2006; Yorimitsu et al., 2006). As with the GST-BHMT assay, 11 μM leupeptin is added at the start of the treatment time to facilitate lysosomal accumulation of GST. Both starvation and tunicamycin treatment are able to induce release and accumulation of GST from the $GST_{LSCS}GFP$-cb5 reporter (Fig. 7.6C), indicating increased delivery of the ER-localized protein to the lysosomal compartment. The lysosome-based proteolysis of the reporter is confirmed by chloroquine treatment, which completely inhibits accumulation of released GST during starvation (Fig. 7.6D). Furthermore, inhibition of GST accumulation by depletion of ULK1 validates that degradation of the cb5-tagged reporter is through macroautophagy of the ER (Fig. 7.6D).

To create an autophagy reporter for mitophagy, 26 amino acids of ActA are fused to the carboxy terminus of $GST_{LSCS}GFP$ to create a $GST_{LSCS}GFP$-ActA reporter (Fig. 7.7A). To confirm mitochondrial localization of the $GST_{LSCS}GFP$-ActA protein, HEK293 and T98G cells, transiently expressing the reporter, are costained with MitoTracker Red (Invitrogen), a fluorescent dye that specifically stains mitochondria. Optimal concentrations of MitoTracker Red and incubation times are determined experimentally. Cells are incubated for 15 min (HEK293 cells) or 30 min (T98G cells) in 50 nM MitoTracker Red in DMEM, after which the live cells are visualized by confocal microscopy. $GST_{LSCS}GFP$-ActA demonstrates a punctate localization pattern throughout the cell, distinct from that observed with $GST_{LSCS}GFP$ and $GST_{LSCS}GFP$-cb5 reporters (Figs. 7.6B, 7.7B and unpublished data). An overlay of the signals from the $GST_{LSCS}GFP$-ActA analysis with those from MitoTracker Red demonstrates strong overlap and confirms that the $GST_{LSCS}GFP$-ActA reporter is localized to the mitochondria (Fig. 7.7B). The results are consistent with what has been published earlier regarding the ability of the ActA sequence to localize fused proteins to the mitochondria (Zhu et al., 1996; Zong et al., 2003).

Increased mitophagy is reported in rat hepatocytes that are nutrient starved in buffer containing glucagon (Lemasters et al., 1998). To test whether nutrient starvation leads to increased accumulation of GST released from the mitochondria-localized $GST_{LSCS}GFP$-ActA reporter, HEK293 cells transfected with the reporter construct are cultured for 13 h in full medium or starvation medium in the presence of leupeptin. $GST_{LSCS}GFP$-ActA protein expression levels are lower than that observed for the parental construct or the cb5-tagged reporter. Therefore, we enrich for mitochondria from cells expressing $GST_{LSCS}GFP$-ActA by collecting the light membrane fraction (LMF), as previously described (Mercer et al., 2008). Extracts from LMFs are analyzed for full-length $GST_{LSCS}GFP$-ActA and released

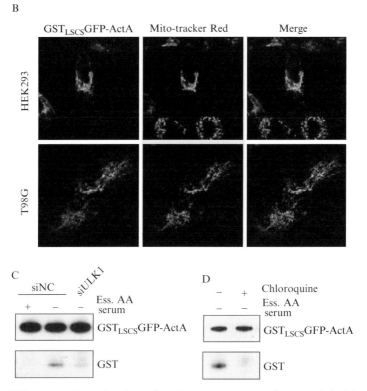

Figure 7.7 Measuring mitophagy by GST$_{LSCS}$GFP-ActA cleavage. (A) Schematic showing the GST$_{LSCS}$GFP reporter with the carboxy-terminal ActA sequence and LSCS (*hatched region*). The primary structure of the ActA sequence is shown. (B) HEK293 and T98G cells were transfected with the GST$_{LSCS}$GFP-ActA construct (*green*) and cultured in full medium. MitoTracker Red was added to the transfected cells, to a final concentration of 50 nM, followed by incubation for 15–30 min at 37 °C before the live cells were analyzed by confocal microscopy. (C) The GST$_{LSCS}$GFP-ActA construct was cotransfected with nonsilencing siRNA (siNC) or siRNA against ULK1 in HEK293 cells that were cultured for 13 h in either full medium (+) or starvation medium (−) in the presence of leupeptin (11 μM). Light membrane fractions were prepared from the treated cells and extracted. Full-length GST$_{LSCS}$GFP-ActA and released GST were analyzed as in Fig. 7.6. (D) HEK293 cells were transfected with GST$_{LSCS}$GFP-cb5 and cultured for 13 h in starvation media. One set of cells was treated with 100 μM chloroquine during the starvation period as indicated. (See Color Insert.)

GST by Western blotting after enrichment on glutathione agarose beads. During nutrient starvation, significantly more released GST accumulates than in the control, indicating an increase in the degradation of mitochondria (Fig. 7.7C). This starvation-induced accumulation is reduced when ULK1 is depleted by siRNA treatment (Fig. 7.7C), implicating a macroautophagic mechanism for degradation. This conclusion is supported by the complete loss of free GST accumulation in starved cells treated with chloroquine (Fig. 7.7D). Together, the data presented for $GST_{LSCS}GFP$-cb5 and $GST_{LSCS}GFP$-ActA demonstrate the versatility of the LSCS-based reporters for studying selective macroautophagy of subcellular compartments.

5. Concluding Remarks

The GST-BHMT assay, presented here as a macroautophagy assay designed to measure autophagic flux, relies on site-specific cleavage of GST-BHMT once delivered to the lysosome by macroautophagy. Cleavage at the L2 loop of GST-BHMT releases the GST tag as well as ~9 kDa of BHMT that remains fused to GST (Mercer et al., 2008). One strength of the assay is that proteolysis of GST-BHMT at its L2 loop seems to occur only in an acidified autolysosome (Mercer et al., 2008). However, it is possible that nonmacroautophagic or nonlysosomal proteolysis occurs under certain experimental conditions. The cleavage of GST-BHMT at its L2 loop and release of the GST-conjugated BHMT fragment depends on the ability of GST-BHMT to form stable oligomers, and therefore depends on the maintenance of quaternary structure immediately after lysosomal delivery (Mercer et al., 2008). The need for stable structure rules out the proteasomal pathway as well as chaperone-mediated autophagy as potential degradation pathways, as both rely on substrate denaturation before proteolysis. The asparaginyl endopeptidase legumain, a protease that is only active in acidified compartments (Chen et al., 2000; Li et al., 2003), cleaves GST-BHMT after two asparagines in BHMT's L2 loop to produce the GST-BHMT fragment (manuscript in preparation). Therefore, drugs that lead to increased lysosomal pH, such as chloroquine and bafilomycin A1, are good tools to validate that cleavage of GST-BHMT occurs in the lysosome. To ensure that GST-BHMT proteolysis occurs downstream of macroautophagy, siRNA-mediated depletion of the macroautophagy proteins Beclin 1 or ULK1 is an effective control.

The GST-BHMT assay has some attractive features over other commonly used autophagy assays. The GFP-Atg8/LC3 assays can used as endpoint biochemical reporters of autophagy, because of the release and accumulation of GFP in the lytic compartment (Hosokawa et al., 2006;

Shintani and Klionsky, 2004). Because Atg8 and LC3 are part of the autophagic machinery required for the formation of an autophagosome, they do not represent cytosolic cargo. Additionally, though LC3-II has generally been thought to be an autophagosome-specific protein, a recent publication has shown that lipidated LC3-II associates with phagosomes in a macrophage cell line (Sanjuan et al., 2007), which suggests that in some circumstances, LC3-II turnover may not be specific for autophagic flux. GST-BHMT represents cytosolic cargo; therefore, it reflects macroautophagy at the level of cytosolic cargo sequestration as well as delivery to the lysosome.

The LC3 assay is also used to measure macroautophagic flux by analyzing the accumulation of autophagosome-associated LC3-II (Kabeya et al., 2000). However, because autophagosomes marked by LC3-II can accumulate under conditions of either increased autophagy or decreased lysosomal function, it is necessary to measure the autophagic flux of LC3-II to avoid ambiguity (Mizushima and Yoshimori, 2007). This requires the measurement of LC3-II at two data points, one without lysosomal inhibition and one with lysosomal inhibitors. The advantage of the GST-BHMT assay is that that the endpoint measurement is of a proteolytic fragment that is only produced after successful delivery of autophagic cargo to an acidified autolysosome; therefore, autophagic flux can be determined from a single data point. Because the GST-BHMT assay does not rely on the accumulation of the substrate, it does not suffer from problems of interpretation inherent with the LC3-II assay.

The LDH assay measures the sequestration of the full length enzyme in sedimented vacuoles (Kopitz et al., 1990). Like the GST-BHMT assay, protease inhibitors are necessary in the LDH assay to allow accumulation of the enzyme. Because of the abundance of LDH, it is important to separate sequestered LDH from a large fraction of the unsequestered, cytosolic enzyme (Kopitz et al., 1990). This requires additional fractionation of extracted cells, slowing down the analysis. In contrast, proteolysis of GST-BHMT occurs only at the lysosomal endpoint; therefore, background is low and fragment accumulation can be analyzed from whole cell extracts. In this way, the GST-BHMT assay is similar to the Pho8Δ60p assay developed for yeast (Noda et al., 1995). Although BHMT expression is limited to the kidney and liver, the GST-BHMT assay can be used in cell lines from a variety of different lineages, even those that do not express endogenous BHMT (Chadwick et al., 2000; Mercer et al., 2008). Finally, enrichment of the GST-fused fragment with glutathione affinity purification allows for an increase of signal strength for measurements performed at early time points or under conditions of low reporter expression.

The discovery of the LSCS in an early version of the GST-BHMT reporter has allowed for the expansion of the endpoint assay to the autophagy of subcellular compartments. The advantages of a GST-LSCS reporter are identical to those of the GST-BHMT assay mentioned earlier. Even though the approach presented here uses targeting sequences to localize a reporter to two specific subcellular compartments, the GST and associated LSCS could theoretically be fused to any protein indigenous to the desired subcellular location to be analyzed. In particular, tagging of specific subcellular compartments with a GST-LSCS fusion protein offers the advantage of specificity over the LC3 assay, in that increased turnover of LC3-II does not distinguish which subcellular compartment is being sequestered and degraded in the lysosome. Thus, the autophagy assays presented here can be modified to any number of applications requiring specific, endpoint analysis of autophagic flux.

ACKNOWLEDGMENTS

We thank Alagammai Kaliappan and Dr. Manju Sharma for their expert technical assistance. We also thank Dr. David Plas for helpful conversations and suggestions, Dr. Birgit Ehmer for technical assistance with confocal microscopy, Dr. Wei-Xing Zong for the kind gift of the $BclX_L$-cb5 and $BclX_L$-ActA constructs used for subcloning and Dr. Anja Jaeschke for critically reading the manuscript. This work was funded by a grant from the American Cancer Society, Ohio Division.

REFERENCES

Chadwick, L. H., McCandless, S. E., Silverman, G. L., Schwartz, S., Westaway, D., and Nadeau, J. H. (2000). Betaine-homocysteine methyltransferase-2: cDNA cloning, gene sequence, physical mapping, and expression of the human and mouse genes. *Genomics* **70**, 66–73.

Chen, C. A., and Okayama, H. (1988). Calcium phosphate-mediated gene transfer: A highly efficient transfection system for stably transforming cells with plasmid DNA. *Biotechniques* **6**, 632–638.

Chen, J. M., Dando, P. M., Rawlings, N. D., Brown, M. A., Young, N. E., Stevens, R. A., Hewitt, E., Watts, C., and Barrett, A. J. (1997). Cloning, isolation, and characterization of mammalian legumain, an asparaginyl endopeptidase. *J. Biol. Chem.* **272**, 8090–8098.

Chen, J. M., Fortunato, M., and Barrett, A. J. (2000). Activation of human prolegumain by cleavage at a C-terminal asparagine residue. *Biochem. J.* 352 Pt **2**, 327–334.

Furuya, N., Kanazawa, T., Fujimura, S., Ueno, T., Kominami, E., and Kadowaki, M. (2001). Leupeptin-induced appearance of partial fragment of betaine homocysteine methyltransferase during autophagic maturation in rat hepatocytes. *J. Biochem. (Tokyo)* **129**, 313–320.

Hosokawa, N., Hara, Y., and Mizushima, N. (2006). Generation of cell lines with tetracycline-regulated autophagy and a role for autophagy in controlling cell size. *FEBS Lett.* **580**, 2623–2629.

Kabeya, Y., Mizushima, N., Ueno, T., Yamamoto, A., Kirisako, T., Noda, T., Kominami, E., Ohsumi, Y., and Yoshimori, T. (2000). LC3, a mammalian homologue of yeast Apg8p, is localized in autophagosome membranes after processing. *EMBO J.* **19**, 5720–5728.

Klionsky, D. J. (2005). The molecular machinery of autophagy: Unanswered questions. *J. Cell Sci.* **118,** 7–18.
Kopitz, J., Kisen, G. O., Gordon, P. B., Bohley, P., and Seglen, P. O. (1990). Nonselective autophagy of cytosolic enzymes by isolated rat hepatocytes. *J. Cell Biol.* **111,** 941–953.
Kraft, C., Deplazes, A., Sohrmann, M., and Peter, M. (2008). Mature ribosomes are selectively degraded upon starvation by an autophagy pathway requiring the Ubp3p/Bre5p ubiquitin protease. *Nat. Cell Biol.* **10,** 602–610.
Lee, D. H., and Goldberg, A. L. (1998). Proteasome inhibitors: Valuable new tools for cell biologists. *Trends Cell Biol.* **8,** 397–403.
Lemasters, J. J., Nieminen, A. L., Qian, T., Trost, L. C., Elmore, S. P., Nishimura, Y., Crowe, R. A., Cascio, W. E., Bradham, C. A., Brenner, D. A., and Herman, B. (1998). The mitochondrial permeability transition in cell death: A common mechanism in necrosis, apoptosis and autophagy. *Biochim. Biophys. Acta* **1366,** 177–196.
Li, D. N., Matthews, S. P., Antoniou, A. N., Mazzeo, D., and Watts, C. (2003). Multistep autoactivation of asparaginyl endopeptidase *in vitro* and *in vivo*. *J. Biol. Chem.* **278,** 38980–38990.
Marshansky, V., and Futai, M. (2008). The V-type H(+)-ATPase in vesicular trafficking: Targeting, regulation and function. *Curr. Opin. Cell Biol.* **20,** 415–426.
Mercer, C. A., Kaliappan, A., and Dennis, P. B. (2008). Macroautophagy-dependent, intralysosomal cleavage of a betaine homocysteine methyltransferase fusion protein requires stable multimerization. *Autophagy* **4,** 185–194.
Mizushima, N., and Yoshimori, T. (2007). How to Interpret LC3 Immunoblotting. *Autophagy* 3.
Neermann, J., and Wagner, R. (1996). Comparative analysis of glucose and glutamine metabolism in transformed mammalian cell lines, insect and primary liver cells. *J. Cell Physiol.* **166,** 152–169.
Noda, T., Matsuura, A., Wada, Y., and Ohsumi, Y. (1995). Novel system for monitoring autophagy in the yeast *Saccharomyces cerevisiae*. *Biochem. Biophys. Res. Commun.* **210,** 126–132.
Ogata, M., Hino, S., Saito, A., Morikawa, K., Kondo, S., Kanemoto, S., Murakami, T., Taniguchi, M., Tanii, I., Yoshinaga, K., Shiosaka, S., Hammarback, J. A., Urano, F., and Imaizumi, K. (2006). Autophagy is activated for cell survival after endoplasmic reticulum stress. *Mol. Cell Biol.* **26,** 9220–9231.
Ohsumi, Y., and Mizushima, N. (2004). Two ubiquitin-like conjugation systems essential for autophagy. *Semin. Cell Dev. Biol.* **15,** 231–236.
Overbye, A., Fengsrud, M., and Seglen, P. O. (2007). Proteomic analysis of membrane-associated proteins from rat liver autophagosomes. *Autophagy* **3,** 300–322.
Reitzer, L. J., Wice, B. M., and Kennell, D. (1979). Evidence that glutamine, not sugar, is the major energy source for cultured HeLa cells. *J. Biol. Chem.* **254,** 2669–2676.
Sanjuan, M. A., Dillon, C. P., Tait, S. W., Moshiach, S., Dorsey, F., Connell, S., Komatsu, M., Tanaka, K., Cleveland, J. L., Withoff, S., and Green, D. R. (2007). Toll-like receptor signalling in macrophages links the autophagy pathway to phagocytosis. *Nature* **450,** 1253–1257.
Schellens, J. P., and Meijer, A. J. (1991). Energy depletion and autophagy. Cytochemical and biochemical studies in isolated rat hepatocytes. *Histochem. J.* **23,** 460–466.
Shintani, T., and Klionsky, D. J. (2004). Cargo proteins facilitate the formation of transport vesicles in the cytoplasm to vacuole targeting pathway. *J. Biol. Chem.* **279,** 29889–29894.
Ueno, T., Ishidoh, K., Mineki, R., Tanida, I., Murayama, K., Kadowaki, M., and Kominami, E. (1999). Autolysosomal membrane-associated betaine homocysteine methyltransferase. Limited degradation fragment of a sequestered cytosolic enzyme monitoring autophagy. *J. Biol. Chem.* **274,** 15222–15229.

Yorimitsu, T., Nair, U., Yang, Z., and Klionsky, D. J. (2006). Endoplasmic reticulum stress triggers autophagy. *J. Biol. Chem.* **281,** 30299–30304.

Zhu, W., Cowie, A., Wasfy, G. W., Penn, L. Z., Leber, B., and Andrews, D. W. (1996). Bcl-2 mutants with restricted subcellular location reveal spatially distinct pathways for apoptosis in different cell types. *EMBO J.* **15,** 4130–4141.

Zong, W. X., Li, C., Hatzivassiliou, G., Lindsten, T., Yu, Q. C., Yuan, J., and Thompson, C. B. (2003). Bax and Bak can localize to the endoplasmic reticulum to initiate apoptosis. *J. Cell Biol.* **162,** 59–69.

CHAPTER EIGHT

Monitoring Starvation-Induced Reactive Oxygen Species Formation

Ruth Scherz-Shouval* and Zvulun Elazar*

Contents

1. Introduction	120
2. Detection of ROS in Living Cells Using Fluorescent Dyes	121
2.1. Qualitative determination of ROS formation using live, single-cell microscopy analysis	121
2.2. Quantitative and comparative determination of ROS levels using fluorimetric analysis	123
3. Interfering with ROS Formation: The Use of Antioxidants to Inhibit Autophagy	126
3.1. N-acetyl-L-cysteine	126
3.2. Catalase	128
4. Concluding Remarks	128
Acknowledgments	128
References	129

Abstract

Reactive oxygen species (ROS) are potentially harmful to cells because of their ability to oxidize cell constituents such as DNA, proteins, and lipids. However, at low levels, and under tight control, this feature makes them excellent modifiers in a variety of signal transduction pathways, including autophagy. Autophagy was traditionally associated with oxidative stress, acting in the degradation of oxidized proteins and organelles. Recently, a signaling role was suggested for ROS in the regulation of autophagy, leading, under different circumstances, either to survival or to death. To study the effects of ROS on this pathway, one must determine the localization, intensity, kinetics, and essentiality of the oxidative signal in autophagy. Moreover, once characterized, detection and manipulation of ROS formation could be used to monitor and control autophagic activity. In this chapter we discuss methods to examine ROS in the context of autophagy.

* Department of Biological Chemistry, Weizmann Institute of Science, Rehovot, Israel

1. Introduction

High levels of reactive oxygen species (ROS) can oxidize many cell constituents, including lipids, proteins, and DNA, and thus impose a threat to cell integrity. Cells have evolved various defense mechanisms to cope with oxidative stress, among which autophagy plays a major role (Kiffin et al., 2006). Though autophagy is largely considered nonselective, there is accumulating evidence for specific autophagic processes in response to ROS. These include (1) the selective degradation of mitochondria, termed *mitophagy*, proposed to decrease the potential oxidative damage caused by defective mitochondria (Kissova et al., 2004); (2) chaperone-mediated autophagy (CMA) (Cuervo and Dice, 1996; Dice, 1990), suggested to exhibit higher efficiency in degrading oxidized substrate proteins than their unaltered counterparts (Finn and Dice, 2005; Kiffin et al., 2004); and (3) macroautophagy in plants, shown recently to act in the degradation of oxidized proteins following severe oxidative stress (Xiong et al., 2007). Once oxidative damage exceeds repair, autophagic cell death is activated as an avenue of the cell's death programs (Gomez-Santos et al., 2003).

A signaling role for ROS in autophagy and in autophagic cell death was shown in different cell lines in response to various stimuli, such as nerve growth factor (NGF) deprivation (Kirkland et al., 2002; Xue et al., 1999), TNF-α treatment to sarcoma cells (Djavaheri-Mergny et al., 2006), lipopolysaccharides (LPSs) and caspase inhibitor Z-VAD treatment to macrophages (Xu et al., 2006), and rapamycin treatment to yeast (Reef et al., 2006). Furthermore, selective autophagy of the ROS scavenger catalase occurs in response to caspase inhibition, thereby promoting accumulation of ROS in mitochondria and, consequently, cell death (Yu et al., 2006). Finally, starvation-induced autophagy requires H_2O_2 as a signaling molecule (Scherz-Shouval and Elazar, 2007).

To distinguish between a local oxidative signal and a general oxidative stress, intracellular ROS levels and the kinetics of the oxidative signal must be measured. Moreover, ROS stands for a variety of molecules with some common but many distinct functions. Some reports implicate H_2O_2 as a signaling ROS in autophagy (Djavaheri-Mergny et al., 2006; Scherz-Shouval et al., 2007; Yu et al., 2006). But is that the only ROS involved in autophagy? To determine the identity of the oxidative molecule in one's system, several probes with different ROS-specificity should be tested. Finally, intracellular ROS have different sources; they can enter the cell through the plasma membrane, via NADPH oxidase (NOX) and dual oxidase (DUOX) (Bedard and Krause, 2007; Edens et al., 2001; Lambeth, 2004; Suh et al., 1999) or accumulate in the mitochondria and exit to the cytoplasm (Adam-Vizi and Chinopoulos, 2006). Thus far, mitochondria are the main ROS source suggested in the context of autophagy (Chen et al., 2007; Djavaheri-Mergny et al., 2006; Gomez-Santos et al., 2003; Kissova et al., 2006; Scherz-Shouval and

Elazar, 2007), though other sources cannot be precluded. Microscopy analysis using specific ROS-sensitive reagents is required to determine the localization of the oxidative signal. In this chapter, we describe several assays for analysis of intracellular ROS in the context of autophagy, using ROS-sensitive fluorescent probes and antioxidants. By means of these assays, one can attempt to answer the questions raised above and characterize the source, levels, kinetics and identity of the autophagy-related ROS signal in question.

2. Detection of ROS in Living Cells Using Fluorescent Dyes

Several ROS-sensitive fluorescent probes are available, which display different specificities and mechanisms of action. Here we describe two commonly used probes, $2',7'$-dichlorofluorescin (DCF) and dihydroethidium (DHE). As $2',7'$-dichlorofluorescin diacetate (DCFH-DA) enters the cell, the acetate group is cleaved and nonfluorescent $2',7'$-dichlorofluorescin (DCFH) is trapped. Upon oxidation by intracellular oxidants, the fluorescent product DCF is formed. DHE, another cell-permeable, ROS-sensitive probe, fluoresces upon DNA binding. The absorption wavelength depends on the redox state of this compound, so that interaction with ROS, which oxidize DHE to ethidium, leads to a change in fluorescence once the compound binds DNA in the nucleus or mitochondria (Vanden Hoek et al., 1997). The two dyes exhibit different substrate specificity: DHE is particularly sensitive to superoxide, whereas DCFH is more sensitive to H_2O_2 and hydroxyl radicals than to superoxide (Cathcart et al., 1983; Vanden Hoek et al., 1997; Zhu et al., 1994). Another important difference between the dyes is that DCF fluoresces at the site of ROS formation, whereas ethidium fluoresces mainly on DNA binding; therefore, DCF, but not DHE, can be used to characterize the site of ROS formation.

2.1. Qualitative determination of ROS formation using live, single-cell microscopy analysis

2.1.1. Method

Both DCF and DHE can be visualized in living cells by fluorescence microscopy. In our experimental setup, we found DCF more sensitive to changes in starvation-induced ROS than DHE (Scherz-Shouval and Elazar, 2007); nevertheless, the conditions required for the use of both dyes are detailed subsequently:

1. Prepare stock solutions of DCF or DHE. Both can be ordered as dry powders from Invitrogen (Cat. Nos. D-399 and D-1168, respectively).

Notably, both are air and light sensitive and should therefore be kept in the dark, under nitrogen or argon.

- DCF: dissolve DCF in 100% ethanol, make 5-mg aliquots and dry them in a speed vacuum machine in the dark. Once the aliquots are completely dry, pour argon into the each tube, seal with Parafilm, and store in a dark container filled with desiccating material, at $-20\,°C$. Before use, dissolve each aliquot in 100 μl of high-quality anhydrous dimethylsulfoxide (DMSO) to obtain a stock concentration of 100 mM. Dissolved aliquots can usually be stored up to 2–3 months at $-20\,°C$.
- DHE: dissolve DHE in high quality anhydrous dimethylsulfoxide (DMSO) to a stock concentration of 100 mM, and make aliquots. Pour argon into each tube, seal with Parafilm, and store in a dark container filled with desiccating material at $-20\,°C$. Caution should be taken when working with DHE, as with any other DNA-binding material.

2. Plate cells (HeLa, CHO, COS, and MEF cells were tested) on coverslips compatible with a growth-supporting chamber for the microscope (e.g., Lab-Tek from Nunc) and grow them in α−MEM (modified eagle's medium) with serum to 50% confluence.
3. Transfer coverslips to a growth-supporting chamber, wash, and replace the medium with serum-free, amino acid–free medium such as Earle's balanced salts solution (EBSS; see the first note herein) for at least 5 min (longer periods of starvation may be used [see the third note below]. Under our conditions, the ROS signal does not increase and cell death is not observed over a period of 5 h [Scherz-Shouval and Elazar, 2007]).
4. Add 30 μM DCF or 50 μM DHE (from a 100 mM stock in DMSO) to the cells, in the dark, and incubate at $37\,°C$ for 10 min.
5. Wash the cells twice with PBS in the dark and add fresh EBSS.
6. View immediately under the microscope using minimal intensity and time of exposure (focus on the cells using transmission, not fluorescence). DCF is excited at 492 nm and emits at 517–527 nm. In its reduced state, DHE is excited at 355 nm and emits at 420 nm. In its oxidized, DNA-bound state, ethidium is excited at 518 nm and emits at 605 nm. The DCF signal peaks at 10 min; the DHE signal peaks at 30 min. For calibration of the optimal exposure, images should be taken every 1 or 3 min over 10 or 30 min for DCF and DHE, respectively, in a confocal microscope, using very low laser intensity. This is most critical for DCF, which is more sensitive to light (see second note).

2.1.2. Notes

- Nonstarved cells should be tested as a control for the effect of starvation. We tested several types of control: full medium (α−MEM + fetal calf serum [FCS]), serum-starvation (α−MEM), and EBSS supplemented with amino acids [according to Biological Industries' [Israel] formulation sheet for Cat. No. 01-042-1, with the following exceptions: double concentrations of arginine, leucine, lysine, and methionine, shown to be essential and sufficient for inhibition of autophagy [Shvets et al., 2008], were used). We found that full medium completely inhibits the DCF signal, whereas serum-starvation (α–MEM or EBSS + amino acids) induces a DCF signal, which is, however, substantially lower than the signal obtained in the presence of EBSS alone (Scherz-Shouval et al., 2007) (Fig. 8.1).

- Both dyes are light sensitive, as are most fluorescent dyes, and DCF is particularly sensitive. We found that high laser intensities tend to increase the DCF signal (Ichimura et al., 2003; Peng and Jou, 2004), possibly due to photodynamic effects on the Fenton reaction (Zorov et al., 2000). Hence, microscopy inspection should be calibrated to the minimal time and intensity of exposure, and controls must be used to ensure that the observed signal is not an artifact.
- In a basic amino acid starvation assay, the duration of starvation does not affect the ROS signal (i.e. the DCF signal measured after 5 min to 3 h of starvation is similar; see Scherz-Shouval et al., 2007). Nevertheless, different treatments might affect ROS detoxification, resulting in increased accumulation of ROS over time. Therefore, we recommend calibration of the period of starvation/treatment for each experimental setup.

2.1.3. Application

This assay allows visualization of ROS in live cells. Using DCF, this method detects the intracellular site of ROS accumulation and, in combination with other dyes (e.g., MitoTracker; Scherz-Shouval et al., 2007) or fluorescently tagged proteins, identifies this site. Notably, DCF's spectrum overlaps with that of GFP, thus affecting the choice of dyes or tagged proteins. There are two main disadvantages to this method. First, the number of cells monitored is limited, as high magnification (X40) is required to identify the intracellular localization of the signal; also, there is a limited number of exposures possible before photo-oxidation starts affecting the system. Finally, the results obtained are images, which are more difficult to analyze than numbers, especially in high throughput.

2.2. Quantitative and comparative determination of ROS levels using fluorimetric analysis

2.2.1. Method

1. Plate cells in multiwell plates in α-MEM medium with serum, so that the cells reach 50%–80% confluence at the day of measurement.

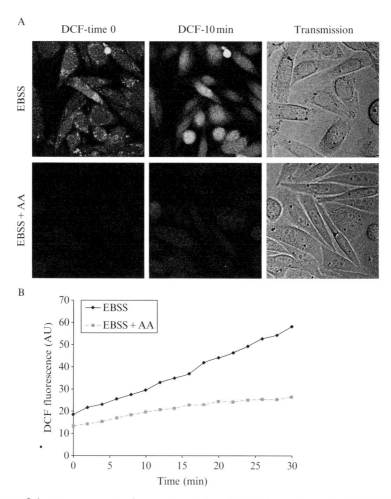

Figure 8.1 Measurement of starvation-induced ROS formation using DCF. Cells grown (A) on 18-mm coverslips or (B) in a 96-well plate were incubated in starvation medium (EBSS) or in EBSS supplemented with amino acids for 2 h, after which they were treated with DCF according to the protocol described in this chapter. (A) Cells were visualized using an Olympus IX-70 confocal microscope equipped with a 37°C heating chamber, and images were taken every 1 min for 10 min. The first and last DCF images and a transmission image are shown for each treatment. (B) DCF fluorescence was measured every 2 min for 30 min by a SPECTRAmax Gemini fluorimeter equipped with a 37 °C heating chamber. The raw measurements of a typical experiment are presented.

a. Higher or lower confluences are not recommended. If, however, these confluences are obtained, the dye concentration should be increased by 20% for higher densities and decreased by 20% for lower densities.

b. For multiwell plates of up to 96 wells, transparent plates can be used. For 384-well plates, we recommend using black plates with a clear bottom (Cat. No. 781091, Greiner).
2. Preheat the chamber of a fluorimetric plate reader to 37 °C.
3. Wash and replace the medium to serum-free, amino acid–free medium, such as Earle's balanced salts solution (EBSS) for at least 5 min (longer periods of starvation can be used; over a period of 5 h the ROS signal does not increase and no cell death is observed; Scherz-Shouval and Elazar, 2007).
4. Add to the cells 30 μM DCF (from a 100 mM stock in DMSO, prepared according to the manufacturer's protocol), in the dark, and incubate at 37° for 10 min.
5. Wash the cells twice with PBS, in the dark, and add fresh EBSS.
6. Read immediately in a fluorimetric plate reader, using 492/517–527 nm excitation/emission, for 5–30 min, using a 2–3 min interval (2–10 recordings). The fluorescent signal emitted by DCF increases with each excitation, resulting in amplification of the signal. Hence, the longer one measures, the greater the differences become. For high-throughput experiments, where time is a limiting factor, short measurements (5 min) are preferable. To detect minor changes or modest effects, longer measurements (30–45 min) might be required. Importantly, even 1 h of measurement (30 excitations) did not result in excitation of control samples, indicating that long measurements can be safely used.

Note: The signal obtained in this assay is DCF fluorescence/well. The DCF signal is proportional to the number of emitting cells (though the correlation is not 1:1). Hence, to accurately transform this readout to DCF signal/cell or to compare DCF signals between different wells, the number of cells per well should either be measured or assumed to be equal. Short (2–3 h) starvation does not affect cell viability; therefore, as long as the wells were plated in equal density, the increase in DCF fluorescence/well indicates increase in DCF fluorescence/cell. However, longer periods of starvation (overnight) or different treatments (e.g., transfections, drugs) might affect cell viability, in which case adding a viability dye to the measurements is recommended. We recommend using Hoechst 33342 (Cat. No. H3570, Invitrogen), which is actively transported into cells, and therefore serves as a marker for live cells.

2.2.2. Application

This method allows a quantitative measurement of the average ROS production in a given cell population. Using a multiwell plate reader, one can easily and accurately compare ROS production in hundreds and thousands of different samples. The main disadvantage of this method is the inability to discern differences in ROS localization.

Note: DCF and DHE fluorescence can also be measured using FACS flow cytometry (Hafer *et al.*, 2008; Iwai *et al.*, 2003).

3. Interfering with ROS Formation: The Use of Antioxidants to Inhibit Autophagy

Several antioxidants were shown to inhibit autophagy when induced by various signals. A list of these reagents and the conditions used for the inhibition of autophagy is described in Table 8.1 Here we explain in some detail the use of two antioxidants, N-acetyl-L-cysteine (NAC) and catalase, as inhibitors of starvation-induced autophagy. It should be noted that measuring the expression levels and activity of antioxidation enzymes, such as catalase and superoxide dismutase (SOD), as well as manipulating the levels of these enzymes using overexpression or knockdown, could also be used to monitor ROS (see, e.g., Chen *et al.*, 2007; Yu *et al.*, 2006).

3.1. N-acetyl-L-cysteine

N-acetyl-L-cysteine (NAC) is a derivative of the amino acid cysteine. It serves as a general antioxidant and is widely used in the pharmaceutical industry.

1. Dissolve NAC (Cat. No. A7250, Sigma) to a 100 mM solution in double distilled water. NAC is easily oxidized by air in its aqueous form, and therefore should be freshly prepared. Vortex vigorously then incubate at 37 °C until a clear solution is obtained.
2. Add 1–10 mM NAC to the control medium (α–MEM) of the cells and mix. Note that the color of the medium becomes yellow.
3. Incubate the cells with NAC for 10 min at 37 °C.
4. Wash and replace the medium with EBSS containing NAC.
5. Incubate the cells with NAC for the desired period of starvation (we tested up to 4-h treatment) at 37 °C.

Note: To test whether NAC worked, it is best to measure DCF fluorescence as compared to starved cells that were not exposed to NAC. DCF can be measured after 5 min of starvation. To monitor autophagy, one can look at autophagosome formation and Atg8-family protein lipidation (best after 2–3 h of starvation), protein degradation (after 4 h of starvation), or other autophagic assays (Tasdemir *et al.*, 2008).

Table 8.1 Antioxidants shown to inhibit autophagy

Reagent	Autophagy inducer	Autophagy outcome	Conditions	Reference
Ascorbic acid	Dopamine	Cell death	2.5 mM	(Gomez-Santos et al., 2003)
Butylated hydroxyanisole (BHA)	TNF-α	Cell death	100 μM, 2 h prior to TNF-α treatment	(Djavaheri-Mergny et al., 2006)
Catalase	Starvation	Survival	1000 u/ml, 10 min or 14 h prior to starvation	(Scherz-Shouval et al., 2007a)
Lipoic acid	Arsenic trioxide	Cell death	50 μM in combination with 5 μM arsenic trioxide for 24 h	(Cheng et al., 2007)[a]
N-acetyl-L-cysteine	Starvation	Survival	10 mM, 10 min prior to starvation	(Scherz-Shouval et al., 2007a)
N-acetyl-L-cysteine	Dopamine	Cell death	2.5 mM	(Gomez-Santos et al., 2003)
N-butyl-α–phenyl-nitrone	TNF-α	Cell death	50 μM, 2 h prior to TNF-α treatment	(Djavaheri-Mergny et al., 2006)
Tiron	Mitochondrial electron complex I and II inhibitors	Cell death	1 mM	(Chen et al., 2007)

[a] Cheng, T. J., Wang, Y. J., Kao, W. W., Chen, R. J., and Ho, Y. S. (2007). Protection against arsenic trioxide-induced autophagic cell death in U118 human glioma cells by use of lipoic acid. *Food Chem. Toxicol.* **45**, 1027–1038.

3.2. Catalase

Catalase catalyzes the decomposition of H_2O_2 to H_2O and O_2. Extracellularly added catalase affects intracellular levels of H_2O_2 because it serves as a sink. H_2O_2 can diffuse freely through the cell membrane; therefore, once it is decomposed in the extracellular medium, intracellular levels of H_2O_2 are affected as well, as additional H_2O_2 will exit the cell moving down the concentration gradient (Preston et al., 2001; Sakurai and Cederbaum, 1998; Xu et al., 2003).

1. Add 1000–4000 units/ml of freshly prepared (according to manufacturer's protocol) bovine brain catalase (Cat. No. C1345, Sigma) to the control medium (α–MEM) of the cells and mix.
2. Incubate the cells with catalase for 10 min (see note) at 37 °C.
3. Wash the cells and replace the medium with EBSS containing catalase.
4. Incubate the cells with catalase for the desired period of starvation at 37 °C.

Note: Similar to NAC, to test whether catalase worked, it is best to measure DCF fluorescence as compared to starved cells that were not exposed to catalase. DCF can be measured after 5 min of starvation. To monitor autophagy, one can look for autophagosome formation and Atg8 lipidation (best after 2–3 h of starvation). To measure long-lived protein degradation in the presence of catalase, it is preferable to preincubate the cells with catalase for 14 h in control medium prior to starvation (Scherz-Shouval and Elazar, 2007).

4. Concluding Remarks

Numerous studies have revealed the involvement of autophagy in a wide range of processes, varying from development and differentiation, through housekeeping degradation of pathogens, damaged organelles and proteins, to pathologies and cell death. ROS are involved in virtually all of these processes. Therefore, studying the involvement of ROS in autophagy is important for our fundamental understanding of this process. Moreover, once this involvement has been established in a certain experimental setup, monitoring ROS could be used as a tool to detect and manipulate autophagy.

ACKNOWLEDGMENTS

This work was supported in part by the Israel Science Foundation, the Binational Science Foundation, and the Weizmann Institute's Minerva Center.

REFERENCES

Adam-Vizi, V., and Chinopoulos, C. (2006). Bioenergetics and the formation of mitochondrial reactive oxygen species. *Trends Pharmacol. Sci.* **27,** 639–645.
Bedard, K., and Krause, K. H. (2007). The NOX family of ROS-generating NADPH oxidases: Physiology and pathophysiology. *Physiol. Rev.* **87,** 245–313.
Cathcart, R., Schwiers, E., and Ames, B. N. (1983). Detection of picomole levels of hydroperoxides using a fluorescent dichlorofluorescein assay. *Anal. Biochem.* **134,** 111–116.
Chen, Y., McMillan-Ward, E., Kong, J., Israels, S. J., and Gibson, S. B. (2007). Mitochondrial electron-transport-chain inhibitors of complexes I and II induce autophagic cell death mediated by reactive oxygen species. *J. Cell Sci.* **120,** 4155–4166.
Cuervo, A. M., and Dice, J. F. (1996). A receptor for the selective uptake and degradation of proteins by lysosomes. *Science* **273,** 501–503.
Dice, J. F. (1990). Peptide sequences that target cytosolic proteins for lysosomal proteolysis. *Trends Biochem. Sci.* **15,** 305–309.
Djavaheri-Mergny, M., Amelotti, M., Mathieu, J., Besancon, F., Bauvy, C., Souquere, S., Pierron, G., and Codogno, P. (2006). NF-kB activation represses tumor necrosis factor-alpha-induced autophagy. *J. Biol. Chem.* **281,** 30373–30382.
Edens, W. A., Sharling, L., Cheng, G., Shapira, R., Kinkade, J. M., Lee, T., Edens, H. A., Tang, X., Sullards, C., Flaherty, D. B., Benian, G. M., and Lambeth, J. D. (2001). Tyrosine cross-linking of extracellular matrix is catalyzed by Duox, a multidomain oxidase/peroxidase with homology to the phagocyte oxidase subunit gp91phox. *J. Cell Biol.* **154,** 879–891.
Finn, P. F., and Dice, J. F. (2005). Ketone bodies stimulate chaperone-mediated autophagy. *J. Biol. Chem.* **280,** 25864–25870.
Gomez-Santos, C., Ferrer, I., Santidrian, A. F., Barrachina, M., Gil, J., and Ambrosio, S. (2003). Dopamine induces autophagic cell death and α-synuclein increase in human neuroblastoma SH-SY5Y cells. *J. Neurosci. Res.* **73,** 341–350.
Hafer, K., Iwamoto, K. S., and Schiestl, R. H. (2008). Refinement of the dichlorofluorescein assay for flow cytometric measurement of reactive oxygen species in irradiated and bystander cell populations. *Radiat. Res.* **169,** 460–468.
Ichimura, H., Parthasarathi, K., Quadri, S., Issekutz, A. C., and Bhattacharya, J. (2003). Mechano-oxidative coupling by mitochondria induces proinflammatory responses in lung venular capillaries. *J. Clin. Invest.* **111,** 691–699.
Iwai, K., Kondo, T., Watanabe, M., Yabu, T., Kitano, T., Taguchi, Y., Umehara, H., Takahashi, A., Uchiyama, T., and Okazaki, T. (2003). Ceramide increases oxidative damage due to inhibition of catalase by caspase-3-dependent proteolysis in HL-60 cell apoptosis. *J. Biol. Chem.* **278,** 9813–9822.
Kiffin, R., Bandyopadhyay, U., and Cuervo, A. M. (2006). Oxidative stress and autophagy. *Antioxid. Redox. Signal* **8,** 152–162.
Kiffin, R., Christian, C., Knecht, E., and Cuervo, A. M. (2004). Activation of chaperone-mediated autophagy during oxidative stress. *Mol. Biol. Cell* **15,** 4829–4840.
Kirkland, R. A., Adibhatla, R. M., Hatcher, J. F., and Franklin, J. L. (2002). Loss of cardiolipin and mitochondria during programmed neuronal death: Evidence of a role for lipid peroxidation and autophagy. *Neuroscience* **115,** 587–602.
Kissova, I., Deffieu, M., Manon, S., and Camougrand, N. (2004). Uth1p is involved in the autophagic degradation of mitochondria. *J. Biol. Chem.* **279,** 39068–39074.
Kissova, I., Deffieu, M., Samokhvalov, V., Velours, G., Bessoule, J. J., Manon, S., and Camougrand, N. (2006). Lipid oxidation and autophagy in yeast. *Free Radic. Biol. Med.* **41,** 1655–1661.

Lambeth, J. D. (2004). NOX enzymes and the biology of reactive oxygen. *Nat. Rev. Immunol.* **4,** 181–189.

Peng, T. I,, and Jou, M. J. (2004). Mitochondrial swelling and generation of reactive oxygen species induced by photoirradiation are heterogeneously distributed. *Ann. N.Y. Acad. Sci.* **1011,** 112–122.

Preston, T. J., Müller, W. J., and Singh, G. (2001). Scavenging of extracellular H2O2 by catalase inhibits the proliferation of HER-2/Neu-transformed rat-1 fibroblasts through the induction of a stress response. *J. Biol. Chem.* **276,** 9558–9564.

Reef, S., Zalckvar, E., Shifman, O., Bialik, S., Sabanay, H., Oren, M., and Kimchi, A. (2006). A short mitochondrial form of p19ARF induces autophagy and caspase-independent cell death. *Mol. Cell* **22,** 463–475.

Sakurai, K., and Cederbaum, A. I. (1998). Oxidative stress and cytotoxicity induced by ferric-nitrilotriacetate in HepG2 cells that express cytochrome P450 2E1. *Mol. Pharmacol.* **54,** 1024–1035.

Scherz-Shouval, R., and Elazar, Z. (2007). ROS, mitochondria and the regulation of autophagy. *Trends Cell Biol.* **17,** 422–427.

Scherz-Shouval, R., Shvets, E., Fass, E., Shorer, H., Gil, L., and Elazar, Z. (2007). Reactive oxygen species are essential for autophagy and specifically regulate the activity of Atg4. *EMBO J.* **26,** 1749–1760.

Shvets, E., Fass, E., and Elazar, Z. (2008). Utilizing flow cytometry to monitor autophagy in living mammalian cells. *Autophagy* **4,** 621–628.

Suh, Y. A., Arnold, R. S., Lassegue, B., Shi, J., Xu, X., Sorescu, D., Chung, A. B., Griendling, K. K., and Lambeth, J. D. (1999). Cell transformation by the superoxide-generating oxidase Mox1. *Nature* **401,** 79–82.

Tasdemir, E., Galluzzi, L., Maiuri, M. C., Criollo, A., Vitale, I., Hangen, E., Modjtahedi, N., and Kroemer, G. (2008). Methods for assessing autophagy and autophagic cell death. *Methods Mol. Biol.* **445,** 29–76.

Vanden Hoek, T. L., Li, Z., Shao, Z., Schumacker, P. T., and Becker, L. B. (1997). Significant levels of oxidants are generated by isolated cardiomyocytes during ischemia prior to reperfusion. *J. Mol. Cell Cardiol.* **29,** 2571–2583.

Xiong, Y., Contento, A. L., Nguyen, P. Q., and Bassham, D. C. (2007). Degradation of oxidized proteins by autophagy during oxidative stress in Arabidopsis. *Plant. Physiol.* **143,** 291–299.

Xu, J., Yu, S., Sun, A. Y., and Sun, G. Y. (2003). Oxidant-mediated AA release from astrocytes involves cPLA$_2$ and iPLA$_2$. *Free Radic. Biol. Med.* **34,** 1531–1543.

Xu, Y., Kim, S. O., Li, Y., and Han, J. (2006). Autophagy contributes to caspase-independent macrophage cell death. *J. Biol. Chem.* **281,** 19179–19187.

Xue, L., Fletcher, G. C., and Tolkovsky, A. M. (1999). Autophagy is activated by apoptotic signalling in sympathetic neurons: An alternative mechanism of death execution. *Mol. Cell Neurosci.* **14,** 180–198.

Yu, L., Wan, F., Dutta, S., Welsh, S., Liu, Z., Freundt, E., Baehrecke, E. H., and Lenardo, M. (2006). Autophagic programmed cell death by selective catalase degradation. *Proc. Natl. Acad. Sci. USA* **103,** 4952–4957.

Zhu, H., Bannenberg, G. L., Moldeus, P., and Shertzer, H. G. (1994). Oxidation pathways for the intracellular probe 2′,7′-dichlorofluorescein. *Arch. Toxicol.* **68,** 582–587.

Zorov, D. B., Filburn, C. R., Klotz, L. O., Zweier, J. L., and Sollott, S. J. (2000). Reactive oxygen species (ROS)-induced ROS release: A new phenomenon accompanying induction of the mitochondrial permeability transition in cardiac myocytes. *J. Exp. Med.* **192,** 1001–1014.

CHAPTER NINE

FLOW CYTOMETRIC ANALYSIS OF AUTOPHAGY IN LIVING MAMMALIAN CELLS

Elena Shvets* *and* Zvulun Elazar*

Contents

1. Introduction	132
2. Qualitative Analysis of LC3 Levels in Cells	133
3. Establishing Flow Cytometry and FACS Analysis to Quantify GFP-LC3 Levels During Amino Acid Deprivation	134
3.1. Method	134
4. Quantification of Decline in GFP-LC3 Using Flow Cytometry and FACS Analysis	137
4.1. Application	137
5. Using FACS to Quantify Autophagy in Response to Specific Treatments	138
5.1. Method	138
5.2. Application	139
6. Concluding Remarks	139
Acknowledgments	140
References	140

Abstract

Autophagy is a major intracellular catabolic pathway induced in response to amino acid starvation. Recent findings implicate it in diverse physiological/pathophysiological events, such as protein and organelle turnover, development, aging, pathogen infection, cell death, and neurodegeneration. However, experimental methods to monitor this process in mammalian cells are limited because of the deficiency of autophagic markers. Recently, MAP1-LC3 (LC3), a mammalian homolog of the yeast ubiquitin-like (UBL) protein Atg8, has been shown to selectively incorporate into the autophagosomal membrane, thus serving as a unique bona fide marker of autophagosomes in mammals. Thus, the autophagic activity can be largely determined by GFP-LC3/LC3,

* Department of Biological Chemistry, The Weizmann Institute of Science, Rehovot, Israel

predominantly associated with autophagosomes (when LC3 is conjugated to phosphatidylethanolamine), both biochemically and microscopically. However, current methods to quantify autophagic activity using LC3 are time consuming, labor intensive, and require expertise in accurate interpretation. In this chapter we describe the use of flow cytometry and fluorescence-activated cell sorting (FACS) as a new assay designed to quantify autophagy in cells stably expressing GFP-LC3. Flow cytometry is a well-established technique for performing quantitative fluorescence measurements, allowing quick, accurate, and simultaneous determination of many parameters in cell subpopulations. Here flow cytometry and FACS were used to quantify the turnover of GFP-LC3 (reflecting an autophagic flux) as a reliable and simple assay to measure autophagic activity in living mammalian cells.

1. Introduction

Autophagy is a major intracellular catabolic pathway, primarily regulated by amino acids (Blommaart et al., 1997; Mortimore et al., 1987), where long-lived proteins and organelles are delivered by a double-membrane vesicle, called an autophagosome, to the lysosomal/vacuolar system for degradation and consequent recycling. Recent reports show that autophagy plays a role in diverse biological events, including development, aging, immunity, pathogen infection, and cell death, and its dysfunction is associated with cancer, as well as muscular and neurodegenerative diseases (Boland and Nixon, 2006; Huang and Klionsky, 2007; Shintani and Klionsky, 2004). However, methods to monitor and quantify autophagy are limited because of a lack of autophagic markers. The classic direct approaches of measuring autophagy include quantitative electron microscopy (see the chapter by Ylä-Anttila and Eskelinen in this volume), quantification of the degradation rate of long-lived proteins (autophagic protein degradation; see the chapter by Bauvy et al., in this volume), and autophagic sequestration assays (see the chapter by Seglen et al., in this volume). Although electron microscopy is a sensitive and accurate way to detect the induction of autophagy, it is labor intensive and requires well-trained personnel and expensive equipment. Moreover, electron microscopy is limited to morphological observations of fixed cells and is poorly amenable for unbiased quantification, as it often leads to misinterpretation regarding the autophagic activity. Methods of measuring autophagic degradation or sequestration, although used for quantitative functional analyses to evaluate autophagy, are technically demanding, as they generally require prelabeling or injection of radioactive materials into the cells. In addition, these assays may be affected by nonspecific total protein degradation and by processes such as proteosomal degradation.

In recent years, the molecular mechanism underlying autophagy has been extensively researched in yeast, and two evolutionarily conserved ubiquitin-like (UBL) conjugation systems, namely Atg12 and Atg8, were found to play a pivotal role in early stages of autophagosome biogenesis (Ohsumi, 2001). Atg8 and its mammalian homologs, including MAP1-LC3 (LC3), undergo conjugation to a lipid (phosphatidylethanolamine, PE) at their C-terminus, and the resultant lipidated form remains associated with the autophagosomal membrane until fusion with the lysosome, thus serving as a bona fide marker for autophagosomes (Kabeya et al., 2000; Kabeya et al., 2004). The discovery of LC3 as the first marker of autophagosomes *in vivo* provides a new tool to study the molecular basis of autophagy in mammals.

2. Qualitative Analysis of LC3 Levels in Cells

Using LC3 (or GFP-LC3) as a bona fide marker of autophagosomes allows for specific monitoring of the autophagic activity both biochemically and microscopically (Klionsky, 2005; Mizushima et al., 2004). One approach is based on the detection of LC3 lipidation by separating LC3 unconjugated (LC3-I) and conjugated (LC3-II) forms by SDS-PAGE (see the chapter by Kimura et al., in this volume). Another widely used method to determine autophagic activity is to follow the subcellular translocation of LC3/GFP-LC3 from the cytosol to autophagosomes, where it appears as punctate dots. However, several difficulties are encountered in quantifying and interpreting data from this analysis. First, formation of lipidated LC3 does not always correlate with formation of autophagosomes. In addition, this analysis represents the dynamic, steady-state level between formation and degradation of LC3-II-labeled autophagosomes. Therefore, to quantify the level of LC3-II or autophagosomes delivered to lysosomes (autophagic flux), one should compare their amount in the absence or presence of lysosomal protease inhibitors (Tanida et al., 2005). Finally, a recent report (Kuma et al., 2007) shows that GFP-LC3 has a tendency to aggregate, especially as a result of transient transfection, leading to the appearance of LC3 punctate structures that do not represent autophagosomes. Therefore, it is recommended to use stable clones of GFP-LC3 expressing cells with at most a moderate level of overexpression (Fass et al., 2006). To improve the purity of stable transfectants one may consider sorting GFP-LC3-expressing cells by flow cytometry. Although anti-LC3 antibodies can reduce potential artifacts resulting from overexpression, using GFP-LC3 is suitable where the endogenous amount is below the level of detection and exogenous expression is needed. Moreover, the use of GFP-fused proteins confers an advantage for visualization of autophagosomes, which are highly dynamic organelles, in living cells (Bampton et al., 2005; Fass et al., 2006; Mizushima and Kuma, 2008). In addition, GFP, being sensitive to an acidic environment (Campbell *et al.*, 2002), may be used as a reporter of delivery into lysosomes (Kimura *et al.*, 2007; Kneen *et al.*, 1998) (also see the chapter by Kimura *et al.*). We propose to exploit these features of GFP to quantify the delivery of GFP-LC3 into lysosomes as a quantitative assay to monitor autophagy.

3. Establishing Flow Cytometry and FACS Analysis to Quantify GFP-LC3 Levels During Amino Acid Deprivation

3.1. Method

This method is based on the main characteristics of LC3, that is, its ability to undergo ubiquitin-like conjugation to PE on the autophagosomal membrane. Following this conjugation, a significant portion of this protein is delivered to the lysosomes for consequent degradation, resulting in reduction in GFP fluorescence intensity of starved cells (Fig. 9.1) Hence, this method measures not only conjugation of LC3 to the autophagosomal membrane (LC3-II) but also formation of autophagosomes as well as their delivery and fusion with lysosomes. Notably, the reduction in the GFP-LC3 fluorescence reflects its delivery into the lysosomes rather than degradation, as this signal is blocked by low pH of the lysosomal lumen following fusion with the lysosomes.

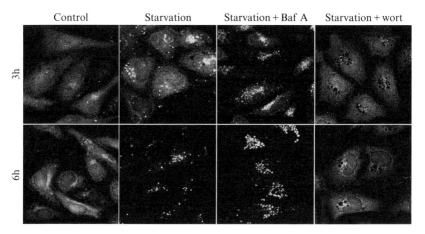

Figure 9.1 Microscopy analysis of fixed CHO cells stably expressing GFP-LC3. Cells incubated for 3 or 6 hours in αMEM medium (control) or in EBSS medium (starvation) showing a significant reduction in the fluorescence intensity, which is inhibited by bafilomycin A_1 (100 nM) or wortmannin (100 nM).

The method of quantifying GFP-LC3 delivery into the lysosomes is based on comparison between the level of the protein (estimated as fluorescence intensity detected by flow cytometry) under control conditions vs. autophagy-induced conditions. Therefore, every sample should be compared (normalized) to its control. For example, the level of GFP-LC3 in cells starved for a certain time period should be compared to its level in cells incubated for the same time period in the presence of amino acids.

1. Prepare a 10-cm dish of confluent GFP-LC3 stably-expressing cells (CHO or HeLa cells are suitable).
2. Twenty-four h prior to the flow cytometry experiment, trypsinize the cells:
   a. Warm trypsin to 37 °C in a water bath.
   b. Aspirate the culture medium; wash cells with phosphate-buffered saline (PBS) twice.
   c. Add 1 ml of trypsin (0.25% Trypsin in EDTA) to a 10-cm dish.
   d. Incubate at 37 °C in 5% CO_2 for about 1–3 min. The cells are ready when they become round.
   e. Resuspend the cells in complete medium (we use αMEM+10% FCS) and count the cells using a hemocytometer.

3. Plate the cells in 3 ml of complete medium into 6-well plates, where each well represents a sample. The cells should be ~80% confluent in the wells prior to the next steps of the experiment, and the number of cells, plated one day earlier, may vary between different cell lines. For CHO cells, ~2.5 × 10^5 cells/well of a 6-well plate is recommended.
4. Culture the cells at 37 °C in 5% CO_2.
5. Plan the starting time-point of the experiment. Take into account that the sample preparation before flow cytometry analysis takes approximately 1 h per 25 samples. To detect an ~50% reduction in GFP-LC3 intensity between control (nonstarved cells) and starvation samples, CHO or HeLa cells should be incubated (starved) for 4–5 h. The period of starvation may vary between different clones or cell lines. Therefore, one should start the experiment (starvation) 5–6 h before the scheduled time of FACS analysis.
6. Start the experiment by discarding the culture medium and washing the cells twice with 1x PBS.
7. Add to each sample 1–2 ml of the appropriate medium. For amino acid starvation, add EBSS or HBSS medium. As a control, use amino acid rich (αMEM, RPMI) medium without serum.

Notes: (a) The nutritional conditions that inhibit autophagy can be tested specifically by incubating cells in medium where only a single/several amino acids are omitted. (b) To ensure that the decay in fluorescence intensity is due to autophagic activity, we recommend adding a control sample of cells starved in the presence of inhibitors such as 3-methyladenine/wortmannin and/or bafilomycin A_1 (Fig. 9.2).

1. At the end of the incubation period, wash the cells with PBS and add 200 µl of trypsin/EDTA.
2. Incubate at 37 °C with 5% CO_2 for about 1–3 min.
3. Resuspend the cells in 5 ml/sample of PBS.
4. Transfer the sample into 5-ml polypropylene FACS tubes (nonsterile) and keep on ice.
5. Centrifuge the cells at 2000 rpm at 4 °C for 5 min.
6. Aspirate the supernatant fraction and resuspend the cells in 5 ml of cold PBS.
7. Centrifuge the cells at 1500 rpm, at 4 °C for 5 min.

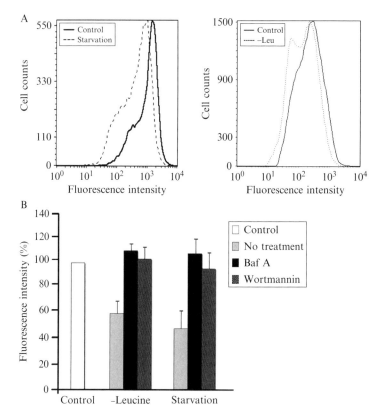

Figure 9.2 (A) Histogram plot presentation of GFP intensity vs. cell count. FACS analysis of CHO cells stably expressing wild-type GFP-LC3 incubated in αMEM (control), EBSS (starvation), or medium deficient in leucine (−Leu) for 5 h. (B) Graph presentation. FACS analysis of CHO cells stably expressing wild-type GFP-LC3 incubated in αMEM (control), EBSS (starvation), or medium lacking leucine (−Leu) for 5 h in the absence or presence of wortmannin (100 nM) or bafilomycin A1 (Baf A; 100 nM).

8. Aspirate the supernatant fraction and resuspend the sample in 0.5 ml of cold PBS.
9. Immediately prior to flow cytometry analysis, vortex the sample for 10 s.
10. Analyze the GFP intensity (wavelength = 488 nm) of every sample by flow cytometry. The number of cells analyzed may vary from 3×10^4 to 1×10^5 cells/sample.

4. Quantification of Decline in GFP-LC3 Using Flow Cytometry and FACS Analysis

1. To observe and analyze the population of viable cells, use CellQuest software to draw dot plots. *Note*: In case the user is not experienced in flow cytometry analysis, we recommend working with someone qualified in the instrumentation and software.
2. Choose cell size (Forward Scatter; FSC) and cell granularity (Side Scatter; SSC) as the X and Y parameters, respectively, to exclude broken cells and cell debris from the cell population being analyzed.
3. You may plot a histogram of GFP (FL1) intensity (X parameter) versus cell count (Y parameter) for a given cell population. To observe the shift in intensity you may overlay one data set over another in a histogram plot (control versus starvation) (Fig. 9.2A).
4. Get various statistics for a given plot.
5. Choose for analysis the mean or geometric mean of GFP fluorescence intensity in a given population.
6. Normalize fluorescence intensity for each treatment to the level of the control sample (cells incubated in serum-free, amino acid–rich medium for similar time periods) as follows:

   a. Set the level of every control sample to 100%.
   b. Calculate the relative level of GFP-LC3 for other treatments incubated for similar time periods according to its control.
   c. Summarize and present the results using column graph in Excel (Fig. 9.2B). *Note*: In case of more than 2–3 treatments/samples, it is recommended to quantify and present results by a column graph (e.g., Excel) rather than by a histogram plot.

4.1. Application

This method quantifies autophagy by measuring the delivery of GFP-LC3 into lysosomes, rendering it more sensitive than other assays, which are based on bulk protein or selective LC3 degradation. Given that the GFP molecule is sensitive to low pH, its fluorescence disappears immediately upon delivery into lysosome. It should be noted that inhibitors of lysosomal degradation that do not affect lysosomal acidification would not prevent the fluorescence drop. Thus, only factors affecting autophagosome formation, delivery, fusion with lysosomes, or lysosomal acidification (Fig. 9.2), but not their proteolytic activity, inhibit the decay in GFP-LC3 fluorescence intensity. It may serve as a quantitative assay to compare lipidation activity of different homologs/mutants; for example, the signal of the mutated form of the protein, GFP-LC3^{G120A}, which is unable to undergo conjugation to the autophagosomal membrane, remains unchanged under starvation conditions (Shvets *et al.*, 2008). Because this analysis is based on disappearance of the GFP signal and not on appearance of autophagosomes labeled by LC3, it cannot be performed on LC3-immunostained cells. Moreover, although reduction in GFP signal in response to amino acids starvation is detectable also in fixed cells (Fig. 9.1), cell fixation may significantly affect quantification, as it may releases part of the cytosolic proteins. We therefore recommend carrying out this assay only in living cells. Finally, this method specifically quantifies the activity level of the autophagic machinery, such as the rate of autophagosome production and delivery into lysosomes, but not bulk protein degradation, and, therefore, this system cannot detect defects in substrate incorporation into autophagosomes.

5. Using FACS to Quantify Autophagy in Response to Specific Treatments

5.1. Method

In addition to measuring autophagy following amino acid starvation, flow cytometry analysis of GFP-LC3 delivery into the lysosomes may serve as an assay to measure autophagy in response to other stimuli that induce autophagy (Fig. 9.3).

1. Prior to flow cytometry analysis (24–48 h), plate the appropriate number (2–3 × 10^5 cells/well) of GFP-LC3-expressing cells into 6-well plates.
2. Culture the cells at 37 °C in 5% CO_2.
3. For long-period incubations, start the experiment 18–16 h before the scheduled time of flow cytometry analysis.
4. To start the experiment, discard the culture medium and wash the cells twice with 1× PBS.
5. Add to each sample 1–2 ml of the appropriate medium in the presence or absence of desired drug. For example, to induce ER stress, add tunicamycin to amino acid–rich (αMEM, RPMI) medium. As a control, use incubation of cells in this medium in the absence of this drug. *Note*: In case of long treatment periods, consider using complete growth medium (containing serum), as long periods of serum withdrawal may both

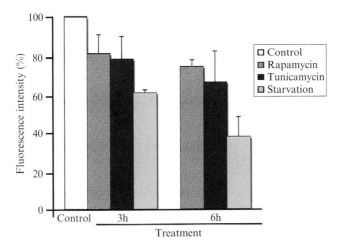

Figure 9.3 Quantification and comparison of autophagic activity induced by different stimuli. CHO cells stably expressing wild-type GFP-LC3 were incubated either in EBSS (starvation) medium or αMEM (control) medium, or in αMEM in the presence of rapamycin (200 nM) or tunicamycin (5 μg/ml) for 3 h or 6 h. The relative level of GFP-LC3 was measured using flow cytometry.

induce autophagy and affect the synthesis of GFP-LC3, leading to its autophagy-unrelated reduction.
6. At the end of the incubation period, harvest cells with trypsin/EDTA and prepare for analysis as described in the previous section.

5.2. Application

FACS analysis of GFP-LC3 turnover, being a quantitative and sensitive method, allows for quantitative screening for drugs/conditions that affect autophagy. However, certain conditions may affect fluorescence intensity and consequently lead to misinterpretation of autophagic activity. For example, conditions/drugs that affect protein expression (serum starvation, protein synthesis inhibitors), membrane permeability, or overall fluorescence signal may lead to autophagy-unrelated reductions in the GFP-LC3 level. Therefore, to avoid such potential shortcomings, it is essential to inspect cells by microscopy prior to flow cytometry analysis.

6. Concluding Remarks

The fact that lipidated GFP-LC3, similar to endogenous LC3, is specifically delivered into lysosomes in response to induction of autophagy is at the core of the methods presented in this chapter. This selective disappearance of GFP-LC3 is readily detectable in living cells using techniques that measure the fluorescence signal, such as flow cytometry analysis. Provided precautions are taken to successfully interpret the flow cytometry data, its application to detect and quantify the selective autophagy-mediated disappearance of GFP-LC3 offers numerous advantages. By allowing semi-automation, flow cytometry analysis may be used to evaluate more samples with larger cell sample sizes (population of 1×10^5 or more) than is practically possible through conventional immunofluorescence microscopy. FACS analysis is also an objective and reproducible methodology, which does not require experience in interpreting the visual appearance of the cells as is the case with electron microscopy. In flow cytometry, different parameters (including size, granularity, and fluorescence intensity) are measured on individual cells. This makes it possible to quantify sample heterogeneity, identify cell subpopulations, detect contaminants, and distinguish between viable and nonviable cells. The method is quantitative, sensitive to various stimuli, such as drugs that induce or block autophagy, and can be used to quantify the level of autophagic activity in response to omitting a single or several amino acids from the medium. Finally, the fact that flow cytometry analysis is performed in living cells can be exploited to sort and collect specific cell subpopulations that are of interest. Importantly, this technique makes it possible to perform large-scale screens for identification of yet unknown factors/signals involved in autophagy.

ACKNOWLEDGMENTS

This work was supported in part by the Israel Science Foundation, the Binational Science Foundation, and the Weizmann Institute's Minerva Center.

REFERENCES

Bampton, E. T., Goemans, C. G., Niranjan, D., Mizushima, N., and Tolkovsky, A. M. (2005). The dynamics of autophagy visualized in live cells: From autophagosome formation to fusion with endo/lysosomes. *Autophagy* **1,** 23–36.

Blommaart, E. F., Krause, U., Schellens, J. P., Vreeling-Sindelarova, H., and Meijer, A. J. (1997). The phosphatidylinositol 3-kinase inhibitors wortmannin and LY294002 inhibit autophagy in isolated rat hepatocytes. *Eur. J. Biochem.* **243,** 240–246.

Boland, B., and Nixon, R. A. (2006). Neuronal macroautophagy: From development to degeneration. *Mol. Aspects. Med.* **27,** 503–519.

Campbell, R. E., Tour, O., Palmer, A. E., Steinbach, P. A., Baird, G. S., Zacharias, D. A., and Tsien, R. Y. (2002). A monomeric red fluorescent protein. *Proc. Natl. Acad. Sci. USA* **99,** 7877–7882.

Fass, E., Shvets, E., Degani, I., Hirschberg, K., and Elazar, Z. (2006). Microtubules support production of starvation-induced autophagosomes but not their targeting and fusion with lysosomes. *J. Biol. Chem.* **281,** 36303–36316.

Huang, J., and Klionsky, D. J. (2007). Autophagy and human disease. *Cell Cycle* **6,** 1837–1849.
Kabeya, Y., Mizushima, N., Ueno, T., Yamamoto, A., Kirisako, T., Noda, T., Kominami, E., Ohsumi, Y., and Yoshimori, T. (2000). LC3, a mammalian homologue of yeast Apg8p, is localized in autophagosome membranes after processing. *EMBO J.* **19,** 5720–5728.
Kabeya, Y., Mizushima, N., Yamamoto, A., Oshitani-Okamoto, S., Ohsumi, Y., and Yoshimori, T. (2004). LC3, GABARAP and GATE16 localize to autophagosomal membra on form-II formation. *J. Cell Sci.* **117,** 2805–2812.
Kimura, S., Noda, T., and Yoshimori, T. (2007). Dissection of the autophagosome maturation process by a novel reporter protein, tandem fluorescent-tagged LC3. *Autophagy* **3,** 452–460.
Klionsky, D. J. (2005). The molecular machinery of autophagy: Unanswered questions. *J. Cell Sci.* **118,** 7–18.
Kneen, M., Farinas, J., Li, Y., and Verkman, A. S. (1998). Green fluorescent protein as a noninvasive intracellular pH indicator. *Biophys. J.* **74,** 1591–1599.
Kuma, A., Matsui, M., and Mizushima, N. (2007). LC3, an autophagosome marker, can be incorporated into protein aggregates independent of autophagy: Caution in the interpretation of LC3 localization. *Autophagy* **3,** 323–328.
Mizushima, A., and Kuma, A. (2008). Autophagosomes in GFP-LC3 Transgenic Mice. *Methods Mol. Biol.* **445,** 119–124.
Mizushima, N., Yamamoto, A., Matsui, M., Yoshimori, T., and Ohsumi, Y. (2004). *In vivo* analysis of autophagy in response to nutrient starvation using transgenic mice expressing a fluorescent autophagosome marker. *Mol. Biol. Cell* **15,** 1101–1111.
Mortimore, G. E., Poso, A. R., Kadowaki, M., and Wert, J. J., Jr., (1987). Multiphasic control of hepatic protein degradation by regulatory amino acids. General features and hormonal modulation. *J. Biol. Chem.* **262,** 16322–16327.
Ohsumi, Y. (2001). Molecular dissection of autophagy: two ubiquitin-like systems. *Nat. Rev. Mol. Cell Biol.* **2,** 211–216.
Shintani, Y., and Klionsky, Y. (2004). Autophagy in health and disease: A double-edged sword. *Science* **306,** 990–995.
Shvets, E., Fass, E., and Elazar, Z. (2008). Utilizing flow cytometry to monitor autophagy in living mammalian cells. *Autophagy* **4,** 621–628.
Tanida, I., Minematsu-Ikeguchi, N., Ueno, T., and Kominami, E. (2005). Lysosomal turnover, but not a cellular level, of endogenous LC3 is a marker for autophagy. *Autophagy* **1,** 84–91.

CHAPTER TEN

Monitoring Autophagy by Electron Microscopy in Mammalian Cells

Päivi Ylä-Anttila,[*] Helena Vihinen,[†] Eija Jokitalo,[†] *and* Eeva-Liisa Eskelinen[*]

Contents

Abbreviations	144
1. Introduction	144
1.1. The autophagic pathway	144
1.2. Identification of autophagic compartments by light and electron microscopy	145
2. Methods	146
2.1. Resin embedding of aldehyde-fixed animal cell pellets using unbuffered osmium tetroxide	146
2.2. Embedding of mouse liver using imidazole-buffered osmium tetroxide	147
2.3. Resin flat embedding of aldehyde-fixed animal cell cultures using reduced osmium tetroxide	148
2.4. Electron tomography	150
2.5. Quantitation of autophagic compartments in thin sections	150
3. Results and Discussion	155
3.1. Fine structure of autophagosomes and autophagic compartments in plastic sections	155
3.2. Tomography of a phagophore	159
3.3. Quantitation of autophagy by electron microscopy	161
4. Concluding Remarks	162
Acknowledgments	163
References	163

[*] Department of Biological and Environmental Sciences, Division of Biochemistry, University of Helsinki, Helsinki, Finland
[†] Institute of Biotechnology, Electron Microscopy Unit, University of Helsinki, Helsinki, Finland

Methods in Enzymology, Volume 452 © 2009 Elsevier Inc.
ISSN 0076-6879, DOI: 10.1016/S0076-6879(08)03610-0 All rights reserved.

Abstract

Electron microscopy remains one of the most accurate methods for the detection of autophagy and quantification of autophagic accumulation. Compared to fluorescence microscopy, the resolution of transmission electron microscopy is superior. In this chapter we describe the fine structure of early and late autophagic compartments in mammalian cells. Instructions are given for the preparation of samples for conventional electron microscopy using three different protocols suitable for cultured cells and animal tissues. We also introduce tomography as a tool to study the three-dimensional morphology of autophagic organelles and show the morphology of a phagophore as an example. Finally, we describe a protocol for the quantification of autophagic compartments by electron microscopy and point counting.

Abbreviations

AP, autophagosome; AC, autophagic compartment; ER, endoplasmic reticulum; NaCac, sodium cacodylate buffer; PBS, phosphate-buffered saline; RT room temperature

1. Introduction

1.1. The autophagic pathway

Autophagy is a lysosomal degradation pathway for cytoplasmic material (Eskelinen, 2008b; Xie and Klionsky, 2007). After an induction signal such as amino acid starvation, autophagy starts when a flat membrane cistern wraps around a portion of cytoplasm and forms a closed double-membraned vacuole containing cytosol and/or organelles (Arstila and Trump, 1968). The membrane cistern is called the *phagophore* or the *isolation membrane*. The sealed vesicle is called the *autophagosome*. Autophagosomes mature by fusing with endosomal and lysosomal vesicles, which also deliver lysosomal membrane proteins and enzymes (Dunn, 1994; Eskelinen, 2005b). Autophagosomes that have fused with an endosome or a lysosome are called *amphisomes* and *autolysosomes*, respectively. The segregated cytoplasm is degraded by lysosomal hydrolases, and the degradation products are transported back to the cytoplasm by pumps located in the lysosomal limiting membrane.

In electron microscopy, autophagic compartments can be identified as membrane-bound vesicles containing cytoplasmic material or organelles (Eskelinen, 2005a, 2008, 2008a; see also chapters by M. Baba and Sigmond *et al.*, in the previous volume). Morphologically, autophagic compartments can be further classified into early or initial autophagic compartments, containing morphologically intact cytosol or organelles, and to late or degradative autophagic compartments, containing partially degraded cytoplasmic material (Dunn, 1990a,b, 1994).

1.2. Identification of autophagic compartments by light and electron microscopy

Several novel yeast genes essential for autophagy (autophagy-related, or *ATG* genes) were recently characterized (Klionsky *et al.*, 2003), and mammalian homologs have been identified for most of these genes. The microtubule-associated protein 1 light chain 3, or LC3, is the mammalian homolog of Atg8 (Kabeya *et al.*, 2000), and it is the only known marker protein for autophagosomes. Anti-LC3 has been used as a marker for autophagosomes especially in light microscopy but also in electron microscopy (Jäger *et al.*, 2004). LC3 is present mainly in phagophores and autophagosomes. Because it is degraded by the incoming lysosomal hydrolases, less LC3 is present in amphisomes and autolysosomes (Jäger *et al.*, 2004).

In transmission electron microscopy, autophagic compartments can be identified by morphology, particularly when samples embedded in conventional plastic resins are used. Electron microscopy has a superior resolution compared to light microscopy, but its disadvantage is that the sample size is very small (see also Chapter by Razi and Tooze in this volume). Instead of whole cells, ultrathin slices of cells, typically of 70–80 nm thickness, can be observed. Therefore, it can be difficult to get an idea of the size and total volume of different compartments inside the cells. However, relatively simple quantification methods have been developed that produce reliable and repeatable estimates of volume and size of cell compartments. For comparison of the autophagic accumulation in different samples, it is recommended to use unbiased stereological methods to quantify the autophagic structures.

This chapter describes the fine structure of autophagic compartments in plastic-embedded samples. We include a short description on the use of tomography to study the three-dimensional structure of autophagosomes. Three sample preparation protocols are presented. The first protocol is suitable for embedding of cell pellets. This procedure is well suited for quantification of autophagic compartments by point counting. The second method uses imidazole-buffered osmium tetroxide. This procedure gives a high contrast to lipids, including the limiting membranes of autophagosomes, helping in their identification even at low magnification. The third protocol is suitable for flat embedding of adherent cultured cells. This protocol uses reduced osmium tetroxide, which gives good contrast to the limiting membranes of autophagosomes. This method is well suitable for tomography. Finally, we describe a simple protocol for the quantification of autophagic compartments by stereology in cell pellets or tissue blocks.

2. Methods

2.1. Resin embedding of aldehyde-fixed animal cell pellets using unbuffered osmium tetroxide

This protocol is suitable for plastic embedding of cell pellets or tissue cubes. The protocol includes block staining with uranyl acetate, which serves two purposes: first, it gives contrast to ribosomes, helping in the identification of the cytoplasmic contents of autophagosomes; second, it enhances the contrast of the autophagosome limiting membranes. Further details on fixatives and embedding media can be found in the literature (Glauert and Lewis, 1998; Hayat, 2000).

1. Use 3- or 6-cm culture dishes, depending on the volume of the cells. The cell culture should be semiconfluent. You need a visible pellet, but the pellet should not be too large, because penetration of the reagents is slow. A suitable pellet size is approximately 10–30 μl. The whole embedding procedure is carried out in a fume hood with gloves because the chemicals are toxic and/or volatile.
2. Fix cells in 2% glutaraldehyde (electron microscopy grade, such as MP Biomedicals Cat. No. 195199) in 0.2 M HEPES, pH 7.4, at RT for 2 h. If you use adherent cells, start the fixation on the culture plate. Scrape the cells from the plate after 30 min fixation, using a razor blade or cell scraper. Pellet the cells at full speed in a microcentrifuge for 5 min. Continue fixation as a pellet. Do not resuspend the pellet during the rest of the embedding procedure but use gentle rotation during incubations to keep the pellet in motion. We use a rocking platform where the tubes are placed in a horizontal position.
3. Wash the pellets in 0.2 M HEPES, pH 7.4. If the pellets are firm, the solutions can be changed by decanting. The pellets can be stored in this buffer at 4 °C for some weeks.
4. Wash in phosphate-buffered saline (PBS) or 0.1 M phosphate buffer, pH 7.4, 3 times.
5. Postfix the pellets in 1% osmium tetroxide in water at RT for 1 h.
6. Wash in water twice.
7. Stain the pellets in 2% uranyl acetate (Agar Scientific, Cat. No. R1260A) in water at RT, in the dark, for 1 h.
8. Because the embedding resin is not water soluble, all water must be removed from the samples. This is done using ethanol at increasing concentrations. Propylene oxide at the end of the dehydration further helps the penetration of the resin. Gentle mixing is important during dehydration.

Dehydration at RT: 70% ethanol, 15 min; 95% ethanol, 15 min; 100% ethanol, twice for 15 min; propylene oxide, 20 min (toxic and volatile; may be replaced with acetone).

9. Infiltration with resin★: Incubate the pellets in a mixture of resin and propylene oxide (1:1) at RT for 2 h, and then in 100% resin at RT overnight.
10. Transfer the pellets to fresh resin in beem capsules (Agar Scientific Cat. No. G362 or G360) (Glauert and Lewis, 1998) and incubate at RT for 4–6 h.
11. Polymerize the blocks at 60 °C for 2 days.
12. Cut 80-nm sections with a diamond knife, pick them up on 200 mesh grids with square or hexagonal openings, and poststain with 2% uranyl acetate in water for 15 min and 0.3% lead citrate (Agar Scientific, Cat. No. R1210) for 3 min, according to standard grid staining procedures (Hayat, 2000).

★Resin: Durcupan mixture (Fluka, catalog numbers for each component are given below):

Mix 10 g Component A (44611), 10 g Component B (44612), 0.3 g Component C (44613), and 0.3 g Component D (44614). The mixture can be stored at -20 °C for several days. Do not open the container before the mixture has reached RT.

Agar 100 mixture (Agar Scientific, Cat. No. R1045) is also suitable. See manufacturer's instructions for the mixing of the components. Use the medium formula. Other Epon equivalent embedding media such as LX-112 (Ladd) or TAAB 812 resin (TAAB) will work as well.

2.2. Embedding of mouse liver using imidazole-buffered osmium tetroxide

Imidazole-buffered osmium tetroxide gives a strong contrast to unsaturated lipids (Angermüller and Fahimi, 1982). We have shown that this protocol gives a very strong contrast to autophagosome membranes (Reunanen et al., 1985), which helps in their identification even at low magnifications. Perfusion fixation is recommended for tissues from larger animals such as mouse or rat. It is important to note that lack of oxygen causes alterations in tissue fine structure as soon as a few minutes after the sacrifice of the animals (Glauert and Lewis, 1998).

1. Anesthesize the mice. After that, everything is done in a fume hood using gloves. Fix the liver by perfusion via the inferior vena cava, below the branching point of the renal veins. We use open, one directional flow of the perfusion solutions. Perfusion can be done using gravity, by placing the container with perfusion solution approximately 120 cm above the bench. The flow should be adjusted to 30–40 ml/min. To wash out the blood, first rinse the liver with 0.1 M sodium cacodylate buffer, pH 7.4 (NaCac), containing 0.8% NaCl, 0.4% lidocaine chloride

(Sigma, Cat. No L5647) and 28 µg/ml heparin (Sigma, Cat. No H7405). Then start the fixation using 2% glutaraldehyde (electron microscopy grade) in the same buffer containing also 4.0% polyvinylpyrrolidone to increase the osmolarity (Sigma, Cat. No. PVP10), and 0.05% $CaCl_2$. Continue perfusion fixation for 4 min. Then cut approximately 1-mm^3 cubes from the liver and immerse them in the fixation solution. Incubate at RT for 1–2 h. Further details on perfusion fixation can be found in the literature (Glauert and Lewis, 1998).
2. Wash the tissue cubes in 0.1 M NaCac 3 times for 5 min each. Usually it is possible to change the solutions by decanting.
3. Imidazole-buffered osmium tetroxide postfixation: Add 0.2 M imidazole, pH 7.4, to the tissue cubes. Then add an equal volume of 4% osmium tetroxide in distilled water to make a solution of 2% osmium tetroxide in 0.1 M imidazole buffer. Incubate at RT in the dark for 30 min.
4. Wash the cubes in 0.1 M NaCac three times for 5 min each.
5. Dehydrate the cubes in ethanol:70% ethanol, 3×5 min; 96% ethanol, 3×5 min; 2% uranyl acetate in absolute ethanol, 30 min; absolute ethanol, 3×5 min.

Uranyl acetate block staining gives contrast to autophagosome membranes and ribosomes.

6. Incubate the cubes in propylene oxide three times for 7 min each.
7. Infiltrate the samples with epoxy resin (Agar100 from Agar Scientific, Cat. No. R1045, LX-112 from Ladd, Cat. No. 21205, or equivalent): First, infiltrate in a mixture of resin and propylene oxide (1:1) for 1 h, and then in pure resin for 1 day. Propylene oxide may be replaced with acetone, which is less harmful.
8. Place the samples in embedding molds (beem or gelatin capsules are fine), fill with resin, and polymerize at 60 °C for two days.
9. Cut 70- to 80-nm sections and poststain them in lead citrate only, or first in uranyl acetate and then in lead citrate as described previously.

2.3. Resin flat embedding of aldehyde-fixed animal cell cultures using reduced osmium tetroxide

This protocol is well suited for adherent cells. In flat embedding, the sections are cut parallel to the culture substrate. This is important in situations where cell processes need to be identified, such as axons and dendrites in neuronal cells. Reduced osmium tetroxide used in this protocol gives good contrast to the limiting membranes of autophagosomes. This is important in applications such as tomography. Ribosomes, which can be used as a marker of the cytoplasmic contents of autophagosomes, have less contrast. This can be improved to some extent by using uranyl acetate block staining.

1. Grow cells on glass cover slips, preferably thickness No 1. From step 2 onward, everything is done in a fume hood with gloves. Do not let the cells dry out at any point during the procedure. This is achieved by replacing the solution quickly after removal of the previous one.
2. Remove the cells from the culture incubator just before fixation. Fix the monolayers in 2% glutaraldehyde (electron microscopy grade) in 0.1 M NaCac, pH 7.4, for 30–60 min at RT. There is no need to wash the cells before fixation. 0.1 M phosphate buffer, pH 7.4, can be used instead of NaCac.
3. Wash with NaCac twice for 3 min.
4. Reduced osmium tetroxide postfixation:

Mix 2%–4% osmium tetroxide in water and NaCac to give final concentrations of 1% osmium tetroxide and 0.1 M NaCac. Then add 15 mg/ml potassium ferrocyanide, $K_4Fe(CN)_6$, (Sigma, Cat. No. P3289). Incubate on the cells for 1 h at RT.

5. Wash with NaCac twice for 3 min, then with water once.
6. Block staining: Incubate the monolayers in 1% uranyl acetate and 0.3 M sucrose in water for 1 h, at 4 °C. This step is recommended as it adds contrast to ribosomes.
7. Dehydration at RT: 70% ethanol, 1 min; 96% ethanol, 1 min; 100% ethanol, 3 × 1 min.
8. Dip the cover slips, one at a time, into acetone and place them on solvent-resistant plates or trays (glass or aluminum). Immediately drop some epoxy resin (TAAB 812 resin, TAAB, UK, or other Epon equivalent embedding resin such as LX-112 from Ladd, or Agar100 from Agar Scientific) to cover the cells, before the acetone has completely evaporated.
9. Fill beem capsules with resin and place them upside down on the cover slips. Incubate at RT for 2 h.
10. Polymerize the resin at 60 °C overnight.
11. Transfer the samples from the oven to a hot plate and carefully remove the cover slips from the resin blocks. Optionally, take the samples from the oven and drop them directly into liquid nitrogen. The cover slips should then easily crack off from the blocks. Make sure all glass has been removed from the block surface before using a diamond knife to cut sections. Glass will destroy the edge of the diamond knife immediately.
12. Trim the block to a pyramid of desired size and directly start cutting 70- to 80-nm thin sections. Remember the sample is only one cell layer thick; therefore, it is not recommended to trim the block face before cutting thin sections.
13. Stain the sections with uranyl acetate and lead citrate as described previously.

2.4. Electron tomography

The reduced osmium tetroxide protocol described previously is suitable for electron microscopy tomography. The exact procedure for tomography depends on the availability of equipment. The tomogram presented in this article was produced as follows:

1. Normal rat kidney cells were incubated in serum and amino acid-free medium for 1 h to induce autophagy and processed as described earlier using the reduced osmium tetroxide protocol.
2. Serial semithin sections (250 nm) were cut from the samples with a diamond knife (35 degrees cutting edge, Diatome, Switzerland).
3. We used single-slot grids (Agar Scientific, Cat. No. G2525C or G2500C) coated with Pioloform (Agar Scientific, R1275) and a thin layer of carbon (Bal-Tec CED030 carbon thread evaporator, Bal-Tec Union, Liechtenstein). Ten-nm gold particles (protein A coated gold particles, available from the University of Utrecht, School of Medicine, Department of Cell Biology, or from commercial suppliers such as British BioCell, Cat. No. EM.AG10) were adhered to the grid surface for alignment of the tilt-series micrographs. Gold particles were diluted in water (1:40) and incubated on the grids for 10 s. The grids were then rinsed in water 3 times.
4. The sections were picked up on the grids and stained with 0.5% uranyl acetate (EMS, Cat. No. 22400) in water for 30 min and with lead citrate (Leica Ultrostain 2 solution, Cat. No. 70553022) for 80 s.
5. Dual axis tilt-series micrographs were acquired at 14,500X primary magnification, providing a pixel size of 1.25 nm, using a Tecnai FEG20 electron microscope (FEI Corp.) at 200 kV, equipped with a 1 k × 1 k Multiscan 794 CCD camera (Gatan Corp.). The angle between the tilts was 1 degree and maximum tilt was ±70 degrees. Three-dimensional reconstruction from the tilt series was done using IMOD software (http://bio3d.colorado.edu/imod/). Membranes of the phagophore and endoplasmic reticulum were traced by hand, and Amira software (TGS Inc.) was used to build a three-dimensional model of these membranes.

2.5. Quantitation of autophagic compartments in thin sections

2.5.1. Sampling

The first crucial step in quantitative microscopy is sampling, which should be unbiased. This is best achieved using so-called uniform random sampling (Howard and Reed, 2005). The principle behind uniform random sampling is that every region of the sample should have an equal opportunity to become included in the counting. To get a representative sample for a whole organ, it is important to sample the whole organ. For whole organs, the use of uniform random sampling is illustrated in Fig. 10.1.

2.5.1.1. Sampling protocol for organs

1. The organ, such as a kidney, is first fixed by perfusion for at least 5–10 min. The organ can then be removed from the animal for the sampling procedure. The organ is first cut into slices of equal thickness. Assume the kidney is longitudinally cut into 12 1-mm-thick slices. Special equipment have been designed for cutting the tissue (Howard and Reed, 2005).
2. Then, we decide to sample every third slice. We take a random number between 1 and 3 (in this example, we end up with the random number 3).
3. The slices are numbered starting from one end of the kidney. The first slice to become sampled is number 3. After that every third slice is sampled, ending up with a sample of 4 slices.
4. To each slice we apply the same procedure, to sample tissue cubes approximately the size of 1 mm^3. Instead of cutting the entire tissue slices into cubes, it is easier to randomly apply a test system of small squares on each slice, and cut the samples according to that, again using uniform random sampling (see Fig. 10.1). The selected 1-mm^3-tissue

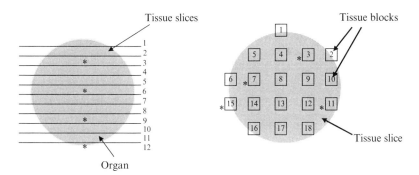

Figure 10.1 The principle of uniform random sampling when applied to an organ such as a kidney. Step 1: the organ is cut into slices of equal thickness. A decision is made on how many of the slices need to be sampled. In this example, it is 4. Twelve divided by 4 is 3, meaning we need to take every third slice. A random number is selected between 1 and 3, in this example we get number 3. The first slice to be sampled is number 3. After slice number 3, every third slice is selected, meaning slices 3, 6, 9, and 12 are sampled. Another uniform random sampling is applied to each tissue slice. In this case, a sampling system is used to ease the procedure. The squares in the sampling system are numbered. We decide to sample every fourth square. Samples are collected starting from square 3, which was given by a random number between 1 and 4. After this first square, every fourth square is selected, and we get squares 3, 7, 11, and 15.

cubes are then immersed in fixation solution for 1–2 h, and embedded in resin as described previously.

2.5.1.2. Sampling of cell pellets Cultured cells are generally easier to sample, as all cells in the culture are similar and have a similar orientation to the culture dish and to one another. The orientation of the cells in the thin sections should be random for quantitative estimation of autophagic accumulation using the approach described here. This is best achieved when using cell pellets. If flat embedded cells are used, the counting should be done using sections cut from more than one plane, which should be located a few micrometers from each other. This is because it is possible that the structures of interest are more concentrated in a certain region of the cytoplasm (e.g., close to the culture surface or above the nucleus). This problem is easily avoided by using sections in which the cells are randomly oriented.

Our experience with different cell lines has shown that when using cell pellets, it is possible to get repeatable results with small variation, with only one step of uniform random sampling. This step is applied to the sections picked up on grids, using the grid itself as a sampling system (Fig. 10.2A). In practice it is adequate to use just one thin section through the cell pellet per sample. This approach makes quantification of autophagic compartments doable, even in experiments that have numerous samples.

2.5.2. Counting

There are two alternatives for the electron microscopy quantification. First, the volume fraction (e.g., the fraction of the cellular volume occupied by autophagic compartments) can be estimated. This can be accurately and efficiently estimated by point counting (Eskelinen *et al.*, 2002a,b) (Fig. 10.2). Second, it is possible to estimate the number of autophagic profiles per cell area on sections (Eskelinen *et al.*, 2004; Jäger *et al.*, 2004). This approach saves a considerable amount of work, as only the cell volume needs to be estimated by point counting, whereas the autophagic compartments are just counted under the microscope.

2.5.2.1. Counting protocol for pellets of cultured cells

1. One thin section covering the pellet is generally needed for the counting. Square or hexagonal grid openings can be used as sampling units in the quantification. The grid openings are selected according to the rules of uniform random sampling (Fig. 10.2A).
2. The whole grid squares are systematically scanned under the microscope for the presence of early and late autophagic compartments (Fig. 10.2B).
3. When estimating the number of compartments per cell area, only the number of the compartments in the grid square is recorded. When estimating volume fraction, each autophagic compartment is

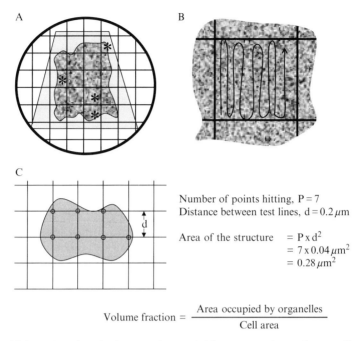

Figure 10.2 Principles of point counting. A. Grid squares can be used as sampling units. It is recommended to count at least 3 squares from each sample. The squares are selected using uniform random sampling. The squares are numbered as explained in Fig. 10.1. A total of 19 grid squares contain cells. We decide to count 4 openings, meaning we need to sample every fifth square. In this example, the random number between 1 and 5 is 3. The first square to be sampled is number 3, and after that every fifth square is sampled. B. Screening of the selected grid squares for the presence of autophagic compartments at higher magnification (12,000X). One photograph is taken from each grid square at low magnification (400X) that covers the entire grid square. C. Point counting. A square lattice is placed on top of a microscopy photograph or photographic negative. The intersections of the lattice lines are used as test points. Points hitting the structure of interest (autophagic compartment, or cell in a cell pellet) are counted. The distance between points (d) is calibrated with the final magnification of the photograph. These two parameters are used to calculate the area of the structures, as indicated next to the drawing. This procedure can be done on the computer screen, if the photographs are digital. The lattice is created using the computer software. If the microscope uses a film camera, point counting can be performed using a light box. The lattice is printed on an overhead foil that is then placed on top of the film sheet. A magnifying loop can be used to help counting. The figure is modified from one published previously (Eskelinen, 2008a).

photographed at 12,000X magnification for point counting of the area (point counting 1; Fig. 10.2C). This can be done directly from photographic negatives if a digital camera is not available in the microscope.
4. The cell area can be measured using approximately 400X magnification, which allows the whole grid squares to fit in one photograph.

5. The area of the cell profiles is then estimated using point counting with a different lattice (point counting 2; Fig. 10.2C).
6. The number of compartments per cell area is calculated by dividing the number of compartments in the grid square with the cell area in the same grid square. Similarly, the volume fraction is calculated by dividing the area occupied by autophagic compartments (from point counting 1) with the cell area in that grid square (from point counting 2). Thus, the ratio of areas directly gives the ratio of volumes (Howard and Reed, 2005).
7. It is recommended to count at least three grid squares for each sample to get a representative estimate. When using cell pellets, these grid squares can all be in one thin section on the same grid (Fig. 10.2A).

In most cases, the number of autophagic compartment profiles per cell area gives a very similar result compared to the stereologically more accurate volume fraction (Eskelinen, 2008c). Because the former saves work, it is better suited for experiments with a high number of samples. It should be kept in mind, however, that the number of profiles per cell area is *not* equal to the number of organelles per cell area. For more exact discussion, please refer to the stereology literature (Howard and Reed, 2005).

2.5.2.2. Counting protocol for more abundant autophagic compartments The procedure described previously is suitable for samples with a relatively low amount of autophagic compartments. In some cell types, such as isolated hepatocytes, autophagic compartments are so frequent that it becomes difficult to count them under the microscope by scanning the grid squares as described earlier (Fig. 10.2B). In these cases, it is better to follow an alternative approach.

1. Grid squares are selected using uniform random sampling as described above. The selected grid openings are then photographed at 12,000X magnification, again using the uniform random sampling principle. The opening is scanned using the systematic approach described in Fig. 10.2B, but now photographs are taken at regular intervals, such as at each half turn of the wheel that moves the sample in the microscope.
2. These photographs are used to point count both the cell area and the area of the autophagic compartments. The two can be point counted using a double lattice, with a tight lattice for autophagic compartments and a loose lattice for the cell area. Details on the double lattices and suitable size of the lattice squares are available in the literature (Howard and Reed, 2005).

2.5.2.3. Counting autophagic compartments in tissue samples The principles described previously for pellets of cultured cells can be used for tissue blocks as well, but the amount of sample blocks needs to be higher to get an estimate for the whole organ. Special arrangements may be necessary for tissues such as skeletal muscle, in cases where the sections need to have a certain orientation such as longitudinal cross-sections through the muscle fibers. For more information, the reader should study the literature on stereology (Howard and Reed, 2005).

2.5.2.4. The reference trap The point-counting protocol described here is based on the amount of autophagic compartments per cell area. It should be kept in mind that if cell area changes (due to cell swelling or shrinkage), then the ratio of organelles per cell area is also changing, although the amount of the organelles actually stays the same (Fig. 10.3). This phenomenon is called the *reference trap*. This should be taken into account if treatments such as viral infection are used that may affect the cell volume. One possibility in such cases would be to measure the cell size and take this into account when calculating the results. However, because cells shrink during the plastic embedding, and the extent of shrinkage depends on the water content of the sample; it may be difficult to get an accurate result using this approach. Another possibility to get around the problem would be to count organelles per cell or nucleus instead of organelles per cell area. Stereological methods for such measurements are described in the literature (Howard and Reed, 2005).

3. Results and Discussion

3.1. Fine structure of autophagosomes and autophagic compartments in plastic sections

By definition, autophagosomes are limited by a double, or occasionally multiple, membrane. They contain undegraded cytoplasm but no lysosomal proteins. In plastic sections, the two limiting membranes can be so close to

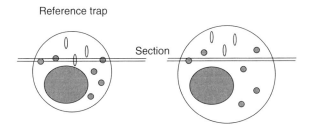

Figure 10.3 Bias caused by cell swelling of shrinking, called the reference trap. Both cells have an equal number of round and oval organelles and a nucleus, all of which have equal size in both cells. The cell on the right has swollen compared to the cell on the left, meaning the reference space is larger in that cell. The swelling has two consequences: first, the thin section will hit fewer organelles because the same amount of organelles is diluted to a larger volume; and second, the ratio of organelle area to cell area is smaller, although the number and size of the organelles has not decreased.

each other that it is not possible to see them as separate membranes (Eskelinen, 2008a). Sometimes the limiting membrane may appear to contain multiple layers. It is possible that this is an artifact caused by the chemical fixation. Sometimes the autophagosome limiting membrane does not have any contrast in thin sections. This is probably caused by extraction of lipids during sample preparation, as lipids are not optimally preserved in conventional aldehyde fixation. In most cases, however, the two limiting membranes are visible in plastic sections. Typically, there is a narrow empty (electron lucent) space between the two membranes (Figs. 10.4, 10.5C), which can be used as one criteria for the identification of autophagosomes at low magnification. Typically, but not always, a cistern of rough endoplasmic reticulum (ER) is located close to the autophagosome or phagophore. Another rough cistern is often located on the other side of the autophagosome or phagophore limiting membrane (Fig. 10.6). Thus the limiting membrane runs in the space between these two rough ER cisterns. The role of this autophagosome-associated ER, if any, is currently unknown.

The second, very important, criterion for autophagic compartments is that they must contain cytoplasmic material. The cytoplasmic contents of autophagosomes include organelles, such as ER membranes and mitochondria (Figs. 10.4, 10.5, 10.6). Ribosomes are frequently seen inside autophagosomes (Fig. 10.4) and serve as a good marker for the cytoplasmic contents. The diameter of autophagosome profiles varies between 300–400 nm and several micrometers. In cultured cells, the average diameter is approximately 600 nm. Autophagosomes are frequently observed in fusion profiles with endosomal or lysosomal vesicles (Eskelinen, 2005b). In the fusion event, the outer limiting membrane fuses with the endo/lysosome limiting membrane. The contents, still surrounded by the inner limiting membrane, are delivered to the endo/lysosome lumen. This membrane must then be degraded, or at least permeabilized, to allow degradation of the cytoplasmic contents. In late autophagic compartments, partially degraded electron-dense ribosomes serve as a criterion for identification of the cytoplasmic contents (Figs. 10.4B and 10.4C).

In samples prepared using cacodylate-buffered or unbuffered osmium tetroxide, autophagosomes can be difficult to identify at low magnification. If large areas of tissue need to be screened, the identification can be improved using imidazole-buffered osmium tetroxide, which stains unsaturated lipids (Angermüller and Fahimi, 1982), including autophagosome membranes (Reunanen et al., 1985). We have successfully used this protocol for mouse liver (Figs. 10.5A and 10.5B) and pancreas (not shown). Interestingly, autophagosome membranes are heavily stained, whereas amphisome and autolysosome limiting membranes only show a normal contrast (Fig. 10.5B). This may suggest that autophagosome membranes have a high content of unsaturated lipids.

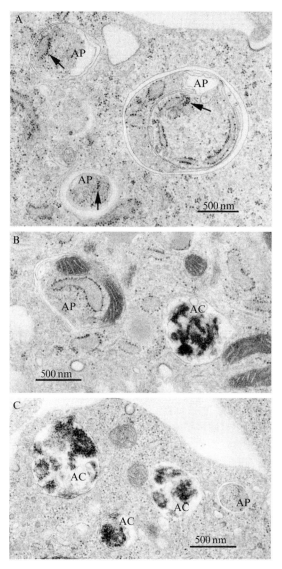

Figure 10.4 Fine structure of autophagosomes (early/initial autophagic compartments) and late autophagic compartments (amphisomes and/or autolysosomes) in plastic embedded mouse embryonic fibroblasts. The cells were incubated in serum and amino acid-free medium to induce autophagy, and prepared according to the unbuffered osmium tetroxide protocol. A. Three autophagosomes (AP) of different sizes. The large autophagosome on the right has formed around another autophagosome that is now visible inside it. Ribosomes (arrows), either free or attached to ER membranes, are visible inside all autophagosomes. B. The autophagosome on the left (AP) contains rough ER and a mitochondrion. The late autophagic compartment (AC) contains partially degraded, electron-dense ribosomes. C. The three late autophagic compartments (AC) contain partially degraded ribosomes. One small autophagosome is also visible (AP).

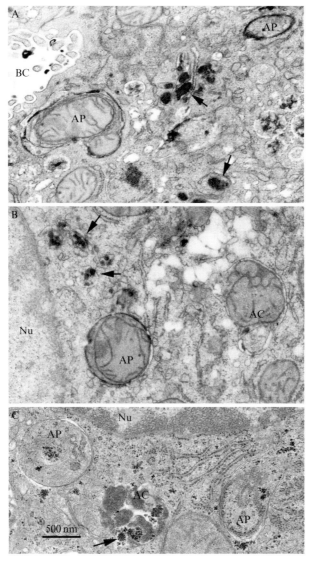

Figure 10.5 Perfusion-fixed mouse liver embedded using imidazole-buffered osmium tetroxide (A, B). Mouse liver embedded using the unbuffered osmium protocol is presented in panel C for comparison. Because hepatocytes are rich in mitochondria, these are frequently seen inside autophagosomes (AP in panels A, B, and AC in panel B). Imidazole-buffered osmium gives a high contrast to autophagosome membranes (AP in panels A and B), whereas later autophagic compartments show less contrast (AC in panel B). Imidazole-buffered osmium also gives high contrast to lipids in general, including lipoprotein particles inside the Golgi cisternae and secretory vesicles (arrows in panels A and B). Note that in hepatocytes processed using unbuffered osmium, the autophagosome membranes show less contrast (AP in panel C). The autophagic compartment in panel C contains glycogen particles (arrow). BC, bile canaliculus, Nu, nucleus.

It should be noted that special attention must be paid when identifying autophagic compartments in conventional transmission electron microscopy (Eskelinen, 2008). There are several examples in the current literature of different organelles, including multilamellar or multivesicular endosomes, being claimed as autophagic compartments. When identification is based on morphology, only vesicles containing cytoplasmic material, in most cases ribosomes, can be claimed as autophagic compartments in mammalian cells.

3.2. Tomography of a phagophore

Using tomography, very thin slices (1–2 nm thick) through the organelles can be viewed. This gives a higher resolution than conventional thin sections that are 50–60 nm thick at best. We used two consecutive

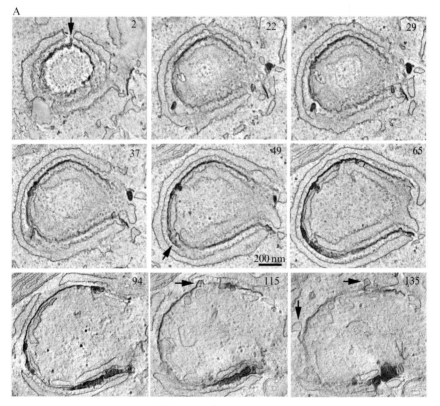

Figure 10.6 (*continued*)

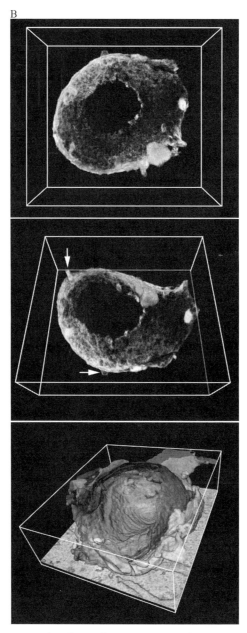

Figure 10.6 A. Tomographic slices of 1.6- to 2-nm thickness, created from the tomographic reconstruction presenting a phagophore in a normal rat kidney cell. A total of 164 images were created from the tilt series collected from two consecutive 250-nm semi-thick sections. The numbers in the panels indicate the number of the slice in the reconstructed image stack. The phagophore membrane is visible as one electron-dense

250-nm semithin sections to produce a three-dimensional tomographic reconstruction of a phagophore lined by cisternae of rough ER on both sides (Figs. 10.6A and 10.6B). The reconstruction contains approximately half of the phagophore. The phagophore membrane is seen as one electron-dense line, instead of a flat membrane cistern. This may be due to imperfect preservation of the membrane in the chemically fixed sample. However, similar findings are described in a careful study that applies both cryofixation and chemical fixation (Kovacs *et al.*, 2007). In fact, Kovacs *et al.* conclude that the empty cleft between the phagophore membranes is an artifact of aldehyde fixation. We observed a few extensions bulging out of the phagophore membrane (Fig. 10.6A, sections 2, 49, 115, and 135; Fig. 10.6B), but no connections between the phagophore and the ER cisternae. This is in agreement with published results (Kovacs *et al.*, 2007). The reconstruction shows that the two ER cisternae tightly line the phagophore membrane on both sides (Figs. 10.6A and 10.6B).

3.3. Quantitation of autophagy by electron microscopy

Even today, quantitative electron microscopy is one of the most sensitive methods to detect the presence of autophagic compartments. In addition, it is possible to detect the nature of the accumulating structures (e.g., whether early or late autophagic compartments accumulate) (Eskelinen, 2005a). The ratio of early to late compartments gives clues on the cause of the accumulation. If autophagosomes predominate, there is probably an increase in their formation or a defect in the maturation of autophagosomes into degradative amphisomes and autolysosomes. On the other hand, large accumulation of late autophagic compartments suggests there may be a defect in degradation in autolysosomes. It should be noted that most of the morphologically identified late autophagic compartments are probably amphisomes. Autolysosomes are short-lived and thus less frequently observed by microscopy, unless lysosomal hydrolases are inhibited using inhibitors such as leupeptin, E64, and pepstatin.

---

line between two ER cisternae. Note the processes (arrows) bulging from the phagophore in slices 2, 49, 115, and 135. B. Three-dimensional models built from the image stack after tracing the phagophore and ER membranes by hand. The upper panel shows the inner surface of the phagophore membrane. The bottom of the phagophore cup is missing, as it was not present in the semithick sections used to collect the tilt series. The middle panel shows the outside of the phagophore membrane, which is now tilted 150 degrees from the upper panel. Note the two extensions bulging from the phagophore (arrows). The lower panel shows the phagophore membrane in purple, the ER cistern outside the phagophore in yellow, and the ER cistern inside the phagophore in red. The orientation of the organelle is such that the reader is looking inside the phagophore from the open end. (See Color Insert.)

The kinetics of the autophagic accumulation can be studied using inhibitors of lysosomal degradation that cause the accumulation of autophagic compartments at the amphisome/autolysosome stage. The buildup of autophagic organelles can thus be estimated by taking samples after several time points during autophagy induction (Klionsky et al., 2008). Similarly, the clearance of autophagic compartments can be followed by taking samples during autophagy induction, as well as after induction followed by a chase period when the induction no longer occurs (Eskelinen et al., 2002a) (Fig. 10.7).

4. Concluding Remarks

Autophagy was originally described using electron microscopy approximately 50 years ago, when electron microscopy and sample preparation methods for biological materials had just emerged. After the discovery of autophagy genes, the field has exponentially expanded and autophagy has gained more and more attention. Despite the enormous developments in different methods for the monitoring of autophagy in cells and animals,

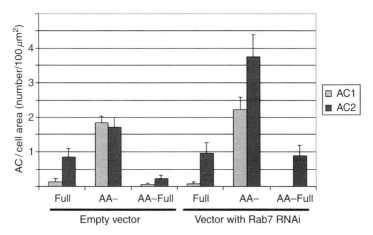

Figure 10.7 Quantification of autophagic compartments in HeLa cells using the point counting protocol. Accumulation and clearance of autophagic compartments was estimated in control cells (empty vector) and in Rab7-depleted (vector with Rab7 RNAi) cells. The cells were fixed without treatment (Full), after a 2-h amino acid starvation (AA-), and after a 2-h amino acid starvation followed by a 2-h chase in full-culture medium (AA-Full). The amounts of autophagosomes (AC1) and late autophagic compartments (AC2) were quantified. Note that in the absence of Rab7, more late autophagic compartments accumulate in amino acid–free medium, and that the clearance of these compartments in full medium is retarded as compared to the control cells. The figure is modified from one published previously (Jäger et al., 2004).

transmission electron microscopy is still needed and can give both qualitative and quantitative information that is not available using any other method. In the future, cryofixation followed by freeze substitution and tomography are likely to provide high resolution three-dimensional images of autophagic compartments that are free of artifacts caused by chemical fixation. This is very likely to aid in finding the answer to the most challenging question in this field: where does the phagophore membrane come from?

ACKNOWLEDGMENTS

We thank Mervi Lindman and Kaisa Happonen for technical help. E.-L.E. would like to thank Pirkko Hirsimäki (Turku, Finland) and Hilkka Reunanen (Jyväskylä, Finland) for the initial introduction to the secrets of electron microscopy and autophagy, and John Lucocq (Dundee, UK) for sharing his knowledge on the quantitative aspects of electron microscopy. The authors were or are supported by University of Helsinki Research Funds, Biocentrum Helsinki, the Academy of Finland (E.-L.E. and E.J.), the Royal Society, and the Magnus Ehrnrooth Foundation (E.-L.E.).

REFERENCES

Angermüller, S., and Fahimi, H. D. (1982). Imidazole-buffered osmium tetroxide: An excellent stain for visualization of lipids in transmission electron microscopy. *Histochem. J.* **14,** 823–835.

Arstila, A. U., and Trump, B. F. (1968). Studies on cellular autophagocytosis. The formation of autophagic vacuoles in the liver after glucagon administration. *Am. J. Pathol.* **53,** 687–733.

Dunn, W. A., Jr. (1990a). Studies on the mechanisms of autophagy: Formation of the autophagic vacuole. *J. Cell Biol.* **110,** 1923–1933.

Dunn, W. A., Jr. (1990b). Studies on the mechanisms of autophagy: Maturation of the autophagic vacuole. *J. Cell Biol.* **110,** 1935–1945.

Dunn, W. A., Jr. (1994). Autophagy and related mechanisms of lysosomal-mediated protein degradation. *Trends Cell Biol.* **4,** 139–143.

Eskelinen, E.-L. (2005a). Autophagy in mammalian cells. In "Lysosomes." (P. Saftig, ed.), pp. 166–180. Landes Bioscience/Eurekah.com, Georgetown.

Eskelinen, E.-L (2005b). Maturation of autophagic vacuoles in mammalian cells. *Autophagy* **1,** 1–10.

Eskelinen, E.-L. (2008). To be or not to be? Examples of incorrect identification of autophagic compartments in conventional transmission electron microscopy of mammalian cells. *Autophagy* **4,** 257–260.

Eskelinen, E.-L. (2008a). Fine structure of the autophagosome. *Methods Mol. Biol.* **445,** 11–28.

Eskelinen, E.-L. (2008b). New insights into the mechanisms of macroautophagy in mammalian cells. *Int. Rev. Cell Mol. Biol.* **266,** 207–247.

Eskelinen, E.-L. (2008c). Fine structure of the autophagosome. *Methods Mol. Biol.* **445,** 11–28.

Eskelinen, E.-L., Illert, A. L., Tanaka, Y., Blanz, J., von Figura, K., and Saftig, P. (2002a). Role of LAMP-2 in lysosome biogenesis and autophagy. *Mol. Biol. Cell* **13,** 3355–3368.

Eskelinen, E.-L., Prescott, A. R., Cooper, J., Brachmann, S. M., Wang, L., Tang, X., Backer, J. M., and Lucocq, J. M. (2002b). Inhibition of autophagy in mitotic animal cells. *Traffic* **3,** 878–893.

Eskelinen, E.-L., Schmidt, C. K., Neu, S., Willenborg, M., Fuertes, G., Salvador, N., Tanaka, Y., Lüllmann-Rauch, R., Hartmann, D., Heeren, J., von Figura, K., Knecht, E., *et al.* (2004). Disturbed cholesterol traffic but normal proteolytic function in LAMP-1/LAMP-2 double deficient fibroblasts. *Mol. Biol. Cell* **15,** 3132–3145.

Glauert, A. M., and Lewis, P. R. (1998). *Biological specimen preparation for transmission electron microscopy*. Portland Press, London.

Hayat, M. A. (2000). "Principles and techniques of electron microscopy: Biological applications." Cambridge University Press, New York.

Howard, C. V., and Reed, M. G. (2005). "Unbiased stereology. Three-dimensional measurement in microscopy." Garland Science/BIOS Scientific Publishers, New York.

Jäger, S., Bucci, C., Tanida, I., Ueno, T., Kominami, E., Saftig, P., and Eskelinen, E.-L. (2004). Role for Rab7 in maturation of late autophagic vacuoles. *J. Cell Sci.* **117,** 4837–4848.

Kabeya, Y., Mizushima, N., Ueno, T., Yamamoto, A., Kirisako, T., Noda, T., Kominami, E., Ohsumi, Y., and Yoshimori, T. (2000). LC3, a mammalian homologue of yeast Apg8p, is localized in autophagosome membranes after processing. *EMBO J.* **19,** 5720–5728.

Klionsky, D. J., Abeliovich, H., Agostinis, P., Agrawal, D. K., Aliev, G., Askew, D. S., Baba, M., Baehrecke, E. H., Bahr, B. A., Ballabio, A., Bamber, B. A., Bassham, D. C., *et al.* (2008). Guidelines for the use and interpretation of assays for monitoring autophagy in higher eukaryotes. *Autophagy* **4,** 151–175.

Klionsky, D. J., Cregg, J. M., Dunn, W. A. Jr., Emr, S. D., Sakai, Y., Sandoval, I. V., Sibirny, A., Subramani, S., Thumm, M., Veenhuis, M., and Ohsumi, Y (2003). A unified nomenclature for yeast autophagy-related genes. *Dev. Cell* **5,** 539–545.

Kovács, A. L., Palfia, Z., Rez, G., Vellai, T., and Kovács, J. (2007). Sequestration revisited: Integrating traditional electron microscopy, de novo assembly and new results. *Autophagy* **3,** 655–662.

Reunanen, H., Punnonen, E.-L., and Hirsimaki, P. (1985). Studies on vinblastine-induced autophagocytosis in mouse liver V. A cytochemical study on the origin of membranes. *Histochemistry* **83,** 513–517.

Xie, Z., and Klionsky, D. J. (2007). Autophagosome formation: Core machinery and adaptations. *Nat. Cell Biol.* **9,** 1102–1109.

CHAPTER ELEVEN

Monitoring Mammalian Target of Rapamycin (mTOR) Activity

Tsuneo Ikenoue,* Sungki Hong,* *and* Ken Inoki*,†

Contents

1. Introduction	166
2. Methods	168
2.1. Cell culture and preparation of cell lysates	168
2.2. Preparation of tissue sample	169
2.3. Substrates of mTOR complexes	169
2.4. Monitoring phosphorylation status of mTOR substrates	170
2.5. mTOR activity *in vitro*	172
3. Concluding Remarks	177
Acknowledgment	178
References	178

Abstract

Mammalian target of rapamycin (mTOR) is an evolutionarily conserved serine/threonine protein kinase implicated in a wide array of cellular processes such as cell growth, proliferation, and survival. Analogous to the situation in yeast, mTOR forms two distinct functional complexes termed mTOR complex 1 and 2 (mTORC1 and mTORC2). mTORC1 activity is inhibited by rapamycin, a specific inhibitor of mTOR, whereas mTORC2 activity is resistant to short-term treatments with rapamycin. In response to growth factors, mTORC2 phosphorylates Akt, an essential kinase involved in cell survival. On the other hand, mTORC1 can be activated by both growth factors and nutrients such as glucose and amino acids. In turn, mTORC1 regulates the activity of the translational machinery by modulating S6 kinase (S6K) activity and eIF4E binding protein 1 (4E-BP1) through direct phosphorylation. Consequently, protein synthesis and cell growth are stimulated in a variety of different cell types. In addition, mTORC1 inhibits autophagy, an essential protein degradation and recycling system, which cells employ to sustain their viability in times of limited availability of nutrients. Recent studies have highlighted the fact that autophagy plays crucial

* Life Sciences Institute, University of Michigan, Ann Arbor, Michigan, USA
† Department of Molecular and Integrative Physiology, University of Michigan Medical School, Ann Arbor, Michigan, USA

roles in many aspects of human health including cancer development, neurodegenerative disease, diabetes, and aging. It is likely that dysregulation of the mTOR-autophagy pathway may contribute at least in part to these human disorders. Therefore, the assessment of mTOR activity is important to understand the status of autophagy in the cells being analyzed and its role in autophagy-related disorders. In this section, we describe methods to monitor mTOR activity both *in vitro* and *in vivo*.

1. INTRODUCTION

Mammalian target of rapamycin (mTOR) is a major constituent of the signaling pathways that regulate cell growth because of the control it exerts on translation, transcription, mRNA turnover, protein stability, actin cytoskeletal organization, and autophagy (Wullschleger *et al.*, 2006). As its name indicates, mTOR is a target of the drug called rapamycin (sirolimus), a macrolide antibiotics from *Streptomyces hygroscopicus*, which is also an FDA-approved immunosuppressant. To confer the effect of rapamycin on mTOR in cells, rapamycin forms a drug-receptor complex with the cellular protein FKBP12 (immunophilin FK506-binding protein 12) and then binds to mTOR, which prevents mTOR from phosphorylating currently known targets (Schmelzle and Hall 2000). mTOR is evolutionarily conserved and it contains a carboxy-terminal amino acid sequence with significant homology to the catalytic domain of phosphoinositide 3-kinase (PI3K) (Abraham 2004). However, similar to other members of the PI3K family such as ATM (ataxia telangiectasia mutated), ATR (ATM- and Rad3-related), and DNA-PKcs (DNA-dependent protein kinase catalytic subunit), mTOR is a genuine serine/threonine protein kinase.

mTOR exists in at least two different complexes, mTOR complex 1 and 2, mTORC1 and mTORC2, respectively (Loewith *et al.*, 2002; Sabatini 2006). These two mTOR complexes have distinct physiological functions and are regulated by different mechanisms. Importantly, mTORC1 activity is highly sensitive to inhibition by rapamycin, whereas mTORC2 activity is resistant at least after a relatively short treatment (Sarbassov *et al.*, 2006). mTORC1 exists as a multiprotein complex containing mTOR, Raptor, PRAS40 and mLST8. The mechanism by which rapamycin dominantly inhibits mTORC1 activity is that the rapamycin-FKBP12 complex preferably disrupts the interaction between mTOR and Raptor, an essential scaffolding protein that recruits mTORC1 substrates into the mTORC1 complex (Kim *et al.*, 2002; Oshiro *et al.*, 2004).

In response to growth stimuli, class I PI3K indirectly activates Akt, also known as PKB (protein kinase B). Stimulation of Akt inhibits the GAP (GTPase activating protein) activity of the TSC1/2 heterodimeric complex by phosphorylating TSC2 of the heterodimeric complex leading to the induction of GTP loading on the Rheb small GTPase (Garami et al., 2003; Inoki et al., 2003; Tee et al., 2003; Zhang et al., 2003). Subsequently, the GTP-Rheb complex activates mTORC1 by an unknown mechanism resulting in the promotion of translation, mRNA maturation, and cell growth via phosphorylating ribosomal S6 Kinase and 4E-BP (Long et al., 2005). Akt also directly phosphorylates PRAS40, a possible negative regulator of the mTORC1 complex and induces the dissociation of PRAS40 from the mTORC1 complex (Sancak et al., 2007; Vander Haar et al., 2007). Recently, it has been proposed that in response to amino acids, the Rag small GTPase complex binds to Raptor, a component of mTORC1, which stimulates mTOR by changing the localization of mTORC1 to endomembrane structures in the perinuclear region within cells where mTORC1 can be activated by Rheb (Sancak et al., 2008). However, whether the Rag-Raptor complex exists as an endogenous preformed complex needs to be further determined.

In spite of the body of the evidence regarding the regulation of the mTORC1 pathway, the regulation of the mTORC2 pathway is still not well understood. mTORC2 consists of mTOR, Rictor, mSin1, mLST8, and PRR5 and phosphorylates Akt Ser473 both *in vivo* and *in vitro* (Sarbassov et al., 2005). A series of the genetic studies convincingly demonstrates that loss of mTOR, Rictor, mSin1, or mLST8 abolishes Akt Ser473 phosphorylation in embryonic fibroblasts, indicating that in addition to mTOR kinase activity, mTORC2 integrity is essential for Akt S473 phosphorylation (Guertin et al., 2006; Jacinto et al., 2006; Shiota et al., 2006). Interestingly, loss of function of mTORC2 selectively attenuates FoxO phosphorylation among the Akt substrates (Jacinto et al., 2006). Although mTORC2 activity is enhanced by serum or growth factors, the molecular mechanism by which mTORC2 is activated by these stimuli remains elusive (Sarbassov et al., 2005).

Analogous to the situation in yeast, rapamycin or nutrient starvation contributes to a stimulation of autophagy in mammalian cells (Mizushima et al., 2008). This indicates that mTORC1 has a negative effect on the autophagic pathway, and when there is a shortage of nutrients, cells stimulate autophagy possibly through inhibiting mTORC1 activity to stay alive. Therefore, the activation of the lysosomal autophagic pathway in response to inhibition of mTORC1 or starvation conditions helps in the production of amino acids and other elements needed for biosynthetic pathways. Interestingly, a recent study has shown that rapamycin-insensitive mTORC2 has a role in the regulation of autophagy through the Akt-FOXO axis (Mammucari et al., 2007); FoxO3 is necessary and sufficient for the induction of autophagy in skeletal muscle. These studies indicate that both mTORC1 and mTORC2 might be involved in the regulation of autophagy. Therefore, the assessment of mTOR activity is critical to determine its role in the autophagy pathway.

2. METHODS

2.1. Cell culture and preparation of cell lysates

The best method to monitor mTOR activity in cells is Western blot analysis using phospho-specific antibodies (antibodies specific to phosphorylated residues of the target proteins) against the mTOR substrates. Growth factors and nutrients such as glucose and amino acids stimulate mTOR activity promoting substrate phosphorylation. Hence, to study the signaling events that are supposed to stimulate mTOR activity, the following general protocol to determine mTOR activity in HEK293 cells is provided (for additional details, see the subsequent protocol for rapamycin treatment):

1. Cells are generally grown in DMEM containing 10% fetal bovine serum (FBS). For serum starvation, cells are washed by exchanging the medium with PBS and then cultured in serum free media for 16 h.
2. Cells are then treated with the appropriate stimuli such as insulin (100 nM to 400 nM) for 15 min.
3. After stimulation, the cells are lysed for 10 min on ice with occasional shaking by hand.

Note: To avoid additional inputs for mTOR activity during harvesting of the cells, we normally put lysis buffer directly into the well without washing the cells with PBS.

4. The soluble components are then collected into 1.5-ml tubes using a pipette and centrifuged at 13,000 rpm for 10 min at 4 °C.
5. After centrifugation, the supernatant fractions are transferred into new 1.5-ml tubes. If the protein extracts will be resolved by SDS PAGE, the collected supernatant fractions need to be denatured immediately by boiling at 100 °C for 5 min in 1X SDS sample loading buffer containing a reducing reagent such as β-mercaptoethanol or DTT. Otherwise, extracts do not need to be denatured at once and can be used for other experiments such as immunoprecipitation (IP) and pull-down (CoIP) assays.

Following is the detailed protocol for protein extraction from cell culture for determining mTOR activity in the presence or absence of rapamycin treatment.

1. Plate 1×10^6 HEK293T cells into two 3.5-cm dishes a day before rapamycin treatment: one dish will be used for a no treatment control and the other dish will be treated with 5–20 nM rapamycin.
2. On the second day, treat cells with or without rapamycin (we generally use 20 nM) and incubate them at 37 °C in a humidified 5% CO_2

incubator for 30 min. Rapamycin is available from Sigma, Cell Signaling, or LC laboratory.
3. Place the cell culture dish on ice and completely remove the medium with mild aspiration, then add in 300 µl of cold lysis buffer (40 mM HEPES, pH 7.5, 120 mM NaCl, 1 mM EDTA, 10 mM pyrophosphate, 10 mM glycerophosphate, 50 mM NaF, 1.5 mM Na_3VO_4, 1% Triton X-100, and EDTA-free protease inhibitors (Roche, 14132300)). Leave on ice for 10 min with occasional shaking by hand.
4. Collect the supernatant fraction using a pipette and transfer into 1.5-ml tubes.
5. Centrifuge at 13,000 rpm for 10 min at 4 °C.
6. Remove the supernatant fractions and transfer into new 1.5-ml tubes. Add SDS sample loading buffer (stock 4X) to 1X working concentration.
7. Denature the proteins by incubating tubes containing the samples at 100 °C for 5 min.
8. Briefly centrifuge the tubes (13,000 rpm, 10 s) and either load the samples onto an SDS-PAGE gel or store at −20 °C for later analysis.

2.2. Preparation of tissue sample

To monitor the mTOR activity in animal tissues by Western blotting, we use the following protocol:

1. The tissue should be immediately lysed in lysis buffer (see section 2.1) to avoid the decrease of mTOR activity during the process of protein extraction from the tissue. The isolated tissues are transferred into cold lysis buffer (200 mg tissue/500 µl lysis buffer).
2. The tissues are homogenized in a Dounce homogenizer by repeated strokes (approximately 20 strokes) until the tissue is torn apart into very small pieces.
3. Centrifuge at 13,000 rpm for 15 min at 4 °C.
4. Transfer the supernatant fractions to new 1.5-ml tubes. It is recommended to avoid any contamination with debris, and especially fat, when the supernatant fractions are transferred into new tubes. Supernatant fractions containing extracted proteins from the tissues can be used for analyses or snap-frozen using liquid nitrogen followed by storage at −80 °C.

2.3. Substrates of mTOR complexes

p70S6K (Thr389), 4E-BP1 (Thr37, Thr46, Ser65, Thr70), and PRAS40 (Ser183) are known to be convincing direct substrates of mTORC1 (Brunn et al., 1997; Burnett et al., 1998; Gingras et al., 1999; Fonseca et al., 2007; Oshiro et al., 2007). The phosphorylation of S6K1 on Thr389 and 4E-BP1 on Thr37, Thr46, and Ser65 are often used as functional readouts of mTORC1 activity, as phosphorylation of these sites by mTORC1 has been confirmed both *in vitro* and *in vivo* and is inhibited by rapamycin treatment. Recently, it has been also reported that Ser183 and Ser221 of PRAS40 are also directly phosphorylated by mTORC1 (Fonseca et al., 2007; Oshiro et al., 2007; Wang et al., 2008). The phosphorylation of 4E-BP1 upon stimulation (growth factors, mitogens, and hormones) occurs at multiple sites in a hierarchical manner (first Thr37 and Thr46, then Thr70, and finally Ser65) (Gingras et al., 2001).

In contrast, Akt (Ser473) is the only substrate shown to be directly phosphorylated by mTORC2 (Sarbassov et al., 2005). The phosphorylation of Ser473 in a hydrophobic motif of Akt has been shown to be mediated by mTORC2 both *in vitro* and *in vivo* (Sarbassov et al., 2005). Although debates still exist as to whether mTORC2 is the sole kinase for Akt on Ser473, the observation that the levels of Akt Ser473 phosphorylation are indeed abolished in Rictor or hSin1 knockout cells allows this site to be employed for monitoring mTORC2 activation. PKCα (S657) is also reported to be regulated by mTORC2. However, the regulation seems to be rather indirect, as direct phosphorylation of PKCα (S657) by mTORC2 has not been successfully demonstrated (Sarbassov et al., 2004).

2.4. Monitoring phosphorylation status of mTOR substrates

To analyze the phosphorylation status of mTORC1 substrates, Western blot analysis using phospho-specific antibodies is performed. Phospho-S6K1 Thr389 (#9234) antibody and phospho-4E-BP1 Thr37/Thr46 (#9459), Ser65 (#9451), and Thr70 (#9455) antibodies are available from Cell Signaling (Fig. 11.1).

To examine the activation status of mTORC2 *in vivo*, Western blot analysis using phospho-Akt Ser473 antibody (#4058, #9271, Cell Signaling) is usually performed. Although mTORC2 enhances phosphorylation of PKCα Ser657 possibly in an indirect manner, it is not ideal to use this phosphorylation as a functional readout for mTORC2 activation, as phosphorylation of the site is quite stable and resistant to at least short-term treatment with stimuli and inhibitors (Ikenoue et al., 2008).

In addition, monitoring reduced mobility of the proteins by SDS PAGE analysis also can be an alternative to phospho-specific antibodies, due to the decrease in protein mobility associated with their posttranslational modifications, especially phosphorylation. To detect the mobility shift of S6K1 (70 kDa) and 4E-BP1 (22 kDa), it is ideal to use 10% and 15% SDS PAGE gels, respectively (Fig. 11.1). Optimal resolution can be achieved by using gels containing a lower concentration of methylene *bis*-acrylamide. The phosphorylation of 4E-BP1 can be often more clearly determined by mobility shift rather than by using a phospho-specific antibody.

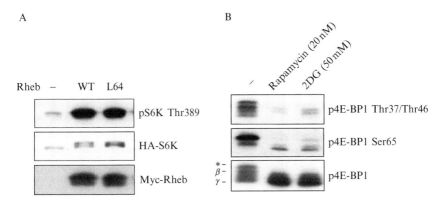

Figure 11.1 Monitoring of mTORC1 activity using phospho-specific antibodies for S6K1 and 4E-BP1 and mobility shifts of S6K1 and 4E-BP1 proteins. (A) HEK293 cells were transfected with HA-S6K1 together with Myc-tagged empty vector, wild-type Rheb (WT), or mutant Rheb (L64). After serum starvation, the cell lysates were analyzed by Western blot analysis using the indicated antibodies. (B) HEK293 cells were treated with the indicated concentration of rapamycin or 2-deoxyglucose (2DG), a nonmetabolizable d-glucose, for 15 min. Phosphorylation of endogenous 4E-BP1 proteins were determined by Western blot analysis using the indicated antibodies. $\star$, β, and γ denote hyper-, less-, and nonphosphorylated forms of 4E-BP1 proteins, respectively.

The 4E-BP1 proteins can be resolved into three bands (usually termed $\star$, β, and γ) with different mobilities that correspond to hyper-, less-, and nonphosphorylated forms, respectively (Fig. 11.1B).

To examine the effect of drugs on mTORC1 or mTORC2, phosphorylation of endogenous substrates for either mTOR complex can be used. In contrast, to analyze the effect of proteins of interest on mTORC1 and mTORC2, cotransfection of the plasmids encoding genes of interest and the substrates of mTOR complexes is frequently required, especially in cells with rather low transfection efficiency. For example, cotransfection of Myc-Rheb and HA-S6K1 can be used for assessing the effect of Rheb on mTORC1 activity (Fig. 11.1A).

The following is an example of how to monitor exogenous phospho-S6K1 Thr389 in HEK293T cells.

1. Plate HEK293T cells into a 6-well plate (30-50% confluency) a day before the transfection.
2. For the transfection, add 100 ng of pRK5 HA-S6K1 in a microcentrifuge tube and mix with 150 μl of Opti-MEM I (GIBCO) by vortexing briefly (2 s). Add 3 μl of lipofectamine (Invitrogen) in the tube and mix it with brief vortexing and incubate at room temperature (RT) for 30 min. Meanwhile, add 850 μl of Opti-MEM I after a wash with this medium in a 6-well plate.
3. Add transfection mixture into the well containing 850 μl of Opti-MEM I.

4. Incubate for 4–6 h at 37 °C in a humidified 5% CO_2 incubator and remove Opti-MEM I with the transfection mixture by aspiration. Then, add cell culture medium (10% FBS in DMEM with 1X penicillin/streptomycin).
5. 48-h posttransfection, the transfected cells can be treated with rapamycin (20 nM) to inhibit phosphorylation of S6K1 on Thr389. Cells cultured in the 6-well plate are lysed in 300 µl of mild lysis buffer (10 mM Tris·HCl, pH 7.5, 100 mM NaCl, 1% Nonidet P-40, 50 mM NaF, 2 mM EDTA, 1 mM PMSF, 10 µg/ml leupeptin, 10 µg/ml aprotinin) on ice for 10 min with occasional shaking.
6. Collect the supernatant fraction and centrifuge at 13,000 rpm for 10 min at 4 °C.
7. Transfer the clear supernatant fraction into new tubes and denature the proteins with SDS sample loading buffer.
8. Samples are resolved by 10% SDS PAGE and transferred onto PVDF membrane at 350 mA for 2 h at 4 °C. Before transfer, the PVDF membrane is activated by soaking it in methanol and washed in water briefly, and then washed in transfer buffer for 5 min.
9. Block the membrane with 5% nonfat dry milk in Tris-buffered saline (100 mM Tris-HCl, pH 7.5, 150 mM NaCl) containing 0.05% Tween 20 (TBST) for 1 h and incubate the membrane with phospho-S6K1 Thr389 antibody (#9234, Cell Signaling) overnight at 4 °C with gentle shaking, or use antibody against HA to see the mobility shift due to phosphorylation.
10. After several washes in TBST for a total of 30 min, the membrane is incubated with secondary antibody conjugated with horseradish peroxidase (NA934, GE Healthcare) for 1 h at RT with gentle shaking.
11. Following several washes with TBST for a total of 40 min, the membrane is treated briefly with ECL reagent and exposed to X-ray film until clear bands are observed after developing the film.

Rapamycin treatment is supposed to inhibit phosphorylation of S6K1 on Thr389 even in the presence of serum. Therefore, it is rare to see phospho-S6K1 bands in rapamycin-treated samples. If there are residual bands corresponding to S6K1 or phospho-S6K1 even after rapamycin treatment, approximately 2 h of treatment with rapamycin might be necessary to obtain complete inhibition of mTOR.

2.5. mTOR activity *in vitro*

2.5.1. Preparation of substrates for *in vitro* kinase assay

Recombinant S6K1 and 4E-BP1 proteins can be used as substrates for an *in vitro* mTORC1 kinase assay, whereas recombinant Akt can be used for the mTORC2 kinase assay. Recombinant proteins produced from bacteria, baculovirus, or mammalian cells can be used as substrates; however, those from baculovirus or mammalian cells work better as substrates in our experience. To prepare the substrates for mTORC1 or mTORC2 from mammalian cells, we usually transfect GST-S6K1 or GST-Akt expression vectors into HEK293T cells, treat the cells with rapamycin or PI3 kinase inhibitors such as LY294002 (Sigma) for 30 min to induce dephosphorylation of S6K1 on Thr398 or Akt on Ser473.

The following is a standard procedure to purify the GST-fusion proteins from cultured mammalian cells in a 10-cm dish.

1. HEK293T cells (50% confluence) are transfected with a mammalian expression GST-S6K1 or GST-Akt construct (2 μg DNA/10 cm dish) using Lipofectamine (Invitrogen).
2. 48 h posttransfection, the transfected cells are treated with rapamycin (20 nM for S6K1) or LY294002 (20 μM for S6K1 and Akt) for 30 min to induce dephosphorylation of the substrate proteins.
3. Cells are washed with ice-cold PBS and lysed with 1 ml of PBST buffer (PBS containing 1 mM EDTA, 0.1% β-mercaptoethanol, 1% TritonX-100, protease inhibitor mixture; Roche).
4. Collect the lysates into 1.5-ml tubes and centrifuge at 13,000 rpm for 15 min at 4 °C.
5. Transfer the clear supernatant fraction into new tubes and rotate with 50 μl of glutathione sepharose beads (Amersham Biosciences) for 4 h at 4 °C.
6. After incubation, GST-protein bound beads are washed 3 times with 1 ml of PBST and then once with glutathione-free GST buffer (50 mM Tris, pH 8.0, 150 mM NaCl, 1 mM DTT, 5 mM MgCl$_2$).
7. Elute the GST-proteins with 50 μl of GST buffer containing 10 mM reduced glutathione (Sigma) for 20 min at 4 °C. Repeat the elution at least twice.
8. Collected eluents are dialysed in at least 100 times volume of dialysis buffer (10 mM Tris, pH 7.5, 50 mM NaCl, 0.05% β-mercaptorthanol, 0.5 mM EDTA) for 4 h at 4 °C. The dialyzed solution is concentrated by microcon centrifugation (YM100, Millipore).
9. The purified protein and different concentrations of bovine serum albumin (BSA) are subjected to SDS-PAGE and stained with coomassie blue to determine the concentration of GST-protein. To obtain a sufficient amount of purified GST-fusion proteins for the mTOR kinase assay, we normally harvest the cells from at least ten 10-cm dishes.

2.5.2. Purification of mTOR complexes

Recent, landmark studies reveal that Raptor plays a critical role in the mTORC1 complex for recruiting mTOR substrates such as S6K1 and 4E-BP1 to mTORC1 (Hara *et al.*, 2002; Kim *et al.*, 2002; Schalm and

Blenis 2002; Schalm et al., 2003). The interaction between mTOR and Raptor is detergent sensitive. For instance, nonionic detergents such as NP-40 or Triton-X100 dissociate the mTOR-Raptor complex suggesting that hydrophobicity of the residues at the interface of these proteins plays a role in complex formation. Consistent with this idea, the interaction of these proteins can be well sustained in ionic detergent buffers such as CHAPS (3-[(3-cholamidopropyl) dimethylammonio]-1-propanesulfonate)-containing buffer (Hara et al., 2002; Kim et al., 2002). Interestingly, variations in the association of Raptor and mTOR can be seen when using CHAPS buffer for lysis and wash. For example, amino acid withdrawal (nutrient deprivation) enhances Raptor's binding to mTOR, whereas rapamycin treatment reduces the binding (Kim et al., 2002). Considering that both conditions impair phosphorylation of downstream substrates such as S6K1 on Thr389, it is notable that rapamycin and nutrient deprivation inhibit mTORC1 by distinct mechanisms.

In addition to the detergent conditions, the salt concentration in the wash buffer is also important. High-salt washes during mTORC1 purification increase the basal activity of mTORC1 (Sancak et al., 2007). In addition, PRAS40 can be dissociated from TORC1 by high-salt washes leading to an increase in the activity of mTORC1. However, recent studies have shown that PRAS40 can be directly phosphorylated by mTOR, suggestive of PRAS40 as a substrate of mTORC1 (Fonseca et al., 2007; Oshiro et al., 2007). This evokes the question of whether it is appropriate to retain PRAS40 in the purified mTORC1 complex to measure native mTORC1 kinase activity.

The mTOR-Rictor interaction in the mTORC2 complex is also diminished in 1% NP-40 containing buffer, but not in 0.3% CHAPS-containing buffer (Sarbassov et al., 2004). Although Rictor was originally identified as a component of the rapamycin-insensitive mTOR complex, long-term rapamycin treatment disrupts the binding of Rictor to mTOR in a cell type-specific manner (Sarbassov et al., 2006).

For the purification of mTORC1 and mTORC2, we use the following general protocol (see subsequently for further details):

1. Subconfluent cells are lysed in 1 ml of CHAPS lysis buffer (40 mM HEPES, pH 7.4, 120 mM NaCl, 2 mM EDTA, 0.3% CHAPS, 10 mM pyrophosphate, 10 mM glycerophosphate, 50 mM NaF, EDTA-free protease) containing protease and phosphatase inhibitors (Kim et al., 2002; Sarbassov and Sabatini 2005).
2. After collecting the lysates, 30 μl is saved for Western blot analysis as input controls.
3. The remaining lysates are used for the immunoprecipitation (IP). We usually incubate the lysates with antibody for 1–2 h at 4 °C with rocking. Anti-FRAP (mTOR) antibody (N-19; Santa Cruz) can be used for

immunoprecipitation of endogenous mTOR complexes and does not interfere with the kinase reaction.
4. To recover the antibody-antigen complex, 10 μl of a 50% slurry of protein G-sepharose (17–0618-01 GE Healthcare) are added and incubated for 2 h at 4 °C with rocking.
5. Immunoprecipitates are centrifuged at 10,000 rpm for 5 s and then washed with CHAPS lysis buffer 3 times without protease and phosphatase inhibitors.
6. After the final wash, the supernatant fractions are completely removed and the immunoprecipitates are subjected to the kinase reaction.

2.5.3. Reaction of mTOR kinase assay

The kinase assays are performed according to a modified protocol based on previous reports (Kim *et al.*, 2002; Sarbassov *et al.*, 2005; Yang *et al.*, 2006) described briefly here and in detail subsequently.

1. For kinase reactions, 200 ng of GST-S6K1 and 50 ng of His-Akt (#14-279, Upstate) can be used for the mTORC1 and mTORC2 kinase assay, respectively.
2. 250-500 μM ATP with or without 1 μCi of $[\gamma-^{32}P]$ ATP (for detection by autoradiography) is added in the kinase assay buffer (described subsequently) 10 min prior to the start of the kinase reaction.
3. The kinase reactions are stopped by adding 5 μl of 4X SDS sample buffer and resolved in 10% and 15% SDS-PAGE gels for S6K1 and Akt and for 4E-BP1, respectively, and transferred to PVDF membranes followed by autoradiography.

Immunoblotting using phospho-specific antibodies for mTOR-dependent sites in S6K1, 4E-BP1, and Akt described earlier can be performed to determine the kinase activity of mTOR in each immunoprecipitate instead of using radioactivity (Fig. 11.2). A summary of the purification of the TOR complexes with low-salt washes (using exogenously expressed mTOR components) and kinase assays is the following:

1. HEK293T cells are cultured in 10-cm plates.
2. When the cell density reaches 3×10^6 cells per plate, cells are transfected with either mTORC1 components (MYC-mTOR, HA-Raptor, and/or MYC-mLST8) or mTORC2 components (MYC-mTOR, HA-Rictor, MYC-Sin1, and/or MYC-mLST8). After 48 h transfection, the cells are lysed in CHAPS lysis buffer (40 mM HEPES, pH 7.4, 120 mM NaCl, 2 mM EDTA, 0.3% CHAPS, 10 mM pyrophosphate, 10 mM glycerophosphate, 50 mM NaF, EDTA-free protease) on ice.
3. To immunoprecipitate mTORC1 or mTORC2, one μg of anti-HA (for immunoprecipitation of HA-Raptor or HA-Rictor) antibody is added to

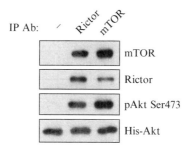

Figure 11.2 Determination of mTORC2 activity *in vitro*. HEK293 cells were grown in 10-cm dishes and lysed using CHAPS Lysis buffer (containing 0.3% CHAPS). The lysates were immunoprecipitated with Rictor or mTOR antibody or without antibody. The immunoprecipitates were subjected to an *in vitro* kinase assay using His-Akt (full-length) as a substrate. Phosphorylation of Akt Ser473 was analyzed using phospho-specific antibody. The amounts of mTOR and Rictor in each immunoprecipitate were also determined by immunoblotting.

each of the cellular lysates and incubated at 4 °C for 120 min with gentle rocking.
4. 10 μl of protein G Sepharose slurry (50%) are added to the lysates and incubated for another hour.
5. The sepharose beads are centrifuged for 15 s at 10,000 rpm. The immunoprecipitates are washed 3 times in low-salt wash buffer (40 mM HEPES, pH 7.4, 150 mM NaCl, 2 mM EDTA, 0.3% CHAPS, 10 mM pyrophosphate, 10 mM glycerophosphate, 50 mM NaF).
6. The immunoprecipitates are washed twice in kinase wash buffer (25 mM HEPES-KOH, pH 7.4, 20 mM KCl). The mTORC1 kinase assays, are performed for 30 min at 30 °C in a final volume of 15 μl consisting of the mTORC1 kinase buffer (25 mM HEPES-KOH, pH 7.4, 50 mM KCl, 10 mM MgCl$_2$, and 250 μM ATP). For the mTORC2 kinase assay, reactions are performed in the mTORC2 Kinase Buffer (25 mM HEPES-KOH, pH 7.4, 100 mM potassium acetate, 1 mM MgCl$_2$, and 500 μM ATP).
7. To stop the reaction, 5 μl of 4x SDS sample buffer are added to each reaction, which are then boiled for 5 min and the reactions are resolved by SDS PAGE and then visualized by autoradiography or Western blot analysis with phospho-specific antibodies for substrates used in the kinase assay.

2.5.4. Monitoring mTOR activity in tissues

To study mTOR activity in animal tissues by immunohistochemistry (IHC), the tissue samples are prepared as either a frozen or a paraffinized section. Because the mTOR activity is very sensitive to the availabilities of both growth factors and nutrients, the preparation for tissue samples should be done as promptly as possible. To obtain a paraffinized sample, the animals are first perfused with 4% paraformaldehyde, and fixed tissue are paraffinized, and embedded in paraffin. Sections of 2–10-μm thickness of sections can be cut using a microtome and stained with antibodies against phospho-S6K1, S6, Akt and mTOR. The following is a standard procedure for immunostaining the paraffin embedded sample using DAB (diaminobenzidine) detection.

1. Deparaffinize and hydrate sections by sequential washes in the following: xylene 3 min (3 times), 100% ethanol 1 min (twice), 95% and 80% ethanol 1 min each, and then wash in distilled water 2 min once. Rock gently in each solution.
2. Warm 300 ml of 10 mM sodium citrate, pH 6.0 to 95 °C–100 °C in a Coplin staining jar.
3. Quickly immerse slides into the hot citrate buffer and incubate for 20–40 min. The optimal incubation time needs to be determined empirically.
4. Cool the slides on the bench in citrate solution at RT for 30 min.
5. Rinse sections in PBS for 2 min twice.
6. Using a hydrophobic pen (S2002, Dako), surround the area to be stained.
7. Block sections in 2% BSA in PBS for 60 min at RT in a humidified chamber to prevent drying.
8. Remove the blocking solution.
9. Add primary antibody (IHC available p-S6 [#4857 Cell Signaling], p-Akt [#3787, #9266 Cell Signaling], or p-mTOR [#2976 Cell Signaling]) diluted in 1% BSA, 30–50 μl per section at the appropriate dilution as indicated by the supplier and incubate for 1 h at RT or overnight at 4 °C in a humidified chamber with gentle agitation.
10. Rinse with PBS for 3 min twice.
11. Block sections with peroxidase blocking solution (3% hydrogen peroxide) for 10 min.
12. Rinse in PBS for 5 min 3 times.
13. Add 20 μl of HRP-conjugated antirabbit secondary antibody (K4002 Dako) per section and incubate 1 h at RT.
14. Rinse in PBS for 10 min 3 times.
15. Add 20 μl of DAB solution (substrate buffer:chromogen = 50:1 [K3466, Dako]). DAB is oxidized and forms a stable brown end product at the site of the target antigen.

3. Concluding Remarks

The mTOR complexes are part of a major pathway that transduces growth factor signaling, leading to the induction of anabolic biosynthesis. Furthermore, regulation of mTOR also affects the autophagic pathway, which plays a major role in cell maintenance in the absence of nutrients. Recent studies have elucidated that both mTORC1 and mTORC2 are involved in the regulation of autophagy. We have introduced methods to evaluate the activity of mTORC1 and mTORC2 *in vivo* and *in vitro*. *In vivo* activities of mTORC1 and mTORC2 can be measured by the phosphorylation status of their downstream targets using Western blot analysis with cell lysates. On the other hand, *in vitro* kinase assays using immunoprecipitated mTORC1 and mTORC2 and their substrates are used for assessing their direct activities on targets *in vitro*. Given that mTOR is the critical regulator of the autophagic pathway, it is useful to establish the methods to measure the activity of mTOR in response to various stimuli, which possibly affect the autophagic pathway. This approach for monitoring mTOR activity will be helpful to resolve the role of proteins that might be involved in the mTOR-involved autophagic pathway. Although the mTOR pathway plays a critical role in the regulation of autophagy in a wide array of organisms, it is noteworthy that mTOR-independent regulation of the autophagy pathway has been postulated (Sarkar *et al.*, 2007; Williams *et al.*, 2008). Furthermore, recent studies also suggest that autophagy can be upstream of mTORC1 activity. For instance, the autophagy-related gene 1 (*ATG1*), an essential kinase for vesicle formation during autophagy, inhibits mTOR1 activity, indicative of another direction in the cross talk between mTORC1 and autophagy (Neufeld 2007). It is certain that current studies that address the molecular mechanism of regulation between the mTOR pathway and autophagy will reveal the detailed mechanism that controls this complex pathway in the mammalian system.

ACKNOWLEDGMENT

We thank Dr. Jennifer Aurandt for critical reading and discussion.

REFERENCES

Abraham, R. T. (2004). PI 3-kinase related kinases: 'big' players in stress-induced signaling pathways. *DNA Repair (Amst.)* **3**(8–9), 883–887.

Brunn, G. J., Fadden, P., Haystead, T. A., and Lawrence, J. C. Jr. (1997). The mammalian target of rapamycin phosphorylates sites having a (Ser/Thr)-Pro motif and is activated by antibodies to a region near its COOH terminus. *J. Biol. Chem.* **272**(51), 32547–32550.

Burnett, P. E., Barrow, R. K., Cohen, N. A., Snyder, S. H., and Sabatini, D. M. (1998). RAFT1 phosphorylation of the translational regulators p70 S6 kinase and 4E-BP1. *Proc. Natl. Acad. Sci. USA* **95**(4), 1432–1437.

Fonseca, B. D., Smith, E. M., Lee, V. H., MacKintosh, C., and Proud, C. G. (2007). PRAS40 is a target for mammalian target of rapamycin complex 1 and is required for signaling downstream of this complex. *J. Biol. Chem.* **282**(34), 24514–24524.

Garami, A., Zwartkruis, F. J., Nobukuni, T., Joaquin, M., Roccio, M., Stocker, H., Kozma, S. C., Hafen, E., Bos, J. L., and Thomas, G. (2003). Insulin activation of Rheb, a mediator of mTOR/S6K/4E-BP signaling, is inhibited by TSC1 and 2. *Mol. Cell* **11**(6), 1457–1466.

Gingras, A. C., Gygi, S. P., Raught, B., Polakiewicz, R. D., Abraham, R. T., Hoekstra, M. F., Aebersold, R., and Sonenberg, N. (1999). Regulation of 4E-BP1 phosphorylation: A novel two-step mechanism. *Genes Dev.* **13**(11), 1422–1437.

Gingras, A. C., Raught, B., Gygi, S. P., Niedzwiecka, A., Miron, M., Burley, S. K., Polakiewicz, R. D., Wyslouch-Cieszynska, A., Aebersold, R., and Sonenberg, N. (2001). Hierarchical phosphorylation of the translation inhibitor 4E-BP1. *Genes Dev.* **15**(21), 2852–2864.

Guertin, D. A., Stevens, D. M., Thoreen, C. C., Burds, A. A., Kalaany, N. Y., Moffat, J., Brown, M., Fitzgerald, K. J., and Sabatini, D. M. (2006). Ablation in mice of the mTORC components *raptor*, *rictor*, or *mLST8* reveals that mTORC2 is required for signaling to Akt-FOXO and PKCalpha, but not S6K1. *Dev. Cell* **11**(6), 859–871.

Hara, K., Maruki, Y., Long, X., Yoshino, K., Oshiro, N., Hidayat, S., Tokunaga, C., Avruch, J., and Yonezawa, K. (2002). Raptor, a binding partner of target of rapamycin (TOR), mediates TOR action. *Cell* **110**(2), 177–189.

Ikenoue, T., Inoki, K., Yang, Q., Zhou, X., and Guan, K.-L. (2008). Essential function of TORC2 in PKC and Akt turn motif phosphorylation, maturation and signalling. *EMBO J.* **27**(14), 1919–1931.

Inoki, K., Li, Y., Xu, T., and Guan, K.-L. (2003). Rheb GTPase is a direct target of TSC2 GAP activity and regulates mTOR signaling. *Genes Dev.* **17**(15), 1829–1834.

Jacinto, E., Facchinetti, V., Liu, D., Soto, N., Wei, S., Jung, S. Y., Huang, Q., Qin, J., and Su, B. (2006). SIN1/MIP1 maintains rictor-mTOR complex integrity and regulates Akt phosphorylation and substrate specificity. *Cell* **127**(1), 125–137.

Kim, D. H., Sarbassov, D. D., Ali, S. M., King, J. E., Latek, R. R., Erdjument-Bromage, H., Tempst, P., and Sabatini, D. M. (2002). mTOR interacts with raptor to form a nutrient-sensitive complex that signals to the cell growth machinery. *Cell* **110**(2), 163–175.

Loewith, R., Jacinto, E., Wullschleger, S., Lorberg, A., Crespo, J. L., Bonenfant, D., Oppliger, W., Jenoe, P., and Hall, M. N. (2002). Two TOR complexes, only one of which is rapamycin sensitive, have distinct roles in cell growth control. *Mol. Cell* **10**(3), 457–468.

Long, X., Lin, Y., Ortiz-Vega, S., Yonezawa, K., and Avruch, J. (2005). Rheb Binds and Regulates the mTOR Kinase. *Curr. Biol.* **15**(8), 702–713.

Mammucari, C., Milan, G., Romanello, V., Masiero, E., Rudolf, R., Del Piccolo, P., Burden, S. J., Di Lisi, R., Sandri, C., Zhao, J., Goldberg, A. L., Schiaffino, S., *et al.* (2007). FoxO3 controls autophagy in skeletal muscle *in vivo*. *Cell. Metab.* **6**(6), 458–471.

Mizushima, N., Levine, B., Cuervo, A. M., and Klionsky, D. J. (2008). Autophagy fights disease through cellular self-digestion. *Nature* **451**(7182), 1069–1075.

Neufeld, T. P. (2007). Contribution of Atg1-dependent autophagy to TOR-mediated cell growth and survival. *Autophagy* **3**(5), 477–479.

Oshiro, N., Takahashi, R., Yoshino, K., Tanimura, K., Nakashima, A., Eguchi, S., Miyamoto, T., Hara, K., Takehana, K., Avruch, J., Kikkawa, U., and Yonezawa, K. (2007). The proline-rich Akt substrate of 40 kDa (PRAS40) is a physiological substrate of mammalian target of rapamycin complex 1. *J. Biol. Chem.* **282**(28), 20329–20339.

Oshiro, N., Yoshino, K., Hidayat, S., Tokunaga, C., Hara, K., Eguchi, S., Avruch, J., and Yonezawa, K. (2004). Dissociation of raptor from mTOR is a mechanism of rapamycin-induced inhibition of mTOR function. *Genes Cells* **9**(4), 359–366.

Sabatini, D. M. (2006). mTOR and cancer: Insights into a complex relationship. *Nat. Rev. Cancer* **6**(9), 729–734.

Sancak, Y., Peterson, T. R., Shaul, Y. D., Lindquist, R. A., Thoreen, C. C., Bar-Peled, L., and Sabatini, D. M. (2008). The Rag GTPases Bind Raptor and Mediate Amino Acid Signaling to mTORC1. *Science* **320**, 1496–1501.

Sancak, Y., Thoreen, C. C., Peterson, T. R., Lindquist, R. A., Kang, S. A., Spooner, E., Carr, S. A., and Sabatini, D. M. (2007). PRAS40 is an insulin-regulated inhibitor of the mTORC1 protein kinase. *Mol. Cell* **25**(6), 903–915.

Sarbassov, D. D., Ali, S. M., Kim, D. H., Guertin, D. A., Latek, R. R., Erdjument-Bromage, H., Tempst, P., and Sabatini, D. M. (2004). Rictor, a novel binding partner of mTOR, defines a rapamycin-insensitive and raptor-independent pathway that regulates the cytoskeleton. *Curr. Biol.* **14**(14), 1296–1302.

Sarbassov, D. D., Ali, S. M., Sengupta, S., Sheen, J. H., Hsu, P. P., Bagley, A. F., Markhard, A. L., and Sabatini, D. M. (2006). Prolonged rapamycin treatment inhibits mTORC2 assembly and Akt/PKB. *Mol. Cell* **22**(2), 159–168.

Sarbassov, D. D., Guertin, D. A., Ali, S. M., and Sabatini, D. M. (2005). Phosphorylation and regulation of Akt/PKB by the rictor-mTOR complex. *Science* **307**(5712), 1098–1101.

Sarbassov, D. D., and Sabatini, D. M. (2005). Redox regulation of the nutrient-sensitive raptor-mTOR pathway and complex. *J. Biol. Chem.* **280**(47), 39505–39509.

Sarbassov, D. D., Ali, S. M., Sengupta, S., Sheen, J. H., Hsu, P. P., Bagley, A. F., Markhard, A. L., and Sabatini, D. M. (2006). Prolonged rapamycin treatment inhibits mTORC2 assembly and Akt/PKB. *Mol. Cell* **22**(2), 159–168.

Sarkar, S., Davies, J. E., Huang, Z., Tunnacliffe, A., and Rubinsztein, D. C. (2007). Trehalose, a novel mTOR-independent autophagy enhancer, accelerates the clearance of mutant huntingtin and α-synuclein. *J. Biol. Chem.* **282**(8), 5641–5652.

Schalm, S. S., and Blenis, J. (2002). Identification of a conserved motif required for mTOR signaling. *Curr. Biol.* **12**(8), 632–639.

Schalm, S. S., Fingar, D. C., Sabatini, D. M., and Blenis, J. (2003). TOS motif-mediated raptor binding regulates 4E-BP1 multisite phosphorylation and function. *Curr. Biol.* **13**(10), 797–806.

Schmelzle, T., and Hall, M. N. (2000). TOR, a central controller of cell growth. *Cell* **103**(2), 253–262.

Shiota, C., Woo, J. T., Lindner, J., Shelton, K. D., and Magnuson, M. A. (2006). Multiallelic disruption of the rictor gene in mice reveals that mTOR complex 2 is essential for fetal growth and viability. *Dev. Cell* **11**(4), 583–589.

Tee, A. R., Manning, B. D., Roux, P. P., Cantley, L. C., and Blenis, J. (2003). Tuberous sclerosis complex gene products, tuberin and hamartin, control mTOR signaling by acting as a GTPase-activating protein complex toward Rheb. *Curr. Biol.* **13**(15), 1259–1268.

Vander Haar, E., Lee, S. I., Bandhakavi, S., Griffin, T. J., and Kim, D. H. (2007). Insulin signalling to mTOR mediated by the Akt/PKB substrate PRAS40. *Nat. Cell Biol.* **9**(3), 316–323.

Wang, L., Harris, T. E., and Lawrence, J. C. Jr. (2008). Regulation of Proline-rich Akt Substrate of 40 kDa (PRAS40) Function by mammalian Target of Rapamycin Complex 1 (mTORC1)-mediated phosphorylation. *J. Biol. Chem.* **283**(23), 15619–15627.

Williams, A., Sarkar, S., Cuddon, P., Ttofi, E. K., Saiki, S., Siddiqi, F. H., Jahreiss, L., Fleming, A., Pask, D., Goldsmith, P., O'Kane, C. J., Floto, R. A., *et al.* (2008). Novel targets for Huntington's disease in an mTOR-independent autophagy pathway. *Nat. Chem. Biol.* **4**(5), 295–305.

Wullschleger, S., Loewith, R., and Hall, M. N. (2006). TOR signaling in growth and metabolism. *Cell* **124**(3), 471–484.

Yang, Q., Inoki, K., Kim, E., and Guan, K.-L. (2006). TSC1/TSC2 and Rheb have different effects on TORC1 and TORC2 activity. *Proc. Natl. Acad. Sci. USA* **103**(18), 6811–6816.

Zhang, Y., Gao, X., Saucedo, L. J., Ru, B., Edgar, B. A., and Pan, D. (2003). Rheb is a direct target of the tuberous sclerosis tumour suppressor proteins. *Nat. Cell. Biol.* **5**(6), 578–581.

CHAPTER TWELVE

Monitoring Autophagic Degradation of p62/SQSTM1

Geir Bjørkøy,* Trond Lamark,* Serhiy Pankiv,* Aud Øvervatn,* Andreas Brech,[†] *and* Terje Johansen*

Contents

1. Introduction	182
2. Monitoring Autophagy-Mediated Degradation of Endogenous p62/SQSTM1	183
2.1. Species-specificity and epitope-mapping of commercially available anti-p62 antibodies	183
2.2. Monitoring the level of p62 by Western blot	185
2.3. Pulse-chase measurement of p62 half-life	187
2.4. Immunofluorescence analysis of p62	190
2.5. Immuno-electron microscopy of p62	191
3. Monitoring the Autophagic Degradation of p62/SQSTM1 by Live Cell Imaging	192
3.1. Ectopic overexpression of p62	192
3.2. Live cell imaging of p62 using the pH-sensitive mCherry-GFP double tag	193
4. Concluding Remarks	194
Acknowledgments	195
References	195

Abstract

The p62 protein, also called sequestosome 1 (SQSTM1), is a ubiquitin-binding scaffold protein that colocalizes with ubiquitinated protein aggregates in many neurodegenerative diseases and proteinopathies of the liver. The protein is able to polymerize via an N-terminal PB1 domain and can interact with ubiquitinated proteins via the C-terminal UBA domain. Also, p62/SQSTM1 binds directly to LC3 and GABARAP family proteins via a specific sequence motif. The protein is itself degraded by autophagy and may serve to link ubiquitinated proteins to the autophagic machinery to enable their degradation in the lysosome.

* Biochemistry Department, Institute of Medical Biology, University of Tromsø, Norway
[†] Department of Biochemistry, Institute for Cancer Research, Norwegian Radium Hospital, Oslo, Norway

Since p62 accumulates when autophagy is inhibited, and decreased levels can be observed when autophagy is induced, p62 may be used as a marker to study autophagic flux. Here, we present several protocols for monitoring autophagy-mediated degradation of p62 using Western blots, pulse-chase measurement of p62 half-life, immunofluorescence and immuno-electron microscopy, as well as live cell imaging with a pH-sensitive mCherry-GFP double tag. We also present data on species-specificity and map the epitopes recognized by several commercially available anti-p62 antibodies.

1. Introduction

p62, also called sequestosome 1 (SQSTM1) in humans, A170 in mouse, ZIP in rats, and Ref 2(p) in *Drosophila melanogaster*, accumulates in cytoplasmic and nuclear ubiquitinated protein aggregates formed in various neurodegenerative diseases, liver diseases, and myofibrillar myopathies (Kuusisto *et al*., 2001, 2007; Olive *et al*., 2008; Zatloukal *et al*., 2002). p62 possesses a short LC3 interaction region (LIR) that facilitates direct interaction with LC3- and GABARAP-family members and causes p62 to be specifically degraded by autophagy (Komatsu *et al*., 2007a; Pankiv *et al*., 2007). The protein is expressed in most cells and tissues and serves as a scaffold or adapter protein in several signaling pathways (reviewed in Seibenhener *et al*., 2007). Because its degradation is dependent on autophagy, the level of p62 increases in response to inhibition of autophagy (Bjørkøy *et al*., 2005). p62 contains an N-terminal PB1 domain that enables efficient formation of p62 polymers where an acidic surface of one p62 PB1 domain binds to a basic surface in the next p62 PB1 domain (Lamark *et al*., 2003; Wilson *et al*., 2003). The PB1 domain seems to be needed for the autophagic degradation of p62. The presence of a C-terminal ubiquitin-associated domain (UBA) that binds mono- and polyubiquitin noncovalently enables p62 to act as a cargo adapter for ubiquitinated proteins that can be degraded by autophagy. In addition, p62 may ferry ubiquitinated substrates to the proteasome (Seibenhener *et al*., 2004). The level and subcellular localization pattern of p62 can be easily monitored by Western blotting or immunostaining. The protein accumulates in cells and tissues from autophagy-deficient mice (Komatsu *et al*., 2007a; Nakai *et al*., 2007; Wang *et al*., 2006). Accumulation of p62 has been used as a marker for inhibition of autophagy or defects in autophagic degradation (Bjørkøy *et al*., 2005; Filimonenko *et al*., 2007; Gal *et al*., 2007; Hara *et al*., 2008; Komatsu *et al*., 2007a,b; Lee *et al*., 2008; Maclean *et al*., 2008; Mizushima and Yoshimori, 2007; Settembre *et al*., 2008; Wang *et al*., 2006). It is also possible to use p62 as a marker for autophagy induction (Marino *et al*., 2008; Mizushima and Yoshimori, 2007; Tasdemir *et al*., 2008).

Here we describe methods to evaluate the localization and level of endogenous and ectopically expressed p62 as a function of autophagic activity in fixed or live cells and in cell lysates using different commercially available p62 antibodies. Autophagy of ectopically expressed p62 can be visualized by live cell imaging using a tandem fusion tag of mCherry and EGFP whose fluorescent properties change as a result of changes in pH (see also the chapter by Kimura *et al.*, in this volume).

2. MONITORING AUTOPHAGY-MEDIATED DEGRADATION OF ENDOGENOUS P62/SQSTM1

The subcellular localization and level of endogenous p62 can easily be determined by immunostaining of cells and Western blotting of cell lysates. There are good commercially available antibodies that work well in standard protocols for Western blotting, immunofluorescence, and immunoprecipitation. Knockdown of endogenous p62 using siRNA provides a nice control to evaluate the specificity of the antibodies. We have used the following siRNA 5′-GCATTGAAGTTGATATCGAT-3′ (Dharmacon ON-TARGET plus) at 20 nM transfected 2 times with 24-h intervals in HeLa cells using Lipofectamine 2000 (Invitrogen). When we analyze 24 h after the last transfection, we routinely observe efficient knockdown of the protein judged as loss of fluorescent signal in more than 80% of the cells and a more than 80% reduction of the p62 band on Western blots of cell lysates compared to transfection with control siRNA. As control siRNA we have used 20 nM nontargeting siRNA (Dharmacon, cat. no. 001210-01).

2.1. Species-specificity and epitope-mapping of commercially available anti-p62 antibodies

The amino acid sequence of p62 is highly conserved among the higher eukaryotes. We have evaluated some of the commercially available mono- and polyclonal antibodies raised against human p62 for their ability to detect mouse p62. We expressed GFP-tagged human p62 in HeLa cells and mouse embryonic fibroblasts and GFP-tagged mouse p62 in HeLa cells. The monoclonal p62 antibody from BD Transduction Laboratories (clone 3, cat. no. 610832) as well as the monoclonal antibody from Santa Cruz Biotechnology (clone D-3, cat. no. sc-28359) efficiently recognize human p62 (Fig. 12.1A). However, neither of the two monoclonal antibodies recognizes mouse p62; neither the endogenous protein in mouse fibroblasts nor mouse p62 overexpressed in HeLa cells (Fig. 12.1A). On the other hand, the monoclonal antibody from Abnova (clone 2C11, cat. no. H00008878-M01) efficiently recognizes both human and mouse p62.

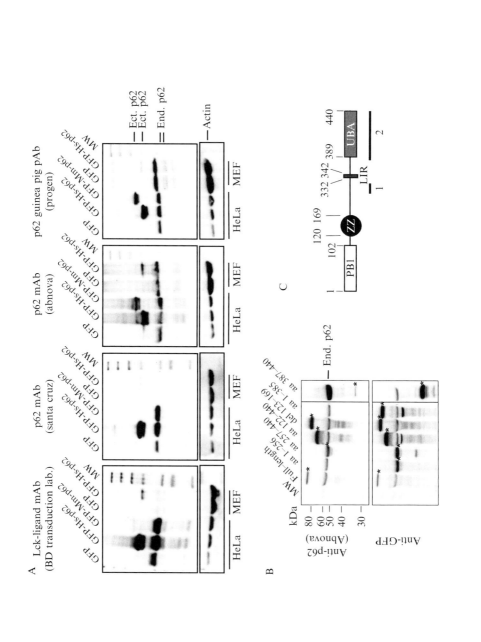

As expected, the guinea pig polyclonal antibody (Progen Biotechnik, cat. no. GP62-C) raised against the highly conserved C-terminal 20 amino acids of human p62 recognizes both human and mouse p62.

The two monoclonal anti-p62 antibodies, clone 3 (BD Transduction Laboratories) and clone D-3 (Santa Cruz Biotechnology), are raised against human p62 amino acids 257–437 and 151–440, respectively. By expressing deletion constructs of human p62 we find the epitope for both of these antibodies to be located to amino acids 303–320 of human p62 (Pankiv et al., in prep). Within this amino acid stretch there are 4 amino acids that differ between mouse and human p62. The monoclonal anti-p62 antibody clone 2C11 from Abnova is raised against recombinant full-length human p62 protein. Expression of deletion constructs of p62 identifies the C-terminal 54 amino acids containing the UBA domain as necessary and sufficient for antibody recognition (Fig. 12.1B). Thus, the Abnova monoclonal antibody clone 2C11 does not discriminate between human and mouse p62 and can be used to detect expression constructs that contain the UBA domain of p62 from these species. The dilutions of the antibodies used for western blot and immunostaining of cells are given below in sections 2.2 and 2.4, respectively.

2.2. Monitoring the level of p62 by Western blot

For estimation of p62 levels in cell extracts, we use the following protocol:

1. Approximately 10^5 cells are lysed directly in 100 μl of 2xSDS-PAGE gel loading buffer (125 mM Tris-HCl, pH 7.4, 4% SDS, 0.04% bromophenol blue, 8% sucrose and 30 mg per ml DTT, added immediately before use).

Figure 12.1 Determination of species-specificity and epitope mapping of commercially available anti-p62 antibodies. (A) Human and mouse p62 cDNA fused to the C terminus of GFP expressed in HeLa cells and mouse embryonic fibroblasts (MEFs). The indicated GFP-p62 fusions of human (Hs-p62) and mouse p62 (Mm-p62) constructs were transiently expressed in HeLa and MEFs. The Western blots were developed with the indicated p62 antibodies (all diluted 1:1000) and reprobed with an anti-actin antibody (1:1000 dilution). (B) Detection of the indicated deletion constructs of GFP-p62 to map the epitope of the Abnova (clone 2C11) anti-p62 monoclonal antibody (diluted 1:1000). The membrane was stripped and reprobed with an anti-GFP antibody (Abcam, ab 290, 1:2000 dilution) to control for the expression of the different constructs. Note that the upper right insert is from a longer exposure due to the low expression level of GFP-p62 amino acids 385-440. (C) Schematic illustration showing the location of the PB1, ZZ zinc-finger and UBA domains, as well as the LC3-interacting region (LIR) of human p62. The amino acid positions defining the borders of the domains are indicated. The location of the epitope (amino acid positions 303–320) recognized by the human-specific monoclonal anti-p62 antibodies from BD Transduction Laboratories and Santa Cruz Biotechnology is indicated (1). The monoclonal anti-p62 antibody from Abnova recognizes the UBA domain of both human and mouse p62 (2).

2. The lysates are sonicated briefly 3 times for 5 s using output 5 on a W-370 sonicator equipped with a minitip (Heat systems, Ultrasonics) followed by boiling for 5 min.
3. The lysate (20 μl) is separated by electrophoresis using an 8% polyacrylamide gel and electrotransferred to a Hybond-ECL nitrocellulose membrane (Amersham Biosciences).
4. Nonspecific labeling is blocked by incubating the membrane in TBS-T (150 mM NaCl, 10 mM Tris-HCl, pH 7.4, 0.1% Tween-20) containing 5% (weight/volume) nonfat dry milk powder for 1 h.
5. To reduce the amount of primary and secondary antibody used, the membranes are placed in a 50-ml capped plastic tube and 1.5-ml antibody solution. We routinely use the following dilutions of primary anti-p62 antibodies; 1:2500 dilution of the monoclonal antibody from Abnova, 1:1000 dilution of the monoclonal antibodies from BD Transduction Laboratories or Santa Cruz Biotechnology, or 1:5000 dilution of the polyclonal antibody from Progen Biotechnik. The appropriate primary antibody is diluted in TBS-T and incubated on a vertical rotating wheel for 1 h at room temperature or overnight at 4 °C.
6. Unbound primary antibodies are removed by washing 5 times for 5 min with TBS-T.
7. Horseradish peroxidase-conjugated secondary antibodies (goat-anti-mouse IgG$_1$ from BD Pharmingen, cat. no. 554002 and goat anti-guinea pig IgG$_1$ Santa Cruz Biotechnology, cat no. SC-2438) are diluted 1:1000 in TBS-T, incubated for 1 h at room temperature and unbound antibodies removed by washing as previously.
8. The membrane is developed using chemiluminescence following the supplier's instructions and imaged using a Lumi-Imager F1 (Roche Molecular Biology).

To normalize for protein loading, the membrane is redeveloped using a rabbit polyclonal anti-β-actin antibody (Sigma, cat. no. A2066). Relative band intensities in the 16-bit gray-scale image is estimated using the Lumi-Imager software. The image is converted to an 8-bit gray-scale image and processed using Canvas software.

To monitor the rate of autophagic degradation of p62, we perform Western blotting with extracts isolated from cells treated or not treated with the lysosomal inhibitor bafilomycin A$_1$ at 200 nM for 2–18 h. Bafilomycin A$_1$ causes an increase in the relative p62 level compared to loading controls such as β-actin, as the inhibitor prevents degradation of p62 molecules that are recruited into autophagic structures. The difference in p62 level between cells treated with bafilomycin A$_1$ and untreated cells represents the amount of p62 that has been recruited into autophagic vesicles during the period of treatment. Although this method is less reliable than a pulse-chase measurement (see subsequent sections), the assay is faster and easier to perform. Generally we have observed a good correlation between the half-life of p62 measured by pulse-chase measurement and the accumulation of p62 seen by Western blotting after treatment of cells with bafilomycin A_1.

2.3. Pulse-chase measurement of p62 half-life

It is difficult to definitively determine the rate of autophagy by counting the number of autophagic structures in cells, or the levels of autophagic substrates, such as p62, in cell lysates. These methods do not discriminate between changes due to altered rates of synthesis or degradation. Increased levels of p62 may formally be caused by either increased synthesis or decreased degradation. However, pulse-chase labeling of cellular proteins with radioactive amino acids allows the determination of their half-lives and how their rates of degradation change with time depending on the experimental conditions. We have previously found that p62 has a half-life of approximately 6 h in HeLa cells under normal growth conditions (Bjørkøy et al., 2005). Here, we compared the degradation of p62 under normal growth conditions and after amino acid starvation (Fig. 12.2).

1. The cells are plated at 2.5×10^5 cells per 6-cm diameter culture dish approximately 24-h before labeling.
2. To label the cells, the normal medium is removed by aspiration and the cultures washed 2 times in PBS (37 °C) followed by a 1-h incubation in the CO_2-incubator in methionine-free labeling medium (Sigma cat. no. D0422) containing 10% dialyzed fetal calf serum (stagger the labeling times to allow simultaneous processing of all samples; see the note below).
3. The medium is removed and replaced with 0.5 ml of labeling medium containing 10% dialyzed fetal calf serum and ^{35}S-methionine (50 μCi ^{35}S-methionine per dish).
4. The cells are labeled in the CO_2-incubator placed in a separate partially open container for 30 min. The dishes are carefully swirled every 5 min during labeling.
5. Following labeling, the medium is removed and the dishes are carefully washed twice in PBS (37 °C), containing 2 ml of normal medium (MEM, GIBCO cat. no. 31095) and 10% fetal calf serum, or 2 ml of HANKS medium (to induce autophagy) with dialyzed 10% fetal calf serum and incubated for different times.
6. Following appropriate incubation times, the cell cultures are washed twice in PBS, and trypsinized (0.1 ml of 0.25% trypsin from Sigma and 0.005% EDTA in PBS, pH 7.5, per dish) for 5 min.
7. Next, 2 ml of MEM with 10% fetal calf serum is added and the cell suspension is transferred to centrifugation tubes.

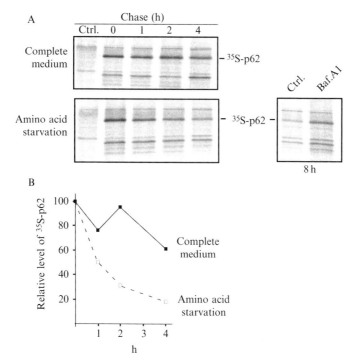

Figure 12.2 Amino acid starvation speeds up the autophagic degradation of p62 in HeLa cells. (A) Autoradiography images of electro-blotted membranes following SDS-PAGE of immunopurified ^{35}S-methionine pulse-labeled p62 from extracts of cells incubated in either complete MEM medium or HANKS medium (amino acid starvation) for the indicated times. Immunoprecipitation of p62 was performed using 1 μg/lysate of the monoclonal anti-p62 antibody from BD Transduction Laboratories. As a negative control (Ctrl.), an extract made immediately after labeling was processed as for p62 but using 1 μg of the anti-myc tag antibody (clone 9E10, Santa Cruz Biotechnology). Two dishes were pulse-labeled and left for 8 h in amino acid–free medium either with or without Bafilomycin A_1 (200 nM) to control for the lysosomal dependence of the p62 degradation (lower right panel). (B) Relative levels of immunopurified p62 (from A). Following quantification of the radioactive signals, the membranes were processed for Western blotting to verify that equal amounts of total p62 had been immunopurified from the cell extracts.

8. The cells are collected by centrifugation at 1200 rpm (200×g) for 5 min using a swing-out rotor for 15-ml tubes (MSE minor centrifuge).
9. The supernatant fraction is discarded and the cells are lysed by resuspending the pellet in 400 μl of ice-cold lysis buffer (50 mM Tris-HCl, pH 7.5, 150 mM NaCl, 2 mM EDTA, 1 mM EGTA, 1% Triton X-100) and incubated on ice for 15 min. The lysis buffer contains protease inhibitor, one complete minitablet (Roche, cat. no. 11836170001) per 10 ml of lysis buffer, and phosphatase inhibitor cocktail set II (Calbiochem, cat. no. 524625) to 1% final concentration.

10. The lysates are precleared by centrifugation for 15 min at 13,000 rpm in a microcentrifuge at 4 °C (Biofuge 13, Heraeus).
11. The supernatant fraction (380 µl) is transferred to new 1.5-ml microcentrifuge tubes and incubated for 20 min on a rotating wheel at 4 °C with 20 µl of a 50% slurry of protein G-coupled agarose beads (Santa Cruz Biotechnology cat. no. sc-2002) to remove proteins that bind unspecifically to the beads.
12. Following centrifugation (30 s, 13,000 rpm), 380 µl of the supernatant is carefully transferred to new 1.5-ml microcentrifuge tubes and 1 µg of the monoclonal p62 antibody from BD Transduction Laboratories is added followed by incubation on the rotating wheel overnight at 4 °C. As a negative control, a lysate from cells harvested immediately after labeling is immunoprecipitated using 1 µg anti myc-tag antibody (clone 9E10, Santa Cruz Biotechnology, cat. no. sc-40) that is of the same isotype (IgG1) as the monoclonal p62 antibody used. As a control for the requirement of lysosomal activity for degradation of p62 in response to amino acid starvation, two 6-cm dishes with cells are incubated for 8 h after labeling in amino acid–free medium in the presence or absence of bafilomycin A_1 (200 nM).
13. After incubation with p62 or control antibodies, the immunocomplexes are collected by adding 20 µl of a 50% slurry of protein G-coupled agarose beads, and incubating for 1 h at 4 °C on the rotating wheel. The beads are collected by centrifugation (30 s at 13,000 rpm in the microcentrifuge).
14. The supernatant fraction is carefully removed and the beads are washed by resuspension in 1 ml of ice-cold lysis buffer containing protease and phosphatase inhibitors before the beads are pelleted by centrifugation as previously. This washing procedure is repeated 5 times.
15. Immunocomplexes are eluted from the beads by adding 20 µl of 2X-concentrated SDS-PAGE gel loading buffer to the final bead pellet. The samples are boiled for 5 min in tubes where the cap is perforated using a syringe tip. Beads are removed by centrifugation and 15 µl of each supernatant fraction is subjected to SDS-PAGE on an 8% polyacrylamide gel.
16. Following electrophoresis, the proteins are electrotransferred to a Hybond-ECL nitrocellulose membrane (Amersham Biosciences) and radioactive bands on the membrane are detected using an autoradiograpy screen reader (Fujifilm BAS-5000 phosphoimager). The membrane is subsequently developed for chemiluminescent detection of the total amount of purified p62 (see previous sections). The images of the radioactive proteins from the phosphoimager and the Lumi-Imager F1 are digitally superimposed to confirm that the radioactive band comigrated with p62. Developing the membrane with anti-p62 antibody also provides a control for the total amount of p62 immunoprecipitated from the different samples.

Note that the labeling for the different time points is started at different times to allow simultaneous lysis and immunoprecipitation. Materials and solutions are prechilled and all procedures performed at 4 °C or on ice starting from lysis of the cells until boiling of the immunocomplexes. Special care must be taken to avoid skin and oral exposure to the radioactive material and all materials and solutions containing radioactivity must be collected in separate containers and disposed of following the local guidelines for radioactive waste.

Immunoprecipitation using the p62 antibody isolates a radioactive labeled protein that migrates as approximately 60 kDa, whereas the isotype control antibody does not purify a similar protein. The degradation of the pulse-labeled 60 kDa protein is faster in cells placed in amino acid–free medium than in normal medium and the degradation is blocked by inhibiting lysosomal activity. Chemiluminescence detection after staining the membrane with p62 antibody demonstrates that the radioactive 60 kDa protein comigrates with p62. By quantification of the pulse-labeled p62 we estimated the half-life of p62 in HeLa cells to be reduced from approximately 6 h under basal conditions to between 1–2 h under amino acid starvation conditions (Fig. 12.2).

2.4. Immunofluorescence analysis of p62

All the antibodies tested here efficiently stain human p62, whereas the p62 antibody from BD Transduction Laboratories as well as the monoclonal antibody from Santa Cruz Biotechnology give no or very weak signal in mouse cells. This correlates with the test of species-specificity on immunoblotting (see previous sections). As a routine protocol, we do all the immunofluorescence on cell culture chamber slides with confocal microscope glass bottoms (8-well Lab-Tek chambered coverglass, Nunc cat. no. 155411). This allows sequential staining and reduces hands-on time and cell detachment associated with mounting.

1. HeLa cells are plated at 8000 cells and HEK293 at 12,000 cells per well approximately 16 h before fixation.
2. The 8-well coverglasses containing 200 µl of medium per well are placed on ice and 200 µl of cold, freshly made 8% weight/volume paraformaldehyde in PBS (pH adjusted to 7.0–7.4) are added.
3. The cells are fixed for 10 min on ice and then washed twice in PBS.
4. Depending on what fraction of p62 one wants to efficiently visualize, permeabilization can either be performed using 0.1% Triton X-100 in PBS for 5 min at room temperature, ice-cold methanol for 10 min in ice or 40 µg/ml digitonin in PBS for 5 min at room temperature. To efficiently visualize p62 that has entered into the vesicular fraction including autophagosomes and lysosomes, we recommend using

digitonin permeabilization. For staining of aggregates, both Triton X-100 and methanol permabilization work well and the choice of protocol may depend on the other proteins that are to be stained in the same experiment.
5. Nonspecific binding is blocked by a 30 min incubation in PBS containing 3% (v/v) serum from the species where the secondary antibodies are raised (usually goat serum).
6. The blocking solution is removed and replaced with a 150 μl solution of primary anti-p62 antibody (1:200 dilution of the monoclonal antibodies and 1:2000 dilutions of the guinea pig polyclonal antibody) in PBS containing 1% serum. Primary antibodies are incubated for one hour at room temperature or overnight at 4 °C.
7. Excess primary antibody is removed by 3 times 5 min incubations in 0.5 ml of PBS.
8. Bound anti-p62 antibodies are visualized by incubation with a 1:500 dilution of the appropriate Alexa 488, 555, or 647 conjugated secondary antibodies in PBS containing 1% serum. Secondary antibodies are incubated for 30 min at room temperature and excess antibodies removed by 3 times 5 min incubation in PBS.
9. Finally, the cells are left in 200 μl of PBS and imaged using a water immersion objective (40x) on an inverted microscope equipped with a confocal laser-scanning unit.

For prolonged storage (up to several weeks) of the stained cells, 0.02% thimerosal is added to the PBS, and the stained cells are stored at 4 °C in the dark. Alternatively, fixed cells can be stored in PBS containing 0.02% thimerosal for several weeks before refixation, permeabilization, and staining. The siRNA-mediated knockdown of p62 followed by staining using the different p62 antibodies is the best negative control for immunostaining using the p62 antibodies. Alternatively, isotype controls for primary antibodies at the same final concentration can be used.

2.5. Immuno-electron microscopy of p62

For preparation of cells for immuno-electron microscopy, we usually grow $1-5\times10^5$ cells in either 3- or 6-cm dishes. Cells are fixed in either 4% formaldehyde alone or a mixture of 4% formaldehyde and 0.15% glutaraldehyde in phosphate buffer (pH 7.4). Both fixations allow detection of endogenous p62 in human cells with monoclonal (clone 3, BD Transduction Laboratories) and the guinea pig polyclonal (Progen Biotechnik, GP62-C) antibodies. Cells are further processed according to the following protocol (Peters et al., 1991):

1. After fixation, the cells are scraped up in 10%–12% gelatin/PBS, and pelleted at 13,000 rpm for 5 min in 1.5 ml of microcentrifuge tubes.
2. The samples are transferred onto ice, and blocks of approximately 1 mm³ are prepared from the pellet.
3. The blocks are infused with 2.3 M sucrose for 1 h in the cold, further reduced to 0.2–0.3 mm side length and mounted on silver pins and frozen in liquid nitrogen.
4. Ultrathin cryosections are cut at -110 °C on a cryo microtome and collected with a 1:1 mixture of 2% methyl cellulose and 2.3 M sucrose.
5. Sections are transferred to formvar/carbon-coated grids and labeled with primary antibodies for 30 min at room temperature, followed by incubation with protein A-gold conjugates (Slot *et al.*, 1991), which should not exceed 20 min. In the case of monoclonal antibodies we use bridging secondary antibody. Primary antibodies are used at 1:100 dilution and protein A-gold conjugates are used at the manufacturer's recommendations (ECM, Utrecht). For double labeling experiments, we include a blocking step between the first protein A-gold and the second primary antibody (15 min incubation in 0.1% glutaraldehyde in 0.1 M PBS).
6. After embedding in 2% methyl cellulose/0.4% uranyl acetate, sections are observed at 60–80 kV in a TEM.

3. Monitoring the Autophagic Degradation of p62/SQSTM1 by Live Cell Imaging

3.1. Ectopic overexpression of p62

Expression of fluorescent proteins fused to p62 is an informative approach to study autophagy of p62 in living cells. Both endogenous and ectopically expressed p62 often localize to cytoplasmic bodies with a diameter of more than 0.5 μm (Bjørkøy *et al.*, 2005). Transiently expressed p62 has a strong tendency to be located in such bodies dependent on PB1 domain-mediated homo-polymerization and on an intact UBA domain (Lamark *et al.*, 2003). The p62 structures are reduced both in number and size from 24–48 h after transfection. The localization pattern of transiently expressed enhanced green fluorescent protein-tagged p62 (GFP-p62) 2 days after transfection is very similar to what we observe by staining endogenous p62. Expression of fluorescent protein-tagged p62 also allows a nice control for p62 immunofluorescence. Larger GFP-p62-positive cytoplasmic bodies with homogenous green intensity will appear as a positive ring at the outside of the structure when stained with anti-GFP or anti-p62 antibodies. This demonstrates that the antibodies do not efficiently penetrate the core of these structures.

3.2. Live cell imaging of p62 using the pH-sensitive mCherry-GFP double tag

A number of different fluorescent proteins are now available as fusion tags. These have several biochemical and photometric differences. To generate a probe that can be followed into the acidic environment of the lysosomes we took advantage of the very different pH stabilities of GFP (enhanced green fluorescent protein) and mCherry. GFP is found to be quite pH labile with a pKa of 6.15, whereas mCherry is much more pH stable with a pKa lower than 4.0 (Llopis et al., 1998; Shaner et al., 2004). To confirm the pKa values of GFP and mCherry and to determine if the fusion between these two proteins would behave as the two isolated proteins, we expressed and purified from E. coli the tandem mCherry-GFP tag fused to GST (glutathione S-transferase). The purified GST-mCherry-GFP protein was diluted to 20 μg/ml in citrate phosphate buffer. The pH range of 2.5 to 7.75 was studied using increments of 0.25 pH units. Fluorescent intensity was estimated at the different pH values in triplicate wells in a 96-well plate using a monochromator fluorescence plate reader (Spectramax Gemini, Molecular Devices). GFP fluorescence was measured using a 470 ± 9 nm excitation filter and a 525 ± 9 nm emission filter. The mCherry fluorescence was determined using a 544 ± 9 nm excitation filter and a 612 ± 9 nm emission filter. Using these settings, we found the absolute emission of mCherry to be 2 times less intense than GFP. The well with the maximum intensity at the respective emissions was set to 1.0 and relative intensities estimated as average relative intensity in the triplicates with standard deviation (Fig. 12.3). We found that the green fluorescence was rapidly lost as the pH dropped from pH 7.0 with a pKa of approximately 6.2. The red fluorescence of the fusion protein was clearly much more pH stable with an estimated pKa of 3.8. This is in excellent agreement with the pKa values of the isolated enhanced GFP and mCherry. Hence, this shows that the 2 fluorescent proteins retained their fluorescent features when fused in tandem.

The pH-sensitive mCherry-GFP double tag can be observed as both red and green in a neutral microenvironment and as red only in an acidic milieu. Expression of mCherry-GFP-p62 is analyzed in live cells 48 h after transient transfection. The live cell imaging is performed at 37 °C using normal medium when using a microscope stage with CO_2 control (5% CO_2) or in HANKS medium (Gibco, cat. no. 14025) containing 10% serum for up to 1 h without CO_2 control. HANKS medium is used because of a higher buffering capacity than normal cell culture medium to avoid acidification of the medium in the absence of a CO_2 atmosphere. Neutral p62 bodies are both green and red and represent mainly cytoplasmic aggregates (Pankiv et al., 2007). In addition, we observe at this time a high number of red-only structures. These structures are generally smaller in size (0.4–0.8 μm in diameter) and are often much more mobile. It is important to express the

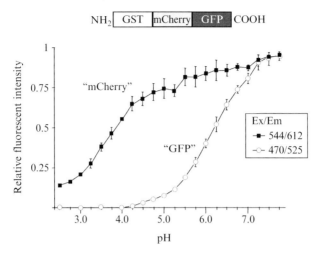

Figure 12.3 The pH dependence of fluorescence intensity of the mCherry-GFP tandem fusion protein. The fluorescent intensity of the recombinant purified mCherry-GFP fusion protein was measured in a microplate reader using 470 nm and 544 nm excitation for GFP (= enhanced GFP) and mCherry, respectively. The emission of GFP (closed squares) was detected using a 525-nm emission filter and that of mCherry (open circles) measured using a 612-nm emission filter. All filters had a 9-nm-band pass width. Triplicate wells of the indicated that pH values were measured and the relative average values with standard deviation are shown.

mCherry-GFP alone as a control of red fluorescence enriched in acidic structures due to the stability of the mCherry fluorophore in acidic vesicles. We find this enrichment to be very weak compared to the high number of intensely fluorescing red-only structures observed when the mCherry-GFP double tag is fused to p62. Using sequential scans it is important to reduce the laser power output of the 488-nm laser line of the Argon laser as much as possible to avoid photo-bleaching. Normally, live cell imaging will work well using 2%–4% laser output and 40%–50% power of the 30 mW Argon laser. To control for possible bleed-through between the green and red channel, the 488 nm and the 543 nm or 563 nm laser line is switched off individually and no differences should be observed in the other channel.

4. Concluding Remarks

Given the complexity of both the autophagy process itself and its regulation, there is presently no single marker or assay that can be used as a stand-alone assay to monitor autophagy in mammalian cells (Klionsky *et al.*, 2008). The realization that p62 binds directly to LC3, resulting in specific degradation by autophagy makes p62 a valuable marker of autophagy that potentially will be interesting in further studies of the regulation of this fundamental process. The level of p62 is clearly dependent on the rate of autophagy. However, p62 has also been found to be induced at the transcriptional level by reactive oxygen species directly or by cellular stress conditions that cause oxidative stress. Therefore, as for punctate GFP-LC3 structures and LC3-II levels in cell extracts, caution should be taken when evaluating the rate of autophagy based on the level of p62 alone. However, in combination with other assays and autophagy markers, the methods described here to measure the protein levels and autophagic degradation of p62 provide a valuable addition to the toolbox used to study autophagy in mammalian cells.

ACKNOWLEDGMENTS

This work was supported by grants from the FUGE and "Top Research Programme" of the Norwegian Research Council, the Norwegian Cancer Society, the Aakre Foundation, Simon Fougner Hartmanns Familiefond and the Blix Foundation to T.J. A.B. is recipient of a career fellowship from the FUGE programme of the Norwegian Research Council.

REFERENCES

Bjørkøy, G., Lamark, T., Brech, A., Outzen, H., Perander, M., Øvervatn, A., Stenmark, H., and Johansen, T. (2005). p62/SQSTM1 forms protein aggregates degraded by autophagy and has a protective effect on huntingtin-induced cell death. *J. Cell Biol.* **171**, 603–614.

Filimonenko, M., Stuffers, S., Raiborg, C., Yamamoto, A., Malerod, L., Fisher, E. M., Isaacs, A., Brech, A., Stenmark, H., and Simonsen, A. (2007). Functional multivesicular bodies are required for autophagic clearance of protein aggregates associated with neurodegenerative disease. *J. Cell Biol.* **179**, 485–500.

Gal, J., Strom, A. L., Kilty, R., Zhang, F., and Zhu, H. (2007). p62 accumulates and enhances aggregate formation in model systems of familial amyotrophic lateral sclerosis. *J. Biol. Chem.* **282**, 11068–11077.

Hara, T., Takamura, A., Kishi, C., Iemura, S., Natsume, T., Guan, J. L., and Mizushima, N. (2008). FIP200, a ULK-interacting protein, is required for autophagosome formation in mammalian cells. *J. Cell Biol.* **181**, 497–510.

Klionsky, D. J., Abeliovich, H., Agostinis, P., Agrawal, D. K., Aliev, G., Askew, D. S., Baba, M., Baehrecke, E. H., Bahr, B. A., Ballabio, A., Bamber, B. A., Bassham, D. C., *et al.* (2008). Guidelines for the use and interpretation of assays for monitoring autophagy in higher eukaryotes. *Autophagy* **4**, 151–175.

Komatsu, M., Waguri, S., Koike, M., Sou, Y. S., Ueno, T., Hara, T., Mizushima, N., Iwata, J., Ezaki, J., Murata, S., Hamazaki, J., Nishito, Y., *et al.* (2007a). Homeostatic levels of p62 control cytoplasmic inclusion body formation in autophagy-deficient mice. *Cell* **131**, 1149–1163.

Komatsu, M., Wang, Q. J., Holstein, G. R., Friedrich, V. L., Jr., Iwata, J., Kominami, E., Chait, B. T., Tanaka, K., and Yue, Z. (2007b). Essential role for autophagy protein Atg7 in the maintenance of axonal homeostasis and the prevention of axonal degeneration. *Proc. Natl. Acad. Sci. USA* **104**, 14489–14494.

Kuusisto, E., Kauppinen, T., and Alafuzoff, I. (2008). Use of p62/SQSTM1 antibodies for neuropathological diagnosis. *Neuropathol. Appl. Neurobiol.* **34,** 169–180.

Kuusisto, E., Salminen, A., and Alafuzoff, I. (2001). Ubiquitin-binding protein p62 is present in neuronal and glial inclusions in human tauopathies and synucleinopathies. *Neuroreport* **12,** 2085–2090.

Lamark, T., Perander, M., Outzen, H., Kristiansen, K., Øvervatn, A., Michaelsen, E., Bjørkøy, G., and Johansen, T. (2003). Interaction codes within the family of mammalian Phox and Bem1p domain-containing proteins. *J. Biol. Chem.* **278,** 34568–34581.

Lee, I. H., Cao, L., Mostoslavsky, R., Lombard, D. B., Liu, J., Bruns, N. E., Tsokos, M., Alt, F. W., and Finkel, T. (2008). A role for the NAD-dependent deacetylase Sirt1 in the regulation of autophagy. *Proc. Natl. Acad. Sci. USA* **105,** 3374–3379.

Llopis, J., McCaffery, J. M., Miyawaki, A., Farquhar, M. G., and Tsien, R. Y. (1998). Measurement of cytosolic, mitochondrial, and Golgi pH in single living cells with green fluorescent proteins. *Proc. Natl. Acad. Sci. USA* **95,** 6803–6808.

Maclean, K. H., Dorsey, F. C., Cleveland, J. L., and Kastan, M. B. (2008). Targeting lysosomal degradation induces p53-dependent cell death and prevents cancer in mouse models of lymphomagenesis. *J. Clin. Invest.* **118,** 79–88.

Marino, G., Ugalde, A. P., Salvador-Montoliu, N., Varela, I., Quiros, P. M., Cadinanos, J., van der Pluijm, I., Freije, J. M., and Lopez-Otin, C. (2008). Premature aging in mice activates a systemic metabolic response involving autophagy induction. *Hum. Mol. Genet.* **17,** 2196–2211.

Mizushima, N., and Yoshimori, T. (2007). How to interpret LC3 immunoblotting. *Autophagy* **3,** 542–545.

Nakai, A., Yamaguchi, O., Takeda, T., Higuchi, Y., Hikoso, S., Taniike, M., Omiya, S., Mizote, I., Matsumura, Y., Asahi, M., Nishida, K., Hori, M., et al. (2007). The role of autophagy in cardiomyocytes in the basal state and in response to hemodynamic stress. *Nat. Med.* **13,** 619–624.

Olive, M., van Leeuwen, F. W., Janue, A., Moreno, D., Torrejon-Escribano, B., and Ferrer, I. (2008). Expression of mutant ubiquitin (UBB+1) and p62 in myotilinopathies and desminopathies. *Neuropathol. Appl. Neurobiol.* **34,** 76–87.

Pankiv, S., Clausen, T. H., Lamark, T., Brech, A., Bruun, J. A., Outzen, H., Overvatn, A., Bjorkoy, G., and Johansen, T. (2007). p62/SQSTM1 binds directly to Atg8/LC3 to facilitate degradation of ubiquitinated protein aggregates by autophagy. *J. Biol. Chem.* **282,** 24131–24145.

Peters, P. J., Neefjes, J. J., Oorschot, V., Ploegh, H. L., and Geuze, H. J. (1991). Segregation of MHC class II molecules from MHC class I molecules in the Golgi complex for transport to lysosomal compartments. *Nature* **349,** 669–676.

Seibenhener, M. L., Babu, J. R., Geetha, T., Wong, H. C., Krishna, N. R., and Wooten, M. W. (2004). Sequestosome 1/p62 is a polyubiquitin chain binding protein involved in ubiquitin proteasome degradation. *Mol. Cell Biol.* **24,** 8055–8068.

Seibenhener, M. L., Geetha, T., and Wooten, M. W. (2007). Sequestosome 1/p62: more than just a scaffold. *FEBS Lett.* **581,** 175–179.

Settembre, C., Fraldi, A., Jahreiss, L., Spampanato, C., Venturi, C., Medina, D., de Pablo, R., Tacchetti, C., Rubinsztein, D. C., and Ballabio, A. (2008). A block of autophagy in lysosomal storage disorders. *Hum. Mol. Genet.* **17,** 119–129.

Shaner, N. C., Campbell, R. E., Steinbach, P. A., Giepmans, B. N., Palmer, A. E., and Tsien, R. Y. (2004). Improved monomeric red, orange and yellow fluorescent proteins derived from Discosoma sp. red fluorescent protein. *Nat. Biotechnol.* **22,** 1567–1572.

Slot, J. W., Geuze, H. J., Gigengack, S., Lienhard, G. E., and James, D. E. (1991). Immunolocalization of the insulin regulatable glucose transporter in brown adipose tissue of the rat. *J. Cell Biol.* **113,** 123–135.

Tasdemir, E., Maiuri, M. C., Galluzzi, L., Vitale, I., Djavaheri-Mergny, M., D'Amelio, M., Criollo, A., Morselli, E., Zhu, C., Harper, F., Nannmark, U., Samara, C., et al. (2008). Regulation of autophagy by cytoplasmic p53. *Nat. Cell Biol.* **10,** 676–687.

Wang, Q. J., Ding, Y., Kohtz, D. S., Mizushima, N., Cristea, I. M., Rout, M. P., Chait, B. T., Zhong, Y., Heintz, N., and Yue, Z. (2006). Induction of autophagy in axonal dystrophy and degeneration. *J. Neurosci.* **26,** 8057–8068.

Wilson, M. I., Gill, D. J., Perisic, O., Quinn, M. T., and Williams, R. L. (2003). PB1 domain-mediated heterodimerization in NADPH oxidase and signaling complexes of atypical protein kinase C with Par6 and p62. *Mol. Cell* **12,** 39–50.

Zatloukal, K., Stumptner, C., Fuchsbichler, A., Heid, H., Schnoelzer, M., Kenner, L., Kleinert, R., Prinz, M., Aguzzi, A., and Denk, H. (2002). p62 Is a common component of cytoplasmic inclusions in protein aggregation diseases. *Am. J. Pathol.* **160,** 255–263.

CHAPTER THIRTEEN

Cytosolic LC3 Ratio as a Quantitative Index of Macroautophagy

Motoni Kadowaki* *and* Md. Razaul Karim*

Contents

1. Introduction	200
2. Measurement of Proteolysis	201
2.1. Proteolysis measurement in fresh rat hepatocytes	201
2.2. Proteolysis measurement in the H4-II-E cell line	203
3. Measurement of the Cytosolic LC3 Ratio	203
3.1. Subcellular fractionation	203
3.2. Characterization of cytosolic LC3-II (LC3-IIs)	205
3.3. Calculation of the cytosolic LC3 ratio	207
3.4. Evaluation of macroautophagic flux as a quantitative index	208
4. Concluding Remarks	211
References	211

Abstract

Macroautophagy, an intracellular bulk degradation process and a typical form of autophagy in eukaryotes, is sensitive to physiological regulation, such as the supply and deprivation of nutrients. Microtubule-associated protein 1 light chain 3 (LC3), a mammalian homologue of yeast Atg8, plays a critical role in macroautophagy formation and is considered a suitable marker for this process. In mammalian cells, there is a limitation for biochemical and morphological methods to monitor autophagy within a short period of time. During analysis of the subcellular distribution of LC3, we found that the cytosolic fraction contains not only a precursor form (LC3-I), but also an apparently active form, denoted as LC3-IIs. Both LC3-I and LC3-IIs in the cytosolic fraction, and thus the LC3-IIs/I ratio (designated the cytosolic LC3 ratio), were more responsive to amino acids than monitoring LC3-II or the LC3-II/I ratio in the total homogenate, and remarkably reflected the total proteolytic flux in fresh rat hepatocytes and the cultured H4-II-E cell line. Thus, in addition to representing a sensitive index

---

* Department of Applied Biological Chemistry, Faculty of Agriculture, Graduate School of Science and Technology, Niigata University, Niigata, Japan

of macroautophagy, examining the cytosolic LC3 ratio is an easy and quick quantitative method for monitoring the regulation of this process in hepatocytes and H4-II-E cells.

1. Introduction

Macroautophagy constitutes a cellular self-restructuring system through degradation of intracellular components (e.g., organelles, proteins, RNA) and is actively controlled by nutritional, physiological, and pathological processes (Mortimore and Kadowaki, 2001; Meijer and Codogno, 2004; Cuervo, 2004; Yorimitsu and Klionsky, 2005; Kadowaki *et al.*, 2006). Although several biochemical and morphological methods have been developed for autophagy assessment, they are associated with limited and somewhat unsatisfactory results in mammalian cells (Mizushima, 2004; Klionsky *et al.*, 2007). In general, biochemical methods are quantitative. However, it is difficult to distinguish autophagy from other intracellular degradation pathways such as the ubiquitin-proteasome system or endocytosis, although autophagy is believed to account for the majority of long-lived proteolysis (Mortimore and Kadowaki, 2001). Until recently, morphological and morphometric methods utilizing electron microscopy have been the most reliable methods for monitoring autophagy, but they are time-consuming and require advanced skills (Schwörer *et al.*, 1981; Mizushima, 2004; Kawai *et al.*, 2006; Klionsky *et al.*, 2007). A fluorescence method based on overexpression of GFP-LC3 was recently introduced as a relatively simple and specific marker, but it does not provide a convenient measure for assessing autophagic flux (Mizushima, 2004).

Although LC3-II is a very specific marker of autophagosomes/autolysosomes, detection of its steady-state levels is not appropriate for estimating cellular autophagic flux (Tanida *et al.*, 2005) (see also the chapter by Kimura *et al.* in this volume). For this reason, we calculated the ratio between LC3-II and LC3-I to estimate the activation step of autophagy formation that could be compared with the bulk proteolytic flux in isolated rat hepatocytes and rat hepatoma H4-II-E cells. During our analysis of the subcellular localization of LC3, we discovered a soluble delipidated form of LC3-II that was localized in the cytosolic fraction. This soluble form, termed LC3-IIs, was different from the typical form found in membrane fractions (LC3-IIm). When the ratio between LC3-IIs and LC3-I in the cytosolic fraction was assessed, it was found to reflect quantitative changes in the proteolytic flux much better than that of the total homogenate. Thus, in this chapter, we describe the cytosolic LC3 ratio (LC3-IIs/LC3-I) as a novel sensitive index of macroautophagy in hepatocytes and H4-II-E cells.

2. Measurement of Proteolysis

We describe proteolysis measurement techniques for comparison with the LC3 method because it is still a reliable method for quantifying bulk proteolysis including autophagy. Here, two different proteolytic methods are described. For the study of fresh hepatocytes, we employ the combination of monitoring valine (Val) release from the hepatocytes (within an hour) in the presence of cycloheximide, relying on the fluorescent detection of Val by HPLC, because this method is more sensitive to physiological stimuli such as amino acids at a low cell density of 2×10^6/mL than the C^{14}-Val release method (Venerando et al., 1994). In addition, the latter method needs to use a radioactive tracer to label protein in intact animals, which may cause problems in terms of the waste disposal for radioactively contaminated animals. For the cultured cell line study, we employ the C^{14}-Val release method, with which it is easier to detect the label with a longer incubation period of proteolysis (several hours), and which causes less trouble with regard to radioactive waste. It is not possible to detect cold (i.e., nonradioactive) Val release using the HPLC method.

2.1. Proteolysis measurement in fresh rat hepatocytes

Liver parenchymal cells from male Wistar/ST rats weighing approximately 200–400 g are isolated using a previously described collagenase method (Seglen et al., 1976). Our protocol is slightly modified to obtain cells that are more sensitive to amino acids (Venerando et al., 1994) as follows:

1. After separation, the hepatocytes are resuspended in 80 mL of Krebs-Ringer bicarbonate (KRB; final concentration 118.5 mM NaCl, 4.74 mM KCl, 2.5 mM CaCl$_2$ · 2H$_2$O, 1.18 mM KH$_2$PO$_4$, 1.18 mM MgSO$_4$ · 7H$_2$O, 23.5 mM NaHCO$_3$) buffer (pH 7.4) containing 6 mM glucose and 0.5% BSA oxygenated with O$_2$:CO$_2$ (95:5, v/v) gas, and adjusted to a cell density of 2×10^6/mL using a hematocytometer to determine the cell count.
2. Aliquots of the cell suspension (3 ml) are placed in 10-ml conical flasks and incubated in a rotary shaker with holders to accommodate 4 columns of water-bath flasks, with six 10-ml flasks per row at 37 °C in a closed-transparent plastic box with gas.

Maintaining the cell density at approximately 2×10^6 cells/mL during the incubation helps to avoid the influence of amino acids that spontaneously accumulate in the medium at higher cell densities and the resulting, and perhaps unexpected, feedback inhibition (Venerando et al., 1994). When necessary, regulatory amino acids (RegAA), which exert direct inhibitory effects on autophagic proteolysis and show an equal effect to complete amino acids (CAA), are added at concentrations that are several fold higher than their normal portal plasma concentrations. The normal (1x) portal plasma concentrations of RegAA, including the coregulatory Ala, are set as follows (μM): Leu, 204; Tyr, 98; Pro, 437; Met, 60; His, 92; Trp, 93; Ala, 475 (Niioka et al., 1998). For the composition of CAA, refer to Niioka et al. (1998).

Proteolysis can be measured by Val release from hepatocytes in the presence of 20 μM cycloheximide (Kanazawa et al., 2004). Because cycloheximide may inhibit not only protein synthesis but also autophagic proteolysis (Khairallah and Mortimore, 1976; Abeliovich et al., 2000), we employ a short exposure method; within this time frame the rate is not affected by this agent (Khairallah and Mortimore, 1976).

1. After starting the incubation using the cell suspensions above, cycloheximide (final concentration, 20 μM) is added at 30 min to block protein synthesis. The suspensions (350 μL) for Val analysis are sampled at 37 min and 47 min (i.e., a 10-min interval) to calculate the Val release rate per min as a function of proteolysis.
2. Samples are deproteinized using ice-cold perchloric acid (PCA; 6% final concentration) in microcentrifuge tubes, and kept on ice for 15–20 min. The samples are then centrifuged at 12,000 rpm for 5 min. The supernatant fractions are collected and transferred to new tubes and stored at $-20\,^\circ C$ until analysis; the pellets are discarded.
3. Ten and 20 μL of 200 μM Val are used as a standard and made up to 100 μL with distilled water (final concentration at 20 and 40 μM in 100 μL). Add 40 μL of 100 μM norvaline as an internal standard. Add 100 μL of 10% PCA.
4. Frozen samples (100 μL) are mixed with 40 μL of 100 μM norvaline and the volume was adjusted to 240 μL with 100 μL of distilled water.
5. Thirty-five μL of 3 N KOH and 25 μL of 0.5 M Na_2CO_3 buffer (pH 9.5) are added to the samples, vortexed and kept on ice for 15 min, and then the pH is adjusted to 9.60–9.90 by titrating with 1 N KOH or 6% PCA.
6. Samples are centrifuged at 12,000 rpm for 5 min at room temperature, and the supernatant fractions (200 μL) are collected into new microcentrifuge tubes.
7. Dansyl chloride (100 μL diluted to 20 $\mu L/mL$ acetonitrile from a stock solution of 100 mg dansyl chloride/mL acetone) is added for derivatization, vortexed and incubated for at least 45 min at 4 $^\circ C$. After incubation, the samples are centrifuged again at 12,000 rpm for 15 min, and finally aliquots (150–200 μL) of the supernatant fractions are used for HPLC analysis.
8. Val as well as norvaline are separated by reverse phase-HPLC using a Supelcosil LC-18 column (4.6 × 150 mm) as follows: Samples are eluted at room temperature at a flow rate of 0.8 ml/min with a gradient

of eluent A and B, in which eluent B is increased from 40% to 50% during the first 10 min, maintained at the 50% level until 25 min, then increased to 100%.

Eluent A: distilled water 880 mL; acetonitrile 120 mL; acetic acid 3 mL; triethylamine 0.35 mL. Filter through a 0.45 μ filter and degas for 10–15 min.
Eluent B: 100% methanol.

9. The retention time of Val and norvaline usually corresponds to 20 min and 21 min, respectively. The Val concentration can be calculated using norvaline as an internal standard.

2.2. Proteolysis measurement in the H4-II-E cell line

Proteolysis in the H4-II-E cell line is analyzed according to Lavieu *et al.* (2006) (also see the chapter by Bauvy *et al.*, in this volume). After 18 h of radiolabeling with ^{14}C-Val (0.2 μCi/ml), cells are incubated for 1 h in chase medium (nutrient-free Hanks' Balanced Salt Solution (HBSS) containing 10 mM cold Val) to abolish reutilization of ^{14}C-Val derived from short-lived degradation. Next, the medium is replaced with fresh chase medium and the cells are incubated for another 3 h. When required, a mixture of amino acids or Leu alone is added to the chase medium as indicated in the figure legends (Lavieu *et al.* 2006). The radioactivity present in the medium and cell lysate are determined by liquid scintillation counting. Long-lived proteolysis is calculated by dividing the acid-soluble radioactivity in the medium by the total radioactivity of the whole cell lysate and the medium, and expressed as a percentage of the control rate (HBSS), which represents the most accelerated proteolytic rate.

3. MEASUREMENT OF THE CYTOSOLIC LC3 RATIO

3.1. Subcellular fractionation

Because this method is rather simple but requires obtaining the cytosolic fraction, we describe the subcellular fractionation technique (Furuya *et al.*, 2001).

1. To collect hepatocytes, 1.2 mL of cell suspensions (2 × 10^6 cells/mL) in the KRB buffer are centrifuged at 1,000 rpm for 50 s at 4 °C in a Kubota 6930 centrifuge with an RA-300F rotor.
2. The supernatant fraction is discarded and the cell pellet is resuspended in 800 μL of 0.25 M sucrose/1 mM EDTA, pH 7.4, and homogenized by 120 strokes with a tightly fitting Dounce homogenizer (size, 7 ml; Wheaton Science Products, NJ, USA, Cat. No.357542) on ice. In order

to homogenize the cells, it is critical to change from the KRB buffer to the non-buffered sucrose/EDTA solution. This mechanical homogenization is preferable to maintain the organelles and membranes in their intact forms as much as possible (Niioka et al., 1998). The broken whole cell lysate after 120 strokes is considered the total homogenate (TH).
3. The TH is transferred into a microcentrifuge tube and centrifuged at $700 \times g$ for 10 min at 4 °C to separate the nuclei and unbroken cells from the post-nuclear supernatant (PNS) fraction. The PNS fraction is transferred to a new tube. The pellet is re-suspended in sucrose/EDTA, rehomogenized and centrifuged again to wash away any contaminants. The $700 \times g$ pellet is denoted as the nuclear (N) fraction.
4. The PNS fraction is centrifuged at $17,500 \times g$ for 10 min at 4 °C to obtain a mitochondrial-lysosomal (ML) fraction as a pellet, which includes mitochondria and autophagy-related vacuoles (i.e., autophagosomes, autolysosomes, and lysosomes); the supernatant fraction is transferred to a new tube. The pellet is rehomogenized, washed again using the same solution to remove contaminants, and centrifuged as before to get the ML pellet fraction.
5. The $17,500 \times g$ supernatant fraction from step 4 is finally centrifuged at $100,000 \times g$ for 60 min at 4 °C in a Hitachi CS 100 GXL ultracentrifuge with an S 100 AT4-557 rotor. The pellet is a microsomal (pellet, P) fraction, while the supernatant comprises the cytosolic (supernatant, S) fraction. Finally keep all of the separated fractions at -40 °C until ready for analysis by Western blotting.

When you need only the cytosolic fraction for routine assay, you can go to the final step directly after the initial homogenization.

Western blotting of LC3
1. The fractionated samples are boiled for 3 min in SDS-PAGE sample buffer (final concentrations: 12.5 mM Tris-Hcl, pH 6.8, 2% (v/v) glycerol, 0.4% (w/v) SDS, 1% (v/v) β-mercaptoethanol and 0.01% (w/v) bromophenol blue; prepare as 5X stock).
2. The proteins (15 μg for hepatocytes and 25 mg for cell lines) are separated by SDS-PAGE using a 15% gel (100 V; 24 mA) according to the Laemmli method (Laemmli, 1970). To load an equal amount of protein in each lane, the protein concentrations are determined by the Lowry method (Lowry et al., 1951) using BSA as a standard.
3. The separated proteins are transferred from a gel to a PVDF membrane using a semi-dry transfer unit (15 V; 30 min, Atto Corporation, Japan).
4. The membrane is blocked immediately with 7.5% skim milk in PBS containing 0.1% Tween 20 (PBS-T) by slow shaking at 12–14/min for 1 h at room temperature. It is then washed five times with the same buffer at room temperature.

5. The membrane is incubated with an LC3 antiserum (a gift from Juntendo University School of Medicine, Japan, 1.37 mg/ml; 1:666 dilution in TBS containing 2% BSA and 0.1% NaN$_3$) for 1 h at room temperature, followed by incubation with HRP-conjugated donkey anti-rabbit IgG (Amersham Biosciences UK limited; Cat. No. NA934V; 1:4,000 dilution in PBS-T) overnight at 4 °C.
6. After 5 washes with PBS-T, an ECL Western blot detection kit is used as the substrate for detection of the bound HRP-conjugated secondary antibody, followed by exposure of the membrane to Hyperfilm ECL (Amersham Biosciences). Finally, LC3 is quantified by densitometric analysis (Scion Image 1.63.1; NIH, MD, USA).

3.2. Characterization of cytosolic LC3-II (LC3-IIs)

As shown in Fig. 13.1A, LC3-I and LC3-II are both visualized in the TH of isolated hepatocytes. After the subcellular fractionation, LC3-I is only detected in the S fraction, whereas LC3-II is localized not only in the N, ML and P fractions, but also the S fraction. It is well recognized that LC3-I is a precursor and exists as a soluble form in the cytosol, whereas LC3-II is an active membrane-bound form found in membrane fractions (Kabeya *et al.*, 2000, 2004; Ichimura *et al.*, 2000; Mizushima *et al.*, 2004; Tanida *et al.*, 2004b). In contrast, a soluble form of LC3-II (LC3-IIs) is observed in the S fraction. The reason for this unusual finding is not clear, but may be due to the different methods used for disrupting the cells, namely cell lysis with Triton X-100 and mechanical homogenization.

To examine the properties of LC3-II in the S fraction, Triton X-114 (TX-114), a nonionic detergent for phase partition of membrane proteins, is used (Bruska and Radolf, 1994; Chen *et al.*, 2003).

1. Briefly, subcellular fractions are solubilized in 1 ml of PBS(−), pH 7.4, containing 0.5% (v/v) TX-114 and a protease inhibitor cocktail (SIGMA-ALDRICH Inc. USA; Cat No. P8340) (3.2 µl), and incubated at 4 °C overnight for complete extraction.
2. The detergent-insoluble debris is removed by centrifugation at 13,000×g for 10 min at 4 °C; the pellet fraction is discarded.
3. The supernatant fraction is then incubated at 37 °C for 10 min followed by centrifugation at 13,000×g for 10 min at room temperature, which leads to the formation of a hydrophilic aqueous and hydrophobic detergent (cloudy) phase.
4. After separation, the aqueous phase is transferred to a new tube and TX-114 is added at a final concentration of 2%. The detergent phase is mixed thoroughly with 1 ml of PBS(−) at 0 °C, and the separation step is repeated twice.

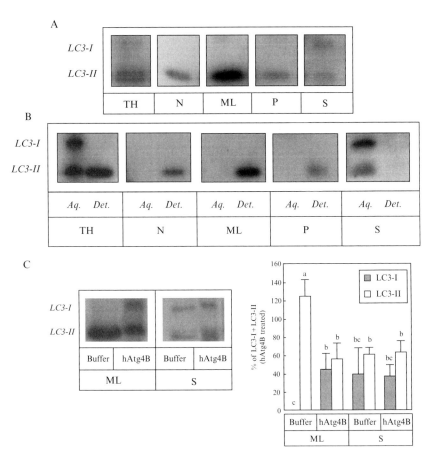

Figure 13.1 Subcellular distribution of LC3 in rat hepatocytes: characterization of cytosolic LC3-II by phase partitioning and hAtg4B treatment. (A) Fresh hepatocytes were isolated from male Wistar/ST rats. Total homogenates (TH) of the hepatocytes were fractionated into N (nuclear), ML (mitochondrial–lysosomal), P (microsomal pellet) and S (cytosolic supernatant) fractions. LC3 proteins were separated by SDS-PAGE and assayed by immunoblotting using an anti-LC3 antibody. Equal amounts of protein (15 μg) were loaded in each lane. (B) All the fractions were separated into aqueous (Aq) and detergent (Det) phases by phase partitioning with Triton X-114 as described in the text. C) Assay for delipidation of LC3-II by hAtg4B. The ML and S fractions (80 μg protein) were incubated with purified hAtg4B (33 μg) at 37 °C for 30 min. The delipidated (LC3-I) and lipidated (LC3-II) forms were assayed by immunoblotting using an anti-LC3 antibody (*left*). *Buffer*: PBS(-) was added instead of hAtg4B. The relative densities of each LC3 relative to the total LC3 proteins after hAtg4B treatment were calculated by densitometric analysis using Scion Image 1.63 (*right*). Data are means ± SEM (n=3-4). [a-c]Values without common letters differ significantly at $p < 0.05$. This figure is reproduced from Karim *et al.* 2007 with permission of the publisher.

The results clearly indicate that LC3-II exists as two different forms, namely the typical membrane-bound hydrophobic form in the N, ML and P fractions and a soluble hydrophilic form in the S fraction (Fig. 13.1B).

The possibility that the soluble LC3-II contained a phospholipid moiety was tested using hAtg4B, an enzyme specific for delipidation of LC3-II (Tanida et al., 2004a; Kabeya et al., 2004). Enzyme activity was confirmed in the ML fraction by *in vitro* experiments (Fig. 13.1C, *left*). After hAtg4B treatment, one-half of the LC3-II in the ML fraction was shifted to LC3-I. On the other hand, LC3-II in the S fraction was not altered by the same treatment (Fig. 13.1C, *right*), which excluded the possibility of lipidation. Therefore, the novel soluble form of LC3-II, denoted as LC3-IIs, may be a transient intermediate form of the typical membrane-bound LC3-II (LC3-IIm). To date, molecular differences between LC3-I and LC3-II have only been indicated before and after the addition of phosphatidylethanolamine, with the exception of intermediate conjugates to hAtg7 or hAtg3 (Kabeya et al., 2004; Sou et al., 2006).

3.3. Calculation of the cytosolic LC3 ratio

To determine whether the balance of LC3-I and LC3-IIs reflects physiological stimuli, we investigated the response of the cytosolic LC3 ratio to amino acids. In Fig. 13.2A, the LC3 protein bands were visualized by immunoblotting analysis (*top panel*). The densitometric data were calculated after scans of the protein blots were obtained on X-ray plates. The data for a 45-min incubation were calculated as the percentage of those for 0 min for both LC3-I and LC3-IIs (*middle panel*). In the bottom panel, the LC3-IIs/I ratio in each condition was calculated from the values obtained in the middle panel, and finally the percentage relative to the control was obtained. A ten-fold plasma level of RegAA, which is known to inhibit autophagy quite effectively, reduced the concentration of LC3-IIs and concomitantly increased the concentration of LC3-I. Leu, the strongest single amino acid regulator of autophagy, produced similar responses. These findings indicate that the transformation of LC3-I to LC3-IIs is reduced and effectively suppressed by RegAA and Leu. We also examined the time course of the LC3 transformation. When hepatocytes were exposed to $10\times$ RegAA (Fig. 13.2B), the LC3-IIs/I ratio began to decrease within 5 min and reached a minimal plateau level by 10 min. This time course was faster than the kinetics of a regression of proteolysis (the final step of autophagy) and the volume of autophagic vacuoles (an earlier step of autophagy) that occur after the addition of amino acids, and display regression with a half-life of 8 min (Schwörer et al., 1981). Thus, the time course we determined agrees with the model that the LC3 transformation step from LC3-I to LC3-IIs may precede the modification step of LC3-IIm (i.e., delipidation).

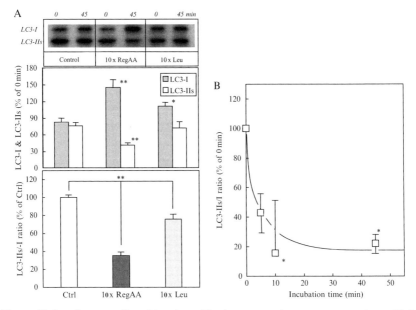

Figure 13.2 The cytosolic LC3 ratio and its time course in response to amino acids in rat hepatocytes. (A) Hepatocytes were incubated with 10× RegAA or 10× Leu for 45 min. The S fractions at 0 and 45 min were obtained and separated by SDS-PAGE. The LC3 proteins were visualized by Western blotting (*top*). Densitometric data of the LC3 bands were calculated, and LC3-I and LC3-IIs at 45 min were individually represented as their percentages relative to 0 min (*middle*). The cytosolic LC3 ratios (LC3-IIs/I) were further calculated as percentages of the control (*bottom*). ★$p < 0.05$, ★★$p < 0.01$, versus the respective control (Ctrl). (B) Time course responses of the LC3-IIs/LC3-I ratio to 10 × RegAA. ★$p < 0.05$, significant difference from 0 min of incubation. This figure is reproduced from Karim *et al.* 2007 with permission of the publisher.

3.4. Evaluation of macroautophagic flux as a quantitative index

Tanida et al. (2005) indicate that the intracellular level of LC3-II, mostly LC3-IIm, increases during macroautophagy induction. However, the total amount of this protein within the cell does not represent the autophagic flux. LC3-IIm subsequently becomes detached from the autolysosomal membrane through delipidation or is digested within autolysosomes. Therefore, the amount of LC3-IIm is a function of multiple and sequential steps, and it is difficult to quantify autophagic flux from typical LC3-IIm levels alone. Since we detected a novel form of LC3, namely LC3-IIs, a new index for autophagy assessment was examined. Initially, isolated hepatocytes were used, since they are considered to be very sensitive to physiological signals, such as amino acids. The LC3 bands in the TH and S fraction after 0 and 45 min of incubation (Fig. 13.3A) were quantified by densitometric

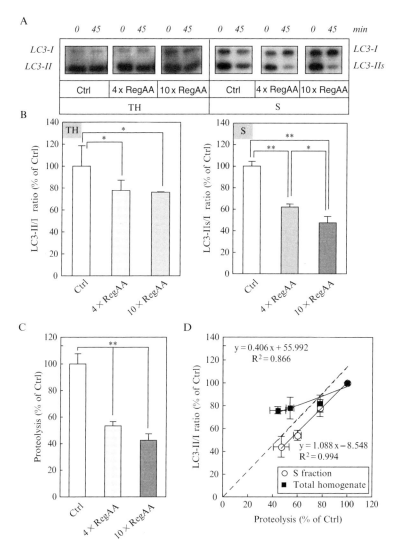

Figure 13.3 Comparison of the LC3 ratio and autophagic proteolysis between the TH and S fraction of rat hepatocytes. (A) Hepatocytes were maintained with or without 4× and 10× RegAA for 45 min. The TH and S fractions were obtained after homogenization and the LC3 bands were visualized by Western blotting. (B) Densitometric data of the LC3 bands were calculated and the LC3-II/I ratios were represented as percentages of the control. ★$p < 0.05$, ★★$p < 0.01$. (C) Proteolysis was measured in the same hepatocytes using the Val release rate as described in the text. ★★$p < 0.01$, versus the control. (D) Relationship of the LC3 ratio and proteolysis in the TH (*squares*) and S fraction (*circles*). The dotted line represents the 1:1 proportional relationship. This figure is reproduced from Karim *et al.* 2007 with permission of the publisher.

analysis and calculated as the LC3 ratios (Fig. 13.3B). The index values in the TH and S fraction were calculated as the percentages of the control (no amino acids) and found to be 73% and 60% for 4× RegAA and 72% and 45% for 10× RegAA treatment, respectively. Fig. 13.3C illustrates the proteolysis data obtained from the same cell population, which were suppressed to 54% and 42% of the control by 4× and 10× RegAA, respectively. The proteolytic response to amino acids was then compared with the ratios of the TH and S fraction. The LC3-IIs/I ratio exhibited a proportion of almost 1:1 with proteolysis (slope = 1.088; R^2= 0.994), whereas the response of the LC3-II/I ratio in the TH (slope = 0.406; R^2= 0.866) was one-half relative to that in the S fraction (Fig. 13.3D). Thus, the LC3 ratio in the S fraction is preferable over that in the TH.

Since the proportions of LC3-I and LC3-II may differ according to the cell type (Kabeya et al., 2000; Tanida et al., 2005), the rat hepatoma H4-II-E cell line was tested to ascertain whether this index is applicable to other cell types (Fig. 13.4). Again, the ratio in the S fraction was more sensitive to amino acids than that in the TH. The data for the LC3 ratios of the TH and S fraction were plotted and high correlations were obtained between the proteolysis and the LC3 ratios of the S fraction (slope = 0.812; R^2 = 0.954) and the TH (slope = 0.419; R^2 = 0.932). Consistent with the earlier results,

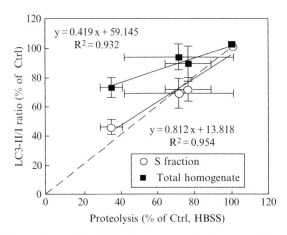

Figure 13.4 Comparison of the LC3 ratio and autophagic proteolysis between the TH and S fraction of the H4-II-E cell line. Confluent H4-II-E cells were maintained with HBSS (Control), 4× CAA, 4× RegAA, 4× Leu and DMEM/F12 (Dulbecco's modified Eagle's medium nutrient mixture F-12 Ham; SIGMA-ALDRICH Co. Ltd., UK; Cat. No. D8437) +10% FBS for 1 h. The TH and S fraction were prepared after homogenization and the LC3 bands were visualized by Western blotting. The LC3-II/I ratio was calculated from densitometric data. Proteolysis was measured by ^{14}C-Val release as described in the text. The relationship of the LC3 ratio and proteolysis in the TH (*squares*) and S fraction (*circles*) is shown. The dotted line represents the 1:1 proportional relationship. This figure is reproduced from Karim et al. 2007 with permission of the publisher.

the ratio in the S fraction was more indicative of autophagy. In the case of the cell line study, we did not take samples at 0 min because the cell numbers were much more stable. The results were in good agreement with the relationship depicted in Fig. 13.3D and strongly support the hypothesis that autophagy kinetics are reflected by the cytosolic LC3 ratio (Karim *et al.*, 2007; Klionsky *et al.*, 2008).

4. Concluding Remarks

In conclusion, we identified a cytosolic subpopulation of LC3-II, designated LC3-IIs, which is hydrophilic and not lipidated, in fresh rat hepatocytes and the H-4-II-E cell line. The cytosolic LC3 ratio derived from LC3-IIs was found to represent a reliable index of autophagy formation and was much more sensitive and proportional to changes in the proteolytic rates than the ratio from the total homogenate. The use of this index should lead to a better understanding of the mechanisms of macroautophagy, which will hopefully lead to the discovery of new participants in the control of this process.

REFERENCES

Abeliovich, H., Dunn, W. A., Jr., Kim, J., and Klionsky, D. J. (2000). Dissection of autophagosome biogenesis into distinct nucleation and expansion steps. *J. Cell Biol.* **151,** 1025–1033.

Bruska, J. S., and Radolf, J. D. (1994). Isolation of integral membrane proteins by phase partitioning with Triton X-114. *Methods Enzymol.* **228,** 182–193.

Chen, X., Ernst, S. A., and Williams, J. A. (2003). Dominant negative Rab3D mutants reduce GTP-bound endogenous Rab3D in pancreatic acini. *J. Biol. Chem.* **278,** 50053–50060.

Cuervo, A. M. (2004). Autophagy: many paths to the same end. *Mol. Cell. Biochem.* **263,** 55–72.

Furuya, N., Kanazawa, T., Fujimura, S., Ueno, T., Kominami, E., and Kadowaki, M. (2001). Leupeptin-induced appearance of partial fragment of betaine homocysteine methyltransferase during autophagic maturation in rat hepatocytes. *J. Biochem. (Tokyo).* **129,** 313–320.

Ichimura, Y., Kirisako, T., Takao, T., Satomi, Y., Shimonishi, Y., Ishihara, N., Mizushima, N., Tanida, I., Kominami, E., Ohsumi, M., Noda, T., and Ohsumi, Y. (2000). A ubiquitin-like system mediates protein lipidation. *Nature* **408,** 488–492.

Kabeya, Y., Mizushima, N., Ueno, T., Yamamoto, A., Kirisako, T., Noda, T., Kominami, E., Ohsumi, Y., and Yoshimori, T. (2000). LC3, a mammalian homologue of yeast Apg8p, is localized in autophagosome membranes after processing. *EMBO J.* **19,** 5720–5728.

Kabeya, Y., Mizushima, N., Yamamoto, A., Oshitani-Okamoto, S., Ohsumi, Y., and Yoshimori, T. (2004). LC3, GABARAP and GATE16 localize to autophagosomal membrane depending on form-II formation. *J. Cell Sci.* **117,** 2805–2812.

Kadowaki, M., Karim, M. R., Carpi, A., and Miotto, G. (2006). Nutrient control of macroautophagy in mammalian cells. *Mol. Aspects Med.* **27,** 426–443.
Kanazawa, T., Taneike, I., Akaishi, R., Yoshizawa, F., Furuya, N., Fujimura, S., and Kadowaki, M. (2004). Amino acids and insulin control autophagic proteolysis through different signaling pathways in relation to mTOR in isolated rat hepatocytes. *J. Biol. Chem.* **279,** 8452–8459.
Karim, M. R., Kanazawa, T., Daigaku, Y., Fujimura, S., Miotto, G., and Kadowaki, M. (2007). Cytosolic LC3 ratio as a sensitive index of macroautophagy in isolated rat hepatocytes and H4-II-E cells. *Autophagy* **3,** 553–560.
Kawai, A., Takano, S., Nakamura, N., and Ohkuma, S. (2006). Quantitative monitoring of autophagic degradation. *Biochem. Biophys. Res. Commun.* **351,** 71–77.
Khairallah, E. A., and Mortimore, G. E. (1976). Assessment of protein turnover in perfumed rat liver: evidence for amino acid compartmentation from differential labeling of free and tRNA-bound valine. *J. Biol.Chem.* **251,** 1375–1384.
Klionsky, D. J., Cuervo, A. M., and Seglen, P. O. (2007). Methods for monitoring autophagy from yeast to human. *Autophagy* **3,** 181–206.
Klionsky, D. J., Ableovich, H., Agostinis, P., Agarwal, D. K., Aliev, G., Askew, D. S., Baba, M., Baehrecke, E. H., Bahr, B. A., Ballabio, A., Bamber, B. A., Bassham, D. C., *et al.* (2008). Guidelines for the use and interpretation of assays for monitoring autophagy in higher eukaryotes. *Autophagy* **4,** 151–175.
Laemmli, U. K. (1970). Cleavage of structural proteins during the assembly of the head of bacteriophage T4. *Nature* **227,** 680–685.
Lavieu, G., Scarlatti, F., Sala, G., Carpentier, S., Levade, T., Ghidoni, R., Botti, J., and Codogno, P. (2006). Regulation of autophagy by sphingosine kinase 1 and its role in cell survival during nutrient starvation. *J. Biol. Chem.* **281,** 8518–8527.
Lowry, O., Rosebrough, N., Farr, A., and Randall, R. (1951). Protein measurement with the folin phenol reagent. *J. Biol. Chem.* **193,** 265–275.
Meijer, A. J., and Codogno, P. (2004). Regulation and role of autophagy in mammalian cells. *Int. J. Biochem. Cell Biol.* **36,** 2445–2462.
Mizushima, N., Yamamoto, A., Matsui, M., Yoshimori, T., and Ohsumi, Y. (2004). In vivo analysis of autophagy in response to nutrient starvation using transgenic mice expressing a fluorescent autophagosome marker. *Mol. Biol. Cell* **15,** 1101–1111.
Mizushima, N. (2004). Methods for monitoring autophagy. *Int. J. Biochem. Cell Biol.* **36,** 2491–2502.
Mortimore, G. E., and Kadowaki, M. (2001). *Regulation of protein metabolism in liver* in Handbook of Physiology (L. S. Jefferson and A. D. Sherrington, eds.), Vol. II, pp. 553–577, American Physiological Society, Oxford University Press, New York.
Niioka, S., Goto, M., Ishibashi, T., and Kadowaki, M. (1998). Identification of autolysosomes directly associated with proteolysis on the density gradients in isolated rat hepatocytes. *J.Biochem. (Tokyo).* **124,** 1086–1093.
Schwörer, C. M., Shiffer, K. A., and Mortimore, G. E. (1981). Quantitative relationship between autophagy and proteolysis during graded amino acid deprivation in perfused rat liver. *J. Biol. Chem.* **256,** 7652–7658.
Seglen, P. O. (1976). *In* "Methodsin Cell Biology" (Prescott, D. M., ed), Academic Press, New York **13,** 29–83.
Sou, Y. S., Tanida, I., Komatsu, M., Ueno, T., and Kominami, E. (2006). Phosphatidylserine in addition to phosphatidylethanolamine is an *in vitro* target of the mammalian Atg8 modifiers, LC3, GABARAP, and GATE-16. *J. Biol. Chem.* **281,** 3017–3024.
Tanida, I., Minematsu-Ikeguchi, N., Ueno, T., and Kominami, E. (2005). Lysosomal turnover, but not a cellular level, of endogenous LC3 is a marker for autophagy. *Autophagy* **2,** 84–91.

Tanida, I., Sou, Y. S., Ezaki, J., Minematsu-Ikeguchi, N., Ueno, T., and Kominami, E. (2004a). HsAtg4B/HsApg4B/autophagin-1 cleaves the carboxyl termini of three human Atg8 homologues and delipidates microtubule-associated protein light chain 3- and GABA receptor-associated protein-phospholipid conjugates. *J. Biol. Chem.* **279,** 36268–36276.

Tanida, I., Ueno, T., and Kominami, E. (2004b). LC3 conjugation system in mammalian autophagy. *Int. J. Biochem. Cell Biol.* **36,** 2503–2518.

Venerando, R., Miotto, G., Kadowaki, M., Siliprandi, N., and Mortimore, G. E. (1994). Multiphasic control of proteolysis by leucine and alanine in the isolated rat hepatocyte. *Am. J. Physiol.* **266,** C455–C461.

Yorimitsu, T., and Klionsky, D. J. (2005). Autophagy: molecular machinery for self-eating. *Cell Death Differ.* **12,** 1542–1552.

CHAPTER FOURTEEN

METHOD FOR MONITORING PEXOPHAGY IN MAMMALIAN CELLS

Junji Ezaki,* Masaaki Komatsu,* Sadaki Yokota,[†] Takashi Ueno,* *and* Eiki Kominami*

Contents

1. Introduction	216
2. Experimental Models for the Study of Pexophagy in Mammals	216
2.1. Induction and accumulation of peroxisomes	216
2.2. Administration protocol of peroxisome proliferation and degradation	217
3. Monitoring Degradation of Excess Peroxisomes	217
3.1. Induction of peroxisome degradation	217
3.2. Measurement of peroxisome degradation by immunoblot analysis	218
3.3. Immunofluorescence analysis of peroxisome degradation	220
3.4. Electron microscopy analyses of peroxisome degradation	220
4. Concluding Remarks	223
Acknowledgment	225
References	225

Abstract

The abundance of peroxisomes within a cell is rapidly controlled depending on environmental changes and physiological conditions. It is well established that phthalate esters can cause a marked proliferation of peroxisomes (Yokota, 1986). Following induction of peroxisomes by a 2-week treatment with phthalate esters in mouse livers, peroxisomal degradation via autophagy can be induced for the subsequent week after discontinuation of the phthalate esters. Autophagic degradation of peroxisomes can be monitored by electron microscopy as well as biochemical assay for some peroxisome markers. Although most of the excess peroxisomes in the liver are selectively degraded within one week, this rapid removal is exclusively impaired in the autophagy-deficient liver.

* Department of Biochemistry, Juntendo University School of Medicine, Hongo, Tokyo, Japan
† Faculty of Pharmaceutical Science, Nagasaki International University, Sasebo, Nagasaki, Japan

1. Introduction

Cell organelles in eukaryotic cells exist in a constitutive flow of biogenesis and degradation. In addition, a dynamic reorganization of organelles occurs as an adaptation to environmental changes that accompany the cell cycle, development, and differentiation (Lazarow and Fujiki, 1985).

The peroxisome, a single membrane-bound organelle ubiquitously distributes in eukaryotic cells, has diverse functions, including decomposition of hydrogen peroxide and oxidation of fatty acids. It is well established that a group of hypolipidemic drugs (Fahimi *et al.*, 1982) and other chemicals (Reddy and Krishnakantha, 1975; Yokota, 1986) can cause the specific proliferation of peroxisomes. A potential mechanism for the proliferation involves stimulation of PPAR-α and its downstream regulators (Guo *et al.*, 2007; Lee *et al.*, 1995; Yu *et al.*, 2001). As a consequence, peroxisome proliferators increase the size and number of peroxisomes, and enzymes involved in fatty acid metabolism (e.g., peroxisomal thiolase, peroxisomal bifunctional protein, and those involved in peroxisomal fatty acid β-oxidation) (Subramani, 1998; Yokota, 1993).

Recent studies on autophagy have thrown new light on the molecular mechanism for the degradation of peroxisomes. In yeast species such as *Pichia pastoris*, *Hansenula polymorpha*, and *Saccharomyces cerevisiae*, proliferating peroxisomes are degraded by an autophagy-related process named *pexophagy*, as an adaptation to changes of carbon sources in the growth medium (methanol or oleic acid to glucose or ethanol). Many *ATG* (autophagy-related) genes play indispensable roles in this selective degradation of peroxisomes (Dunn *et al.*, 2005). An important point is that this process occurs selectively toward peroxisomes and is distinct from nonselective autophagy (that can also degrade peroxisomes), which is generally induced by nutrient starvation (Sakai *et al.*, 2006). Autophagy is also essential for the degradation of accumulated peroxisomes in the mouse liver (Iwata *et al.*, 2006; Yokota, 1993). Thus, the drug-induced proliferation of liver peroxisomes and subsequent degradation upon withdrawal of the drug is a convenient experimental model for studying mammalian pexophagy. In this article, we describe procedures for *in vivo* induction of peroxisomes in mouse livers using phthalate, and methods for monitoring the degradation of peroxisomes after removal of this drug.

 ## 2. Experimental Models for the Study of Pexophagy in Mammals

2.1. Induction and accumulation of peroxisomes

A plasticizer, di-(2-ethylhexyl) phthalate (DEHP) and its active metabolite monoethylhexyl phtalate (MEHP) can cause marked increase in both the size and the number of peroxisomes as well as induction of peroxisomal enzymes

(Lake *et al.*, 1975). After DEHP administration for 2 weeks, the ratio of liver weight to body weight became 7.2%, whereas the ratio of control mice was 4.8% (unpublished data). By the DEHP treatment, the amount of total protein of mitochondrial/lysosomal/peroxisomal fractions increased approximately 2-fold (Iwata *et al.*, 2006). The numerical density and the volume density of peroxisomes increased approximately 3-fold, respectively (Iwata *et al.*, 2006; Yokota, 1986). This is accompanied with a marked increase in the activities of peroxisomal β-oxidation enzymes (Hashimoto, 1982).

2.2. Administration protocol of peroxisome proliferation and degradation

Male C57B6J mice are fed on a standard pellet laboratory diet and tap water *ad libitum*. For the experiments in which peroxisomes are accumulated by phthalate treatment, mice receive DEHP (Sigma D201154) (115 mg/100 g/day) or vehicle (corn oil; Sigma C8267) by gastric intubation via tube daily for 2 weeks, because some of the mice evade eating DEHP-containing laboratory diet. DEHP solution for the administration is prepared as a mixture of DEHP and corn oil (1:3.28) and is administered at a volume of 0.5 ml/100 g/day. Control mice receive the corresponding quantity (0.5 ml/100 g/day) of corn oil only. In practice, the mice are divided into five experimental groups:

Group 1: Untreated control

Group 2: Vehicle (corn oil, 0.5 ml/100 g/day) is administered daily for 2 weeks. This group is used as a control for peroxisome induction.

Group 3: DEHP (115 mg/100 g/day) is administered daily for 2 weeks. This group is used for monitoring peroxisome induction.

Group 4: Vehicle (corn oil, 0.5 ml/100 g/day) is administered daily for 2 weeks. Subsequently, mice are fed a conventional laboratory diet for one additional week. This group is used as a control for peroxisome induction and degradation.

Group 5: DEHP (115 mg/100 g/day) is administered daily for 2 weeks. Subsequently, mice are fed a conventional laboratory diet for one additional week. This group is used for monitoring peroxisome induction and degradation.

3. Monitoring Degradation of Excess Peroxisomes

3.1. Induction of peroxisome degradation

After peroxisomes are induced and accumulate by DEHP administration for 2 weeks, the proliferated peroxisomes diminish rapidly and markedly to basal levels when mice are fed a normal laboratory diet for 1 week.

Concomitantly, a parallel decrease in the levels of peroxisomal enzymes, including peroxisomal thiolase (PT) and peroxisomal bifunctional protein (BF), occur. This recovery process, which can be examined morphologically and biochemically, is a good model for analyzing the autophagic degradation of peroxisomes.

3.2. Measurement of peroxisome degradation by immunoblot analysis

1. Livers are removed from the 5 groups of mice. For the experiments, at least 3 mice are used for each group.
2. The excised livers are homogenized with 4 volumes of 0.25 M sucrose containing protease inhibitor cocktail (Roche Diagnostics 11701000) and 10 mM HEPES-NaOH, pH 7.4, using a motor-driven glass/Teflon homogenizer (5 up-down strokes at 800 rpm) (SHINTO Scientific Co., Three-One Motor BL600).
3. Subfractionation of liver homogenates is achieved by differential centrifugation according to the method of de Duve*et al.* (1971). In brief, the 20% homogenates of the livers are centrifuged at $650 \times g$ for 5 min to remove nuclei and unbroken cells. The pellet fractions are resuspended in the same volume of homogenizing buffer and are recentrifuged. The supernatant fractions from these two centrifugations are combined and used as postnuclear supernatants (PNS).
4. The PNS are solubilized with a {1/3} volume of lysis buffer (4% SDS, 4 mM EDTA, 40% glycerol, .004% bromophenol blue (BPB), 120 mM Tris-HCl, pH 6.8) to give a final concentration of 1% SDS, 1 mM EDTA, 10% glycerol, 0.001% BPB, 30 mM Tris-HCl, pH 6.8, boiled for 3 min, and subjected to SDS-PAGE.
5. Immunoblotting analyses are performed according to the method of Towbin *et al.* (1979) except that Super Signal West Pico Chemiluminescent Substrate or Supersignal West Dura Extended Duration Substrate (Pierce Biotechnology) is used as the substrate for the horseradish peroxidase conjugate of the secondary antibodies. Peroxisomal thiolase (PT) and bifunctional protein (BF) are used for the marker proteins of the peroxisome. The levels of mitochondrial proteins, β-subunit of ATP synthase and manganese superoxide dismutase (Mn^{2+}-SOD), endoplasmic reticulum marker (BiP), and cytosolic protein, tubulin, are also examined for comparison. The antibodies against Bip and tubulin are purchased from Affinity Bioreagent and Chemicon International, respectively. The antibodies for Atg7 (Tanida *et al.*, 1999), LC3 (Komatsu *et al.*, 2005), BF (Usuda *et al.*, 1991), PT (Tsukamoto *et al.*, 1990), Mn^{2+}-SOD (Matsuda *et al.*, 1990) and the β subunit of ATP synthase (Ezaki *et al.*, 1995) are prepared as described previously.

6. The immunoblotted patterns are scanned and analyzed with a calibrated densitometer (Bio-Rad, GS-800) with 200-μm resolution and with the highest sensitivity setting. Scanned raw data are processed with the gel analysis software Quantity One (Bio-Rad).

In practice, following the induction of peroxisomes by a 2-week treatment with DEHP, peroxisomal degradation is monitored after a 1-week discontinuation of phthalate ester. Fig. 14.1 shows representative data from the measurement of peroxisome degradation. PT and BF, marker proteins of peroxisomes, increased significantly after administration of DEHP but not the vehicle, and both diminished significantly to basal levels at 1 week after DEHP discontinuation. In comparison, the levels of β-subunit, Mn^{2+}-SOD, Bip and tubulin remain unchanged during the same manipulations. Catalase might not be an adequate marker for peroxisome proliferation,

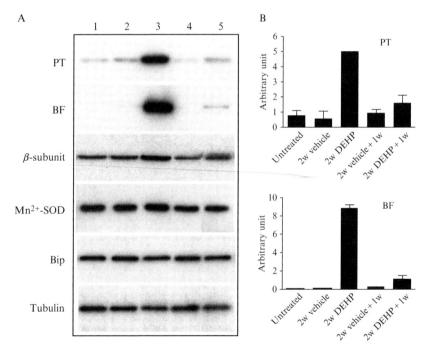

Figure 14.1 The recovery process of excess peroxisomes induced by DEHP treatment. (A) Lane 1: Untreated. Lane 2: Vehicle was administered for 2 weeks. Lane 3: DEHP was administered for 2 weeks. Lane 4: Vehicle was administered for 2 weeks and the mice were kept alive for 1 more week. Lane 5: DEHP was administered and the mice were kept alive for 1 more week. PNS of mouse livers were analyzed by immunoblotting analysis with anti-thiolase (PT), bifunctional protein (BF), β-subunit of ATP synthase (β-subunit), Mn^{2+}-SOD, binding protein (Bip), and tubulin antibodies. (B) Quantitative densitometry of immunoblotting data determined with PT and BF in three different liver PNS samples in each of five experimental groups.

because a high quantity of catalase exists even in the liver of untreated control mice (group 1) and it is difficult to distinguish the differences in cellular levels of this enzyme among different experimental groups by immunoblotting analysis.

3.3. Immunofluorescence analysis of peroxisome degradation

For the histological analysis, we use the following protocol:

1. DEHP-treated and untreated mice are anesthetized with diethyl ether and fixed by perfusion through the left ventricle with saline followed by 4% paraformaldehyde containing 0.1 M sodium-phosphate buffer, pH 7.4. *In situ* fixation is necessary to prevent induction of autophagy during tissue preparation.
2. Livers are excised, dissected into small pieces and further fixed overnight with the same fixative.
3. The fixed samples are then treated with 15% sucrose in phosphate-buffered saline (PBS) for 4 h and then a 30% sucrose solution overnight.
4. The tissues are embedded in Tissue-Tek OCT (Sakura Finetek, 45833) compound and frozen in dry ice/isopentane.
5. The samples are sectioned at 5–7 μm thickness with a cryostat, air dried for 30 min, and stored at $-20\ ^\circ$C until use.
6. For immunohistochemical analysis, the sections are blocked with 5% normal goat serum in PBS containing 0.2% Triton X-100 for 1 h at room temperature.
7. The sections are incubated with the primary antibody against PT or BF for 1 h at room temperature. They are then washed 4 times with PBS and incubated with fluorophore-conjugated second antibody such as Alexa Fluor 488 (Molecular Probes/Invitrogen).
8. Fluorescence images are obtained using a fluorescence microscope equipped with a cooled charge-coupled device camera. We find that immunofluorescence analysis using anti-PT or anti-BF antibody reveals that a 2-week administration of DEHP (Fig. 14.2), but not vehicle, results in the appearance of numerous dots representing peroxisomes, and most of these dots disappear at 1 week after discontinuation of DEHP (Iwata *et al.*, 2006).

3.4. Electron microscopy analyses of peroxisome degradation

3.4.1. Fixation protocol of leupeptin-treated liver specimens

Intra-autophagosomal components and autophagosomal membranes are rapidly degraded by lysosomal hydrolases following autophagosome fusion with the lysosome to mature into an autolysosome. Due to the rapid

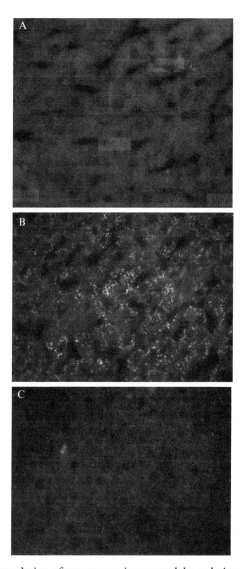

Figure 14.2 Accumulation of excess peroxisomes and degradation. Mice were treated with DEHP for 2 weeks and then chased for 1 week. The frozen sections of livers were stained with anti-PT antibody to detect peroxisomes. Magnification , X400. (A) Untreated mouse. (B) Mouse was treated with DEHP for 2 weeks. (C) Mouse was treated with DEHP for 2 weeks and then subjected to a chase without DEHP for 1 week.

turnover of autophagosomes, inhibition of lysosomal degradation is required to conduct quantitative analyses of autophagosomes by electron microscopy (see also the chapter by Ylä-Anttila *et al.*, in this volume).

1. Usually, mice are injected with leupeptin (2 mg/100 g of body weight) intraperitoneally 1 h prior to isolation of liver specimens.
2. Then, livers are perfusion-fixed with fixative containing 2% paraformaldehyde, 1% glutaraldehyde, and 0.1 M HEPES-KOH buffer, pH 7.4, through the left ventricle for 10 min.
3. Isolated livers are sliced to the appropriate size (2 mm × 2 mm) for postfixation.
4. The slices are divided into 2 groups; one for conventional transmission electron microscopy and the other for peroxidase visualization. Slices for conventional electron microscopy are post fixed with 1% reduced osmium tetroxide. Slices for visualizing peroxisomes (peroxidase) are incubated with alkaline 3, 3-diaminobenzidine (DAB) medium consisting of 2 mg/ml 3, 3-diaminobenzidine tetrahydrochloride (Nacalai Tesque, 11009-41), 0.02% hydrogen peroxide, and 0.1 M glycine-NaOH, pH 10.0, for 1 h at room temperature to visualize peroxisomes. Then, they are postfixed with 1% reduced osmium tetroxide for 1 h. Peroxidase of the peroxisomes catalyzes the conversion of the chromogenic substrate DAB into a dark-stained reaction product that is visible at both light and electron microscopy levels.
5. All liver slices are dehydrated in a graded series of ethanol (50% ethanol, 15 min; 70% ethanol, 15 min; 80% ethanol, 15 min; 90% ethanol, 15 min; 95% ethanol, 15 min; 100% ethanol, 2 × 15 min; propylene oxide (PO), 2 × 15 min; 50% Epon/50% PO, overnight) and embedded in Epon.
6. Thin sections are cut with a diamond knife using an ultramictrotome and contrasted with 40 mM lead citrate for 5 min.
7. The samples are then examined with an electron microscope. We detect the increase of peroxisomes in hepatocytes following a 2-week DEHP treatment, and most of these structures disappear after 1 week of discontinuation of DEHP (Fig. 14.3).

3.4.2. Quantification

For each liver slice, 20 digital electron micrographs for peroxidase visualization are acquired at 5000x magnification, enlarged 2.7-fold, and printed by a laser printer. Using the printed figures, we measure the area of peroxisomes and that of the cytoplasmic area of hepatocytes using a SigmaScan scientific measurement system equipped with a computer (Jandel Scientific, San Rafael, CA). The relative total area of the peroxisomes is calculated using the following formula: (number of peroxisomes in the average area of peroxisome/cytoplasmic area) and expressed in $\mu m^2/100\ \mu m^2$ of cytoplasmic area. After 2-week DEHP treatment, the relative total area of peroxisomes increases but then decreases again to the basal level at 1 week after DEHP withdrawal (Fig. 14.3).

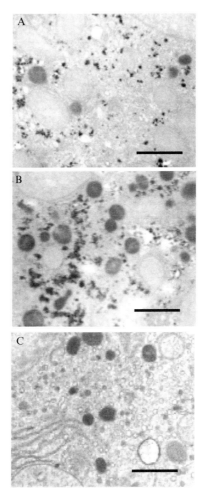

Figure 14.3 Electron microscopy of DEHP-treated mouse livers. Electron micrographs of DEHP-treated and untreated mouse livers. (A) Untreated control liver. (B) Liver of the mouse treated with DEHP for 2 weeks. (C) Liver of the mouse treated with DEHP for 2 weeks and then subjected to a chase without DEHP for 1 week.

4. Concluding Remarks

Orally administered DEHP for 2 weeks successfully and specifically induces peroxisomes in the livers of wild-type mice. No increase in mitochondria and endoplasmic reticulum is induced, so this experimental model is useful for investigating the selective degradation of proliferating peroxisomes via autophagy after discontinuation of DEHP for a subsequent

1 week. During the recovery process, induced peroxisomes are degraded rapidly to below one-tenth of the maximally induced levels. Essentially the same induction of peroxisomes can be reproduced in conditional autophagy-deficient ($Atg7^{-/-}$) livers (Komatsu et al., 2005), but in this case the accumulated peroxisomes remain completely in the liver during the recovery after withdrawal of DEHP from the diet. Peroxisome degradation can be quantitatively monitored by immunoblotting and immunofluorescence in ordinary laboratory manipulations, if appropriate antibodies to peroxisomal marker enzymes are available (Fig. 14.4). Also, quantitative morphometry of peroxisomes using electron microscopy can be also considered as a tool for direct observation of peroxisomes.

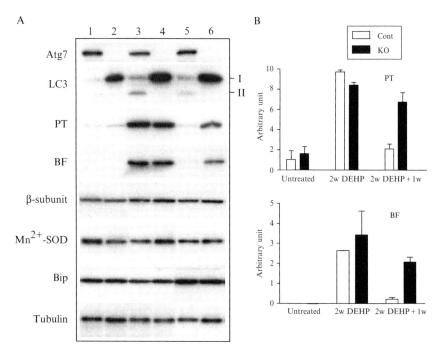

Figure 14.4 The recovery process of excess peroxisomes was impaired in $Atg7$-deficient liver. (A) Immunoblot analyses of liver PNS from DEHP-treated control wild-type and liver-specific $Atg7$-deficient mice. Lane 1: Untreated control mouse. Lane 2: Untreated $Atg7^{F/F}$:Mx1 mouse. Lane 3: Control mouse treated with DEHP for 2 weeks. Lane 4: $Atg7^{F/F}$:Mx1 mouse treated with DEHP for 2 weeks. Lane 5: Control mouse treated with DEHP for 2 weeks and then subjected to a chase without DEHP for 1 week. Lane 6: $Atg7^{F/F}$:Mx1 mouse treated with DEHP for 2 weeks and then subjected to a chased without DEHP for 1 week. Liver PNS were analyzed by immunoblotting analysis using anti-PT, BF, β-subunit of ATP synthase, Mn^{2+}-SOD, Bip, and tubulin antibodies. (B) Quantitative densitometry of immunoblotting data was performed for PT and BF.

We also consider it very useful to establish a procedure to trace peroxisome degradation in cell culture to allow the application of molecular biological techniques. We have attempted to induce peroxisomes in several cultured cell lines, but have so far been unable to do this. The establishment of the method for monitoring pexophagy in cultured cells is important for future analyses.

ACKNOWLEDGMENT

This work was supported by grants-in-aid 15032263, 16790195, 15590254, 09680629, and 1270040 from the Ministry of Education, Culture, Sports, Science and Technology of Japan.

REFERENCES

de Duve, C. (1971). Tissue fractionation. Past and present. *J. Cell Biol.* **50**, 20d–55d.
Dunn, W. A. Jr., Cregg, J. M., Kiel, J. A. W. K., van der Klei, I. J., Oku, M., Sakai, Y., Sibirny, A. A., Stasyk, O. V., and Veenhuis, M. (2005). Pexophagy: The selective autophagy of peroxisomes. *Autophagy* **1**, 75–83.
Ezaki, J., Wolfe, L. S., Higuti, T., Ishidoh, K., and Kominami, E. (1995). Specific delay of degradation of mitochondrial ATP synthase subunit c in late infantile neuronal ceroid lipofuscinosis (Batten disease). *J. Neurochem.* **64**, 733–741.
Fahimi, H. D., Reinicke, A., Sujatta, M., Yokota, S., Ozel, M., Hartig, F., and Stegmeier, K. (1982). The short- and long-term effects of bezafibrate in the rat. *Ann. N.Y. Acad. Sci.* **386**, 111–135.
Guo, D., Sarkar, J., Suino-Powell, K., Xu, Y., Matsumoto, K., Jia, Y., Yu, S., Khare, S., Haldar, K., Rao, M. S., Foreman, J. E., Monga, S. P., Peters, J. M., Xu, H. E., and Reddy, J. K. (2007). Induction of nuclear translocation of constitutive androstane receptor by peroxisome proliferator-activated receptor alpha synthetic ligands in mouse liver. *J. Biol. Chem.* **282**, 36766–36776.
Hashimoto, T. (1982). Individual peroxisomal beta-oxidation enzymes. *Ann. N.Y. Acad. Sci.* **386**, 5–12.
Iwata, J., Ezaki, J., Komatsu, M., Yokota, S., Ueno, T., Tanida, I., Chiba, T., Tanaka, K., and Kominami, E. (2006). Excess peroxisomes are degraded by autophagic machinery in mammals. *J. Biol. Chem.* **281**, 4035–4041.
Komatsu, M., Waguri, S., Ueno, T., Iwata, J., Murata, S., Tanida, I., Ezaki, J., Mizushima, N., Ohsumi, Y., Uchiyama, Y., Kominami, E., Tanaka, K., and Chiba, T. (2005). Impairment of starvation-induced and constitutive autophagy in Atg7-deficient mice. *J. Cell Biol.* **169**, 425–434.
Lake, B. G., Gangolli, S. D., Grasso, P., and Lloyd, A. G. (1975). Studies on the hepatic effects of orally administered di-(2-ethylhexyl) phthalate in the rat. *Toxicol. Appl. Pharmacol.* **32**, 355–367.
Lazarow, P. B., and Fujiki, Y. (1985). Biogenesis of peroxisomes. *Annu. Rev. Cell Biol.* **1**, 489–530.
Lee, S. S., Pineau, T., Drago, J., Lee, E. J., Owens, J. W., Kroetz, D. L., Fernandez-Salguero, P. M., Westphal, H., and Gonzalez, F. J. (1995). Targeted disruption of the α isoform of the peroxisome proliferator-activated receptor gene in mice results in abolishment of the pleiotropic effects of peroxisome proliferators. *Mol. Cell Biol.* **15**, 3012–3022.

Matsuda, Y., Higashiyama, S., Kijima, Y., Suzuki, K., Kawano, K., Akiyama, M., Kawata, S., Tarui, S., Deutsch, H. F., and Taniguchi, N. (1990). Human liver manganese superoxide dismutase. Purification and crystallization, subunit association and sulfhydryl reactivity. *Eur. J. Biochem.* **194,** 713–720.

Reddy, J. K., and Krishnakantha, T. P. (1975). Hepatic peroxisome proliferation: Induction by two novel compounds structurally unrelated to clofibrate. *Science* **190,** 787–789.

Sakai, Y., Oku, M., van der Klei, I. J., and Kiel, J. A. W. K. (2006). Pexophagy: Autophagic degradation of peroxisomes. *Biochim. Biophys. Acta* **1763,** 1767–1775.

Subramani, S. (1998). Components involved in peroxisome import, biogenesis, proliferation, turnover, and movement. *Physiol. Rev.* **78,** 171–188.

Tanida, I., Mizushima, N., Kiyooka, M., Ohsumi, M., Ueno, T., Ohsumi, Y., and Kominami, E. (1999). Apg7p/Cvt2p: A novel protein-activating enzyme essential for autophagy. *Mol. Biol. Cell* **10,** 1367–1379.

Towbin, H., Staehelin, T., and Gordon, J. (1979). Electrophoretic transfer of proteins from polyacrylamide gels to nitrocellulose sheets: Procedure and some applications. *Proc. Natl. Acad. Sci. USA* **76,** 4350–4354.

Tsukamoto, T., Yokota, S., and Fujiki, Y. (1990). Isolation and characterization of Chinese hamster ovary cell mutants defective in assembly of peroxisomes. *J. Cell Biol.* **110,** 651–660.

Usuda, N., Yokota, S., Ichikawa, R., Hashimoto, T., and Nagata, T. (1991). Immunoelectron microscopic study of a new D-amino acid oxidase-immunoreactive subcompartment in rat liver peroxisomes. *J. Histochem. Cytochem.* **39,** 95–102.

Yokota, S. (1986). Quantitative immunocytochemical studies on differential induction of serine:pyruvate aminotransferase in mitochondria and peroxisomes of rat liver cells by administration of glucagon or di-(2-ethylhexyl)phthalate. *Histochem.* **85,** 145–155.

Yokota, S. (1993). Formation of autophagosomes during degradation of excess peroxisomes induced by administration of dioctyl phthalate. *Eur. J. Cell Biol.* **61,** 67–80.

Yu, S., Cao, W. Q., Kashireddy, P., Meyer, K., Jia, Y., Hughes, D. E., Tan, Y., Feng, J., Yeldandi, A. V., Rao, M. S., Costa, R. H., Gonzalez, F. J., and Reddy, J. K. (2001). Human peroxisome proliferator-activated receptor α (PPARα) supports the induction of peroxisome proliferation in PPARα-deficient mouse liver. *J. Biol. Chem.* **276,** 42485–42491.

CHAPTER FIFTEEN

MITOPHAGY IN MAMMALIAN CELLS: THE RETICULOCYTE MODEL

Ji Zhang,[*] Mondira Kundu,[†,1] and Paul A. Ney[*,1]

Contents

1. Introduction	228
2. Reticulocyte Production and Maturation	229
3. Methods	230
3.1. Monitoring mitochondrial clearance in reticulocytes	230
3.2. Monitoring mitochondrial depolarization	234
3.3. Ultrastructural evaluation of mitophagy	234
3.4. Expression of autophagy genes	236
3.5. Immunoblotting of mitochondrial and autophagy proteins	236
4. Mouse Models of Mitophagy	239
4.1. Nix and the role of Bcl-2-related proteins	239
4.2. Ulk1 and autophagy initiation	239
5. Conclusions	240
Acknowledgments	241
References	241

Abstract

Mitochondria are the site of oxidative phosphorylation in animal cells and a primary target of reactive oxygen species-mediated damage. To prevent the accumulation of damaged mitochondria, mammalian cells have evolved strategies for their elimination. Autophagy is one means for the controlled elimination of mitochondria; however, although there has been considerable progress in defining the requirements for nonselective autophagy, relatively little is known about the genes that regulate selective autophagy of organelles. To improve our understanding of mitochondrial autophagy in mammals, we have undertaken a genetic analysis of mitochondrial clearance in murine reticulocytes. Reticulocytes provide an ideal model to study this process, because mitochondria are rapidly cleared from reticulocytes during normal development through an

---

[*] Department of Biochemistry, St. Jude Children's Research Hospital, Memphis, Tennessee, USA
[†] Department of Pathology and Laboratory Medicine, Abramson Family Cancer Research Institute, University of Pennsylvania, Philadelphia, Pennsylvania, USA
[1] These two authors contributed equally

autophagy-related process. Here we describe several methods for monitoring mitochondrial clearance and autophagy in reticulocytes, and show that in reticulocytes these processes require genes involved in both nonselective and selective autophagy.

1. INTRODUCTION

Mitochondria are the site of oxidative phosphorylation and energy production in respiring animal cells, and a major source of reactive oxygen species (ROS) (Pieczenik and Neustadt, 2007; Wallace, 2005). Mitochondria generate ROS as a result of the inherent inefficiency of the electron transport chain, and in addition are a primary target of ROS-mediated damage. ROS damage mitochondrial proteins; for example, iron-sulfur center proteins of the electron transport chain. The iron in these proteins is capable of catalyzing the Fenton reaction and generating additional ROS in the form of highly reactive hydroxyl radicals. ROS also damage mitochondrial DNA, due to its proximity to the site of ROS production and lack of protective histones. The end result of ROS-mediated damage to mitochondrial proteins and DNA is a decrease in mitochondrial function, electron transport, and energy production, and an increase in ROS production. Mitochondrial damage accumulates with age, and it has been hypothesized that decreased mitochondrial function and increased ROS production underlies age-related conditions including degenerative diseases and cancer (Pieczenik and Neustadt, 2007; Wallace, 2005).

The clearance of defective mitochondria is central to the maintenance of cellular homeostasis and is primarily accomplished through selective mitochondrial autophagy (*mitophagy*, hereafter). Mitochondria are constantly undergoing cycles of fusion and fission (Twig *et al.*, 2008). Mitochondrial fusion triggers fission, and fission can generate daughter mitochondria with varying degrees of membrane polarization. Depolarized mitochondria are less likely to undergo subsequent fusion, and more likely to be targeted by mitophagy. Thus, cyclical mitochondrial fusion and fission may be one mechanism for the sorting and targeting of defective mitochondria for clearance. Caloric restriction is another; caloric restriction increases life span and is associated with increased mitochondrial biogenesis (Guarente, 2008). Caloric restriction decreases mammalian target of rapamycin (mTOR) activity and increases mitochondrial recycling by increasing autophagy and mitochondrial clearance. This clearance coupled with increased mitochondrial biogenesis has beneficial effects on aging, including increased energy production and decreased ROS production.

Clearance of mitochondria by autophagy is itself an imperfect process. Autophagy delivers mitochondria to lysosomes, where they are exposed to lytic enzymes and degraded. However, due in part to the effects of ROS, lysosomes accumulate a nondegradable polymeric substance, known as lipofuscin (Terman *et al.*, 2007). Lipofuscin accumulation decreases the degradative capacity of aged cells, and correlates with the appearance of large, abnormal mitochondria. Inhibition of autophagy leads to the reappearance of small mitochondria, suggesting that small but not large mitochondria are actively recycled (Terman *et al.*, 2003). Thus, a deleterious positive feedback loop is established with age, whereby ROS impair the degradative capacity of lysosomes, and this in turn leads to the accumulation of large, defective mitochondria and increased production of ROS.

In light of the role of senescent mitochondria in the generation of ROS (Sohal and Sohal, 1991), programmed mitochondrial clearance is likely to be important for the health of long-lived postmitotic cells, and defects of this process may contribute to age-related disease. Studies in yeast led to the identification of two genes involved in mitophagy and to an appreciation of the role of mitochondrial bioenergetic status in mitophagy induction (Kissova *et al.*, 2007; Priault *et al.*, 2005; Tal *et al.*, 2007). Studies of mitophagy in higher eukaryotes have suggested roles for lipid degradation, ubiquitination, oxidative damage, mitochondrial depolarization, and mitochondrial fission (Chen *et al.*, 2007; Elmore *et al.*, 2001; Sutovsky *et al.*, 1999; Twig *et al.*, 2008; van Leyen *et al.*, 1998). Recent studies in erythroid cells have begun to address the genetic requirements for mitophagy (Kundu *et al.*, 2008; Matsui *et al.*, 2006; Sandoval *et al.*, 2008; Schweers *et al.*, 2007). During development, newly enucleated erythroid cells, also known as reticulocytes, eliminate all of their mitochondria, as they mature into fully differentiated erythrocytes. In this chapter, we discuss the use of reticulocytes as a model to study the mechanism of mitophagy in mammalian cells.

2. Reticulocyte Production and Maturation

Erythroid maturation in humans and mice takes place in the bone marrow, in a 3-dimensional structure known as the erythroblastic island (Allen and Dexter, 1982; Mohandas and Prenant, 1978). Erythroblastic islands are composed of a central macrophage surrounded by erythroblasts at various stages of maturation. As erythroblasts mature in the bone marrow, the erythroid nucleus undergoes marked condensation. At the orthochromatic erythroblast stage, the nucleus is extruded, with its surrounding membrane, leaving behind a cell with organelles but no nucleus called a reticulocyte (Palis, 2008). Several primary cell models recapitulate terminal erythroid differentiation: these include *in vivo* expansion of splenic proerythroblasts by the anemia-inducing strain of Friend virus (FVA), or by treatment with thiamphenicol, followed by *in vitro* culture (Koury *et al.*, 1984; Nijhof and Wierenga, 1983); and the *in vitro* expansion of Ter119-negative fetal liver erythroblasts in growth factors and dexamethasone, followed by culture under differentiation conditions (Dolznig et al., 2001).

Newly formed reticulocytes remain in the bone marrow for approximately 24 h, then are released into circulation (Ganzoni et al., 1969; Tarbutt, 1969). During that time, and in the first 2 days in circulation, reticulocytes change from large motile cells to small biconcave discoid cells, lose surface area and volume, undergo cytoskeletal reorganization, and eliminate ribosomes and all membrane-bound organelles (Chasis et al., 1989; Gronowicz et al., 1984; Koury et al., 2005; Mel et al., 1977; Waugh et al., 1997). The clearance of mitochondria from reticulocytes may be regulated by 15-lipoxygenase (van Leyen et al., 1998). Alternatively, ultrastructural studies indicate that mitochondrial clearance from reticulocytes is mediated by an autophagy-related process (Gronowicz et al., 1984; Heynen et al., 1985). Besides lens epithelial cells (Bassnett, 2002), reticulocytes are one of the few cell types in the human body to undergo programmed mitochondrial clearance during development. Reticulocytes can be generated in large numbers, and are easily obtained in a single cell suspension; thus, reticulocytes provide an ideal mammalian model to examine the genetic requirements of mitophagy.

3. METHODS

3.1. Monitoring mitochondrial clearance in reticulocytes

3.1.1. Flow cytometry

To monitor mitochondrial clearance in reticulocytes, we stain mitochondria with the fluorescent dyes MitoTracker Red CMXRos (MTR), MitoTracker Red 580, or MitoTracker Green FM (MTG) (Molecular Probes, Eugene, OR; cat. nos. M7512, M22425, and M7514) (Table 15.1). MTR yields excellent analytical separation between mitochondria-negative and positive cell populations, and can be used in conjunction with, or without, thiazole orange (TO) (Fig. 15.1). TO stains nucleic acids and provides a measure of ribosomal content. Initial uptake of MTR is sensitive to the polarization state of mitochondria; in contrast, MTG is less effective in separating the 2 populations but stains both polarized and depolarized mitochondria. Therefore, to achieve a complete picture, both dyes should be used as well as other independent approaches. In this regard, MTG works well with immunofluorescence microscopy (see Sections 3.1.2 and 3.2).

1. To induce reticulocytosis, phlebotomize a 10- to 16-week-old mouse, 0.35 ml daily, for 4 days, by retro-orbital puncture with a 100-μl disposable micropipette (Drummond Scientific, Broomall, PA; cat. no. 2-000-100), under general anesthesia, and inject an equal volume of phosphate-buffered saline (PBS), intraperitoneally, for fluid replacement; discard the blood at this stage. The final hematocrit (ratio of

Table 15.1 Physical and spectral properties of fluorescent mitochondrial dyes

Fluorescent dye	Molecular weight (grams/mole)	Ex[a] (nm)	Em[b] (nm)	Potentiometric?	Resistant to fixation?
MitoTracker Red CMXRos	532	578	599	One-way[c]	Yes
MitoTracker Red 580	724	588	644	One-way[c]	Yes
MitoTracker Green FM	672	490	516	No	No
Tetramethylrhodamine methylester	501	549	573	Yes	No

[a] Fluorescence excitation maximum.
[b] Fluorescence emission maximum.
[c] These dyes bind covalently to mitochondrial proteins.

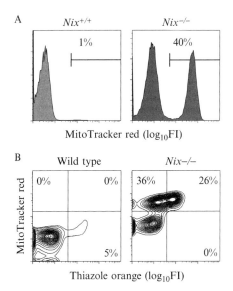

Figure 15.1 Analysis of mitochondrial and ribosomal content of red blood cells by flow cytometry. (A) Single-color flow cytometry of MTR-stained red blood cells, showing mitochondrial retention in $Nix^{-/-}$ red blood cells. (B) Two-color flow cytometry of MTR and TO-stained red blood cells. $Nix^{-/-}$ red blood cells exhibit a population of early reticulocytes (ribosome-positive/mitochondria-positive cells, upper right quadrant), and a selective defect of mitochondrial clearance (ribosome-negative/mitochondriapositive cells, upper left quadrant). Control red blood cells, from wild-type mice, exhibit a small population of late reticulocytes (ribosome-positive/mitochondria-negative cells, lower right quadrant) (Schweers et al., 2007).

packed erythrocyte volume to whole blood volume) should be 20%–25%, which is well tolerated by mice, in our experience. On the fifth day, phlebotomize the mouse 0.35 ml of whole blood (or less, depending on the number of assays to be performed, and the volume of cells required), at which point reticulocytes usually range between 25% to 50% of the red blood cells, and proceed to step 3.
2. Alternatively, induce reticulocytosis by injecting phenylhydrazine-HCl (Sigma, St. Louis, MO; cat. no. P6926), 40 mg/kg in PBS, intraperitoneally, on days 0, 1, and 3. On the seventh day, phlebotomize the mouse 0.35 ml of whole blood, at which point reticulocytes reportedly range between 85% and 95% of the red blood cells (Han et al., 2001). One potential problem with this approach is an increase in autofluorescence due to precipitated hemoglobin.
3. Dilute whole blood from a phlebotomized or phenylhydrazine-treated mouse, 1:500, in complete medium: 30% fetal bovine serum, 1% deionized bovine serum albumin (BSA), 0.001% monothioglycerol (Sigma, St. Louis, MO; cat. no. M6145), 2 mM glutamine, and penicillin-streptomycin in Iscove's Modified Dulbecco's Medium (IMDM) (Invitrogen, Carlsbad, CA; cat. nos. 15140, 25030, and 12440). Aliquot 1.5 ml of cells into the desired number of wells of a 24-well tissue culture plate. Culture cells in a humidified incubator with 5% CO_2, at 37 °C, for 3 days. To prepare deionized BSA stir 1 liter of 30% BSA (Serologicals, Norcross, GA; cat. no. 3415-01) with 50 g of AG 501-X8 resin (Biorad, Hercules, CA; cat. no. 143-6424) for 1 h at 4 °C, decant, repeat with 50 g fresh resin, filter through sterile gauze, render isotonic with 10X PBS (Invitrogen, Carlsbad, CA; cat. no. 70011), dilute to a final concentration of 10% BSA with IMDM, and sterilize by filtering through a 0.22-μ tissue culture filter flask. Transfer to sterile glass bottles and store frozen at −20 °C.
4. To stain reticulocyte mitochondria, add MTR (200 nM) to the cells in the individual wells, and incubate in a humidified incubator with 5% CO_2, at 37 °C, for 30 min. Transfer the cells from a single well to a microcentrifuge tube, centrifuge at 1500×g for 30 s, wash once in cold PBS, and resuspend in 1 ml of PBS for analysis.
5. To monitor both mitochondrial and ribosomal clearance, stain reticulocytes with MTR and TO (Sigma, St. Louis, MO; cat. no. 390062). Start with cells that have been stained with MTR, as described in step 4, only instead of washing, centrifuge and resuspend the cells in 1 ml of TO (2 μg/ml) in PBS at room temperature. Cover from light, and incubate the cells at room temperature for 40 min. Wash the cells twice in cold PBS, and resuspend in 1 ml of PBS for analysis. TO stock (1 mg/ml in methanol) is stored at −20 °C, and is stable for at least 1 year.
6. Measure MTR fluorescence emission at 600 to 620 nm, after 562-nm laser excitation, with a BD LSR II flow cytometry analyzer. Measure TO fluorescence at 500 to 520 nm, after excitation with a 488-nm laser. One potential problem with this approach is an increase in autofluorescence due to precipitated hemoglobin.

3.1.2. Immunofluorescence microscopy

Another approach to monitoring mitochondrial clearance is immunofluorescence microscopy (Fig. 15.2). As noted previously, MTR does not stain depolarized mitochondria; therefore, results obtained with this dye need to be confirmed by another approach, for example transmission electron microscopy (Section 3.3).

1. Stain reticulocytes with MTR (200 nM) or MTG (500 nM), as described in Section 3.1.1, except use 1 ml of cold complete medium (Section 3.1.1) for the wash, resuspend in 0.2 ml of complete medium, and transfer 0.1 ml of cells to a 35-mm, No. 1.5, poly-d-lysine-coated, glass-bottomed, Petri dish (MatTek Corp., Ashland, MA; cat. no. P35GC-1.5-10-C).
2. Perform laser-scanning confocal microscopy with a Nikon TE2000-E inverted microscope equipped with a C1Si confocal system (Nikon, Melville, NY), 488-nm argon ion laser, and 561-nm diode-pumped solid state laser (Melles Griot, Carlsbad, CA). Maintain cells in an environmental control chamber (*In Vivo* Scientific, St. Louis, MO) with 5% CO_2, at 37 °C. We recommend a Nikon Plan Apochromatic 60X oil-immersion objective (N.A. 1.45) with DIC optics. To visualize the mitochondria, excite reticulocytes with the 561-nm (for MTR) or 488-nm (for MTG) laser, and direct the emission to a 605/75 nm or

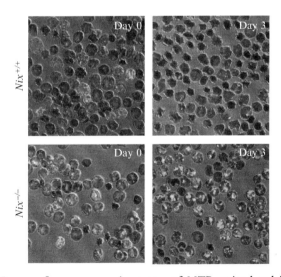

Figure 15.2 Immunofluorescence microscopy of MTR-stained red blood cells in culture. Images show overlay of the red channel (561-nm laser) and transmitted light in the DIC mode. $Nix^{+/+}$ reticulocytes clear their mitochondria in culture over 3 days (upper panels). In contrast, $Nix^{-/-}$ reticulocytes are unable to clear their mitochondria over the same period of time *in vitro* (Schweers et al., 2007). (See Color Insert.)

515/30 nm photomultiplier, respectively. To visualize the cells, use transmitted light in the DIC mode. Adjust power to the 561-nm and 488-nm lasers to a point below the saturation level of the strongest red and green signals, respectively. Set pixel dwell to 10.08 µs, and pinhole to 33 µm. Acquire images at a resolution of 1024×1024 pixels.

3.2. Monitoring mitochondrial depolarization

To assess mitochondrial depolarization, stain mitochondria with MTG and a potentiometric quenching dye, such as tetramethylrhodamine methylester (TMRM) (Table 15.1). Polarized mitochondria, excited with a 488-nm laser, fluoresce red due to fluorescent resonance energy transfer from MTG to TMRM, whereas depolarized mitochondria fluoresce green due to loss of TMRM and retention of MTG (Elmore *et al.*, 2001).

1. Stain reticulocytes in complete medium (Section 3.1.1) with MTG (500 nM), in a humidified incubator with 5% CO_2, at 37 °C, for 30 min. Next, add TMRM (1 µM), incubate for another 30 min, wash the cells once in 1 ml of cold complete medium, resuspend in 0.2 ml of complete medium, transfer 0.1 ml of the stained cells to a glass-bottomed dish, and image the mitochondria by confocal microscopy, as described in Section 3.1.2.
2. It is important to include controls that are stained only with MTG or TMRM to demonstrate that MTG and TMRM signals are not appearing in the red and green channels, respectively, at the chosen laser intensity. If this is the case, then power to the laser should be decreased until the signal disappears.

3.3. Ultrastructural evaluation of mitophagy

Examination of reticulocyte ultrastructure provides several types of information. First, it provides confirmation of mitochondrial autophagy and clearance. Second, under conditions where mitochondrial clearance is defective, it may provide insight into the cause of the defect. Third, it allows a general assessment of cellular remodeling during reticulocyte maturation, for example the clearance of other membrane-bound organelles and ribosomes. To examine the ultrastructure of reticulocytes, we employ a standard protocol of sample preparation and transmission electron microscopy (Hayat, 1989).

1. Begin with whole blood diluted in complete medium, as described in Section 3.1.1, and aliquot approximately 50 ml of cells into each of four 15-cm tissue culture dishes (one for each day of the 3-day culture

period). On each day, harvest the cells from one plate, transfer to a 50-ml conical test tube, and centrifuge at $300 \times g$ for 10 min, at 4 °C.
2. Resuspend the pellet in 10 ml of cold PBS, transfer to a 15-ml conical test tube, and centrifuge at $300 \times g$ for 5 min, at 4 °C. Aspirate and discard PBS, slowly add 1 ml of cold 2.5% glutaraldehyde in PBS to the cell pellet (Tousimis, Rockville, MD; cat. no. 1051), and fix the cells for at least 24 h , at 4 °C.
3. Aspirate and discard glutaraldehyde, and remove traces of glutaraldehyde with a tissue. To immobilize the pellet, add 50 µl of warm 2% low melting point agarose (Sigma, St. Louis, MO; cat. no. A9414). Postfix in osmium tetroxide, and embed in Eponate-Araldite, in a Lynx II tissue processor (Emgrid, Australia), according to the following protocol (all steps are at 20 °C, unless otherwise specified): 0.1 M sodium cacodylate [pH, 7.4] for 10 min (twice); 1:1 mixture of 4% osmium tetroxide (Electron Microscopy Sciences, Hatfield, PA; cat. no. 19190) and 0.1 M sodium cacodylate [pH 7.4] for 2 h; 0.1 M sodium cacodylate [pH, 7.4] for 10 min (twice); 70% ethanol for 15 min, 90% ethanol for 10 min, absolute ethanol for 20 min (this step twice); propylene oxide (Electron Microscopy Sciences, Hatfield, PA; cat. no. 20412) for 15 min (twice); 1:1 mixture of propylene oxide and Eponate-Araldite for 45 min, at 30 °C; and Eponate-Araldite (under vacuum) for 90 min at 30 °C, 90 min at 40 °C, and 1 h at 50 °C. Polymerize the Eponate-Araldite overnight at 70 °C. To prepare the Eponate-Araldite, combine 12.5 ml of Eponate-12 resin, 10 ml of Araldite-502, 31 ml of Dodecenyl Succinic Anhydride (DDSA), and 1.2 ml of Tris-2,4,6-(dimethylaminomethyl)phenol (DMP) (Ted Pella, Redding, CA; cat. nos. 18005, 18060, 18022, and 18042), and mix for 1 h.
4. To locate the cells in the block, cut thick sections (500 nm) on a Leica UCT ultramicrotome (Leica Microsystems, Bannockburn, IL) with a glass knife. Mount thick sections on glass slides, stain with 0.5% toluidene blue (Electron Microscopy Sciences, Hatfield, PA; cat. no. 22050) with 1% sodium borate for 1 min, rinse with water, air-dry, attach coverslip, and examine by light microscopy.
5. After locating the cells, cut thin sections (80 nm) on a Leica UCT ultramicrotome with an ultra diamond knife 45° (Diatome, Hatfield, PA), and mount on copper grids. Stain thin sections with 8% uranyl acetate (Electron Microscopy Sciences, Hatfield, PA; cat. no. 22400) for 15 min, rinse with water (immerse 20 times per cycle, repeat 5 cycles with fresh water), air-dry, then stain with Reynold's lead citrate for 1 min, rinse with water, as earlier, and air-dry. To prepare Reynold's lead citrate, dissolve 1.33 g of lead nitrate in 30 ml of double-distilled water, then add 1.76 g of sodium citrate (Electron Microscopy Sciences, Hatfield, PA; cat. nos. 17900 and 21140) (solution turns cloudy), mix for 30 min, add 8 ml of 1 N NaOH (solution turns clear), add 50 ml of doubly-distilled water, and briefly mix again.

6. Examine thin sections with a Jeol JEM-1200EX II electron microscope, at 60 kV, and archive the images with an Advanced Microscopy Techniques XR111, bottom-mount, CCD digital camera.

3.4. Expression of autophagy genes

To monitor expression of autophagy genes during erythroid maturation, we employ the Friend virus model. Splenic erythroblasts isolated from FVA-infected mice (FVA cells) are partially synchronized at the proerythroblast stage, and undergo terminal differentiation over 48 h, in culture, with erythropoietin. Details of this protocol are provided elsewhere (Koury *et al.*, 1984). We examined several autophagy genes, including the murine homologs of *ATG1*, *Ulk1* and *Ulk2*, and one of the four murine homologs of *ATG8*, *Map1lc3b* (He *et al.*, 2003; Kabeya *et al.*, 2000; Yan *et al.*, 1998; Yan *et al.*, 1999). We found that Ulk1 and Map1lc3b expression increase during erythroid differentiation, but the increase is gradual and delayed relative to stimulation of the erythropoietin receptor (Kundu *et al.*, 2008).

1. Extract RNA from FVA cells with TRIzol (Invitrogen, Carlsbad, CA), at serial time points during the 48-h culture period.
2. To make cDNA, use 1 μg of RNA, and SuperScript II reverse transcriptase, in accordance with the manufacturer's instructions (Invitrogen, Carlsbad, CA; cat. no. 18064).
3. Measure expression by quantitative real-time PCR, using murine *Ulk1* (assay ID, Mm00437238_m1), *Ulk2* (assay ID, Mm00497023_m1), and *Map1lc3b* (assay ID, Mm00782868_sH) primers and probes (Applied Biosystems, Foster, CA), and a 7900HT Sequence Detection System. Analyze results with SDS 2.1 software, and normalize to 18s RNA transcript levels (assay ID, Hs99999901_s1).

3.5. Immunoblotting of mitochondrial and autophagy proteins

Immunoblotting allows the assessment of autophagy induction, mitochondrial clearance, and proteins involved in mitophagy. In yeast, growth of autophagy membranes depends on dual ubiquitin-like conjugation pathways, which produce a protein-protein conjugate, Atg12-Atg5, and a protein-phospholipid conjugate, Atg8–phosphatidylethanolamine (PE) (Ichimura *et al.*, 2000; Mizushima *et al.*, 1998). With regard to Atg8, there are four known mammalian homologs, LC3, GABARAP, GATE-16, and ATG8L (Hemelaar *et al.*, 2003; Kabeya *et al.*, 2004; Kabeya *et al.*, 2000), and we have used modification of three of these, LC3, GABARAP, and GATE-16, to monitor autophagic activity in erythroid cells (see also the chapter by Kimura *et al.*, in this volume). Mitochondrial clearance can be monitored by depletion of integral mitochondrial proteins, such as Cox IV or succinate dehydrogenase subunit B (SDHB). Finally, expression of autophagy proteins, as well as members of the Bcl-2 family, can be assessed by immunoblotting during terminal erythroid differentiation. For these studies we use FVA cells, freshly isolated E14.5 fetal liver cells, or expanded fetal liver cells (Dolznig et al., 2001), differentiated in culture. To evaluate protein expression in reticulocytes, we obtain reticulocyte-enriched red blood cells from phlebotomized or phenylhydrazine-treated mice, as described in Section 3.1.1 (reticulocytes can be further enriched by selection for the transferrin receptor [CD71] through the use of magnetic beads).

1. Whole-cell extracts of FVA cells, fetal liver cells, or red blood cells are made with RIPA buffer (150 mM NaCl, 1.0% Nonidet P40, 0.5% sodium deoxycholate, 0.1% sodium dodecyl sulfate, 1 mM EDTA, and 50 mM Tris-HCl [pH, 8.0], with protease and phosphatase inhibitors: 5 μg/ml Aprotinin, 5 μg/ml Leupeptin, 1 μg/ml Pepstatin (Roche Applied Science, Indianapolis, IN; cat. nos. 10-236-624-001, 11-017-101-001, and 10-253-286-001), 1 mM dithiothreitol, 1 mM PMSF, and 1 mM sodium orthovanadate (Sigma, St. Louis, MO; cat. nos. P7626 and S6508)). Typically, we use 2–5×10^7 FVA or fetal liver cells, or 2–5×10^8 red blood cells, for this purpose. Harvest cells, transfer to a 50-ml conical test tube, and centrifuge at $300 \times g$ for 10 min, at 4 °C. Resuspend the cells in 1 ml of cold PBS, transfer to a microcentrifuge tube, and centrifuge at $1500 \times g$ for 30 s. Resuspend the cells in a 5-fold excess volume of cold RIPA buffer, and mix by rotation, at 4 °C, for 30 min. Centrifuge the lysate at $20,000 \times g$ for 10 min, at 4 °C, transfer the supernatant to a new tube, and discard the pellet. Determine the protein concentration of the lysate with the Bio-Rad protein assay (Bio-Rad, Hercules, CA; cat. no. 500-0006).

2. To perform immunoblotting, resolve the protein (30 μg) on a 12% NuPAGE Bis-Tris gel, with MOPS-SDS running buffer (Invitrogen, Carlsbad, CA; cat. nos. NP0341BOX and NP0001), in an XCell Sure Lock apparatus, 200 V, for approximately 1 h. Next, transfer the protein to PVDF membrane with a Mini-Protean II Cell apparatus (Bio-Rad, Hercules, CA), and NuPAGE transfer buffer (with 10% methanol and 0.1% NuPAGE antioxidant), 16 V, overnight, at 4 °C. Block the membrane in TBST buffer (0.15 M NaCl, 10 mM Tris [pH, 7.4], 0.05% Tween) containing 5% nonfat dry milk for 1 h. Incubations are at room temperature, unless otherwise stated. Wash the membrane with TBST for 5 min, twice. Probe the membrane with primary antibody, diluted in TBST containing 1% milk, overnight, at 4 °C. Wash with the membrane with TBST for 10 min, 3 times. Probe the membrane with secondary antibody, diluted in TBST containing 1% milk, for 1 h. Secondary antibodies are donkey antirabbit IgG-HRP conjugate, and

sheep antimouse IgG–horseradish peroxidase conjugate (1:2,000) (GE Healthcare Bio-sciences, Piscataway, NJ; cat. nos. NA934V and NXA931). Wash the membrane with TBST for 10 min, 3 times. Develop the signal with ECL Western blotting detection reagents, in accordance with the manufacturer's instructions. Exceptions to this protocol are noted subsequently.

3. For LC3, resolve the protein on a 12% NuPAGE Bis-Tris gel, with MES-SDS running buffer (Invitrogen, Carlsbad, CA; cat. no. NP0002). If the LC3 signal is weak, it may be necessary to load more protein, and to use a more sensitive detection method (SuperSignal West Dura Extended Duration Substrate, Pierce Biotechnology, Rockford, IL). Also, use LC3 mouse IgG1 monoclonal primary antibody (1 μg/ml) (Medical and Biological Laboratories, Woburn, MA; cat. no. M115-3) and include controls for identification of both the unconjugated and the phospholipid-modified (faster migrating) forms of LC3. For this purpose, we use extracts of wild-type and $Atg7^{-/-}$ murine embryonic fibroblasts (MEF) (Komatsu et al., 2005), treated in culture with 50 μM chloroquine for 24 h. Chloroquine treatment increases accumulation of the conjugated form of LC3; wild-type MEFs express both forms, and $Atg7^{-/-}$ MEFs express only the unconjugated form of LC3.

4. For GABARAP and GATE-16, resolve the protein on a 16% Tris-Glycine gel (Invitrogen, Carlsbad, CA; cat. no. EC6495BOX), with Tris-Glycine running buffer. This gel-buffer combination can also used for LC3. For GABARAP, use mouse IgG1 monoclonal primary antibody (1 μg/ml) (Medical and Biological Laboratories, Woburn, MA; cat. no. M135-3), and for GATE-16, use rabbit polyclonal primary antibody (1:1000) (Medical and Biological Laboratories, Woburn, MA; cat. no. PM038).

5. For Beclin-1 and Atg7, resolve the protein on a 4%–12% gradient NuPAGE Bis-Tris gel (Invitrogen, Carlsbad, CA; cat. no. NP0321BOX), with MOPS-SDS running buffer. For Beclin 1, use rabbit polyclonal primary antibody (0.1 μg/ml) (Santa Cruz Biotechnology, Santa Cruz, CA; cat. no. sc-11427), and for Atg7, use rabbit polyclonal primary antibody (1 μg/ml) (ProSci, Poway, CA; cat. no. 3617).

6. For Nix, transfer the protein to nitrocellulose membrane, rather than PVDF, and use rabbit polyclonal primary antibody (0.5 μg/ml) (Exalpha Biologicals, Watertown, MA; cat. no. X1120P).

7. Other primary antibodies we use to study mitophagy in reticulocytes are Cox IV mouse IgG2a monoclonal antibody (0.25 μg/ml) (Abcam, Cambridge, MA; cat. no. ab14744); SDHB, goat polyclonal antibody (0.1 μg/ml) (Santa Cruz Biotechnology, Santa Cruz, CA; cat. no. sc-34150); Bcl-X_L mouse IgG2b monoclonal antibody (0.5 μg/ml) (BD Biosciences, San Jose, CA; cat. no. 610210); Mcl1 rabbit polyclonal

antibody (0.1 µg/ml) (Rockland Immunochemicals, Gilbertsville, PA; cat. no. 600-401-394); Bak rabbit polyclonal antibody (0.5 µg/ml) (Upstate Cell Signaling Solutions, Lake Placid, NY; cat. no. 06-536); and β-actin mouse IgG1 monoclonal antibody (1:10,000) (Sigma, St. Louis, MO; cat. no. A5441). Bax rabbit polyclonal antibody (1:3000) is a gift of Joseph Opferman (St. Jude Children's Research Hospital, Memphis, TN).

4. Mouse Models of Mitophagy

4.1. Nix and the role of Bcl-2-related proteins

Our studies, and those of others, identified *Nix* as a gene required for mitophagy in higher eukaryotes (Sandoval *et al.*, 2008; Schweers *et al.*, 2007). Interestingly, Nix is not part of a canonical yeast autophagy pathway, but instead is a Bcl-2-related protein, specifically a member of the BH3-only subgroup (Boyd *et al.*, 1994; Chen *et al.*, 1999; Matsushima *et al.*, 1998; Yasuda *et al.*, 1999). $Nix^{-/-}$ blood contains an abnormal population of erythrocytes that have cleared their ribosomes but retained their mitochondria (Fig. 15.1). Additionally, $Nix^{-/-}$ reticulocytes exhibit defective mitochondrial clearance *in vitro* (Fig. 15.2). Nix is not required for nonselective autophagy, but is essential for the selective targeting of mitochondria to autophagosomes. It is suggested that Nix causes mitochondrial depolarization, and thereby triggers mitophagy (Sandoval *et al.*, 2008); however, Nix does not require Bax or Bak for its activity, nor does its function depend on activation of the mitochondrial permeability transition pore (MPTP) (Schweers *et al.*, 2007). Therefore, any effect of Nix on mitochondrial membrane potential is independent of an effect on these established proapoptotic pathways. Indeed, although mitochondrial depolarization precedes elimination (Sandoval *et al.*, 2008; Zhang and Ney, 2008), it is still unresolved whether it precedes or follows autophagosome formation. Future experiments with autophagy-deficient reticulocytes will help resolve this question.

4.2. Ulk1 and autophagy initiation

Unc51-like kinase (Ulk1) is a serine-threonine kinase that shares homology with the yeast autophagy regulator, Atg1. Atg1 acts downstream of Tor kinase to regulate the size and content of autophagosomal structures, on the basis of its differential interaction with components of the autophagic machinery (Cheong *et al.*, 2005, 2008; Kabeya *et al.*, 2005). Atg1 also regulates the subcellular localization of Atg9, a function that has been linked to autophagosome membrane recycling (Reggiori et al., 2004). To investigate the role of Ulk1 in reticulocyte maturation, we generated $Ulk1^{-/-}$ mice (Kundu et al., 2008). The transferrin receptor, CD71, is downregulated during reticulocyte maturation as the need for iron for hemoglobin synthesis declines. $Ulk1^{-/-}$ reticulocytes exhibited dysynchronous maturation, with the appearance of a population of erythrocytes that downregulated CD71, but retained their ribosomes and mitochondria. This is the first genetic evidence that autophagy plays a role in mitochondrial clearance in reticulocytes. Notably, mitochondria in these CD71-negative cells maintain their membrane potential, suggesting that loss of mitochondrial membrane potential does not occur prior to Ulk1-mediated autophagic clearance of mitochondria. One implication of this result is that genes that are upstream of Ulk1, for example, potentially Nix, do not function by causing mitochondrial depolarization. Atg5 and Atg7 are components of the basal autophagy machinery and germ-line deletion of *Atg5* or *Atg7* leads to perinatal starvation (Komatsu et al., 2005; Kuma et al., 2004). By contrast, $Ulk1^{-/-}$ mice are viable, suggesting that Ulk1 function is redundant, possibly shared with Ulk2 (Hara et al., 2008), and that tissue-restricted Ulk1 may regulate the efficiency or specificity of autophagy in addition to its role as a core component of the autophagy machinery.

5. Conclusions

In conclusion, the dependence of mitophagy on Nix, which is not generally required for autophagy, shows the importance of selective autophagy in mitochondrial clearance. The mechanism of Nix-dependent clearance may involve mitochondrial depolarization or, alternatively, a localized effect on the induction of autophagy. In that regard, there is a growing appreciation of the role of the Bcl-2 family in autophagy regulation (Maiuri et al., 2007; Pattingre et al., 2005). The requirement for Ulk1 suggests that canonical autophagy pathways are also important, although this awaits further characterization of autophagy pathways in $Ulk1^{-/-}$ erythroid cells. In another line of evidence, our studies indicate that another canonical autophagy component, Atg7, is also required for mitochondrial clearance (Zhang and Ney, unpublished results). Still, the defect caused by Ulk1 or Atg7 deficiency is less severe than that caused by loss of Nix, which raises the possibility that there are redundant or autophagy-independent pathways of mitochondrial clearance.

An important issue in interpreting our findings is the applicability of results obtained in the reticulocyte model to mitochondrial clearance in other settings. Initial observations suggested that mitophagy is triggered through mitochondrial depolarization and the MPTP (Elmore et al.,

2001), which would contradict some of our findings in reticulocytes. However, this view has been refined, and there is now evidence for at least two distinct mechanisms for the induction of mitophagy. First, mitochondrial depolarization per se appears to be sufficient to trigger mitophagy, and mitochondrial damage can cause depolarization and mitophagy in a phosphoinositide 3-kinase (PI3-K)-independent manner. Second, other stimuli, for example starvation, can cause the clearance of polarized mitochondria in a PI3-K-dependent manner (Kim *et al.*, 2007). Third, our studies suggest that regulation of Nix is a potential mechanism for controlling the induction of mitophagy. Indeed, Bnip3, which is closely related to Nix, is required for mitophagy during hypoxia (Zhang *et al.*, 2008), suggesting a broader role for these proteins in homeostasis.

Finally, it should be noted that mitochondrial clearance in reticulocytes involves, at least to some extent, the extrusion of membrane-bound cytoplasmic contents (Gasko and Danon, 1972; Noble, 1989), and that this occurs in other cell types as well (Lyamzaev *et al.*, 2008). While this resembles an autophagy-related process in that mitochondria are sequestered in double-membraned vesicles, and in that it requires components of the autophagy machinery, such as Ulk1 and Atg7, it differs from autophagy in the ultimate fate of these structures: extrusion and degradation by macrophages, instead of internal degradation in the cell's own lysosomes.

ACKNOWLEDGMENTS

The authors acknowledge the support of the Animal Resource Center, Flow Cytometry, and Cellular Imaging facilities of SJCRH. This research was supported by grants from the NIH to P.A.N. and M.K., from the Burroughs Wellcome Fund and the American Society of Hematology to M.K., and by the American, Lebanese, and Syrian Associated Charities.

REFERENCES

Allen, T. D., and Dexter, T. M. (1982). Ultrastructural aspects of erythropoietic differentiation in long-term bone marrow culture. *Differentiation*. **21,** 86–94.

Bassnett, S. (2002). Lens organelle degradation. *Exp. Eye Res*. **74,** 1–6.

Boyd, J. M., Malstrom, S., Subramanian, T., Venkatesh, L. K., Schaeper, U., Elangovan, B., Sa-Eipper, C., and Chinnadurai, G. (1994). Adenovirus E1B 19 kDa and Bcl-2 proteins interact with a common set of cellular proteins. *Cell* **79,** 341–351.

Chasis, J. A., Prenant, M., Leung, A., and Mohandas, N. (1989). Membrane assembly and remodeling during reticulocyte maturation. *Blood* **74,** 1112–1120.

Chen, G., Cizeau, J., Vande, V. C., Park, J. H., Bozek, G., Bolton, J., Shi, L., Dubik, D., and Greenberg, A. (1999). Nix and Nip3 form a subfamily of pro-apoptotic mitochondrial proteins. *J. Biol. Chem*. **274,** 7–10.

Chen, Y., Millan-Ward, E., Kong, J., Israels, S. J., and Gibson, S. B. (2007). Mitochondrial electron-transport-chain inhibitors of complexes I and II induce autophagic cell death mediated by reactive oxygen species. *J. Cell Sci*. **120,** 4155–4166.

Cheong, H., Nair, U., Geng, J., and Klionsky, D. J. (2008). The Atg1 kinase complex is involved in the regulation of protein recruitment to initiate sequestering vesicle formation for nonspecific autophagy in *Saccharomyces cerevisiae*. *Mol. Biol. Cell* **19**, 668–681.

Cheong, H., Yorimitsu, T., Reggiori, F., Legakis, J. E., Wang, C-W., and Klionsky, D. J. (2005). Atg17 regulates the magnitude of the autophagic response. *Mol. Biol. Cell* **16**, 3438–3453.

Dolznig, H., Boulme, F., Stangl, K., Deiner, E. M., Mikulits, W., Beug, H., and Mullner, E. W. (2001). Establishment of normal, terminally differentiating mouse erythroid progenitors: molecular characterization by cDNA arrays. *FASEB J.* **15**, 1442–1444.

Elmore, S. P., Qian, T., Grissom, S. F., and Lemasters, J. J. (2001). The mitochondrial permeability transition initiates autophagy in rat hepatocytes. *FASEB J.* **15**, 2286–2287.

Ganzoni, A., Hillman, R. S., and Finch, C. A. (1969). Maturation of the macroreticulocyte. *Br. J. Haematol.* **16**, 119–135.

Gasko, O., and Danon, D. (1972). Deterioration and disappearance of mitochondria during reticulocyte maturation. *Exp. Cell Res.* **75**, 159–169.

Gronowicz, G., Swift, H., and Steck, T. L. (1984). Maturation of the reticulocyte in vitro. *J. Cell Sci.* **71**, 177–197.

Guarente, L. (2008). Mitochondria: a nexus for aging, calorie restriction, and sirtuins? *Cell.* **132**, 171–176.

Han, A. P., Yu, C., Lu, L., Fujiwara, Y., Browne, C., Chin, G., Fleming, M., Leboulch, P., Orkin, S. H., and Chen, J.-J. (2001). Heme-regulated eIF2α kinase (HRI) is required for translational regulation and survival of erythroid precursors in iron deficiency. *EMBO J.* **20**, 6909–6918.

Hara, T., Takamura, A., Kishi, C., Iemura, S., Natsume, T., Guan, J. L., and Mizushima, N. (2008). FIP200, a ULK-interacting protein, is required for autophagosome formation in mammalian cells. *J. Cell Biol.* **181**, 497–510.

Hayat, M. A. (1989). Principles and Techniques of Electron Microscopy Boca Raton, FL: CRC Press, Boca Raton, FL.

He, H., Dang, Y., Dai, F., Guo, Z., Wu, J., She, X., Pei, Y., Chen, Y., Ling, W., Wu, C., Zhao, S., Liu, J. O., and Yu, L. (2003). Post-translational modifications of three members of the human MAP1LC3 family and detection of a novel type of modification for MAP1LC3B. *J. Biol. Chem.* **278**, 29278–29287.

Hemelaar, J., Lelyveld, V. S., Kessler, B. M., and Ploegh, H. L. (2003). A single protease, Apg4B, is specific for the autophagy-related ubiquitin-like proteins GATE-16, MAP1-LC3, GABARAP, and Apg8L. *J. Biol. Chem.* **278**, 51841–51850.

Heynen, M. J., Tricot, G., and Verwilghen, R. L. (1985). Autophagy of mitochondria in rat bone marrow erythroid cells. Relation to nuclear extrusion. *Cell Tissue Res.* **239**, 235–239.

Ichimura, Y., Kirisako, T., Takao, T., Satomi, Y., Shimonishi, Y., Ishihara, N., Mizushima, N., Tanida, I., Kominami, E., Ohsumi, M., Noda, T., and Ohsumi, Y. (2000). A ubiquitin-like system mediates protein lipidation. *Nature* **408**, 488–492.

Kabeya, Y., Kamada, Y., Baba, M., Takikawa, H., Sasaki, M., and Ohsumi, Y. (2005). Atg17 functions in cooperation with Atg1 and Atg13 in yeast autophagy. *Mol. Biol. Cell* **16**, 2544–2553.

Kabeya, Y., Mizushima, N., Ueno, T., Yamamoto, A., Kirisako, T., Noda, T., Kominami, E., Ohsumi, Y., and Yoshimori, T. (2000). LC3, a mammalian homologue of yeast Apg8p, is localized in autophagosome membranes after processing. *EMBO J.* **19**, 5720–5728.

Kabeya, Y., Mizushima, N., Yamamoto, A., Oshitani-Okamoto, S., Ohsumi, Y., and Yoshimori, T. (2004). LC3, GABARAP and GATE16 localize to autophagosomal membrane depending on form-II formation. *J. Cell Sci.* **117**, 2805–2812.

Kim, I., Rodriguez-Enriquez, S., and Lemasters, J. J. (2007). Selective degradation of mitochondria by mitophagy. *Arch. Biochem. Biophys.* **462,** 245–253.

Kissova, I., Salin, B., Schaeffer, J., Bhatia, S., Manon, S., and Camougrand, N. (2007). Selective and non-selective autophagic degradation of mitochondria in yeast. *Autophagy* **3,** 329–336.

Komatsu, M., Waguri, S., Ueno, T., Iwata, J., Murata, S., Tanida, I., Ezaki, J., Mizushima, N., Ohsumi, Y., Uchiyama, Y., Kominami, E., Tanaka, K., and Chiba, T. (2005). Impairment of starvation-induced and constitutive autophagy in Atg7-deficient mice. *J. Cell Biol.* **169,** 425–434.

Koury, M. J., Koury, S. T., Kopsombut, P., and Bondurant, M. C. (2005). In vitro maturation of nascent reticulocytes to erythrocytes. *Blood* **105,** 2168–2174.

Koury, M. J., Sawyer, S. T., and Bondurant, M. C. (1984). Splenic erythroblasts in anemia-inducing Friend disease: a source of cells for studies of erythropoietin-mediated differentiation. *J. Cell Physiol.* **121,** 526–532.

Kuma, A., Hatano, M., Matsui, M., Yamamoto, A., Nakaya, H., Yoshimori, T., Ohsumi, Y., Tokuhisa, T., and Mizushima, N. (2004). The role of autophagy during the early neonatal starvation period. *Nature* **432,** 1032–1036.

Kundu, M., Lindsten, T., Yang, C., Wu, J., Zhao, F., Zhang, J., Selak, M. A., Ney, P. A., and Thompson, C. B. (2008). Ulk1 plays a critical role in the autophagic clearance of mitochondria and ribosomes during reticulocyte maturation. *Blood* **112,** 1493–1502.

Lyamzaev, K. G., Nepryakhina, O. K., Saprunova, V. B., Bakeeva, L. E., Pletjushkina, O. Y., Chernyak, B. V., and Skulachev, V. P. (2008). Novel mechanism of elimination of malfunctioning mitochondria (mitoptosis): Formation of mitoptotic bodies and extrusion of mitochondrial material from the cell. *Biochim. Biophys. Acta.* **1777,** 817–825.

Maiuri, M. C., Le, T. G., Criollo, A., Rain, J. C., Gautier, F., Juin, P., Tasdemir, E., Pierron, G., Troulinaki, K., Tavernarakis, N., Hickman, J. A., Geneste, O., and Kroemer, G. (2007). Functional and physical interaction between Bcl-X_L and a BH3-like domain in Beclin-1. *EMBO J.* **26,** 2527–2539.

Matsui, M., Yamamoto, A., Kuma, A., Ohsumi, Y., and Mizushima, N. (2006). Organelle degradation during the lens and erythroid differentiation is independent of autophagy. *Biochem. Biophys. Res. Commun.* **339,** 485–489.

Matsushima, M., Fujiwara, T., Takahashi, E., Minaguchi, T., Eguchi, Y., Tsujimoto, Y., Suzumori, K., and Nakamura, Y. (1998). Isolation, mapping, and functional analysis of a novel human cDNA (BNIP3L) encoding a protein homologous to human NIP3. *Genes Chromosomes. Cancer* **21,** 230–235.

Mel, H. C., Prenant, M., and Mohandas, N. (1977). Reticulocyte motility and form: Studies on maturation and classification. *Blood* **49,** 1001–1009.

Mizushima, N., Noda, T., Yoshimori, T., Tanaka, Y., Ishii, T., George, M. D., Klionsky, D. J., Ohsumi, M., and Ohsumi, Y. (1998). A protein conjugation system essential for autophagy. *Nature* **395,** 395–398.

Mohandas, N., and Prenant, M. (1978). Three-dimensional model of bone marrow. *Blood* **51,** 633–643.

Nijhof, W., and Wierenga, P. K. (1983). Isolation and characterization of the erythroid progenitor cell: CFU-E. *J. Cell Biol.* **96,** 386–392.

Noble, N. A. (1989). Extrusion of partially degraded mitochondria during reticulocyte maturation. *Prog. Clin. Biol. Res.* **319,** 275–287.

Palis, J. (2008). Ontogeny of erythropoiesis. *Curr. Opin. Hematol.* **15,** 155–161.

Pattingre, S., Tassa, A., Qu, X., Garuti, R., Liang, X. H., Mizushima, N., Packer, M., Schneider, M. D., and Levine, B. (2005). Bcl-2 antiapoptotic proteins inhibit Beclin 1-dependent autophagy. *Cell* **122,** 927–939.

Pieczenik, S. R., and Neustadt, J. (2007). Mitochondrial dysfunction and molecular pathways of disease. *Exp. Mol. Pathol.* **83,** 84–92.

Priault, M., Salin, B., Schaeffer, J., Vallette, F. M., di Rago, J. P., and Martinou, J. C. (2005). Impairing the bioenergetic status and the biogenesis of mitochondria triggers mitophagy in yeast. *Cell Death Differ.* **12,** 1613–1621.

Reggiori, F., Tucker, K. A., Stromhaug, P. E., and Klionsky, D. J. (2004). The Atg1-Atg13 complex regulates Atg9 and Atg23 retrieval transport from the pre-autophagosomal structure. *Dev. Cell* **6,** 79–90.

Sandoval, H., Thiagarajan, P., Dasgupta, S. K., Schumacher, A., Prchal, J. T., Chen, M., and Wang, J. (2008). Essential role for Nix in autophagic maturation of erythroid cells. *Nature* **434,** 232–235.

Schweers, R. L., Zhang, J., Randall, M. S., Loyd, M. R., Li, W., Dorsey, F. C., Kundu, M., Opferman, J. T., Cleveland, J. L., Miller, J. L., and Ney, P. A. (2007). NIX is required for programmed mitochondrial clearance during reticulocyte maturation. *Proc. Natl. Acad. Sci. USA.* **104,** 19500–19505.

Sohal, R. S., and Sohal, B. H. (1991). Hydrogen peroxide release by mitochondria increases during aging. *Mech. Ageing Dev.* **57,** 187–202.

Sutovsky, P., Moreno, R. D., Ramalho-Santos, J., Dominko, T., Simerly, C., and Schatten, G. (1999). Ubiquitin tag for sperm mitochondria. *Nature* **402,** 371–372.

Tal, R., Winter, G., Ecker, N., Klionsky, D. J., and Abeliovich, H. (2007). Aup1p, a yeast mitochondrial protein phosphatase homolog, is required for efficient stationary phase mitophagy and cell survival. *J. Biol. Chem.* **282,** 5617–5624.

Tarbutt, R. G. (1969). Cell population kinetics of the erythroid system in the rat. The response to protracted anaemia and to continuous gamma-irradiation. *Br. J. Haematol.* **16,** 9–24.

Terman, A., Dalen, H., Eaton, J. W., Neuzil, J., and Brunk, U. T. (2003). Mitochondrial recycling and aging of cardiac myocytes: The role of autophagocytosis. *Exp. Gerontol.* **38,** 863–876.

Terman, A., Gustafsson, B., and Brunk, U. T. (2007). Autophagy, organelles and ageing. *J. Pathol.* **211,** 134–143.

Twig, G., Elorza, A., Molina, A. J., Mohamed, H., Wikstrom, J. D., Walzer, G., Stiles, L., Haigh, S. E., Katz, S., Las, G., Alroy, J., Wu, M., et al. (2008). Fission and selective fusion govern mitochondrial segregation and elimination by autophagy. *EMBO J.* **27,** 433–446.

van Leyen, K., Duvoisin, R. M., Engelhardt, H., and Wiedmann, M. (1998). A function for lipoxygenase in programmed organelle degradation. *Nature* **395,** 392–395.

Wallace, D. C. (2005). A mitochondrial paradigm of metabolic and degenerative diseases, aging, and cancer: A dawn for evolutionary medicine. *Annu. Rev. Genet.* **39,** 359–407.

Waugh, R. E., McKenney, J. B., Bauserman, R. G., Brooks, D. M., Valeri, C. R., and Snyder, L. M. (1997). Surface area and volume changes during maturation of reticulocytes in the circulation of the baboon. *J. Lab. Clin. Med.* **129,** 527–535.

Yan, J., Kuroyanagi, H., Kuroiwa, A., Matsuda, Y., Tokumitsu, H., Tomoda, T., Shirasawa, T., and Muramatsu, M. (1998). Identification of mouse ULK1, a novel protein kinase structurally related to C. elegans UNC-51. *Biochem. Biophys. Res. Commun.* **246,** 222–227.

Yan, J., Kuroyanagi, H., Tomemori, T., Okazaki, N., Asato, K., Matsuda, Y., Suzuki, Y., Ohshima, Y., Mitani, S., Masuho, Y., Shirasawa, T., and Muramatsu, M. (1999). Mouse ULK2, a novel member of the UNC-51-like protein kinases: unique features of functional domains. *Oncogene* **18,** 5850–5859.

Yasuda, M., Han, J. W., Dionne, C. A., Boyd, J. M., and Chinnadurai, G. (1999). BNIP3alpha: A human homolog of mitochondrial proapoptotic protein BNIP3. *Cancer Res.* **59,** 533–537.

Zhang, H., Bosch-Marce, M., Shimoda, L. A., Tan, Y. S., Baek, J. H., Wesley, J. B., Gonzalez, F. J., and Semenza, G. L. (2008). Mitochondrial autophagy is a HIF-1-dependent adaptive metabolic response to hypoxia. *J. Biol. Chem.* **283,** 10892–10903.

Zhang, J., and Ney, P. A. (2008). NIX induces mitochondrial autophagy in reticulocytes. *Autophagy* **4,** 354–356.

CHAPTER SIXTEEN

Assessing Mammalian Autophagy by WIPI-1/Atg18 Puncta Formation

Tassula Proikas-Cezanne[*,1] and Simon G. Pfisterer[*]

Contents

1. The Human WIPI Protein Family	248
2. WIPI-1 Puncta-Formation Assay	250
3. Indirect Immunofluorescence to Visualize Endogenous WIPI-1 by Confocal Microscopy	251
3.1. Cell culture conditions	251
3.2. Autophagy assay conditions	252
3.3. Immunostaining of endogenous WIPI-1 to assay for autophagy	254
3.4. Confocal microscopy to assay for autophagy	254
3.5. Quantification of WIPI-1 puncta formation	255
4. Direct Fluorescence of Transiently Expressed GFP-WIPI-1	255
4.1. Transient transfection and autophagy assays	255
4.2. Quantification of GFP-WIPI-1 puncta formation	256
5. Generation of Stable GFP-WIPI-1 Cell Lines	256
6. Live-Cell Imaging of GFP-WIPI-1	257
7. WIPI-1 Protein-Lipid Overlay Assay	258
8. Concluding Remarks	259
Acknowledgments	259
References	259

Abstract

Macroautophagy (autophagy) is an evolutionarily highly conserved self-digestive mechanism that secures eukaryotic cellular homeostasis. Importantly, this process of intracellular bulk degradation is tightly regulated to prevent pathological consequences of disbalanced autophagic activity such as tumor development, neurodegeneration, myopathies and heart disease. A hallmark of the process of autophagy is the generation of autophagosomes, unique vesicles with double membranes of unknown origin and composition. Required for autophagosome formation is the delivery of phospholipids, such as the

---

[*] Autophagy Laboratory, Department of Molecular Biology, Interfaculty Institute for Cell Biology, University of Tübingen, Tübingen, Germany
[1] Corresponding author: Tassula.proikas-cezanne@uni-tuebingen.de

phosphoinositide phsophatidylinositol-3-phosphate (PI(3)P) to phagophore assembly sites (PAS). We identified the human WIPI family of 7-bladed beta-propeller proteins and found that WIPI-1 (Atg18) functions as a novel PI(3)P scaffold at the onset of autophagy. Upon binding to PI(3)P, the WIPI-1 protein accumulation at the phagophore (WIPI-1 puncta formation) can be visualized by fluorescence microscopy. Quantification of WIPI-1 puncta formation is suitable to analyze basal and induced or inhibited levels of phagophore formation, thereby assessing mammalian autophagy. Here we present an experimental step-to-step guide for assaying WIPI-1 puncta formation in human cells by confocal microscopy, live-cell imaging, and phosphoinositide binding.

1. THE HUMAN WIPI PROTEIN FAMILY

Originally, we identified human WIPI-1 as a 5′-truncated cDNA fragment by screening for novel inhibitors of human p53 (Waddell *et al.*, 2001). We isolated additional human WIPI cDNAs by BLAST searching the human genome sequence database and RT-PCR cloning, and defined four human WIPI genes (Proikas-Cezanne *et al.*, 2004). Our bioinformatic analysis shows that WIPI proteins form an evolutionarily highly conserved WD-repeat protein family divided into two large paralogous groups, one containing human WIPI-1 and WIPI-2, and the other human WIPI-3 and WIPI-4 (Table 16.1). By homology modeling we predict that WIPI proteins fold into 7-bladed beta-propellers and that the prototypical WIPI protein contains fully differentiated blades, suggesting that WIPI function is likely to be represented by ancestral properties (Proikas-Cezanne *et al.*, 2004). One ancestral function should be the involvement in macroautophagy, represented by yeast Atg18 (Guan *et al.*, 2001; Barth *et al.*, 2001) and human WIPI-1 (Proikas-Cezanne 2004; Proikas-Cezanne *et al.*, 2007), and should be closely related to the evolutionarily conserved PI(3)P-binding specificity (see Table 16.1). In yeast, PI(3)P-binding of Atg18 promotes localization to the phagophore assembly site (PAS), where Atg18 interacts with Atg2 and Atg9 (Strømhaug *et al.*, 2004; Xie and Klionsky, 2007). This interaction has not yet been demonstrated for human WIPI-1; likewise, depending on PI(3)P-binding, WIPI-1 localizes to the phagophore as demonstrated by electron microscopy and colocalization studies with GFP-LC3 (Proikas-Cezanne *et al.*, 2004, 2007). On the basis of the localization of WIPI-1 to the phagophore, we demonstrated that quantitative fluorescence microscopy of endogenous or overexpressed WIPI-1 represents a novel monitoring system for mammalian autophagy (WIPI-1 puncta formation) (Proikas-Cezanne *et al.*, 2007; Klionsky *et al.*, 2008).

Table 16.1 Human WIPI proteins and yeast orthologs

Human protein	Yeast protein	PI(3)P-binding	PI(3,5)P$_2$-binding	Function in human	Function in yeast
WIPI-1[a]	Atg18[b,c]	Yes[a,d,f,g,h]	Yes[a,f,g,j]	Macroautophagy[a], endosomal/MPR pathway[h]	Macroautophagy[b,c], organelle morphology[i]
WIPI-2[a]	Atg21[d,e]	Yes[d,f]	Yes[d,e]	unknown	Cvt pathway[d,e]
WIPI-3[a]	unknown	unknown	unknown	unknown	unknown
WIPI-4[a]	HSV2	Yes	Yes	unknown	unknown

[a] Proikas-Cezanne et al., 2004,
[b] Barth et al., 2001,
[c] Guan et al., 2001,
[d] Strømhaug et al., 2004,
[e] Meiling-Wesse et al., 2004,
[f] Krick et al., 2006,
[g] Proikas-Cezanne et al., 2007,
[h] Jeffries et al., 2004,
[i] Efe et al., 2007,
[j] Dove et al., 2004.

2. WIPI-1 Puncta-Formation Assay

We demonstrated that WIPI-1 puncta formation in dividing human cells reflects basal phagophore/autophagosome content (see Fig. 16.1). Upon further induction of autophagy by starvation (amino acid deprivation, serum starvation) or administration of compounds such as rapamycin, thapsigargin, and tamoxifen, the number of cells that display WIPI-1 puncta substantially increases. In contrast, upon the inhibition of autophagy by application of PI3KCIII inhibitors such as wortmannin or LY249002, WIPI-1 puncta formation is nullified in the majority of cells. WIPI-1 puncta formation depends on both, availability of PI(3)P and capability of WIPI-1 to bind PI(3)P. Hence, WIPI-1 PI(3)P-binding mutants do not localize to the phagophore and stay distributed in the cytoplasm. Quantification of endogenous WIPI-1 and overexpressed GFP-WIPI-1 (or *myc*-tagged WIPI-1) puncta formation is comparable to GFP-LC3 puncta formation and also to accumulated MDC fluorescence upon induction and inhibition of autophagy (Proikas-Cezanne et al., 2007). As previously reported for the increase of LC3-II (Mizushima and Yoshimori, 2007;

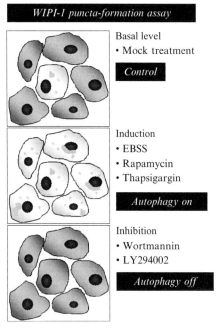

Figure 16.1 Schematic demonstration of the WIPI-1 puncta-formation assay (Proikas-Cezanne *et al.*, 2007)

also see the chapter by Kimura *et al.*, in this volume), the inhibition of autolysosomal cargo breakdown that disturbs the autophagic flux leads to an increase of cells displaying WIPI-1 puncta, probably reflecting an increase of WIPI-1-positive phagophore structures that increase when autophagy is blocked at downstream events (Sabine Ruckerbauer, and Tassula Proikas-Cezanne, unpublished). Therefore, the measurement of both WIPI-1 puncta formation and LC3-II protein monitoring should include controls with protease inhibitors in the culture medium during autophagy assays. Beneficially, in contrast to transiently expressed LC3 that also results in protein aggregations independent of autophagosomal membrane association (Kuma *et al.*, 2007), overexpressed WIPI-1 does not aggregate and is suitable for monitoring autophagy in cells that do not express sufficient levels of endogenous WIPI-1 (Proikas-Cezanne *et al.*, 2007) or where endogenous detection of LC3 is difficult. We have started to employ the WIPI-1 puncta-formation assay in an automated high-throughput procedure and have found that the assay is suitable for drug and siRNA screening approaches as well as for high content analyses of mammalian autophagy by automated fluorescence microscopy (Mario Mauthe and Tassula Proikas-Cezanne, unpublished).

3. Indirect Immunofluorescence to Visualize Endogenous WIPI-1 by Confocal Microscopy

3.1. Cell culture conditions

In general, keeping basal autophagy levels of mammalian cell lines at a constant level is necessary for reproducibility of WIPI-1 puncta-formation results. This can be achieved by performing routine cell culture that includes strict timetables for splitting cells prior to the time they reach confluency and routine screening for mycoplasma infections. Routinely, check for the absence of mycoplasma by PCR analyses and use anti-mycoplasma agents such as Plasmocin (Invivogen) in the culture media. We have successfully used the following human cell lines for autophagy assays followed by WIPI-1 puncta-formation analyses:

1. G-361, human skin, malignant melanoma (ATCC, CRL-1424): detection of endogenous WIPI-1 and/or overexpressed WIPI-1, such as GFP-WIPI-1 (Proikas-Cezanne *et al.*, 2004, available upon request) or c-myc-tagged WIPI-1 (Proikas-Cezanne *et al.*, 2007, available upon request).
2. SK-MEL-28, human skin, malignant melanoma (ATCC, HTB-72): detection of endogenous WIPI-1 and/or overexpressed WIPI-1 (see previous).

3. U-2 OS, human bone, osteosarcoma (ATCC, HTB-96): detection of overexpressed WIPI-1 (see previous).
4. MCF7, human mammary gland, adenocarcinoma (ATCC, HTB-22): detection of overexpressed WIPI-1 (see previous).
5. HeLa, human cervix, adenocarcinoma (ATCC, CCL-2): detection of overexpressed WIPI-1 (see previous).
6. HEK-293, human kidney (ATCC, CRL-1573): detection of overexpressed WIPI-1 (see previous).

We grow all of the above cell lines at 37 °C and 5% CO_2 in Dulbecco's Modified Eagles Medium with 4500 mg/l glucose, GlutaMAX I, pyruvate (DMEM) (Gibco/Invitrogen Corporation), 10% fetal calf serum (FCS) (PAA Laboratories), 100 U/ml penicillin, and 100 μg/ml streptomycin solution (Gibco/Invitrogen Corporation), 5 μg/ml plasmocin ant-mpp (Invivogen), and use 75 cm^2 BD Falcon Culture Flasks with 0.2 μm Vented Blue Plug Seal Cups (BD Europe, France). For long-term storage in liquid nitrogen, we freeze 1×10^7 cells in 1 ml of DMEM, 10% FCS, 10% Dimethylsulfoxide (DMSO).

Generally, we seed cells on coverslips in BD Falcon 6-well tissue culture plates (BD Falcon, Europe, France) as follows:

1. In a tissue culture hood, transfer coverslips into 6-well plates.
2. Add 1 ml of 70% ethanol.
3. Leave at room temperature for a minimum of 30 min.
4. Aspirate off the ethanol.
5. Store dry plates at room temperature until needed.
6. For instant use, wash each well twice with 1 ml of culture medium to take off any residual ethanol.
7. Seed $1-5 \times 10^5$ cells in 2 ml of culture medium in each well.
8. Grow cell monolayers at 37 °C and 5% CO_2.

3.2. Autophagy assay conditions

For autophagy assays use healthy subconfluent cells only and always include mock control treatments. To inhibit basal autophagy by PI3KCIII inhibition, we use wortmannin or the LY294002 compound (both dissolved in DMSO, stored at -20 °C). To induce autophagy above the basal level, we inhibit mTOR by rapamycin (dissolved in DMSO, stored at -20 °C) administration, or by amino acid deprivation using Earle's Balanced Salt Solution (EBSS) (Sigma, catalog number E2888).

Note: Usually, we use the culture medium as mentioned in section 3.1. However, rapamycin administration more profoundly stimulates WIPI-1 puncta formation in the absence of FCS (Gaugel and Proikas-Cezanne, unpublished). Therefore, we recommend omitting FCS when rapamycin is used in autophagy assays.

Routinely, we perform autophagy assays in 6-well plates (see section 3.1) for 1–3 h at 37 °C and 5% CO_2, as follows:

1. Mock control:
   a. Aspirate off the culture medium.
   b. Wash the cells twice with 1 ml of prewarmed culture medium (37 °C).
   c. Add 1 ml of fresh culture medium (37 °C).
2. Rapamycin administration:
   a. Aspirate off the culture medium.
   b. Wash the cells twice with 1 ml of prewarmed culture medium without FCS (37 °C).
   c. Add 1 ml of fresh culture medium (without FCS, 37 °C) containing 100–500 nM rapamycin.
3. Amino acid deprivation:
   a. Aspirate off the culture medium.
   b. Wash the cells twice with 1 ml of prewarmed EBSS (37 °C).
   c. Add 1 ml of fresh EBSS (37 °C).
4. Wortmannin administration:
   a. Aspirate off the culture medium.
   b. Wash the cells twice with 1 ml of prewarmed culture medium (37 °C).
   c. Add 1 ml of fresh culture medium (37 °C) containing 200–250 nM wortmannin.
5. LY294002 treatment:
   a. Aspirate off the culture medium.
   b. Wash the cells twice with 1 ml of prewarmed culture medium (37 °C).
   c. Add 1 ml of fresh culture medium (37 °C) containing 50–100 μM LY294002.
6. Rapamycin/wortmannin administration:
   a. Aspirate off the culture medium.
   b. Wash the cells twice with 1 ml of prewarmed culture medium (37 °C).
   c. Add 1 ml of fresh culture medium (37 °C) containing 200–250 nM wortmannin and incubate the cells at 37 °C, 5% CO_2 for 45-60 minutes (pretreatment).
   d. Aspirate off the culture medium.
   e. Add 1 ml of fresh culture medium (37 °C) containing 200–250 nM wortmannin plus 100–500 nM rapamycin.
7. EBSS/wortmannin treatment:
   a. Aspirate off the culture medium.
   b. Wash the cells twice with 1 ml of prewarmed culture medium (37 °C).
   c. Add 1 ml of fresh, prewarmed culture medium (37 °C) containing 200–250 nM wortmannin and incubate cells at 37 °C, 5% CO_2 for 45–60 min (pretreatment).
   d. Aspirate off the culture medium.
   e. Wash the cells twice with 1 ml of prewarmed (37 °C) EBSS.
   f. Add 1 ml of fresh EBSS plus 200–250 nM wortmannin.

3.3. Immunostaining of endogenous WIPI-1 to assay for autophagy

1. Wash cells (from one or more of the assay conditions in section 3.2) 1–3 times with prewarmed PBS (37 °C).
2. Fix the cells with prewarmed (37 °C) 3.5%–4% paraformaldehyde (dissolved in PBS, pH 7.0, store aliquots at −20 °C for several months) for 10–30 min at room temperature (RT).
3. Wash the cells 1–3 times with PBS.
4. Permeabilize and block the cells with fresh PBS containing 1% BSA, 0.1% Tween-20 for 60 min at RT or overnight at 4 °C.
5. Wash the cells 1–3 times with PBS containing 0.1% Tween-20 and incubate with anti-WIPI-1 antibody (rabbit polyclonal antiserum) (Proikas-Cezanne et al., 2004, available upon request), diluted (1:25–1:250) in PBS containing 0.1% Tween-20 for 30–60 min at 4 °C.
6. Wash the cells 3 times with PBS containing 0.1% Tween-20 and incubate with a secondary antibody, such as Alexa Fluor 488 goat antirabbit IgG (Molecular Probes, catalog number A-11008), diluted (1:250–1:500) in PBS containing 0.1% Tween-20 for 30–60 min at 4 °C. If desired, include a nuclear dye, such as TO-PRO-3 (1:2000, Molecular Probes, catalog number T-3605) during this incubation step.
7. Wash the cells 3 times with PBS containing 0.1% Tween-20 and 1 time with PBS without detergent.
8. Mount the cells using a mounting medium such as Vectashield (Vector Laboratories, catalog number H-1000) or ProLong Antifade Reagent (Molecular Probes, catalog number P-36930) and store the cells at 4 °C in the dark.

Note: Routinely, it is recommended to use staining controls for every set of experiments by including incubation with secondary antibodies alone, incubation with first antibodies alone and unmatched primary and secondary antibodies.

3.4. Confocal microscopy to assay for autophagy

Prepare cells using the assay conditions in section 3.2. For confocal laser microscopy we use an LSM510 microscope (Zeiss) and a 63 × 1.4 DIC Plan-Apochromat oil-immersion objective. Routinely, we use a helium-neon laser for excitations at 543 or 633 nm, and an argon laser for excitations at 488 nm (Proikas-Cezanne et al., 2004; Proikas-Cezanne et al., 2006; Proikas-Cezanne et al., 2007). To acquire images, take a series of 20–30 sections with 0.2- to 0.5-μm spacing along the z-axis. We recommend taking 10 series or a number of fields, enough to document 50–100 cells for each of the assay conditions. We also recommend to take 3–10 representative series with additional zoom to visualize individual cells. Regarding imaging parameters, such as laser intensity, each individual section should be carefully analyzed by using the integrated Range Indicator tool for optimizing confocal settings to avoid oversaturation of images. Imaging results can be represented either as images of individual sections or as projections from confocal stacks created by merging the individual confocal slices. Thereafter, images can be used for quantification of fluorescent intensities with software programs such as ImagePro plus (MediaCybernetics).

3.5. Quantification of WIPI-1 puncta formation

Visualize cells by fluorescence microscopy and group cells into WIPI-1 puncta or WIPI-1 nonpuncta, counting 50–100 cells per assay condition in triplicate sets. Repeat this series of experiments at least 3 times. Count cells positive for WIPI-1 puncta when they display a minimum of 2–3 puncta. Present results as percentage of cells displaying puncta and percentage of cells displaying nonpuncta formation of WIPI-1. If desired, express results as ratio of WIPI-1 puncta:nonpuncta (Proikas-Cezanne *et al.*, 2007).

Note: Counting WIPI-1 puncta per cell as done often for LC3, does not seem to have a value in human cell lines employed in our studies, because the amount of basal and induced WIPI-1 puncta levels greatly fluctuate (see section 3.1) (unpublished observation). However, counting WIPI-1 puncta per cell might be of interest in automated high-throughput approaches.

4. Direct Fluorescence of Transiently Expressed GFP-WIPI-1

4.1. Transient transfection and autophagy assays

1. Seed 5×10^5–1×10^6 cells on sterile coverslips in 6-well plates (see section 3.1) and incubate to 80%–95% confluency at 37 °C and 5% CO_2.
2. Dilute a transfection reagent such as Lipofectamine 2000 (Invitrogen, catalog number 11668019) according to the manufacturer's recommendation in OPTI.MEM (Invitrogen, catalog number 51985042) and incubate for 10 min at RT. Note: We use 10 μl of Lipofectamine 2000 plus 100–150 μl OPTI.MEM.
3. Dilute DNA in OPTI.MEM and combine diluted DNA and diluted Lipofectamine 2000 and incubate 25 min at RT.

Note: We dilute 4 μg DNA in 100–150 μl OPTI.MEM hence use a DNA:Lipofectamine 2000 ratio of 1:2.5.

4. Wash the cells twice with OPTI.MEM, add (0.75–1 ml) OPTI.MEM and slowly apply the DNA/Lipofectamine 2000 mixture (finale volume of 200–300 μl).
5. Incubate for 4–5 h at 37 °C and 5% CO_2.
6. Replace the transfection medium with culture medium containing 10% fetal calf serum (see section 3.1).
7. Incubate for 10–48 h at 37 °C and 5% CO_2 and conduct autophagy assays as described in section 3.2–3.4.

Note: We recommend to conduct autophagy assays 48 h posttransfection for cells that are difficult to transfect.

4.2. Quantification of GFP-WIPI-1 puncta formation

For quantification (see section 3.5) of transiently expressed GFP-tagged WIPI-1, aim for a transfection efficiency of a minimum of 30% and compare only cells displaying similar expression levels, ignoring highly overexpressed GFP signal-intense cells. Routinely, we recommend the use of empty vector controls for every set of experiments. For confocal microscopy (see section 3.4) it is important to keep identical laser intensities and settings. Thereby, we observed that upon the induction of autophagy not only cell numbers displaying GFP-WIPI-1 puncta drastically increase, but that fluorescent intensities of GFP-WIPI-1 puncta were 2- to 3-fold higher in rapamycin-treated cells than in mock treated cells (Sabine Ruckerbauer and Tassula Proikas-Cezanne, unpublished).

5. GENERATION OF STABLE GFP-WIPI-1 CELL LINES

We have generated stable GFP-WIPI-1 expressing human cell lines by initial transfection using Lipofectamine 2000, and we isolated individual clones by standard G418 selection over a period of 3–5 weeks (Anke Jacob and Proikas-Cezanne, unpublished). The cell clones we established using this procedure can be used for at least 3 months for GFP-WIPI-1 puncta-formation assays. However, it is not advisable to use pools of cell clones because this leads to a diverse pattern of GFP-WIPI-1 protein expression, making it more difficult to compare fluorescent intensities in mock treated cells and in cells with induced or inhibited autophagy.

Note: We recommend comparing 2–3 independent clones in subsequent analyses to discard positional insertion artifacts.

6. LIVE-CELL IMAGING OF GFP-WIPI-1

We performed time-lapse microscopy using a Zeiss Cell Observer (Fig. 16.2, left) with an Axiovert 200 inverted microscope equipped with an ApoTome module, an EXFO X-Cite 120 illumination system and a 63 × 1.4 DIC Plan Apochromat oil immersion objective (Carl Zeiss AG, Jena, Germany). The microscope was operated using the Axiovision software, version V.4.5.0.0. To allow live cell imaging the microscope was equipped with a Pecon (Pecon, Erbach, Germany) incubator XL-3, Pecon CO_2 control unit, temperature control unit and heating unit compatible with the XL-3 incubator. Using this setup we either used stable GFP-WIPI-1 cell lines or transiently expressed GFP-WIPI-1 and selected cells displaying similar expression levels for time-lapse microscopy (Simon Pfisterer and Tassula Proikas-Cezanne, unpublished). To prepare time-lapse microscopy, we use the following procedure:

1. Grow cells to 70% confluency in 3.5-cm glass bottom dishes or 6-well plates with a glass bottom and perform autophagy assays (see sections 3.2–3.4).
2. Equilibrate the microscope and incubator to 37 °C and 5% CO_2 before starting the time-lapse microscopy and place cells in an environment-controlled chamber, such as the Pecon heating insert covered with an the Pecon CO_2 lid (Fig. 16.2, right) or a similar device.
3. Acquire images in intervals of 10–15 s. To reduce phototoxic effects and photobleaching, reduce the illuminating light manually at the EXFO light system and Axiovert 200 microscope. We do not recommend using stabilizing compounds such as vitamin C or Trolox to reduce phototoxic

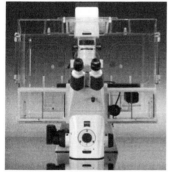

Figure 16.2 Time-lapse microscopy of GFP-WIPI-1 was performed by using the Zeiss Cell Observer (left) equipped with an environment-controlled chamber (Pecon heating insert covered with a CO_2 lid; right) (www.zeiss.de).

effects, as their influence on WIPI-1 puncta formation is not assessed. We recommend using the Apotome module, or any other deconvolution device, for increasing image resolution. For video representation files, display approximately 5 images per second.

7. WIPI-1 Protein-Lipid Overlay Assay

We have shown that WIPI-1 puncta formation strictly correlates with the ability of WIPI-1 to bind PI(3)P (Proikas-Cezanne et al., 2007). We and others suggested, that the major protein pool of WIPI-1/Atg18 that specifically binds PI(3)P is engaged in autophagy, whereas the minor WIPI-1/Atg18 protein pool that specifically binds $PI(3,5)P_2$ is engaged in other vesicle trafficking pathways (Proikas-Cezanne et. al, 2007; Efe et al., 2007; see also Table 16.1). Hence monitoring WIPI-1 puncta formation also indicates the participation of PI(3)P at phagophore formation *in vivo*. Because the majority of WIPI-1 puncta carry PI(3)P, this assay can also be employed to mark PI(3)P *in vivo*, similar to the use of a GFP-fused tandem FYVE domain (GFP-2xFYVE), derived from the endosomal Hrs protein (hepatocyte growth factor regulated tyrosine kinase substrate) (Gillooly et al., 2000). For this purpose, we recommend to control the PI(3)P-binding ability of WIPI-1 by protein-phospholipid protein-overlay assays *in vitro*.

For phospholipid-protein overlay assays, we use the following protocol:

1. Prepare cells using assay conditions in section 3.1., 3.2., or 4.1.
2. Scrape monolayer cells in ice-cold 750 mM aminocaproic acid, 50 mM Bis-Tris, 0.5 mM EDTA, pH 7.0, supplemented with protease inhibitors. Generally, we prepare cell extracts of 1×10^7 to 1×10^8 cells in 1 ml of aminocaproic acid buffer.
3. Generate soluble cellular fractions by repeated cycles (3x) of freezing (dry ice) and thawing (4 °C), and by centrifugation of the extracts at 20,000g and 4 °C.
4. Remove the supernatant fractions to new tubes and determine the total protein concentration.
5. Use 100–1000 μg total cellular extract to overlay 25–100 pmol membrane-immobilized PI(3)P or PI(3,5)P2 (Echelon PIP strips, catalogue number P-6001; Echelon PIP Array, catalog number P-6100).
6. According to the manufacturer's recommendation, incubate the cell extracts with phospholipids overnight at 4 °C and perform a standard ECL detection using anti-WIPI-1 antibodies, or anti-GFP antibodies for detecting GFP-WIPI-1 (Proikas-Cezanne et al., 2007). Binding efficiencies should be normalized over WIPI-1 (or GFP-WIPI-1) protein expression levels for final quantifications.

8. Concluding Remarks

Visualizing and quantifying endogenous WIPI-1/Atg18 accumulation by fluorescence microscopy represents a sensitive novel method to monitor for mammalian autophagy. This approach is employable for screening approaches and for general assessment of mammalian autophagy. Additionally, by overexpressing GFP-WIPI-1 in cells that lack sufficient levels of endogenous WIPI-1/Atg18 protein, active or inactive autophagy can be monitored analogous to the usage of GFP-LC3 or to the detection of LC3/LC3-PE ratio changes in Western blot analyses.

ACKNOWLEDGMENTS

Studies in T. Proikas-Cezanne's laboratory are supported by grants from the Landesstiftung Baden-Wuerttemberg, BioProfile (BMBF), SFB773 (DFG), and a predoctoral fellowship of the Landesgraduiertenförderung Baden-Wuerttemberg to Simon G. Pfisterer.

REFERENCES

Barth, H., Meiling-Wesse, K., Epple, U. D., and Thumm, M. (2001). Autophagy and the cytoplasm to vacuole targeting pathway both require Aut10p. *FEBS Lett.* **508**(1), 23–28.

Barth, H., Meiling-Wesse, K., Epple, U. D., and Thumm, M. (2002). Mai1p is essential for maturation of proaminopeptidase I but not for autophagy. *FEBS Lett.* **512**(1–3), 173–179.

Dove, S. K., Piper, R. C., McEwen, R. K., Yu, J. W., King, M. C., Hughes, D. C., Thuring, J., Holmes, A. B., Cooke, R. H., Michell, R. H., Parker, P. J., and Lemmon, M. A. (2004). Svp1p defines a family of phosphatidylinositol 3,5-bisphosphate effectors. *EMBO J.* **23**(9), 1922–1933.

Efe, J. A., Botelho, R. J., and Emr, S. D. (2007). Atg18 regulates organelle morphology and Fab1 kinase activity independent of its membrane recruitment by phosphatidylinositol 3,5-bisphosphate. *Mol. Biol. Cell* **18**(11), 4232–4244.

Gillooly, D. J., Morrow, I. C., Lindsay, M., Gould, R., Bryant, N. J., Gaullier, J. M., Parton, R. G., and Stenmark, H. (2000). Localization of phosphatidylinositol 3-phosphate in yeast and mammalian cells. *EMBO J.* **19**(17), 4577–4588.

Guan, J., Stromhaug, P. E., George, M. D., Habibzadegah-Tari, P., Bevan, A., Dunn, W. A., Jr., and Klionsky, D. J. (2001). Cvt18/Gsa12 is required for cytoplasm-to-vacuole transport, pexophagy, and autophagy in *Saccharomyces cerevisiae* and *Pichia pastoris*. *Mol. Biol. Cell.* **12**(12), 3821–3838.

Jeffries, T. R., Dove, S. K., Michell, R. H., and Parker, P. J. PtdIns-specific MPR pathway association of a novel WD40 repeat protein, WIPI49. *Mol. Biol. Cell* 15 (6), 2652–2663.

Klionsky, D. J., Abeliovich, H., Agostinis, P., Agrawal, D. K., Aliev, G., Askew, D. S., Baba, M., Baehrecke, E. H., Bahr, B. A., Ballabio, A., Bamber, B. A., Bassham, D. C., *et al.* (2008). Guidelines for the use and Interpretation of assays for monitoring autophagy in higher eukaryotes. *Autophagy* **4**(2),151-175.

Krick, R., Tolstrup, J., Appelles, A., Henke, S., and Thumm, M. (2006). The relevance of the phosphatidylinositolphosphat-binding motif FRRGT of Atg18 and Atg21 for the Cvt pathway and autophagy. *FEBS Lett.* **580**(19), 4632–4638.

Kuma, A., Matsui, M., and Mizushima, N. (2007). LC3, an autophagosome marker, can be incorporated into protein aggregates independent of autophagy: Caution in the interpretation of LC3 localization. *Autophagy* **3**(4), 323–328.

Mizushima, N., and Yoshimori, T. (2007). How to interpret LC3 immunoblotting. *Autophagy* **3**(6), 542–545.

Proikas-Cezanne, T., Waddell, S., Gaugel, A., Frickey, T., Lupas, A., and Nordheim, A. (2004). WIPI-1alpha (WIPI49), a member of the novel 7-bladed WIPI protein family, is aberrantly expressed in human cancer and is linked to starvation-induced autophagy. *Oncogene* **23**(58), 9314–9325.

Proikas-Cezanne, T., Gaugel, A., Frickey, T., and Nordheim, A. (2006). Rab14 is part of the early endosomal clathrin-coated TGN microdomain. *FEBS Lett.* **580**(22), 5241–5246.

Proikas-Cezanne, T., Ruckerbauer, S., Stierhof, Y. D., Berg, C., Nordheim, A., and Human, A. (2007). WIPI-1 puncta-formation: A novel assay to assess mammalian autophagy. *FEBS Lett.* **581**(18), 3396–3404.

Strømhaug, P. E., Reggiori, F., Guan, J., Wang, C.-W., and Klionsky, D. J. (2004). Atg21 is a phosphoinositide binding protein required for efficient lipidation and localization of Atg8 during uptake of aminopeptidase I by selective autophagy. *Mol. Biol. Cell.* **15**(8), 3553–3566.

Waddell, S., Jenkins, J. R., and Proikas-Cezanne, T. A. (2001). "No-hybrids" screen for functional antagonizers of human p53 transactivator function: Dominant negativity in fission yeast. *Oncogene* **20**(42), 6001–6008.

Xie, Z., and Klionsky, D. J. (2007). Autophagosome formation: Core machinery and adaptations. *Nat. Cell Biol.* **9**(10), 1102–1109.

CHAPTER SEVENTEEN

CORRELATIVE LIGHT AND ELECTRON MICROSCOPY

Minoo Razi *and* Sharon A. Tooze

Contents

1. Introduction	262
2. CLEM	263
3. Cell Culture	264
4. Light Microscopy	266
5. Electron Microscopy	268
6. Results, Interpretation, and Presentation	271
7. Tomography and CLEM	273
8. Cryoimmunogold EM and CLEM—the Future?	273
9. Conclusions	274
References	274

Abstract

Autophagy is an intracellular degradative pathway that is essential for cellular homeostasis. Efficient autophagy ultimately relies on the ability of the cell to form autophagosomes, and the efficiency of lysosomal enzymes and lipid hydrolases contained within the autolysosome to degrade sequestered cytosolic material and organelles and recycle these nutrients back to the cytosol. Several assays and techniques to monitor autophagy are available, and these can be quantitative or qualitative, biochemical or morphological. Here we describe a method for monitoring the autophagic process that is based on morphology and the application of both light and electron microscopy, called correlative light and electron microscopy, or CLEM. CLEM provides an advance over either technique (light or electron microscopy) alone and can be performed on any cell or tissue sample, which can be grown or mounted on a gridded coverslip or support compatible with light microscopy. CLEM gives a broad low magnification overview of the cell, allowing an assessment of both spatial and temporal events, as well as providing high-resolution information about individual autophagosomes or single compartments.

London Research Institute, Cancer Research UK, London, UK

1. Introduction

In this chapter we detail how to perform CLEM, correlative light electron microscopy. The application of CLEM to biological samples was first introduced in the mid-1970s and was originally used to correlate conventional light immunohistological staining of cells in tissues with electron microscopy observations, and allowed identification of intracellular vesicles within different cell types (Kobayashi et al., 1978). CLEM can also be adapted to be used in reverse, for example a cytochemical reaction (typically using the enzyme HRP [horseradish peroxidase] followed by DAB [di-aminobenzamidine]) can be carried out on material prepared for electron microscopy (EM), which can then be processed as semi-thin sections and viewed by light microscopy to obtain the gross spatial distribution of the cell within the tissue. CLEM provides an important advance in understanding the properties, topology, and spatial relationship of cells in a tissue (e.g., neurons), in addition to providing important information about the composition, identity and location of subcellular compartments within cells. The purpose of this chapter is to outline a simple application of CLEM to provide an additional tool to understand the process of autophagy in the cell of interest.

Autophagy is an intracellular degradative process, which begins when the nascent cup-shaped double-membrane preautophagosomal membrane (known as the phagophore or isolation membrane; Locke and Sykes, 1975; Seglen et al., 1990) expands, elongates, and encloses cytosol and organelles. Many stress responses can induce autophagy, and one of the most frequently studied is amino acid starvation. Closure of the autophagosome is followed by fusion with endosomal compartments, including late endosomes (Eskelinen, 2005). The final stage of autophagy is the maturation of the autophagosome into an autolysosome in which the degradation of the content occurs. During starvation the degraded content, in the form of amino acids, is recycled back into the cytosol, enabling recovery of amino acid levels.

Many questions remain about autophagy in mammalian cells, and these questions have been discussed in recent reviews (Kundu and Thompson, 2008; Levine and Kroemer, 2008; Mizushima et al., 2008); for example, the source of the isolation membrane and how autophagy-related (Atg) proteins are delivered to the isolation membrane is still poorly understood. The first, and one of the most important techniques used to follow autophagy is EM. EM has been used since the late 1950s to characterize autophagosomes, and more generally autophagic vacuoles (AVs) (Clark, 1957), and it has provided many of the key observations about the autophagic process upon which the recent molecular advances have been placed. A detailed evaluation of how conventional EM can be used to analyze autophagy and autophagosomes will be presented elsewhere in this volume (see the chapter by Ylä-Anttila et al.), and is the subject of recent reviews (Eskelinen, 2008).

We here describe the use of CLEM to characterize the membranous structures labeled by a GFP-tagged (or GFP-derivative-tagged) protein, first at the light level followed by an electron microscopy analysis. Localization of GFP-tagged proteins at the light level allows the broad definition of where the protein is found (cytosol, membrane, nucleus). If membrane associated, further characterization can be achieved by colocalization with well-characterized markers of individual subcellular compartments. However, more detailed information is gained by a higher-resolution analysis of the structures seen in the light microscope by electron microscopy. Fine structure analysis of the shape of the membranous structure, the internal content and immediate environment can all be analyzed by conventional CLEM. From the CLEM analysis one is able to determine whether the GFP-labeled membranes are single- or multilayered, smooth, coated with proteins such as clathrin (responsible for vesicle formation from membranes), or containing lumenal structures. In addition, the nature of the lumenal content, for instance what structures are inside and the morphological state of the content, can be documented, and finally it can be ascertained whether the vesicles are near or associated with other membranes or cytoskeletal elements.

The most basic CLEM experiments correlate the GFP-LC3-labeled structures found in cells undergoing autophagy under the light microscope with morphologically identifiable AVs in the EM, and provide a characterization of the type of AV: if the AV has just formed and is an initial AV (AVi) or has undergone fusion with an acidic, protease-containing endosomal compartment, it is a degradative AV (AVd). However, one should bear in mind that the GFP fluorescence may be diminished in AVds. Therefore, to follow later stages of maturation of the AVd into an autolysosome, it may not be ideal to use GFP-tagged-LC3, as the GFP fluorescence, if diminished, would prevent the AVds from being detected by direct fluorescence in a light microscope. The use of alternative fluorescent proteins such as mRFP or mCherry, or a tandem fluorescent (mRFP-GFP) protein would be advisable (Klionsky *et al.*, 2008), and this adaptation is compatible with CLEM.

2. CLEM

We utilize a cell line, HEK293A, which is stably transfected with GFP-LC3 (Chan *et al.*, 2007), to describe a CLEM approach to verify that structures which are labelled with GFP-LC3 have the morphological appearance of AVs. We will rely on the fluorescence of the GFP-LC3 in a confocal microscope to identify the fluorescent structures, which are formed after induction of autophagy. Autophagy can be induced by a variety of experimental manipulations, including starvation.

CLEM can be used to study cells incubated under other treatments or after transfection with cDNAs, siRNAs, or virus. CLEM can also be used after microinjection of a small number of cells. In the former case, using a tractable cell line, all the cells in the field will have received the treatment, or biological reagent, and thus any cell in the field can be selected. In the latter case, cDNAs, alone or in combination with stable fluorescent tracers (for example Cy3-labeled IgG) or electron-dense tracers such as BSA conjugated to gold (5–15 nm gold) can be microinjected into cells. In this case the gridded coverslip (see step 9 in Cell Culture) is used to record where the microinjected cells are located. The microinjected cells can be identified in the light microscope using the fluorescence of the protein encoded by the cDNA, or the fluorescent tracer.

Finally, CLEM can also be performed on cells expressing an HRP-tagged protein. In this case the location of the labeled protein is revealed by a brown reaction product detectable by phase contrast from the reaction of the HRP with DAB (di-aminobenzamidine) in the presence of H_2O_2, and as a black electron-dense precipitate by EM. Note that the best results are obtained with HRP when the reaction product is contained within the lumen of a compartment (e.g., within the ER or Golgi); otherwise, the diffusion of the reaction product in the cytosol significantly decreases the signal.

3. Cell Culture

1. Warm trypsin (0.25%) and versene (PBS with 0.5 mM EDTA) to 37 °C in a water bath.

   Note: all reagents and materials in step 1–10 are sterile, and the manipulations are performed under sterile conditions. If a method of harvesting and replating cells has already been developed for a particular cell line, start the protocol at step 8 with resuspension in fresh medium.

2. Remove the culture medium (DMEM, 10% FCS) from a 75-cm^2 flask.
3. Wash the HEK293 cells in the flask using 5–10 ml of prewarmed versene.
4. Add 5–10 ml of trypsin diluted in versene and incubate at 37 °C in a humidified, 10% CO_2-containing incubator for 5 min, or until cells detach from the tissue culture flask.
5. Add 5 ml of culture medium to the detached cells in the flask.
6. Collect the suspended cells from the flask and transfer to a 15-ml centrifuge tube.
7. Centrifuge for 5 min at 1200 rpm (1000×g).
8. Aspirate the supernatant fraction. Resuspend the pellet fraction in fresh medium. The volume used to resuspend the pellet will depend on the original cell density, the target cell density, and the number of coverslips being prepared.

9. Prepare glass coverslips containing a grid or pattern that facilitates the identification of the cell of interest during all steps of the procedure. We use a 13-mm round glass coverslip manufactured with a grid containing coordinates (CELLocate, Eppendorf, or Bellco Biotechnology). A pattern can also be scratched manually onto plain 13-mm glass coverslips using a diamond pencil or onto plastic sterile coverslips. A good source of plastic coverslips is Aclar plastic (http://www.tedpella.com). Do not use Thermonox coverslips as they autofluoresce. We suggest a small grid with two different shapes in the upper left- and right-hand corners, for example an open and closed square to simplify orientation. All coverslips should be sterilized before use.

Note: It is advisable to consult the local EM unit for advice about local acceptable practices and for general advice, as at this stage as there are many alternative approaches; for example, some EM units may not work with cells grown on glass coverslips because of the possibility of traces of glass being carried over in the sectioning procedure (see step 15 in Electron Microscopy).

The type of light or confocal microscope available to view the cells on the coverslips will determine how the coverslips are placed into tissue culture dishes. If an upright microscope is available, the coverslips, with the grid facing up, can be placed directly in a 35-mm tissue culture dish. Alternatively, the coverslips can be mounted into a hole drilled in the bottom of a 35-mm tissue culture dish to allow imaging of cells using an inverted microscope. Coverslips can be mounted on the bottom of the dish with grid or etched side facing up into the dish using sterile silicon vacuum grease or hot wax. The wax that is used routinely in our microscopy unit is made of three components, beeswax, paraffin soft yellow and paraffin wax (all from ThermoFisher) in 1:1:1 ratio.

If the cells detach from the substratum easily, it is advisable to precoat the coverslip with poly-D-lysine for 5 min (Sigma, 1 mg/ml in dH_2O, sterilized through a 0.2 μm filter) followed by 3 washes in distilled H_2O at room temperature before plating the cells. This will help the cells adhere to the gridded or scratched coverslips.

The extent of confluency of the cells is important, as the neighboring cells will be used as markers for localizing the cell of interest. Therefore, having a confluent monolayer will make it very difficult to find the cell of interest by EM. The optimal confluency is 50%–60% on the day of the fixation (see below).

10. Plate the cells on the coverslips and incubate until they adhere. Perform any experimental manipulations (e.g., including cDNA or siRNA transfections, drug treatments, microinjection) as required and complete the incubation period at 37 °C in a tissue culture incubator.

4. LIGHT MICROSCOPY

1. After appropriate incubations, wash the cells on the coverslips once with PBS, and fix for 30 min at room temperature using a fixative containing 4% paraformaldehyde and 2% gluteraldehyde in 0.1 M Na cacodylate, pH 7.2, prewarmed to 37 °C in a water bath. The paraformaldehyde and gluteraldhyde are purchased as liquid concentrates at 16% and 25%, respectively from Agar Scientific (Stanstead, England, http://www.agarscientific.com). Prepare 0.1 M Na cacodylate from 125 ml of 0.4 M sodium cacodylate, 20 ml of 0.2 M HCl, 335 ml dH$_2$O (final volume of 480 ml). Use sufficient volume of fixative to ensure a good preservation (i.e., a minimum of 2 ml for a 35-mm dish). It is important to keep the cells on the coverslips covered with liquid at all times to prevent drying out, which will destroy the morphology of the cells. Discard the remaining fixative solution safely.

Note: Do not keep open fixatives or dilute solution in refrigerators or freezers with any biological reagents, as they are very volatile and can fix and damage the biological reagent.

2. Wash the cells 3 times with PBS. The fixed cells remain in PBS while images of the cells are captured. At this stage the labeling should be detectable by light microscopy using either phase contrast (in the case of a detectable reaction product) or by fluorescence (in the case of fluorescent-tagged proteins, or fluorescent dyes to mark the cells of interest).
3. Find a cell of interest and determine and note its position using the marks on the coverslip. The grids or scratch should be visible in the phase contrast images (Fig. 17.1A). Capture both phase contrast and fluorescent images (where appropriate) of the cell of interest (Figs. 17.1B and 17.1C), and the cells in the surrounding area as the shape and arrangement of the neighboring cells will help find the cell of interest under the EM. A good working magnification is a 20X lens for the overview images. The phase contrast images should be focused to give the largest outline of the cells. This focal plane will be used in the next section, at step 20.

    It is useful to select, document and photograph 2 or 3 regions with positive cells because of the possibility that cells can be lost in the EM processing and the orientation may be unclear. Fields containing cells with more distinct features are preferable as it helps in identifying the positive target cells under EM.

4. Next acquire high magnification pictures of the cell of interest using a 60x objective. If a confocal microscope is used to photograph the positive cells, it is important to capture images at different Z positions to help correlate the light and EM images.

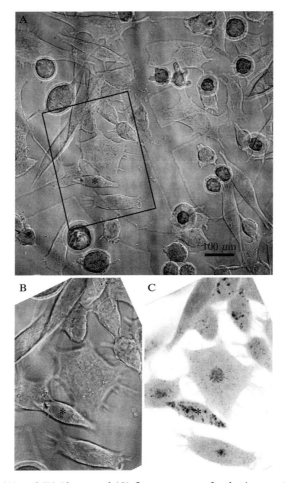

Figure 17.1 (A) and (B) Phase, and (C) fluorescent confocal micrographs of HEK293 cells stably transfected with GFP-LC3 are used to identify a cell of interest in preparation for CLEM. The cells in this experiment were transfected with an siRNA which inhibits AV maturation. In (A) a low magnification image of the GFP-LC3-expressing cell is captured by phase microscopy. The cell of interest is marked by an asterisk and is located on the left side of the letter F on the coverslip grid, on the circle surrounding the F, near an intersecting horizontal line. Note that the cell immediately below the marked cell has well defined filipodia extending to the left, and a larger, flat cell above to the upper right. (B) shows a phase image of the region containing the cell of interest at the same magnification as (C). (C) A single Z-slice confocal image (in reverse contrast) showing the GFP-LC3 fluorescent structures as dark spots. The two cells being used as landmarks are easily seen at this magnification.

5. Electron Microscopy

After capturing images using a light or confocal microscope, the cells on the coverslip are processed for EM using the following protocol. There are many different EM embedding protocols and resins available and any alternative standard protocol is acceptable.

Note: All the reagents used are EM quality. Handling and disposal of the used reagents requires special care, so consult the local EM unit for appropriate handling precautions and read the safety data sheets before starting.

1. Remove all PBS from the dish and add 2% OsO_4 (Agar Scientific) in 0.1 M Na-cacodylate, pH 7.2. Incubate for 1 h on ice in the dark (e.g., using a foil-covered ice bucket) in a fume hood. OsO_4 will fix and stain the lipids in the cell. Note that this reagent must be used with appropriate care and stored as advised by the local EM unit.
2. Remove the OsO_4 and wash the cells twice in 0.1 M Na-cacodylate, pH 7.2.
3. Incubate the cells in freshly made 1% tannic acid (improves contrast of proteins) in 0.05 M Na-cacodylate, pH 7.2, for 40 min at room temperature.
4. Wash twice in 0.05 M Na-cacodylate, pH 7.2.
5. Wash once in double-distilled (dd) H_2O.
6. Dehydrate the cells by incubating with first 70%, then 90% ETOH for a few minutes. It is advisable to keep the 100% ETOH stock solutions stored with desiccant (we use a 3Å molecular sieve) to remove any water. Do not shake the ETOH bottle before removing the stock solution to avoid any particulate material from the desiccant entering the solutions.
7. Incubate with 100% ETOH for 10 min, remove the 100% ETOH, and repeat for an additional 10 min.
8. Mix a 1:1 solution of propylene oxide (from Agar):Epon. Epon is prepared from 20 ml of Araldite cy212 (Agar), 0.8 ml of DMP-30 (TAAB), and 25 ml of DDSA (TAAB).
9. Add the propylene oxide:Epon mix to an aluminium dish (from Agar, see Fig. 17.2A).

Note: Propylene oxide will dissolve most laboratory plastic so EM-grade plasticware or glass should be used for pipetting these solutions.

10. Using fine forceps, remove the coverslip from the 100% ETOH, transfer to the foil dish and incubate for 30 min at room temperature.
11. Remove as much as possible of the 1:1 mixture, replace with fresh 100% Epon, and incubate for 1 h.

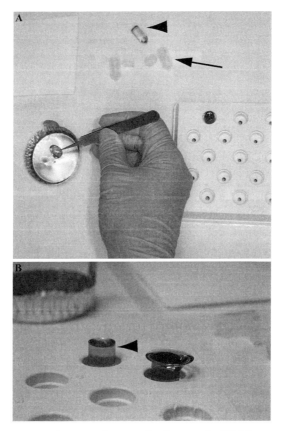

Figure 17.2 Mounting the coverslip onto the Epon stub. (A) The coverslip with the cells facing up is removed after the final incubation from the aluminium tray containing the 100% Epon solution with a fine forceps. Also shown is the empty Beem capsule (arrow), the prepolymerized Epon stub on its side (arrowhead), and the Beem capsule holder. (B) The coverslip is positioned on the flat surface of the Epon stub (arrowhead) using sufficient Epon to ensure there are no air bubbles trapped between the coverslip and the Epon stub.

12. Remove the Epon, replace with fresh Epon and incubate for an additional 1 h.
13. Mount the coverslips cell side down on the flat end of prepolymerized Epon stubs made in a cylindrical capsule (Beem capsules, Agar Scientific) transferring enough liquid Epon to ensure there are no air bubbles between stub and coverslip (Fig. 17.2A). Before positioning the coverslip, gently remove excess Epon from the side of the coverslip without the cells with filter paper to facilitate the removal of the glass coverslip in

step 15. As the diameter of the stubs is smaller than the coverslip, make sure the area placed on the stub has the area of the grid on which the cells of interest are found (Fig. 17.2B).

14. Bake at 60 °C overnight to polymerize the Epon.
15. Remove the glass coverslip from the cells now embedded in Epon by dipping the coverslip-stub in liquid nitrogen for a few seconds. The glass coverslip should break off, and the cells should stay behind on the stub. Make sure all the glass coverslip is removed or any remaining glass fragments may damage the diamond knife used in step 18.
16. Using the eyepiece of a microtome, localize the cells in the region of interest on the gridded coverslip using the coordinates now imprinted on the Epon. Both the cells and the grid should be visible.
17. Carefully trim the block with a single-sided razor blade so that the cells of interest are located in the middle of the trimmed block. If you do not see the cell you wish to study, locate one of the other regions documented in step 2 of the previous section, and position them in the center of the block. It is advisable to trim down the block as much as possible; however, if the block is too small this can cause problems with viewing and photographing in the EM as the edge of the sections can fold as the section is examined under EM and the cells that are too close to the edge will be lost. The optimal size is of the block face is approximately 0.5 mm by 0.2–0.5 mm. It is preferable to trim the block in such a way that the top and bottom of the block are parallel, allowing for the formation of a straight ribbon during sectioning, but the left and right side of the block are trimmed at different angles to make a trapezoid shape with unequal vertical sides. This helps determinate of the orientation of sections in the EM, and allows you to identify the first section cut, which should be at the bottom of the ribbon.
18. Cut serial thin (70 nm) sections for conventional transmission EM, or thick (250 nm) sections for tomography. The location of the structure of interest will be determined from the Z-sections acquired by light microscopy: if the structure is, for example, on the Z-axis in the middle of the cell it is not necessary to thin section through the whole cell. Use thick sections (200 nm thick) to reach the region of interest in the minimum number of sections. Using the known thickness of the section, counting the number of sections cut will allow you to estimate where you are in the cell.
19. Place the ribbon of sections on a slot grid coated with Formvar to support the sections and view with a transmission EM (TEM). Formvar is prepared by dissolving 1.1 g of Formvar (Merck) in 100 ml of chloroform. Leave overnight in the dark at room temperature before coating the grids.

20. To enhance the contrast of the cells on the section, stain with lead citrate. Place a small drop of lead citrate (20 μl) on a clean piece of Parafilm in a 10-cm plastic Petri dish containing 4–5 NaOH pellets, and invert the slot grid on top of the drop for 5–10 min at room temperature. Lead citrate, which stains lipids, forms precipitates with ambient carbon dioxide that can adhere to the sections, and the NaOH pellets absorb the carbon dioxide. Lead citrate solution is prepared from 1.33 g of lead nitrate, 1.76 g of sodium citrate, in 50 ml of ddH_2O, shaken thoroughly for 1 min and then intermittently over 30 mins. Add 8.0 ml of freshly prepared 1 N NaOH.
21. Wash grid thoroughly either on drops of ddH_2O or by dipping into in ddH_2O to remove excess lead citrate. Let the grids dry at room temperature.
22. The section that was closest to the coverslip should be examined first in the electron microscope as it will have the highest similarity to the phase images and this will help to find the cell of interest in the TEM (Fig. 17.3). The grid imprinted in Epon will no longer be visible, which is why the Epon block needs to be trimmed carefully so the region of interest falls in the middle of the sections.
23. Once the region containing the cells is located on the grid in the TEM, photograph the whole region at low magnification (Fig. 17.3B). This allows the comparison of the light and EM images. Next, photograph the whole cell at a higher magnification to make sure no detail is lost (Fig. 17.3C). If necessary, capture serial images of the region of interest from the consecutive serial sections to study individual structures in more detail.

6. Results, Interpretation, and Presentation

1. Prepare a montage of the EM images captured at a high magnification (Fig. 17.3C). Orient the light microscopy and TEM images so they are the same. This simplifies the presentation of the cell of interest. To correlate the fluorescent objects with the TEM images, use different landmarks within the cell such as number and the position of the nucleoli, extensions of the cell of interest, and the shape of the neighboring cells. Compare the light micrographs again with the electron micrographs to verify the position of the structures detected by light microscopy and EM.

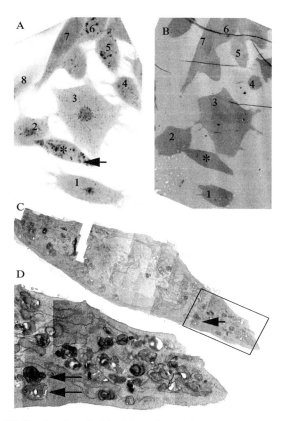

Figure 17.3 (A) Confocal, and (B), (C) and (D) electron micrographs of the GFP-LC3-positive cell. Note (A) is the same image as Fig. 17.1C. In (A) the cell of interest is marked with an asterisk, while the surrounding cells are numbered 1–8. Note the two landmark cells from Fig. 17.1 are numbered 1 and 3. In (B) a low magnification EM image taken of the one of the first sections from the Epon-embedded sample is shown. The cell of interest was found using the landmark cells and the position of the surrounding cells. Note, that cell No. 8 is not visible, either because it has been lost during the preparation or it is above the plane of the section. (C) The EM images of the cell marked with an asterisk have been assembled into a montage to give an overview of the entire cell. In (A) and (C) the two AVs examined at higher magnification in (D) are marked with a single black arrow. (D) A higher magnification of the boxed region in C showing the membrane-bound structures corresponding to the GFP-LC3-labeled structures seen in A. A variety of AVs can be seen at this magnification and further information can be extracted using images taken at higher magnification. The two AVs seen at lower magnification in (A) and (C) are pointed out with two arrows. Under these experimental conditions most of the AVs are classified as late AVds.

2. Note any distinct subcellular structures in the vicinity of the structure or unusual features in the structure. If serial sections were obtained examine the same location on each subsequent EM section again using landmarks, in this case mitochondria, vacuoles, cytoskeletal structures, and so on.

If the structure labeled in the light microscope is expected to correlate with a membrane-bound compartment, then determine the shape and size of the organelle using the serial sections.

7. Tomography and CLEM

More detailed information about the structure or organelle can be obtained using EM tomographs of the cell. Tomography is usually performed after a detailed analysis of the structure done using conventional EM has indicated that the analysis of the structure in 3-D may yield more information. For example, thin membrane cisternae that are perpendicular to the plane of the section can appear as vesicles when cross-sectioned. Tomography can be used to show that the structures are not spherical vesicles but rather tubules and provide additional information about the 3-D shape of the structure. Tomography requires a TEM with a tilting stage to capture images incrementally at closely spaced angles. Customized software is then used to align the images creating a tomogram onto which one can trace and render the structures of interest in 3-D.

1. The same procedure is followed as previously, except at step 18, 250 nm Epon sections are cut, placed on the slotted EM grid, and contrasted with lead citrate.
2. The cell of interest and then the structure of interest are located, and images from the tilt series are captured.
3. Customized software is used to prepare the tomogram.

8. Cryoimmunogold EM and CLEM—the Future?

Whereas conventional CLEM allows the identification of the structures labeled by a fluorescent protein, an important advance in the characterization of the structure would be gained from the colocalization of other proteins to the labeled structure. This could, in principle, be achieved using cryoimmunogold labelling (Slot and Geuze, 2007). The application of cryoimmunogold EM to CLEM is a major technical challenge. The main hurdle is the transfer of the cells of interest (i.e., those photographed by light microscopy) to a holder (a metal pin) for cryosectioning in a way that preserves the orientation of the cells on the coverslip and their subsequent re-identification. For cryoimmunogold labeling the cells are not embedded in Epon but rather gelatin, which does not preserve the imprint of the grid from the coverslip. One possible way to overcome these problems would be to examine thick sections on a light microscope, mounting the sections in the same way histological sections are prepared, then when the region of interest is found, labeling consecutive thin sections using cryoimmunogold techniques.

9. Conclusions

CLEM can be used effectively to characterize structures originally seen in the light microscope at a high resolution in the electron microscope. It is important to be aware that this technique has the usual drawback of any morphological application on fixed cells in that it does not prove that autophagy is induced. We show here the identification of GFP-LC3 containing AVs in a stably transfected HEK293 cell line (Chan et al., 2007) using CLEM. The use of CLEM provides a fine structure analysis of subcellular structures that can be used to (1) identify the structures and (2) to obtain information about their composition. A good example of the use of CLEM is to provide the distinction between different structures that are labeled with a fluorescent protein, which at the low resolution of the light microscope are seemingly identical. An example of this is the identification of the structures labelled by GFP-p62/SQSTM1, called p62 bodies, seen in the confocal microscope as spherical structures (Bjørkøy et al., 2005). Using CLEM, Bjørkøy et al demonstrate that p62 labels both morphologically defined membrane-bound autophagosomes and membrane-less structures called sequestosomes (see Bjørkøy et al., 2005, Fig. 4). Other applications for CLEM to study autophagy would include studying cells undergoing processes such as selective autophagy of, for example, mitochondria (mitophagy) or ribosomes (ribophagy) (Mizushima et al., 2008).

REFERENCES

Bjørkøy, G., Lamark, T., Brech, A., Qutzen, H., Perander, M., Overvatn, A., Stenmark, H., and Johansen, T. (2005). p62/SQSTMI forms protein aggregates degraded by autophagy and has a protective effect on huntingtin-induced cell death. J. Cell Biol. **171**, 603–614.

Chan, E. Y. W., Kir, S., and Tooze, S. A. (2007). siRNA screening of the kinome identifies ULK1 as a multi-domain modulator of autophagy. J. Biol. Chem. M703663200.

Clark, S. L. Jr, (1957). Cellular differentiation in th kidneys of newborn mice studies with the electron microscope. J. Cell Biol. **3**, 349–362.

EsKelinen, E. L. (2005). Maturation of autophagic vacuoles in Mammalian cells. Autophagy **1**, 1–10.

EsKelinen, E. L. (2008). Fine structure of the autophagosome. Methods Mol. Biol. **445**, 11–28.

Klionsky, D. J., Abeliovich, H., Agostinis, P., Agrawal, D. K., Aliev, G., Askew, D. S., Baba, M., Baehrecke, E. H., Bahr, B. A., Ballabio, A., Bamber, B. A., Bassham, D. C., et al. (2008). Guidelines for the use and interpretation of assays for monitoring autophagy in higher eukaryotes. Autophagy **4**, 151–175.

Kobayashi, S., Serizawa, Y., Fujita, T., and Coupland, R. E. (1978). SGC (small granule chromaffin) cells in the mouse adrenal medulla: Light and electron microscopic identification using semi-thin and ultra-thin sections. *Endocrinol. Jpn.* **25,** 467–476.

Kundu, M., and Thompson, C. B. (2008). Autophagy: Basic principles and relevance to disease. *Annu. Rev. Pathol.* **3,** 427–455.

Levine, B., and Kroemer, G. (2008). Autophagy in the pathogenesis of disease. *Cell* **132,** 27–42.

Locke, M., and Sykes, A. K. (1975). The role of the Golgi complex in the isolation and digestion of organelles. *Tissue Cell* **7,** 143–158.

Mizushima, N., Levine, B., Cuervo, A. M., and Klionsky, D. J. (2008). Autophagy fights disease through cellular self-digestion. *Nature* **451,** 1069–1075.

Seglen, P. O., Gordon, P. B., and Holen, I. (1990). Non-selective autophagy. *Semin. Cell Biol.* **1,** 441–448.

Slot, J. W., and Geuze, H. J. (2007). Cryosectioning and immunolabeling. *Nat. Protoc.* **2,** 2480–2491.

CHAPTER EIGHTEEN

Semiconductor Nanocrystals in Autophagy Research: Methodology Improvement at Nanosized Scale

Oleksandr Seleverstov,* James M. Phang,[†] *and* Olga Zabirnyk[†]

Contents

1. Introduction — 278
2. Autophagy Probing in Living Cells Using Different-Sized Fluorescent NP — 279
   2.1. The guidelines for autophagy monitoring in QD-labeled cells — 284
3. Application of Nanotechnology-Based Products for Other Methods in Autophagy Research. The Perspectives — 285
   3.1. Western blotting, protein microarray, and protein kinase assay — 285
   3.2. Single organelle targeting — 286
   3.3. Single molecule tracking within a living object — 286
   3.4. NP-based metabolism studies — 287
   3.5. Bioenergetics of specific metabolic reactions — 288
   3.6. *In situ* hybridization — 288
   3.7. Gene and drug delivery — 289
   3.8. Fluorescence microscopy and flow cytometry — 290
4. Conclusion and Future Perspectives — 290

Acknowledgments — 292
References — 293

Abstract

Our recent findings establish a functional link between foreign nanosized bodies and autophagy. We find that nanoparticles (NP) within a certain size range act as potent autophagy activators, and that autophagic flux is an underlying physiological process of the cellular clearance of the NP. Therefore, NP may be used to study and to monitor autophagy. We provide a detailed description of laboratory protocols designed for studying NP-mediated

---

* Department of Animal Science, College of Agriculture, University of Wyoming, Laramie, Wyoming, USA
[†] Metabolism and Cancer Susceptibility Section, Laboratory of Comparative Carcinogenesis, Center for Cancer Research, National Cancer Institute, Frederick, Maryland, USA

autophagy. In addition, we review available methods of nanotechnology, which may benefit autophagy research.

1. INTRODUCTION

From the first description of the cell's ability to sequester and subsequently degrade its own components within a vacuolar system, nearly 40 years have passed before the recent rapid expansion of knowledge in this field (Deter and De Duve 1967; Pokrovski *et al.*, 1976; Hendy and Grasso 1975). This lag may be mainly due to the methodological difficulties in autophagy research. Autophagy is a highly dynamic process that takes place interchangeably in various cellular compartments. It starts in the cytoplasm with the formation of the phagophore, a membrane that begins to separate a part of this compartment and form a new compartment, the autophagosome. Then the autophagosome matures by fusion with a lysosome to form an autolysosome. Because of the complexity of this process, the usage of standard research analytical techniques is often inadequate. Three main processes have to occur in complete autophagy: (1) targeting of intracellular material (in the case of specific autophagy, e.g., mitophagy), (2) isolation and transport, and (3) digestion of engulfed material. No one single analytical method allows for direct visualization of all three steps. Additionally, autophagy has complex and not well-understood regulatory mechanisms. Last, a big challenge is the exploration of the spatial and dynamic relationships of all of the molecular players. Although several sets of considerably reliable methods (Klionsky *et al.*, 2008) were proposed in an attempt to solve some existing problems, the methodology still remains time-consuming and expensive.

Nanoparticles (NP) are artificial and natural objects ranging in size between 1 nm and 100 nm; however, for practical purposes, some objects beyond that limit are also considered as NP (Stern and McNeil 2008). Although it seems that the definition of NP should be restricted to physical objects with defined chemical composition, physical shape, or origin, currently there are no firm limitations. Therefore, it is not surprising that previously known materials are now defined as NP and, in contrast, nano-sized objects are described as macromolecules. Colloid gold and gold NP may serve as an example of this confusing nomenclature. In the present manuscript, the term NP is applied to all physical objects in the nanometer size range. NP have several unique properties, which distinguish them from molecular and coarse materials of the same chemical composition (Hochella *et al.*, 2008). Therefore the standard experimental methodology needs fine-tuning if one uses nanometer-sized materials (Lewinski *et al.*, 2008). We found recently that certain nanomaterials may serve for the purposes of autophagy induction and monitoring. Furthermore, this work is an attempt to systematize known nanotechnology products, which might be useful in the study of autophagy.

2. AUTOPHAGY PROBING IN LIVING CELLS USING DIFFERENT-SIZED FLUORESCENT NP

We find that semiconductor fluorescent NP widely known as quantum dots (QD) activate autophagy in a size-dependent manner. In cells treated with small-sized QD, transmission electron microscopy reveals the presence of double-membraned vacuoles filled with QD as well as cytoplasmic content consistent with autophagosomes. Further, we observe the fusion of these autophagosomes with lysosomes, suggesting completion of autophagic flux. Additionally, LC3 immunocytochemistry shows accumulation of specific dot-like structures in cells treated with small-sized QD (Seleverstov O *et al.*, 2006; Zabirnyk O, *et al.*, 2007). In combination these findings are sufficient criteria for autophagy detection (Klionsky *et al.*, 2008). Based on our findings, we propose an alternative method for autophagy activation and assessment in living cells.

The unique characteristics of QD are commonly recognized, and they may be briefly described as exceptional brightness and photostability, wide excitation, but narrow, size-dependent emission spectra, opaqueness on electron microscopy, commercial availability of products designed for conjugation to virtually any biological molecule, range in size within that of middle-sized proteins, and the recent introduction of sophisticated, nanometer-sized biologically modifiable complexes (Helmick *et al.*, 2008; Stroh *et al.*, 2005; Agrawal *et al.*, 2008).

The success of a study often depends on the choice of an optimal time frame for steady-state and autophagic flux measurements. We propose to use the set of different-sized NPs in attempt to visualize autophagic flux caused by small-sized green emitting QD. The large-sized NP, which cause no significant autophagic response, may serve as a reference; the decline in their fluorescence signal intensity may be due to other phenomena such as cell division, cytoplasmic flux (secretion, exocytosis), or combinations thereof. Application of these two types of NP allows determination of the autophagic window (AW) which is defined as a time frame of differential clearance autophagy-activating NP and reference nonactivating ones (Fig. 18.1). There are numerous reasons for the loss of NP from metabolically active cells. Whereas some reasons are autophagy-related (e.g., small-sized QD clearance), others are not related to autophagy. Determination of AW would help to classify the NP uptake. The AW is calculated as the time point of signal decay from a reference NP subtracted by the time point of

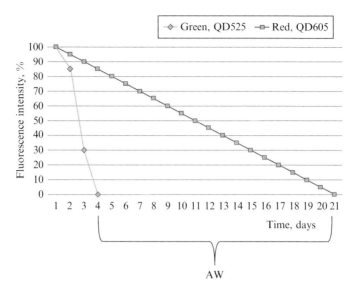

Figure 18.1 Autophagic window. Autophagic window is determined by the fluorescent signal declination. The diamond-marked line represents autophagy-activating QD fluorescence relative to initial intensity, whereas the box-marked line represents fluorescence dynamic for autophagy inert reference QD. Both estimated in vital QD-labeled cells.

signal disappearance from autophagy-activating NP. The examples of AW determination are illustrated in Figs. 18.1 and 18.2.

We propose to subdivide all cell types in accordance to their growth rate, secretion velocity, and autophagic activity into the following main groups:

1. Cells that have long AW: low secretory activity, slow growth mainly as monolayer and high basal activity of autophagy (human mesenchymal stem cells (hMSC)). The signal from large-sized QD is stable and lasts for several weeks, whereas the signal from autophagy-activating small-sized NP is lost within a few days (Fig. 18.2).
2. Cells with moderate AW: tendency to fast growth, and proliferation is not limited when confluency is reached (fibroblasts, cancer cell lines) and/or with significant secretory activity (chondrocytes, keratinocytes, various epithelial cells). Experimental conditions for this cell type should be adjusted in terms of QD labeling concentration and cell seeding density.
3. Cells with fast growth (poorly differentiated cancer cell lines) or very high secretory activity (hepatocytes). Their natural AW is short or absent.

Note: the last statement does not mean autophagy cannot be activated in response to NP in these cell types. It means that due to the natural cellular

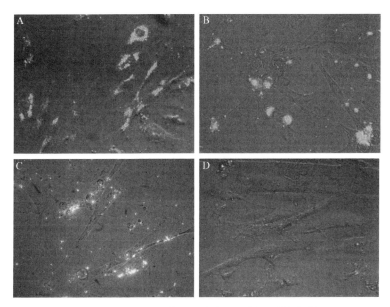

Figure 18.2 Fluorescence signal and morphology of QD-labeled human mesenchymal stem cells. A and C, 24 h after labeling, B and D, 72 h after labeing. 2.5 nM labeling concentration of each (525 [green emitting, small-sized, autophagy activating NP] and 605 [red emitting, large-sized, cytoplasmic flux reference]) QD. Both green and red particles has a diffuse cellular distribution at the beginning, which rapidly changes to a granular one. Green fluorescence disappears from morphologically normal human mesenchymal stem cells at 72 h completely; dead cells may retain well-detectable green fluorescence (D). The red fluorescence is visible at 72 h (B) and up to 52 days (Seleverstov et al., 2006). The AW is calculated as: 52 days (1248 h) − 3 days (72 h) = 49 days (1176 h). Magnification 400x for all. (See Color Insert.)

metabolism, all NP will disappear faster than any measurable autophagy-related changes occur. Moreover, we observe significant LC3-II elevation in such tumor cell lines whereas fluorescence from introduced NP has already become undetectable.

4. Cells manipulated to modulate autophagy or cells with naturally occurring changes in autophagic machinery may have characteristics of several or none of the categories proposed.

There are several aspects of AW practical applicability. Determination of AW may be applied for NP-mediated cell targeting using differences in autophagic activity between targeted and nontargeted cells, for example, for cancer therapy (Zabirnyk et al., 2007). The greater the difference in AW in targeted versus nontargeted cells, the more specific the nanodrug delivery would be. It is essential to determine the best time point for specific methods' application for autophagy analysis (e.g., electron microscopy).

We suggest that the time frame around the start of AW, namely the time of disappearance of autophagy-activating NP, would be optimal. It is extremely difficult to assess true autophagic activity in clinical samples. For example, the tumor samples from a patient before and after treatment may show differences in LC3-II expression, and the reasonable question is whether this is due to autophagy activation or to drug/treatment interference with lysosomal function and subsequent autophagosome accumulation. In this situation, measurement of NP-mediated AW may give a definitive answer. AW prolongation suggests activation of autophagic flux, whereas its shortening would be a sign of negative interference of the applied treatment with the autophagic machinery.

We provide a protocol for QD transfection and autophagy assessment, which is applicable to most cell types. It is important to note that we have chosen an in-suspension transfection procedure for adherent cells. The advantage of this approach is the ability to label relatively large number of cells (up to several millions) using relatively small volumes of media to minimize the amount of QD needed. This optimized protocol also can be used to test an autophagic response in alive cells from clinical samples (e.g., tumors).

1. Wash adherent cells 3 times with PBS to remove medium components. Alternatively wash cells grown in suspension or primary cells by centrifugation.
2. At the same time prepare the QD working solution. When using the Qtracker cell-labeling kit (Invitrogen), mix gently equal amounts of the kit components (component A: Qtracker nanocrystals in borate buffer, pH 8.3; solution B: Qtracker carrier in phosphate-buffered saline, pH 7.2) in a 0.6- or 1.5-ml sterile microcentrifuge tube. Calculate the volume of QD stock based on the final concentration desired in the labeling medium (LM; this can be the regular growth medium for the particular cell type). Allow the mixture to react 5–15 min. We propose to start with 0, 1, 10, and 20 nM working concentration in LM.
3. Detach adherent cells with trypsin solution, using a protocol optimized for that particular cell type. For hMSC we used 0.05% trypsin, 0.53 mM sodium EDTA solution from Invitrogen.
4. Filter the cell suspension through a 40-μm nylon cell strainer (BD Biosciences, Falcon, Cat. 352340) to remove aggregates.
5. Wash the cells with whole medium by centrifugation. For example, hMSC should be washed 3 times by centrifugation at 500×g for 5 min, whereas primary porcine hepatocytes are washed 3 times by centrifugation at 50×g for 5 min.
6. Resuspend the cells in LM (see step 2) in a microcentrifuge tube. The special additives (e.g., 3MA, bafilomycin A_1) might be added to the labeling medium if necessary. To our knowledge, these autophagy-modulating drugs do not affect labeling efficiency. Usage of serum-free medium as LM may dramatically decrease the labeling efficacy.

The labeling volume/cell number relationship should be adjusted for each cell type individually. In our experiments with primary cells, nontransformed and transformed cell lines 200 µl of LM is sufficient to transfect $0.2–1.0 \times 10^6$ cells. The volume can be effectively reduced to 75 µl or expanded as needed.

7. Mix gently.

   Note: If the cell density is overly high, the labeling efficiency will decrease and cell aggregates will promptly form. The rule of thumb is that the cell suspension should be unsaturated (i.e., the medium should contain more nanoparticles than cells can consume during the labeling procedure). The saturation limit can be easily checked by monitoring residual fluorescence in cell-free medium after the labeling procedure is completed. Simply remove labeled cells by centrifugation and illuminate the labeling medium with an ultraviolet light source. Any concentrations of QD above 1–2 nM will provide visible fluorescence.

8. Either incubate the microcentrifuge tube containing cell suspension in LM on a roller or vortex the tube well every 10 min. Usually, an incubation time of 45 min at 37 °C is sufficient to provide good fluorescence in more than 95% of suspended cells.

   Note: Some cells are prone to form aggregate at 37 °C. To avoid that they can be transfected on the roller or orbital shaker at lower (room) temperature.

9. Wash by centrifugation with large volumes of growth medium (see step 5). Alternatively, the cell-containing LM may be diluted by adding a large volume of culture medium. We find that if the labeling medium is diluted more than 100-fold, this wash step is not necessary. At this point, the cells are labeled now.

   Note: If centrifugation is applied for the washing step, the lowest effective centrifuge speed should be used. The QD have significantly higher density than any subcellular components, especially in aggregates (e.g., endosomes). High-speed centrifugation may cause unexpected damage to the NP-bearing cells. The washing step should be established experimentally for each individual cell type. To our knowledge this strongly depends on cytoplasmic viscosity. We advise the use of an immediate cytotoxicity assay (e.g., lactate dehydrogenase (LDH) release) to establish the optimal conditions. To check whether the centrifugation step is damaging or not, the LDH concentration should be compared in cell-free medium before and after centrifugation, with and without QD.

10. The labeling quality can be monitored using the conventional trypan blue exclusion test combined with fluorescence microscopy. We recommend using a 1:1 proportion of cell-containing medium and 0.4% trypan blue dye working solution. First, the percentage of vital

(nonstained) cells should be estimated, and then the proportion of QD-containing cells should be calculated under fluorescence. For that reason the cells in a hemocytometer should be visualized using match (allows QD fluorescent signal passage) and mismatch (to estimate nonspecific fluorescence, e.g., total fluorescent object number) filter sets. The relation between the last two will result in labeling efficacy. The data from the trypan blue exclusion test allows an estimation of the total toxic effect caused by QD during the labeling procedure. Additionally, the fluorescent and light microscopy can be used to make a merged image and the percentage of vital and labeled cells calculated. Alternatively, flow cytometry may be applied.

Note: Polyarginine peptide-mediated intracellular delivery used in the Qtracker cell-labeling kit is considered to be independent of any cell surface molecular targets and is rather related to an enhanced endocytotic process (Futaki 2005). There is little known about the velocity and saturation limits of this process. Our recent findings show QD toxicity depends on concentration rather than duration of exposure for a variety of cells. We note that growth inhibition and metabolic depression are significant at labeling concentrations of 15–20 nM and strongly dependent on the cell type, requiring 40–60 nM for certain cell lines. The nature of this variability is currently under investigation.

2.1. The guidelines for autophagy monitoring in QD-labeled cells

1. Cells can be labeled using the Qtracker cell labeling kit (Invitrogen) with 525 (green fluorescent, small-sized, autophagy activating) and 605 (red fluorescent, large-sized, autophagy inert) QD as described above. However, other NP can provoke autophagy and may be considered as well.

Note: If NP are delivered other than by the polyarginine peptide-mediated method (e.g., by receptor-mediated endocytosis), the application of trypsin or other proteolytic enzymes before transfection should be reconsidered. Labeling of adherent cells would be the best choice in that case. Although a broad spectrum of methods was recently proposed (e.g., lipofection, microinjection, electroporation), there are no data on possible cross talk with autophagic machinery.

2. After the labeling procedure, the cells should be seeded onto culture plates or chambers suitable for fluorescence microscopy of vital cells.
3. Observe the dynamic of the fluorescence signal. The observation time might be 0, 4, 8, 12, 24, 36, 48, and 72 h and then daily. Determine the AW for your cell type.
4. The time point when the signal from autophagy-activating NP (green small-sized QD) begins to decline is the starting point of the AW. The signal from the reference NP (red, large-sized QD) should not change significantly, unless the cells belong to the AW types 3 or 4.

Note: We strongly recommend usage of the fixed photomicroscopy settings throughout the experiment. A higher initial exposure time may be necessary for green emitting QD. This is due to their weaker fluorescence in comparison to red-emitting QD as well as other optical considerations. The initial fluorescent signal from the fixed cells at the initial time point of the experiment may be used as the reference for photobleaching. Although the QD are stable against photobleaching, we recommend restriction of the light exposure of living cells containing these NP. The issue here is that light induces QD to produce cytotoxic reactive oxygen species (Zhang *et al.*, 2006).

5. The time frame around the AW start is the most suitable for the observation of the autophagic flux in most cell types. The earliest autophagic response usually occurs 12–24 h before the AW starts. This lag is due to primary lysosomal processing and possible decoating of NP (Seleverstov *et al.*, 2006; Zabirnyk *et al.*, 2007).

In addition to the NP-based autophagy study method we propose here, the standard autophagy research methods may be applied on the same cells. However, immunocytochemistry applications on the QD-labeled cells or tissues raise several considerations. Simultaneous application of fluorescent NP and organic dyes may lead to fluorescence resonance energy transfer (FRET) (Dennis and Bao 2008). This phenomenon theoretically may be responsible for quenching of the QD signal resulting in a false negative NP and molecular target colocalization; some steps in tissue processing may interfere with the ability of QD to emit fluorescence. Although we have found that conventional fixatives and permeabilizing solutions at concentrations used do not cause any significant loss of fluorescence (Seleverstov *et al.*, 2006), it is highly recommended to check the presence of NP fluorescence after completing every step of the procedure. If necessary, minimize the rinsing steps.

3. Application of Nanotechnology-Based Products for Other Methods in Autophagy Research. The Perspectives

3.1. Western blotting, protein microarray, and protein kinase assay

Although NP is a novel tool in biomedical research, standard lab techniques such as Western immunoblotting may benefit from NP applications. The main advantage of silver and gold NP is improved sensitivity (up to 10,000 times higher) with small proteins (less than 10 kDa) and peptides

(Duchesne and Fernig 2007). Furthermore, usage of fluorescent QD technology makes possible ultrasensitive detection of tracer proteins directly in the cell lysate, avoiding necessary immunoprecipitation and concentration (Bakalova et al., 2005a), which is extremely difficult and in many cases impossible with standard Western blotting technique.

NP-based protein microarrays (Zajac et al., 2007; Wang et al., 2005), highly sensitive NP-based methods detecting functional protein modifications (e.g., activities of kinases and phosphorylation) have been recently described (Kim et al., 2008; Kerman et al., 2007). Although these methods are still in an early testing stage, their application would significantly benefit autophagy research.

3.2. Single organelle targeting

Autophagy is one of the main mechanisms responsible for cellular housekeeping in terms of the removal of damaged organelles (Mathew and White 2007; Colell et al., 2007). Although up to now there is no reliable experimental model of single-organelle damage, recent progress in targeting mitochondria (Yamada et al., 2005) and other subcellular compartments (Hoshino et al., 2004) with the help of NP have made this possible. The usage of NP here has several advantages. In comparison to molecular drugs, which also can damage certain organelles quite specifically, the liberation of drugs packed into NP can be controlled by pH, enzymatic activity (e.g., being digested by a certain enzyme), or may occur as a response to external stimuli (Ganta et al., 2008; Sawant et al., 2006; Kale and Torchilin 2007). The last option is especially attractive for the study of autophagic flux. Certain autophagy modulators packed into the NP may be released at a desired step of the autophagic process within a desired compartment with high precision.

3.3. Single molecule tracking within a living object

Numerous questions in molecular autophagy research require single molecule resolution within a living cell. The possibility for fluorescent tracking of biochemically active enzymes such as lipase (Sonesson et al., 2007) and collagenase (Zhang et al., 2006) was recently shown. Although both enzymes maintain their specific activity, the changes in the kinetics of the biochemical reactions are seen after conjugation with NP. Additionally, NP were used for a long-distance trafficking of proteins in vesicles (Cui et al., 2007), single receptor monitoring (Bouzigues et al., 2007; Lasne et al., 2006), motor protein function and spatial localization assessment (Courty et al., 2006a,b), as well as mRNA probing (Bakalova et al., 2005b). Good correlation is reported between fluorescent NP localization and electron microscopic observation of the same molecular target (Cui et al., 2007).

Specific combined methods are being developed (Sarkar et al., 2004), allowing not only tracking of individual proteins but also on-line visualization of conformational changes and folding process of a specific molecule of interest. In autophagy research this technology may be useful for pH- or enzyme-sensitive constructs with two/multiple-colored NP bioconjugates. Resulting changes in the intensity and spectra of fluorescence signal can be observed from living cells in real-time mode. The technical base for the production of such multicolor hybrids has already been established (Agrawal et al., 2008). The importance of different-sized NP controls in living cell studies should not be underestimated (Seleverstov et al., 2006; Zabirnyk et al., 2007; Lovrić et al., 2005).

3.4. NP-based metabolism studies

The recent revival of research activity on metabolic mechanisms has led to a number of exciting insights in cancer as well as other diseases (Gatenby and Gillies 2004). Not only have investigators clarified the deranged metabolic pathways responsible for the Warburg (1930) effect (i.e., the tumor's reliance on oxidative glycolysis) but also have identified stress substrates and their microenvironmental sources and metabolic pathways (Phang et al., 2008). In this context, NP can be used to augment or to block these metabolic pathways by serving as a delivery vehicle either for inhibitors or for substrates and enzyme systems. Such an approach may be especially important for temporal windows within the microenvironment where vascular delivery is inadequate or has been compromised (e.g., in the developing or metastatic tumor) during wound healing or in the metabolic derangements following cardiac or cerebral vascular occlusion. This type of study would establish a link between autophagy and heterophagy (cellular digestion of an exogenous engulfed substance). Heterophagy includes ecophagy, which is the degradation of substances degraded in the microenvironment serving as a more plausible energy source for prevascularized tumors (Phang et al., 2008). Moreover, heterophagy is a less self-mutilating process than autophagy for individual tumor cells.

An area for exploration would be the proline/hydroxyproline metabolic system in which the degradation of extracellular matrix collagen by matrix metalloproteinases (MMP) provides proline/hydroxyproline as substrate either for the generation of superoxide to initiate apoptosis (Liu et al., 2006) or as a source of energy (Pandhare et al., submitted) for survival. The critical enzymes are MMP, prolidase and proline oxidase, and the substrates would be collagen, imidodipeptides with N-terminal proline or hydroxyproline and free proline or hydroxyproline. Inhibitors include batimastat for MMPs, captopril and carboxybenzylproline for prolidase, and dehydroproline for proline oxidase. To target MMPs and prolidase, the NP could be ideal as these pathways will complement the activity in lysosomes. Proline oxidase, on the other hand, is mitochondrial, and substrates or inhibitors of this pathway can be delivered by NP targeting this organelle.

3.5. Bioenergetics of specific metabolic reactions

The basic physiological questions whether a specific metabolic reaction is energy consuming or producing can be addressed. For this purpose temperature responsive NP can be used. Individual QDs are capable of sensing temperature variations and provide an optical readout (Li *et al.*, 2007). Theoretically, it is possible to sense dynamic temperature variations within individual vacuoles during autophagic flux.

Commonly used detection techniques for proteolytic and other enzymatic activities may be challenging sometimes, especially when the assay has to be performed repeatedly and preferably within the same living organism. Recently, a sophisticated method for detection of proteolytic activity was proposed (Shi *et al.*, 2007; Shi *et al.*, 2006). The principle is based on the conjugation of two fluorophores with different emission peaks—the quantum dots and organic dye residing in an enzyme-sensitive peptide. Upon separation of the two fluorophores with cleavage of the peptide bond, changes in the fluorescent signal may be observed. That phenomenon is called fluorescence resonance energy transfer (FRET). It usually occurs between the quantum dot, serving as energy donor, and an organic dye serving as acceptor, the latter is immobilized on the surface of the quantum dot by the peptide linker. This technique permits on-line measurement of a particular enzymatic activity within the organelle of interest or within a morphological component of the autophagic machinery. Several limitations may interfere here, such as the specificity of the linker molecule as substrate for the enzyme of interest. Even more challenging is a delivery of the intact nanocomplex into the destination compartment.

3.6. *In situ* hybridization

Nanotechnology may revolutionize genetic probing of clinical samples, containing a minute amount of mixed tissue (Jiang *et al.*, 2007). The applicability of NP for *in situ* hybridization for tumor profiling was recently extensively reviewed (Yezhelyev *et al.*, 2006). NP usage will bring new insight into the functional activity of *atg* genes in human disease. Unique optic properties of fluorescent NP are advantageous because a large number of molecular targets can be simultaneously assessed in a single cell (Chan *et al.*, 2005; Bentolila and Weiss 2006). Certain discrepancies in QD and organic probes' distribution are reported (Xiao and Barker 2004) and explained due to different pH and buffer composition during sample processing. However, the same group emphasizes a higher precision of NP-based method.

Modern nanotechnology offers a broad spectrum of custom- and semi-custom engineered diagnostic devices. DNA dendrimers were recently proposed for the detection of small (approximately 1 kb) DNA regions in clinical samples. This method may evolve in the near future as a long-expected technique for detecting small gene deletions/translocations in a single cell or in a relatively small clinical sample. The resolution of dendrimeric probes is significantly higher than that of a conventional fluorescent probes (Mora *et al.*, 2006). Moreover, certain NP allow simultaneous multimodal visualization and tracking of nucleic acid delivery to the cell of interest *in vivo* (Medarova *et al.*, 2007).

3.7. Gene and drug delivery

One of the most effective ways to study the genetic regulation of a physiological process is a knockout or an overexpression of candidate gene. This approach is relatively expensive and time-consuming, conventionally used animal models are not always able to produce a viable offspring (Komatsu *et al.*, 2005). Moreover, virus-based transfection is not devoid of risk for researchers; other pitfalls such as transient or unstable transfection are also present. Although an attractive alternative is a knockdown using siRNA or nonviral gene transfection, other problems exist. Transfection of naked nucleic acids is often very low in efficiency. Electroporation, microinjection, or nuclear transfer yield a very low number of recipient cells, which are often damaged by the procedure. Therefore, the development of more physiological carriers for the delivery of nucleic acids is critical. The most important properties of an effective gene delivery system might be briefly formulated as low side-toxicity (e.g., nonspecific, derived from blank carrier), high efficiency of cargo delivery, sufficient level of subsequent gene expression, tissue high tissue permeability and physiological mode of action (e.g., usage of already active transportation pathways in the targeted tissue and cell). Certain engineered NP have already met most of the requirements for effective gene delivery and are devoid of drawbacks inherent to viral transfection (Wasungu and Hoekstra 2006; Cai *et al.*, 2008; Suzuki *et al.*, 2008).

A number of technical approaches have been successfully applied to deliver both RNA and DNA to the cell. They may be subdivided into 2 groups:

a. Nontargeted delivery, where the NP are used as a carrier that serves mainly for protection from nucleases (Suh *et al.*, 2007; Kaul and Amiji 2002).
b. Targeted delivery, using
   1. Physiological ligand (Kakudo *et al.*, 2004)
   2. Viral peptides (Sasaki *et al.*, 2008; Ferrer-Miralles *et al.*, 2008)

The approaches listed in the group "a" have the common advantages as being simple, low-cost methods for gene/nucleic acid delivery and may be used for *in vitro* studies. However, the low rate of spontaneous endocytosis and subsequent nonsynchronous, poorly controlled transfection may limit their application. Their usage for *in vivo* settings is limited due to nonspecific targeting (retention) in certain organs and tissues restricted mainly to the cells having high phagocytic and endocytotic activity.

Targeted approaches, group "b", seem to be more controllable. The efficacy of transfection may be presented here as a function of receptor number per cell, ligand affinity and NP concentration in the closest proximity of the cell membrane. It is remarkable, that among several hundred publications in the field of gene delivery over the past years there was only one study establishing the functional link between autophagy and NP-based gene therapy (Ohtani *et al.*, 2007).

Although autophagy as a response to intracellular introduced nanoparticles is described in several studies (Zabirnyk *et al.*, 2007; Seleverstov *et al.*, 2006; Yamawaki and Iwai 2006; Harhaji *et al.*, 2007; Stern *et al.*,, 2008), there are no data showing an impact or possible consequences of gene-carrying NP on autophagic machinery.

3.8. Fluorescence microscopy and flow cytometry

Fluorescent NP are found extremely useful for flow cytometry. This method is not established as a standard for autophagy detection and monitoring, but its usage in combination with fluorescent probes is very attractive in addressing specific questions, which are difficult otherwise (e.g., the relationship between autophagy and immunity). The suitability of QD for multicolor flow cytometry (e.g., for simultaneous detection of several antigens) is one of the most important advantages (Chattopadhyay *et al.*, 2006).

The use of NP for conventional fluorescence microscopy has been established in numerous experimental studies and is currently evolving toward clinical application (Xing *et al.*, 2007). QD significantly improve both the resolution and 3-dimensional visualization of individual molecular targets in *vivo* (Smith *et al.*, 2008).

4. Conclusion and Future Perspectives

In this manuscript we demonstrate the applicability of different-sized NP for autophagy activation and monitoring and characterization in various cell types and clinical materials. The proposed methodology is unique in terms of its simplicity and reproducibility. One of the most important and attractive points in QD usage is the fact that autophagy-activating agents

Table 18.1 Nanotechnology Products for Autophagy Research

Autophagy study method	Available nanotechnology-based product	Achieved or potential benefit	Method was tested in autophagy research
Autophagic flux observation	QD	Direct visualization of the cellular traffic by the fluorescence microscopy	Yes
Western blotting	Gold and silver NP	Increase in sensitivity, especially for low-molecular weight proteins	No
Protein microarray	QD, gold NP	Higher sensitivity	No
Protein kinase assay	Gold NP	Higher sensitivity	No
Intracellular compartment and a single-organelle targeting	QD, various liposomes and polymers	Visualization and targeting of a single organelle within a living cell. Minimal impact on entire organism/cell.	No
Single molecule targeting within a living cell	QD, gold NP	Visualization and tracking of a single molecule movements within a living objects, high precision and resolution	No
TEM	QD, gold NP	Correlation study between fluorescent and electron microscopy. Contrasting of vacuole content.	Yes
Enzyme activity estimation	QD	High sensitivity and specificity, applicability within a living object	No

(continued)

Table 18.1 (continued)

Autophagy study method	Available nanotechnology-based product	Achieved or potential benefit	Method was tested in autophagy research
In situ hybridization	QD, dendrimers	Higher sensitivity, small region probing, widening of visualization approaches	No
Gene and drug delivery	Various nanosized formulations	Cell- and tissue-specific targeting, high efficiency, low side effects	Yes
Fluorescence microscopy	QD, gold NP, various organic NP loaded with fluorescent dyes	Improvement of specificity, better signal/noise ratio, stability against photobleaching, multiple antigen detection	Yes
Flow cytometry	QD	Multiple antigen detection	No

may be visualized and tracked within living cells for a long period of time. No additional pharmacological or physical manipulation is required. Beyond QD there are other nanotechnology products available (Table 18.1) and their application may benefit research in this area. Therefore, we expect an explosive increase in the number of studies using these methodologies in the near future.

ACKNOWLEDGMENTS

The preparation of the manuscript was in part (J.M.P. and O.Z.) supported by the Intramural Research Program of the NIH, National Cancer Institute, Center for Cancer Research. The content of this publication does not necessarily reflect the views and policies of the Department of Health and Human Services, nor does the mention of trade names, commercial products, or organizations imply endorsement by the U.S. government.

REFERENCES

Agrawal, A., Deo, R., Wang, G. D., Wang, M. D., and Nie, S. (2008). Nanometer-scale mapping and single-molecule detection with color-coded nanoparticle probes. *Proc. Natl. Acad. Sci. USA* **105**(9), 3298–3303.

Bakalova, R., Zhelev, Z., Ohba, H., and Baba, Y. (2005). Quantum dot-based western blot technology for ultrasensitive detection of tracer proteins. *J. Am. Chem. Soc.* **127**(26), 9328–9329.

Bakalova, R., Zhelev, Z., Ohba, H., and Baba, Y. (2005). Quantum dot-conjugated hybridization probes for preliminary screening of siRNA sequences. *J. Am. Chem. Soc.* **127**(32), 11328–11335.

Bentolila, L. A., and Weiss, S. (2006). Single-step multicolor fluorescence in situ hybridization using semiconductor quantum dot-DNA conjugates. *Cell Biochem. Biophys.* **45**(1), 59–70.

Bouzigues, C., Morel, M., Triller, A., and Dahan, M. (2007). Asymmetric redistribution of GABA receptors during GABA gradient sensing by nerve growth cones analyzed by single quantum dot imaging. *Proc. Natl. Acad. Sci. USA* **104**(27), 11251–11256.

Cai, X., Conley, S., and Naash, M. (2008). Nanoparticle applications in ocular gene therapy. *Vision Res.* **48**(3), 319–324.

Chan, P., Yuen, T., Ruf, F., Gonzalez-Maeso, J., and Sealfon, S. C. (2005). Method for multiplex cellular detection of mRNAs using quantum dot fluorescent in situ hybridization. *Nucleic Acids Res.* **33**(18), e161.

Chattopadhyay, P. K., Price, D. A., Harper, T. F., Betts, M. R., Yu, J., Gostick, E., Perfetto, S. P., Goepfert, P., Koup, R. A., De Rosa, S. C., Bruchez, M. P., and Roederer, M. (2006). Quantum dot semiconductor nanocrystals for immunophenotyping by polychromatic flow cytometry. *Nat. Med.* **12**(8), 972–977.

Colell, A., Ricci, J. E., Tait, S., Milasta, S., Maurer, U., Bouchier-Hayes, L., Fitzgerald, P., Guio-Carrion, A., Waterhouse, N. J., Li, C. W., Mari, B., Barbry, P., *et al.* (2007). GAPDH and autophagy preserve survival after apoptotic cytochrome c release in the absence of caspase activation. *Cell* **129**(5), 983–997.

Courty, S., Bouzigues, C., Luccardini, C., Ehrensperger, M. V., Bonneau, S., and Dahan, M. (2006). Tracking individual proteins in living cells using single quantum dot imaging. *Methods Enzymol.* **414**, 211–228.

Courty, S., Luccardini, C., Bellaiche, Y., Cappello, G., and Dahan, M. (2006). Tracking individual kinesin motors in living cells using single quantum-dot imaging. *Nano Lett.* **6**(7), 1491–1495.

Cui, B., Wu, C., Chen, L., Ramirez, A., Bearer, E. L., Li, W. P., Mobley, W. C., and Chu, S. (2007). One at a time, live tracking of NGF axonal transport using quantum dots. *Proc. Natl. Acad. Sci. USA* **104**(34), 13666–13671.

Dennis, A. M., and Bao, G. (2008). Quantum dot-fluorescent protein pairs as novel fluorescence resonance energy transfer probes. *Nano Lett.* **8**(5), 1439–1445.

Deter, R. L., and de Duve, C. (1967). Influence of glucagon, an inducer of cellular autophagy, on some physical properties of rat liver lysosomes. *J. Cell Biol.* **33**(2), 437–449.

Duchesne, L., and Fernig, D. G. (2007). Silver and gold nanoparticle-coated membranes for femtomole detection of small proteins and peptides by Dot and Western blot. *Anal. Biochem.* **362**(2), 287–289.

Ferrer-Miralles, N., Vázquez, E., and Villaverde, A. (2008). Membrane-active peptides for non-viral gene therapy: making the safest easier. *Trends Biotechnol.* **26**(5), 267–275.

Futaki, S. (2005). Membrane-permeable arginine-rich peptides and the translocation mechanisms. *Adv. Drug Deliv. Rev.* **57**(4), 547–558.

Ganta, S., Devalapally, H., Shahiwala, A., and Amiji, M. (2008). A review of stimuli-responsive nanocarriers for drug and gene delivery. *J. Control Release.* **126**(3), 187–204.

Gatenby, R. A., and Gillies, R. J. (2004). Why do cancers have high aerobic glycolysis? *Nat. Rev. Cancer* **4**(11), 891–899.

Harhaji, L., Isakovic, A., Raicevic, N., Markovic, Z., Todorovic-Markovic, B., Nikolic, N., Vranjes-Djuric, S., Markovic, I., and Trajkovic, V. (2007). Multiple mechanisms underlying the anticancer action of nanocrystalline fullerene. *Eur. J. Pharmacol.* **568**(1–3), 89–98.

Helmick, L., Antúnez de Mayolo, A., Zhang, Y., Cheng, C. M., Watkins, S. C., Wu, C., and Leduc, P. R. (2008). Spatiotemporal response of living cell structures in *Dictyostelium discoideum* with semiconductor quantum dots. *Nano Lett.* **8**(5), 1303–1308.

Hendy, R., and Grasso, P. (1975). Reversibility of lysosomal and glucose 6-phosphatase changes produced in the rat liver by dimethylnitrosamine. *Chem. Biol. Interact.* **10**(6), 395–406.

Hochella, M. F. Jr., Lower, S. K., Maurice, P. A., Penn, R. L., Sahai, N., Sparks, D. L., and Twining, B. S. (2008). Nanominerals, mineral nanoparticles, and Earth systems. *Science* **319**(5870), 1631–1635.

Hoshino, A., Fujioka, K., Oku, T., Nakamura, S., Suga, M., Yamaguchi, Y., Suzuki, K., Yasuhara, M., and Yamamoto, K. (2004). Quantum dots targeted to the assigned organelle in living cells. *Microbiol. Immunol.* **48**(12), 985–994.

Jiang, Z., Li, R., Todd, N. W., Stass, S. A., and Jiang, F (2007). Detecting genomic aberrations by fluorescence in situ hybridization with quantum dots-labeled probes. *J. Nanosci. Nanotechnol.* **7**, 4254–4259.

Kakudo, T., Chaki, S., Futaki, S., Nakase, I., Akaji, K., Kawakami, T., Maruyama, K., Kamiya, H., and Harashima, H. (2004). Transferrin-modified liposomes equipped with a pH-sensitive fusogenic peptide: an artificial viral-like delivery system. *Biochem.* **43**(19), 5618–5628.

Kale, A. A., and Torchilin, V. P. (2007). Design, synthesis, and characterization of pH-sensitive PEG-PE conjugates for stimuli-sensitive pharmaceutical nanocarriers: the effect of substitutes at the hydrazone linkage on the ph stability of PEG-PE conjugates. *Bioconjug. Chem.* **18**(2), 363–370.

Kaul, G., and Amiji, M. (2002). Long-circulating poly(ethylene glycol)-modified gelatin nanoparticles for intracellular delivery. *Pharm. Res.* **19**(7), 1061–1067.

Kerman, K., Chikae, M., Yamamura, S., and Tamiya, E. (2007). Gold nanoparticle-based electrochemical detection of protein phosphorylation. *Anal. Chim. Acta* **588**(1), 26–33.

Kim, Y. P., Oh, Y. H., and Kim, H. S. (2008). Protein kinase assay on peptide-conjugated gold nanoparticles. *Biosens. Bioelectron.* **23**(7), 980–986.

Klionsky, D. J., Abeliovich, H., Agostinis, P., Agrawal, D. K., Aliev, G., Askew, D. S., Baba, M., Baehrecke, E. H., Bahr, B. A., Ballabio, A., *et al.* (2008). Guidelines for the use and interpretation of assays for monitoring autophagy in higher eukaryotes. *Autophagy* **4**(2), 151–175.

Komatsu, M., Waguri, S., Ueno, T., Iwata, J., Murata, S., Tanida, I., Ezaki, J., Mizushima, N., Ohsumi, Y., Uchiyama, Y., Kominami, E., Tanaka, K., *et al.* (2005). Impairment of starvation-induced and constitutive autophagy in Atg7-deficient mice. *J. Cell Biol.* **169**(3), 425–434.

Lasne, D., Blab, G. A., Berciaud, S., Heine, M., Groc, L., Choquet, D., Cognet, L., and Lounis, B. (2006). Single nanoparticle photothermal tracking (SNaPT) of 5-nm gold beads in live cells. *Biophys. J.* **91**(12), 4598–4604.

Lewinski, N., Colvin, V., and Drezek, R. (2008). Cytotoxicity of nanoparticles. *Small* **4**(1), 26–49.

Li, S., Zhang, K., Yang, J. M., Lin, L., and Yang, H. (2007). Single quantum dots as local temperature markers. *Nano. Lett.* **7**(10), 3102–3105.

Liu, Y., Borchert, G. L., Surazynski, A, Hu, C. -A., and Phang, J. M. (2006). *Oncogene* **25** (41), 5640–5647.

Lovrié, J., Bazzi, H. S., Cuie, Y., Fortin, G. R., Winnik, F. M., and Maysinger, D. (2005). Differences in subcellular distribution and toxicity of green and red emitting CdTe quantum dots. *J. Mol. Med.* **83**(5), 377–385.

Mathew, R., and White, E. (2007). Why sick cells produce tumors: the protective role of autophagy. *Autophagy* **3**(5), 502–505.

Medarova, Z., Pham, W., Farrar, C., Petkova, V., and Moore, A. (2007). In vivo imaging of siRNA delivery and silencing in tumors. *Nat. Med.* **13**(3), 372–377.

Mora, J. R., Knoll, J. H., Rogan, P. K., Getts, R. C., and Wilson, G. S. (2006). Dendrimer FISH detection of single-copy intervals in acute promyelocytic leukemia. *Mol. Cell Probes.* **20**(2), 114–120.

Ohtani, S., Iwamaru, A., Deng, W., Ueda, K., Wu, G., Jayachandran, G., Kondo, S., Atkinson, E. N., Minna, J. D., Roth, J. A., and Ji, L. (2007). Tumor suppressor 101F6 and ascorbate synergistically and selectively inhibit non-small cell lung cancer growth by caspase-independent apoptosis and autophagy. *Cancer Res.* **67**(13), 6293–6303.

Phang, J. M., Donald, S. P., Pandhare, J., and Liu, Y. (2008). The metabolism of proline, a stress substrate, modulates carcinogenic pathway. *Amino Acids.* **35**(4), 681–690.

Pokrovski, A. A., Tashev, T. A., Krystev, L. P., Tutel'ian, V. A., and Kravchenko, L. V. (1976). [Ultrastructural changes in the subcellular membranes of hepatocytes during the early periods of starvation]. *Vopr. Pitan.* **2**, 26–31.

Sarkar, A., Robertson, R. B., and Fernandez, J. M. (2004). Simultaneous atomic force microscope and fluorescence measurements of protein unfolding using a calibrated evanescent wave. *Proc. Natl. Acad. Sci. USA* **101**(35), 12882–12886.

Sawant, R. M., Hurley, J. P., Salmaso, S., Kale, A., Tolcheva, E., Levchenko, T. S., and Torchilin, V. P. (2006). "SMART" drug delivery systems: double-targeted pH-responsive pharmaceutical nanocarriers. *Bioconjug Chem.* **17**(4), 943–949.

Sasaki, K., Kogure, K., Chaki, S., Nakamura, Y., Moriguchi, R., Hamada, H., Danev, R., Nagayama, K., Futaki, S., and Harashima, H. (2008). An artificial virus-like nano carrier system: enhanced endosomal escape of nanoparticles via synergistic action of pH-sensitive fusogenic peptide derivatives. *Anal. Bioanal. Chem.* **391**(8), 2717–2727.

Seleverstov, O., Zabirnyk, O., Zscharnack, M., Bulavina, L., Nowicki, M., Heinrich, J. M., Yezhelyev, M., Emmrich, F., O'Regan, R., and Bader, A. (2006). Quantum dots for human mesenchymal stem cells labeling. A size-dependent autophagy activation. *Nano Lett.* **6**(12), 2826–2832.

Shi, L., De Paoli, V., Rosenzweig, N., and Rosenzweig, Z. (2006). Synthesis and application of quantum dots FRET-based protease sensors. *J. Am. Chem. Soc.* **128**(32), 10378–10379.

Shi, L., Rosenzweig, N., and Rosenzweig, Z. (2007). Luminescent quantum dots fluorescence resonance energy transfer-based probes for enzymatic activity and enzyme inhibitors. *Anal. Chem.* **79**(1), 208–214.

Smith, B. R., Cheng, Z., De, A., Koh, A. L., Sinclair, R., and Gambhir, S. S. (2008). Real-time intravital imaging of RGD-quantum dot binding to luminal endothelium in mouse tumor neovasculature. *Nano Lett.* **8**(9), 2599–2606.

Sonesson, A. W., Elofsson, U. M., Callisen, T. H., and Brismar, H. (2007). Tracking single lipase molecules on a trimyristin substrate surface using quantum dots. *Langmuir* **23**(16), 8352–8356.

Stern, S. T., and McNeil, S. E. (2008). Nanotechnology safety concerns revisited. *Toxicol. Sci.* **101**(1), 4–21.

Stern, S. T., Zolnik, B. S., McLeland, C. B., Clogston, J., Zheng, J., and McNeil, S. E. (2008). Induction of autophagy in porcine kidney cells by quantum dots: A common cellular response to nanomaterials? *Toxicol. Sci.* **106**(1), 140–152.

Stroh, M., Zimmer, J. P., Duda, D. G., Levchenko, T. S., Cohen, K. S., Brown, E. B., Scadden, D. T., Torchilin, V. P., Bawendi, M. G., Fukumura, D., and Jain, R. K. (2005). Quantum dots spectrally distinguish multiple species within the tumor milieu *in vivo*. *Nat. Med.* **11**(6), 678–682.

Suh, J., Choy, K. L., Lai, S. K., Suk, J. S., Tang, B. C., Prabhu, S., and Hanes, J. (2007). PEGylation of nanoparticles improves their cytoplasmic transport. *Int. J. Nanomedicine* **2**(4), 735–741.

Suzuki, R., Takizawa, T., Negishi, Y., Utoguchi, N., Sawamura, K., Tanaka, K., Namai, E., Oda, Y., Matsumura, Y., and Maruyama, K. (2008). Tumor specific ultrasound enhanced gene transfer in vivo with novel liposomal bubbles. *J. Control Release* **125**(2), 137–144.

Wang, Z., Lee, J., Cossins, A. R., and Brust, M. (2005). Microarray-based detection of protein binding and functionality by gold nanoparticle probes. *Anal. Chem.* **77**(17), 5770–5774.

Warburg, O. (1930). Ueber den Stoffwechsel der Tumoren London: Constable, London.

Wasungu, L., and Hoekstra, D. (2006). Cationic lipids, lipoplexes and intracellular delivery of genes. *J. Control. Release* **116**(2), 255–264.

Xiao, Y., and Barker, P. E. (2004). Semiconductor nanocrystal probes for human metaphase chromosomes. *Nucleic Acids Res.* **32**(3), e28.

Xing, Y., Chaudry, Q., Shen, C., Kong, K. Y., Zhau, H. E., Chung, L. W., Petros, J. A., O'Regan, R. M., Yezhelyev, M. V., Simons, J. W., Wang, M. D., and Nie, S. (2007). Bioconjugated quantum dots for multiplexed and quantitative immunohistochemistry. *Nat. Protoc.* **2**(5), 1152–1165.

Yamada, Y., Shinohara, Y., Kakudo, T., Chaki, S., Futaki, S., Kamiya, H., and Harashima, H. (2005). Mitochondrial delivery of mastoparan with transferrin liposomes equipped with a pH-sensitive fusogenic peptide for selective cancer therapy. *Int. J. Pharm.* **303**(1–2), 1–7.

Yamawaki, H., and Iwai, N. (2006). Cytotoxicity of water-soluble fullerene in vascular endothelial cells. *Am. J. Physiol. Cell Physiol.* **290**(6), C1 495–502.

Yezhelyev, M. V., Gao, X., Xing, Y., Al-Hajj, A., Nie, S., and O'Regan, R. M. (2006). Emerging use of nanoparticles in diagnosis and treatment of breast cancer. *Lancet Oncol.* **7**(8), 657–667.

Zabirnyk, O., Yezhelyev, M., and Seleverstov, O. (2007). Nanoparticles as a novel class of autophagy activators. *Autophagy* **3**(3), 278–281.

Zajac, A., Song, D., Qian, W., and Zhukov, T. (2007). Protein microarrays and quantum dot probes for early cancer detection. *Colloids Surf. B. Biointerfaces* **58**(2), 309–314.

Zhang, Y., He, J., Wang, P. N., Chen, J. Y., Lu, Z. J., Lu, D. R., Guo, J., Wang, C. C., and Yang, W. L. (2006). Time-dependent photoluminescence blue shift of the quantum dots in living cells: Effect of oxidation by singlet oxygen. *J. Am. Chem. Soc.* **128**(41), 13396–13401.

Zhang, Y., So, M. K., and Rao, J. (2006). Protease-modulated cellular uptake of quantum dots. *Nano Lett.* **6**(9), 1988–1992.

CHAPTER NINETEEN

Methods to Monitor Chaperone-Mediated Autophagy

Susmita Kaushik* *and* Ana Maria Cuervo*

Contents

1. Introduction	298
2. Experimental Models for the Study of CMA	300
3. Properties of CMA Substrates	301
4. Methods to Measure CMA	303
5. Measurement of Protein Degradation Rates	303
5.1. Pulse and chase experiments	303
5.2. Inhibition of different autophagic pathways	306
5.3. Calculations	307
6. Measurement of Levels of Key CMA Components	308
6.1. Isolation of lysosomes	309
6.2. Immunoblot for CMA components	311
7. Analysis of the Subcellular Location of CMA-Active Lysosomes	312
7.1. Immunofluorescence for CMA-active lysosomes	312
7.2. Immunogold and electron microscopy for CMA-active lysosomes	314
8. *In Vitro* Assay to Measure Translocation of CMA Substrates	315
8.1. Radiolabeling of CMA substrates	316
8.2. Protein degradation with isolated lysosomes	317
8.3. Protease protection assay	318
9. Concluding Remarks	320
Acknowledgments	321
References	321

Abstract

Chaperone-mediated autophagy (CMA) is a selective type of autophagy responsible for the lysosomal degradation of soluble cytosolic proteins. In contrast to other forms of autophagy where cargo is sequestered and delivered to lysosomes through membrane fusion/excision, CMA substrates reach the

* Department of Developmental and Molecular Biology, Marion Bessin Liver Research Center, Institute for Aging Research, Albert Einstein College of Medicine, Bronx, New York, USA

lysosomal lumen after direct translocation across the lysosomal membrane. CMA is part of the cellular quality control systems and as such, essential for the cellular response to stress. CMA activity decreases with age, likely contributing to the accumulation of altered proteins characteristic in tissues from old organisms. Furthermore, impairment of CMA underlies the pathogenesis of certain human pathologies such as neurodegenerative disorders. These findings have drawn renewed attention to CMA and a growing interest in the measurement of changes in CMA activity under different physiological and pathological conditions. In this chapter we review the different experimental approaches used to assess CMA activity both in cells in culture and in different organs from animals.

1. Introduction

Chaperone-mediated autophagy (CMA) is a type of autophagy responsible for the degradation of a subset of cytosolic proteins bearing in their amino acid sequence a consensus motif, biochemically related to KFERQ, that targets them for lysosomal degradation (Dice, 1990). This motif is recognized by a cytosolic chaperone, the heat-shock cognate protein of 70 kDa (hsc70), in complex with its cochaperones (Chiang et al., 1989). The substrate/chaperone complex is delivered to the surface of lysosomes, where it binds to a CMA receptor, the lysosome-associated membrane protein type-2A (LAMP-2A) (Cuervo and Dice, 1996). After unfolding, the substrate protein is translocated across the lysosomal membrane in an ATP-dependent manner, assisted by a resident lysosomal chaperone (lys-hsc70) (Agarraberes et al., 1997). Once in the lysosomal lumen, CMA substrates are rapidly degraded (in 5–10 min) by the broad array of lysosomal proteases. These two features, the selectivity toward substrate proteins and their direct translocation across the lysosomal membrane, make CMA distinct from the other types of autophagy in mammalian cells, namely macroautophagy and microautophagy, where cargo is typically delivered in bulk to lysosomes through processes involving vesicular fusion (macroautophagy) and/or membrane excision (microautophagy) (Cuervo, 2004a; Levine and Klionsky, 2004; Shintani and Klionsky, 2004).

Approximately 30% of soluble cytosolic proteins contain the CMA-targeting motif (Dice, 1990). They constitute a heterogeneous pool of intracellular proteins including, among others, some glycolytic enzymes (glyceraldehyde-3-phosphate dehydrogenase, aldolase, phosphoglyceromutase), particular transcription factors and inhibitors of transcription factors (c-fos, the inhibitor of NFκB (IκB)), calcium-binding proteins (Annexins I, II, IV, and VI), vesicular trafficking proteins (α-synuclein), cytosolic forms of secretory proteins (α-2-microglobulin) and even some of the catalytic and regulatory subunits of the proteasome, the major cytosolic protease (reviewed in Dice, 2007; Majeski and Dice, 2004; Massey et al., 2004, 2006b). Given the broad nature of CMA substrate proteins and their participation in many different intracellular processes, it is easy to infer that changes in the activity of this pathway may have major consequences on cell functioning.

Although some level of basal CMA activity is detectable in almost all cells, CMA is maximally up-regulated under stress conditions, such as prolonged nutrient deprivation (serum removal in cultured cells or starvation in rodents) (Cuervo et al., 1995a; Wing et al., 1991), mild oxidative stress (Kiffin et al., 2004), and exposure to toxins (Cuervo et al., 1999). The selectivity that characterizes CMA may be beneficial during prolonged starvation, as it will allow the degradation of nonessential proteins to provide amino acids for the synthesis of proteins required to guarantee cell survival under those stressful circumstances. Activation of CMA in conditions associated with protein damage, such as oxidative stress, may also facilitate the selective removal of altered proteins without disturbing neighboring functional ones.

The signaling mechanisms leading to CMA activation/inactivation are presently unknown, but most of the regulation of this pathway takes place at the lysosomal compartment (Bandyopadhyay et al., 2008; Cuervo and Dice, 2000a,b; Kaushik et al., 2006). Binding of substrate proteins to the CMA receptor, LAMP-2A is rate limiting for this pathway. In fact, levels of LAMP-2A at the lysosomal membrane are tightly regulated and directly correlate with CMA activity (Cuervo and Dice, 2000a,b). In addition, the presence of hsc70 in the lysosomal lumen is also necessary to attain substrate translocation into lysosomes via CMA (Agarraberes et al., 1997). In fact, although all lysosomes contain LAMP-2A in their membrane, only a subset of lysosomes contain hsc70 in their lumen, and they are the only ones competent for CMA (Cuervo et al., 1997). Under conditions such as prolonged starvation, both levels of LAMP-2A at the lysosomal membrane and of hsc70 in the lumen increase gradually, resulting in a progressive increase in CMA activity (Agarraberes et al., 1997; Cuervo et al., 1995a). If starvation persists beyond 3 days in rodents, part of the pool of lysosomes normally unable to perform CMA acquire the lumenal chaperone (hsc70) and become competent for this pathway. During maximal CMA activation, the pool of lysosomes active for this pathway relocates from the cell periphery toward the perinuclear region, although the reasons for this redistribution remain unclear (Cuervo and Dice, 2000b).

A decrease in CMA activity has been reported both in senescent cells in culture and in different organs from old rodents (Cuervo and Dice, 2000c; Dice, 1982). This age-related decline in CMA activity is mainly due to a gradual decrease in the levels of LAMP-2A at the lysosomal membrane because of its increased instability with age (Cuervo and Dice, 2000c; Kiffin et al., 2007). Reduced CMA activity thus contributes to the accumulation of abnormal proteins in the cytosol and it is probably in part responsible for the higher susceptibility to stressors, a characteristic of old organisms (Massey et al., 2006a). Malfunctioning of CMA has also been described in different pathologies such as some lysosomal storage disorders (Cuervo et al., 2003), certain toxic-induced nephropathies (Cuervo et al., 1999), the hypertrophic kidney secondary to diabetes (Sooparb et al., 2004), and in familial forms of Parkinson's disease (Cuervo et al., 2004c). The important roles of CMA as part of the cellular response to stress and the association of its malfunctioning with human pathologies have increased the interest in assessing CMA activity in different physiological and pathological conditions.

Here we describe the experimental models commonly used to study CMA, the characteristics of CMA substrates and the different methods developed by our and other groups to monitor CMA activity: (1) measurement of rates of long-lived protein degradation; (2) monitoring changes in the levels of key CMA components in isolated lysosomes; (3) analysis of the subcellular location of lysosomes active for CMA; and (4) measurement of the translocation of known CMA substrates into isolated lysosomes via an *in vitro* assay.

2. Experimental Models for the Study of CMA

CMA has been identified so far in mammalian cells only. Yeast have a somewhat-related process known as the vacuolar import and degradation pathway in which substrate proteins are first translocated in a chaperone-dependent manner into small vesicles that then fuse with the yeast vacuole where their cargo is degraded (Brown et al., 2003). Although this process resembles a combination of CMA and macroautophagy, the proteins involved in the translocation of substrates to the vesicles are different from those that participate in CMA. In fact, LAMP-2A, the spliced variant of the *lamp-2* gene that is essential for CMA, is not conserved in yeast. In species phylogenetically lower than mammals, such as worms, flies, or fish, a transcript with 40%–50% homology to LAMP-2A has been identified, but this homology is lost in the transmembrane and cytosolic regions of the protein, those that differentiate the three spliced variants of the *lamp-2* gene (Konecki et al., 1995). The LAMP-2 isoform conserved in these species seems to correspond to LAMP-2B, for which a role in macroautophagy but not in CMA is proposed (Eskelinen et al., 2003, 2005). Thus the LAMP-2A variant, required for CMA, appears much later in evolution, being described for the first time in avians and above (Konecki et al., 1995).

Although at different levels, CMA activity has been detected in many different types of transformed cells: NIH3T3 (mouse fibroblasts), 293HEK (human kidney epithelial cells), CHO (Chinese hamster ovary cells), Rat-1 (rat kidney epithelial cells), RALA (rat hepatoblastoma), Huh7 (human hepatoblastoma), astrocytome, several human lung cancer cell lines (H820, A549, H460), and in several primary cells in culture (human skin fibroblasts, dopaminergic neurons, cortical neurons, astrocytes, dendritic cells, macrophages, and CD4+ naive T cells). Among the different tissues in rodents in which CMA activity has been detected (liver, kidney, heart, spleen, lung), liver is by far the tissue in which this pathway has been best characterized (Dice, 2007; Massey *et al.*, 2006b). Although based on the absence of changes in the levels of KFERQ-containing proteins in response to starvation it was initially proposed that CMA is not active in brain (Wing *et al.*, 1991), recent studies with isolated astrocytes and dopaminergic and cortical neurons support the presence of CMA activity in these cells, although it is unresponsive to changes in the nutritional status (Cuervo, 2004c; Martinez-Vicente *et al.*, 2008).

Currently there are no knockout mouse models with impaired CMA. A complete LAMP-2 knockout mouse was developed several years ago (Tanaka *et al.*, 2000). The animals present alterations in macroautophagy that manifest as an accumulation of autophagic vacuoles in different tissues, and inefficient lysosomal biogenesis, that probably contributes to the observed decrease in protein degradation, and abnormal cholesterol metabolism and impaired vesicular trafficking (Eskelinen *et al.*, 2002; Huynh *et al.*, 2007). Surprisingly, mouse embryonic fibroblasts from these animals do not show changes in protein degradation, suggesting possible activation of compensatory mechanisms in undifferentiated cells but not in nondividing differentiated cells (Eskelinen *et al.*, 2004).

Our laboratory has developed a bitransgenic mouse model with regulated expression of LAMP-2A in liver (Zhang and Cuervo, 2008). As in cultured cells, overexpression of LAMP-2A in liver results in an increase in CMA activity. Using this model, we have recently analyzed the consequences of maintaining normal levels of LAMP-2A until advanced ages in liver, by activating the expression of the exogenous form of LAMP-2A once the levels of the endogenous protein start to decrease. We found that livers of old transgenic mice contain lower levels of altered proteins (e.g., oxidized, aggregate) respond more efficiently to different stressors, and show a significant improvement in liver function, supporting the critical role of CMA in maintenance of cellular homeostasis (Zhang and Cuervo, 2008). To determine possible tissue-specific differences in the requirements for functional CMA and whether restoration of CMA in a broad number of tissues will have a positive effect on life span, we are currently developing novel transgenic models with regulatable expression of LAMP-2A in different tissues.

3. Properties of CMA Substrates

An often-asked question when analyzing CMA is whether or not a protein is a substrate for this pathway. Proteins can follow different proteolytic pathways depending on changes in the protein itself (posttranslational modifications) or in the cellular conditions that result in activation/inhibition of particular proteolytic pathways (Cuervo, 2004b). Consequently, whether a protein is a substrate for CMA needs to be experimentally analyzed.

Table 19.1 summarizes the accepted criteria that a protein has to fulfill to be considered a CMA substrate. Briefly, the candidate protein has to bear in its amino acid sequence a CMA-targeting motif. The presence of this motif is necessary and sufficient to target proteins to CMA (Dice, 1990). Thus, proteins that do not carry a CMA-targeting motif can be directed to lysosomes via CMA when the sequence is incorporated as a fusion tag in the proteins. However, the fact a protein contains the motif in its sequence indicates that it *can be* degraded via CMA, but it does not necessarily imply that the protein *is* undergoing degradation through CMA, as often the targeting motifs are only exposed on the surface of the protein after conformational modifications. As with other proteins degraded via lysosomes, CMA substrates have usually long half-lives (ranging from >10 h up to several days), and their half-life changes with changes in CMA activity (increases when CMA is blocked or decreases if CMA is maximally

Table 19.1 Requirements for a protein to be considered as a CMA substrate

Requirement	Assay	References
Presence of KFERQ-like motif in its sequence	Sequence analysis	(Dice, 1990)
Long half-life Increases with CMA blockage Decreases with CMA activation	Metabolic labeling/immunoprecipitation	(Cuervo et al., 1998, 2004c)
Binding to cytosolic hsc70	Co-immunoprecipitation from cytosol	(Cuervo et al., 1998, 1999)
Binding to LAMP-2A at the lysosomal membrane	Co-immunoprecipitation from isolated lysosomes	(Cuervo et al., 2004c)
Translocation into isolated lysosomes ATP/hsc70-dependent Competed by other CMA substrates	In vitro translocation/degradation assays	(Cuervo et al., 1994, 1995b, 1998, 1999; Terlecky and Dice, 1993; Terlecky et al., 1992)

activated). All CMA substrates interact with the two major components of this pathway, hsc70, the chaperone in the cytosol, and the LAMP-2A receptor at the lysosomal membrane. Finally, the ultimate evidence of a protein being a bona fide CMA substrate is if it can be translocated into isolated lysosomes in an ATP- and hsc70-dependent manner (see subsequently). Of the substrate proteins identified until date, most of them are soluble cytosolic proteins. In fact, although it is plausible to think that this could also be a biogenic pathway for the delivery of enzymes into lysosomes, none of the known lysosomal hydrolases contain the CMA-targeting motif.

4. Methods to Measure CMA

The applicability of the methods described in the following sections depends on the experimental model. The four procedures described here can be used to track CMA in cultured cells, though it is true that the number of cells required for the isolation of lysosomes for the *in vitro* assays can be a limitation for some types of primary cells in culture. For animal tissues, such as the liver, where it is relatively easy to prepare a homogenous culture of hepatocytes, all procedures can be applied. However, when cell culture is not possible, measurement of protein degradation is not a straightforward procedure and measurement of CMA relies on the other three procedures.

5. Measurement of Protein Degradation Rates

With some exceptions, proteins degraded in lysosomes have long half-lives (Cuervo, 2004b). Consequently, measurement of the rates of degradation of long-lived protein in cultured cells through metabolic labeling in pulse-chase experiments can be used as a good assessment of lysosomal function (Fig. 19.1) (Auteri *et al.*, 1983). This procedure relies on the incorporation of a radiolabeled amino acid in the proteins synthesized during the labeling period (pulse) and tracking of the release of radiolabeled amino acid into the medium as the labeled proteins undergo degradation (chase). Separation of amino acid and small peptides from intact proteins in the medium is attained by precipitation of the proteins in acid.

5.1. Pulse and chase experiments

1. Plate the cells to approximately 40% of confluence in 12-well plates in the culture medium used for normal maintenance of that particular cell type.

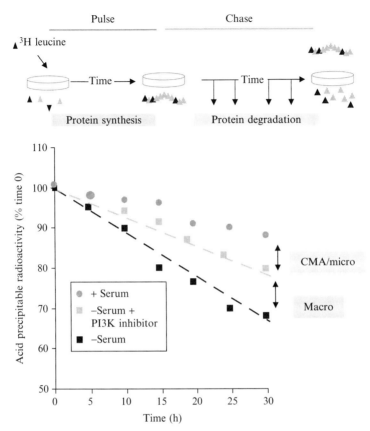

Figure 19.1 Measurement of long-lived protein degradation. Top: Confluent cells in culture are incubated with a radiolabeled amino acid for 48 h and after extensive washing the amount of acid-soluble radioactivity (amino acids and small peptides) released into the medium at different times is determined. Bottom: Typical example of rates of degradation of long-lived proteins in cultured cells due to CMA or micro- or macroautophagy. CMA activity is calculated as the increase in protein degradation during serum deprivation sensitive to lysosomal protease inhibition and insensitive to the effect of PI3-K type III inhibitors.

2. Pulse: When cells reach 60%–70% confluence, label the cells with 2 μCi/ml [^{3}H]leucine or [^{3}H]valine.
3. Incubate the cells at 37 °C for 48 h to maximize labeling of long-lived proteins.
4. Chase: At the end of the labeling, aspirate the medium, wash cells profusely (5 times) with Hanks's solution, and plate the cells in 0.5–0.7 ml of chase medium (chase medium contains 50 times the molar concentration of the unlabeled form of the radiolabeled amino acid to prevent reuse of the labeled amino acid into proteins synthesized during the

chase period). In half of the cells, the serum-free chase medium should be supplemented with serum, to be able to analyze changes in protein degradation in response to serum removal, one of the best characterized stimuli of CMA.

Note: Some caution in the use of leucine for the labeling has been recommended because of the inhibitory effect of this amino acid on macroautophagy in different cell types. However, high specific activity of the radiolabeled amino acid allows for the use of very low concentrations of leucine in the chase medium (30 times below the described inhibitory concentrations), making it feasible for most cell types. [^{3}H]valine has been proposed as a good alternative for labeling. However, the fact that the regulatory effect of amino acids on autophagy is not universal brings about the same concerns regarding the use of valine as a replacement for leucine. Consequently, for cells in which the inhibitory role of amino acids has been studied, it is recommended to use the amino acid with less inhibitory ability, whereas for cells where this effect is unknown it is recommended to at least verify if the result obtained with one radiolabeled amino acid is also reproducible using the other, or to directly assess the effect of both amino acids on macroautophagy in those cells using the procedures described in other chapters (see the chapter by Bauvy *et al.*, in this volume).

5. Collection of samples: To measure the amount of free radiolabeled amino acid released into the medium upon intracellular degradation of the proteins synthesized during the pulse:
   a. Incubate the cells in a CO_2 incubator at 37 °C.
   b. At the desired times (e.g., 0, 4, 12, 20, 24 h) collect aliquots (50–70 µl) of the medium from each well. It is important that only medium without floating cells is collected. If cell detachment is a problem, the 12-well plate should be centrifuged (500g for 5 min) before taking the medium, or the medium aliquots should be collected in separate microcentrifuge tubes, spun down using the same conditions, and the supernatant fractions transferred to a clean tube for precipitation. Otherwise, samples can be directly placed in a 0.45-µm pore filter-bottom 96-well plate (Millipore Multiscreen Assay System, Millipore, Bedford, MA, MSVMHTS00) containing half of the final volume of 20% trichloroacetic acid (TCA) (for a final concentration of 10%).
   c. Precipitation is facilitated by addition of 0.5 mg/ml final concentration of bovine serum albumin (BSA, standard grade powder (protease free)) dissolved in water) followed by incubation at 4 °C for at least 1 h.
   d. After taking the last time point, wash the cells twice with Hanks's solution and add 1–2 ml of solubilization buffer (0.1 N NaOH, 0.1% sodium deoxycholate, Sigma-Aldrich, D6750) per well. These samples

will be used to calculate the total amount of radioactivity incorporated by the cells during the labeling time, as indicated in 6 d.

Incubate the plate at 37 °C for 2–6 h, until cells are solubilized.

6. Sample processing: To count the amount of radiolabeled amino acid released into the medium (following protein degradation) and present inside the cells (incorporated into proteins):

   a. Collect the acid-soluble fraction of the aliquots taken from the medium (containing amino acids and short peptides) by vacuum filtration using a Millipore manifolder into a 96-well plate.
   b. Transfer the collected flow-through (approx. 200 μl) into individual scintillation vials, add scintillation liquid and read in a Beta scintillation counter.
   c. If desired, the filters can be left to air-dry and then punched into scintillation vials to account for the total amount of undegraded protein released into the medium.
   d. Take 50 μl of the solubilized cells and count in scintillation vials (this will be the total radioactivity still present inside cells).
   e. Count all the samples (flow-through, filters, and cells) as disintegrations per minute in a liquid scintillation analyzer by correcting for quenching using an external standard.

5.2. Inhibition of different autophagic pathways

To discriminate the pool of long-lived proteins degraded in lysosomes from those degraded in other proteolytic systems (e.g., ubiquitin/proteasome, calpains) and the contribution of each type of autophagy to the degradation of the long-lived proteins, blockers of lysosomal proteolysis are commonly used in these pulse and chase experiments. Weak bases, such as ammonium chloride or chloroquine, accumulate inside lysosomes neutralizing their intrinsic acid pH, required for maximal activity of the lysosomal proteases (Klionsky et al., 2007). Of the two weak bases, the former is commonly preferred because chloroquine has been shown to also affect protein synthesis. One of the limitations of the use of ammonium chloride is that its neutralizing effect rarely lasts more than 12 h, requiring periodic refreshing of the treatment if the chase is extended beyond this time. To overcome this problem, the group of Erwin Knecht has recently shown that a combination of 20 mM ammonium chloride with 0.1 mM leupeptin results in blocking lysosomal-dependent degradation most effectively without affecting other proteolytic systems (Fuertes et al., 2003). By comparing the degradation in cells supplemented or not with this cocktail it is possible to discriminate lysosomal-dependent degradation from that through other intracellular pathways.

The use in these studies of phosphatidylinositol-3-kinase (PI3K) inhibitors such as 3-methyladenine makes it possible to separate the percentage of lysosomal degradation (sensitive to ammonium chloride/leupeptin) that occurs via macroautophagy (sensitive to 3-methyladenine), as PI3K are required in this type of autophagy. The remaining lysosomal degradation (insensitive to 3-methyladenine) can be attributed in most cell types to the other autophagic pathways (i.e., microautophagy and CMA) (see subsequently). As a side note, 3-methyladenine is used in most cell types at 10 mM final concentration and prepared as a 2x stock in the growing medium, as solubility of this compound is dependent on pH.

5.3. Calculations

Proteolysis is calculated as the amount of acid-precipitable radioactivity (protein) transformed to acid-soluble radioactivity (peptides and amino acids) at each time during the incubation and it is expressed in percentage. To this purpose, the radioactivity in the aliquot collected from the medium needs to be corrected by the final volume of medium remaining in the well (as the amount of chasing medium decreases with each time point). The total amount of radiolabeled protein in each well is calculated by adding the amount of radioactivity present in the solubilized cells, plus the amount of radioactivity taken in each aliquot (acid soluble in the flow-through and acid precipitable in the filters), plus the amount of radioactivity present in the medium in the last time point.

$P =$ acid precipitable radioactivity
$S_t =$ acid soluble radioactivity in aliquot of the medium
$S_0 =$ acid soluble radioactivity in aliquot of the medium at time 0
$V_0 =$ total volume of medium at time 0
$v =$ volume of the aliquot taken from the medium
$T =$ radioactivity in the aliquot taken from the solubilized cells
$VT =$ total volume that cells were solubilized in
$n =$ number of aliquots taken before a given time point
$a =$ volume of aliquot taken from solubilized cells

$$\text{Proteolysis} = \frac{(\text{Asol}_t - \text{Asol}_0)}{\text{Total incorporated radioactivity}} \cdot 100$$
$$\text{Asol}_t = S_t \cdot (V_0 - (v \cdot n))/v$$
$$\text{Asol}_0 = S_0 \cdot (V_0)/v_0$$

Total incorporated radioactivity =
$P + S1 + S_2 + S_3 \ldots + (S_{end} \cdot (V_0 - (v \cdot n))/v) + (T \cdot VT/a)$

Although a certain level of basal CMA is present in all cells, CMA is maximally up-regulated in confluent cultured fibroblasts at approximately 10 h after serum removal. Hence, the inducible form of this pathway can be measured as the percentage of long-lived proteins degraded after removal of serum, inhibited by ammonium chloride (inhibitor of all types of autophagy), but insensitive to phosphatidylinositol-3-kinase inhibitors (Finn and Dice, 2005; Finn et al., 2005; Massey et al., 2006c) (Fig. 19.1). The two major limitations of this approach are that in certain cells there is considerable basal CMA activity, and consequently, considering only the percentage of degradation responsive to serum removal as CMA underestimates the contribution of this autophagic pathway to protein degradation. On the other hand, the lack of methods to quantify microautophagy in mammals or of selective inhibitors for this pathway makes it difficult to separate CMA-dependent degradation from that occurring via microautophagy, as this pathway contributes to lysosomal degradation both in the presence and in the absence of serum. Consequently, measurements of long-lived protein degradation should be complemented with other methods to analyze CMA.

6. Measurement of Levels of Key CMA Components

Levels of hsc70, the cytosolic chaperone responsible for targeting CMA substrates to lysosomes, remain constant in most conditions, as this is the constitutive member of the hsp70 family of molecular chaperones (Cuervo et al., 1995a). However, lysosomal levels of LAMP-2A and lysosomal-hsc70 increase with the increase in CMA activity (Agarraberes et al., 1997; Cuervo and Dice, 2000b). It is thus possible to monitor the increase in CMA activity via immunoblot for LAMP-2A and lys-hsc70 in lysosomes isolated from the tissues/cells of interest. Note that total cellular levels of LAMP-2A and hsc70 often remain constant as most of the changes occur in the particular group of lysosomes involved in CMA (approximately 30%–60% of total lysosomes depending on cell type and cell conditions). Although there are circumstances with extreme changes in CMA activity in which an increase or decrease of total cellular levels of LAMP-2A can be observed, it is advisable, when possible, to analyze the levels of this protein in the lysosomal fraction. As described in more detail subsequently, one of the limitations for the isolation of lysosomes competent for CMA is the large number of cultured cells or starting tissue required. An alternative that has been shown to be valid in some cases is the use of the light mitochondrial/lysosomal fraction obtained by differential centrifugation instead (see subsequently) as this can be prepared from as few as 2×10^6 cells or 0.1 g of tissue. In this case, because both lysosomes and mitochondria are highly enriched in this fraction, it is important to normalize the results to the levels of some abundant mitochondrial protein (e.g., cytochrome c, GRP78) to compensate for changes in the mitochondrial content.

6.1. Isolation of lysosomes

We describe in this section a procedure to purify lysosomes active for CMA from different tissue samples. A detailed protocol for the isolation of lysosomes from cultured cells can be found in the literature (Storrie and Madden, 1990). A critical point in this procedure is the way in which cells are disrupted, because the intrinsic fragility of this organelle requires the use of nitrogen cavitation rather than other common physical or mechanical procedures for cell disruption to get intact lysosomes. The lysosomal fraction obtained through this method is highly enriched in lysosomes active for CMA (those containing hsc70 in their lumen), reaching levels of approximately 80% of total lysosomes in that fraction (Agarraberes et al., 1997). Although other procedures also render a highly purified lysosomal fraction (Marzella et al., 1982), the percentage of CMA-active lysosomes in that fraction is considerably lower (20%–40% of total lysosomes, depending on the cell type).

The following protocol for the isolation of lysosomes with high and low CMA activity from tissue samples (Cuervo et al., 1997) was developed through modification of a previously published method of lysosomal purification from rat liver (Wattiaux et al., 1978). In addition to liver, we have successfully applied this protocol for the purification of lysosomes from kidney, spleen, lung, and different brain regions (gray matter).

1. Extensively wash the tissue of interest with cold 0.25 M sucrose (4–5 times with at least 3 times the volume of the tissue). The amount needed depends on the particular tissue and organism being analyzed; we find it practical to start with amounts of tissue between 1 g and 10 g. After weighing and mincing it, homogenize it in 0.25 M sucrose (3 volumes/g) in a motorized Teflon-glass homogenizer with 8–10 strokes at maximum speed.
2. Filter the homogenates through double gauze (common gauze cheesecloth) (to remove some of the interfering lipids), add 4 volumes of cold 0.25 M sucrose, and centrifuge at $6800g$ for 5 min at 4 °C.
3. Collect the supernatant fraction in another tube (ensure that you do not collect the heavy mitochondria-enriched fraction, the white floating material close to the pellet, which also contains any unbroken cells, nuclei, and red blood cells). Resuspend the pellet fraction with a cold finger (dry glass test tube filled with ice) in the starting volume of 0.25 M sucrose and centrifuge under the same conditions.
4. Pool the two supernatant fractions and centrifuge at $17,000g$ for 10 min at 4 °C.
5. Resuspend the pellet fraction with the cold finger, add 3.5 volumes of 0.25 M sucrose/g tissue, and centrifuge under the same conditions (wash step).

6. Resuspend the pellet (light mitochondria and lysosome-enriched fraction, (ML fraction)) with the cold finger, add 1 ml 0.25 M sucrose/3 g tissue and 2 volumes of 85.6% Metrizamide (AK Scientific, #69696), and mix gently. Load every 10 ml of this final 57% Metrizamide ML fraction at the bottom of an ultracentrifuge tube and layer on top a discontinuous Metrizamide gradient: 6 ml of 32.8% Metrizamide, 10 ml of 26.3% Metrizamide, and 11 ml of 19.8% Metrizamide (all diluted in water, pH 7.3). Fill the tube with 0.25 M sucrose and centrifuge in a SW 28 rotor at 141,000g for 1 h at 4 °C.

Note: The pellet of the second 17,000g, 10 min spin can be used as an alternative to purified lysosomes in conditions in which the amount of starting tissue is limiting.

7. After the centrifugation, white to brownish material (depending on the original tissue) is visible at each of the interphases (*IP*) enriched in the following fractions (from bottom to top): *IP1*—mitochondria; *IP2*—mixture of mitochondria and lysosomes; *IP3*—CMA-active and CMA-inactive lysosomes; *IP4*—CMA-active lysosomes. Collect *IP3* and *IP4* separately with a Pasteur pipette (in approximately 2–4 ml), dilute with at least 5 volumes of 0.25 M sucrose and centrifuge at 37,000g for 15 min at 4 °C.
8. Carefully resuspend the pellet of *IP3* with a blunt Pasteur pipette in 1 ml of 0.25 M sucrose and centrifuge at 10,000g for 5 min at 4 °C. This pellet is enriched in secondary lysosomes with low CMA activity (lacking lys-hsc70).
9. Use the supernatant fraction of this step to resuspend the pellet of the *IP4* fraction to get the fraction enriched in secondary lysosomes with high CMA activity (enriched in lys-hsc70).

Note: To guarantee the reproducibility of any future studies performed with the isolated fractions it is important to systematically evaluate for purity, recovery and enrichment of lysosomal enzymes and for integrity of the lysosomal membrane. Purity can be determined by measuring specific activity of enzyme markers of the main contaminant intracellular components: succinic-dehydrogenase (mitochondria), catalase (peroxisomes) and lactate-dehydrogenase (cytosol) as described before (Storrie and Madden, 1990). Measurement of total and specific activity of β-hexosaminidase or β-N-acetyl-glucosaminidase, two well-characterized lysosomal enzymes, are used routinely to determine the recovery (percentage of total cellular lysosomes recovered in the isolated fraction) and the enrichment (fold increase in specific activity of lysosomal markers in the isolated fraction) of lysosomes in the isolated fraction (Storrie and Madden, 1990). Last, β-hexosaminidase latency (percentage of β-hexosaminidase activity detected in the incubation medium when lysosomes are incubated in an isotonic buffer for increasing periods of time) is used to assess the integrity of the lysosomal membrane (Storrie and Madden, 1990). Preparations of lysosomes with more than 10% broken lysosomes should be discarded.

For studies requiring separate analysis of lysosomal membranes and matrices (lysosomal content), these two fractions can be isolated after disrupting lysosomes with a hypotonic shock. Briefly, collect the isolated lysosomes by centrifugation (25,000g for 10 min), resuspend the pellet fraction in a hypotonic buffer (0.025 M sucrose), and after 30-min incubation on ice, spin the samples at 150,000g for 30 min to recover the membrane fraction in the pellet and the lysosomal content in the supernatant fraction (Ohsumi *et al.*, 1983).

6.2. Immunoblot for CMA components

Because most of the changes in the levels of CMA components that occur with changes in the activity of this pathway are not transcriptionally regulated, direct measurement of the protein of interest by immunoblot is the most used method for analysis. The choice of antibody for detection of CMA components is important as although both hsc70 and hsp70 share high homology and most of the available antibodies recognize both proteins, only hsc70 but not hsp70 participates in CMA (Chiang *et al.*, 1989). Thus, antibodies that recognize both chaperones should be avoided. The group of Fred Dice extensively characterized the mouse monoclonal IgM antibody clone 13D3 (now available from different commercial sources, e.g., Abcam, ab2788) as highly selective for hsc70 (Agarraberes *et al.*, 1997). Regarding LAMP-2A, only levels of the A splice variant of the *lamp2* gene correlates with CMA activity (Cuervo and Dice, 1996, 2000a,b). Most of the commercially available antibodies against LAMP-2 have been developed against the lumenal part of the protein, also shared by the other two isoforms (Carlsson *et al.*, 1988). It is important, thus, to use antibodies against the cytosolic tail that will discriminate between each of these isoforms. We developed an antibody against the cytosolic tail of the rat LAMP-2A that also recognizes the mouse isoform, but it does not cross-react with the human protein (now commercially available through Zymed Laboratories, Invitrogen, 51-2200). To the best of our understanding, antibodies selective against human LAMP-2A are not currently available.

An increase in the levels of lysosomal LAMP-2A is usually observed in conditions when CMA is activated. Likewise, decreased levels of LAMP-2A in lysosomes are a good indication of diminished CMA activity. However, although reduced levels of lys-hsc70 will result in decreased lysosomal ability for CMA, because binding to LAMP-2A is the limiting step of this pathway, increased levels of lys-hsc70, do not necessarily transduce into increased CMA. Thus, in conditions such as aging in which a decrease of lysosomal LAMP-2A has been well reported, levels of lys-hsc70 are higher than in control cells, probably reflecting some type of compensatory mechanism (Cuervo and Dice, 2000c).

7. ANALYSIS OF THE SUBCELLULAR LOCATION OF CMA-ACTIVE LYSOSOMES

As pointed out in the introduction, only a subset of lysosomes is competent for CMA (Cuervo *et al.*, 1997). These are distinguishable because they have LAMP-2A in their membrane (as do most lysosomes) and contain detectable levels of hsc70 in their lumen. This pool of lysosomes can be tracked with the antibodies specific for these two proteins both by immunofluorescence with secondary antibodies conjugated to fluorophores or by the use of two differently sized immunogold-conjugated secondary antibodies in electron microscopy sections of cells or tissues (Cuervo *et al.*, 1997). Using these procedures, it was found that in conditions such as prolonged starvation, the number of lysosomes containing hsc70 in their lumen increases gradually with the increase in CMA activity (Cuervo *et al.*, 1997). Similar procedures were used to identify that the previously described increase in lys-hsc70 in aging originated indeed from an increase in the number of CMA-competent lysosomes rather than a net increase in the amount of hsc70 per lysosome (Cuervo and Dice, 2000c).

Interestingly, for reasons still unknown, activation of CMA is associated with the mobilization of hsc70/LAMP-2A-enriched lysosomes to the perinuclear region of the cells (Cuervo and Dice, 2000b) (Fig. 19.2). Visualizing the increase in the number of hsc70/LAMP-2A lysosomes and their subcellular location are also indirect ways to monitor CMA in cells in culture and in tissue sections.

7.1. Immunofluorescence for CMA-active lysosomes

Colocalization of LAMP-2A and hsc70 by immunofluorescence is used often to identify the CMA-active lysosomes in cultured cells. This procedure can be performed following a standard immunofluorescence protocol, with the exception that fixation of the samples should be done using methanol. Methanol fixation eliminates the soluble form of hsc70 (very abundant in the cytosol) allowing for the detection of the low percentage vesicle-associated hsc70.

1. Grow cells on coverslips at the bottom of 6-well plates in the complete (serum-supplemented) medium that the particular cell type usually requires, until they reach semiconfluence (40%–60% confluence). At this point, replace the medium with serum-free medium in half of

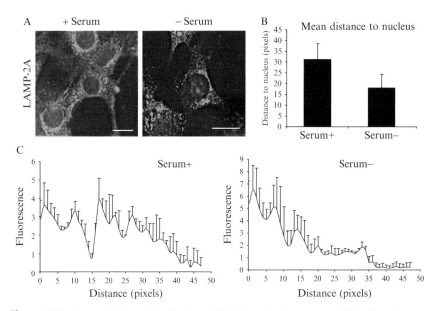

Figure 19.2 Intracellular redistribution of CMA-active lysosomes. (A) Indirect immunofluorescence for LAMP-2A in cultured mouse fibroblasts maintained in the presence/absence of serum. Bar: 10 μm. (B) Mean distance of the fluorescent puncta (lysosomes) to the nucleus. (C) Graph representing the intracellular distribution of fluorescent puncta with respect to the nucleus in the two indicated conditions. Values are mean + standard error of 4 different cells in each condition.

the coverslips and culture in these conditions for 12–20 h to maximally activate CMA in the serum-deprived cells.

2. Wash twice with serum-free medium (to remove all remaining IgGs) and fix in −20 °C cooled methanol for 1 min at room temperature.
3. Add 2 ml of blocking solution (0.2% (w/v) powdered nonfat milk, 2% newborn calf serum, 0.1 M glycine, 1% BSA, and 0.01% Triton X-100 in PBS) per well and incubate at room temperature for 30 min.
4. Aspirate the blocking solution and wash with PBS 3 times.
5. Dilute the primary antibody (rabbit IgG anti-LAMP-2A 1:100 in filtered 0.1% BSA in PBS). Layer a Petri plate with Parafilm, place the coverslips on it (cells facing up) and put 25 μl of diluted primary antibody on top; incubate in a humidified chamber at room temperature for 1 h.
6. Extensively wash the coverslips with PBS (by successive immersion of cover slips held by blunt-end forceps in beakers containing PBS).
7. Incubate the coverslips with 25 μl of the fluorophore-conjugated secondary antibody (diluted in filtered 0.1% BSA in PBS) for 30 min at room temperature, as described in step 5.

8. After extensive washes (as described in step 6), incubate the cover slips in 25 µl of the other primary antibody (mouse IgM anti-hsc70) diluted 1:150 in filtered 0.1% BSA in PBS, under the same conditions as in step 5.
9. After extensive washes (as described in step 6), incubate the coverslips in 25 µl of the fluorophore-conjugated secondary antibody (diluted in filtered 0.1% BSA in PBS) for 30 min at room temperature as described in step 5, wash (as described in step 5) and mount by placing them (cells facing down) on top of 15 µl of DAPI (4′,6-diamidino-2-phenylindole)-containing antifade mounting medium (Invitrogen, P-7481) spotted on a glass slide, and seal with nail polish to prevent drying.
10. Visualize the slides using a fluorescence microscope (Axiovert 200, Carl Zeiss), and deconvolute the captured images using the manufacturer's software.

Note: Standard immunofluorescence control slides (incubated only with secondary antibodies or incubated only with one primary antibody and the secondary for the other primary antibody) should be included.

The following parameters can be analyzed as direct indication of CMA activity in the cultured cells:

a. Colocalization of the two antibodies can be quantified using the Just Another Colocalization Plugin of the ImageJ program (NIH) after setting the appropriate threshold.
b. The mean distance of the vesicles positive for each antibody to the nucleus is calculated with the Straight Line Tool and the Analyze Particles function of the ImageJ program, by drawing straight lines from the most distant vesicle positive for each antibody to the nucleus and computing the particle distribution (distance and density). An average of 6 different radial lines per cell and 20 cells per field is usually calculated to determine changes in the intracellular distribution of CMA-active lysosomes (Fig. 19.2). Although changes in lysosomal localization may be due to many different reasons (e.g., alterations in vesicular trafficking, problems with microtubule polymerization), decreased distance to the nuclear region of the CMA-active lysosomes associated with positive values in any of the other procedures described in this work are good support for CMA activation.

7.2. Immunogold and electron microscopy for CMA-active lysosomes

Immunogold staining for LAMP-2A and hsc70 in tissue sections, cultured cells, or isolated lysosomes can also be used to assess changes in the amount of CMA-active lysosomes, and consequently in CMA activity (Cuervo *et al.*, 1995b, 1997).

1. Fix tissue, cultured cells, or isolated lysosomes in 4% paraformaldehyde, 0.05% glutaraldehyde, and 0.1 M cacodylate in 0.25 M sucrose and process for electron microscopy analysis following standard procedures, with the exception that the sections should be subjected to only one staining step to minimize masking of the gold particles (Cuervo et al., 1995b, 1997) (see also the chapter by Ylä-Anttila et al., in this volume).
2. Perform immunogold staining on ultrathin sections mounted onto copper grids. Incubate the grids with the anti-LAMP-2A and anti-hsc70 antibodies (diluted 1:100) for 8–10 h at room temperature in a humidified chamber, followed by incubation with different-sized gold-conjugated secondary antibodies (Electron Microscopy Sciences, EM grade) (1:100) for 2 h at room temperature.
3. Rinse the grids extensively in water.
4. Negatively stain with 1% uranyl acetate.
5. Capture images using a JEOL 100CX II transmission electron microscope at 80kV.

Note: Required control samples are the same as indicated for immunofluorescence.

Morphometric analysis of digital images of the sections can be done using ImageJ after drawing the lysosomal profiles and after applying the Clearing Outside function determining the number of particles per lysosome for each size gold particle and the number of lysosomes containing both sizes of gold particles.

8. *In Vitro* Assay to Measure Translocation of CMA Substrates

The most unequivocal method to measure CMA activity is by monitoring the direct translocation of known CMA substrates in lysosomes isolated from the tissues or cultured cells of interest. This method requires the isolation of intact highly purified CMA-active lysosomes from the samples (as described in the "Isolation of lysosomes" section). The light mitochondria/lysosomal fraction proposed as an alternative to purified lysosomes for the immunoblot assays is not usually a good replacement for purified lysosomes in these assays due to the high consumption of ATP by the contaminant mitochondria in the fraction and the difficulty to normalize for mitochondria content in these assays. The ideal fraction for the assays is that of lysosomes enriched in hsc70 in their lumen (CMA^+ lysosomes), though a pool of secondary lysosomes with differing hsc70 content can also be used for transport assays (Cuervo et al., 1997).

The purity of the lysosomal fraction and the integrity of the lysosomal membrane are both critical for proper assessment of CMA substrate translocation (Storrie and Madden, 1990). Thus, if lysosomal enzymes leak from the lumen during the experiment, they could degrade the substrate proteins outside lysosomes and provide the erroneous idea that the degradation is taking place after lysosomal translocation. Accurate measurements can be attained only with tight control of lysosomal membrane integrity and of contaminant components in the fraction as described earlier ("Isolation of lysosomes" section). Despite the exorbitant cost of Metrizamide, the density medium used for the isolation of CMA-competent lysosomes, we still strongly recommend its use, as this is the density medium that guarantees higher purity and lower lysosomal breakage during isolation (researchers are strongly discouraged from using sucrose as a density medium, as it is actively transported into lysosomes and results in a high percentage of lysosomal breakage by hyperosmotic shock). Likewise, a rigorous control of the osmolarity of the incubation medium and all the additions to the medium is required to preserve lysosomal integrity during the incubation.

There are two procedures to track CMA substrate translocation into lysosomes. We routinely apply both methods, as they allow analyzing different steps in CMA. Measurement of the degradation of radiolabeled CMA substrates (e.g., [^{14}C]GAPDH) by isolated lysosomes recapitulates the three main lysosomal steps of CMA: binding, translocation, and proteolysis (Terlecky and Dice, 1993) (Fig. 19.3A). Parallel experiments with lysosomes disrupted by hypotonic shock (to allow free access of the enzymes to the CMA substrate) permits determining possible changes in proteolysis independent of binding/uptake. This method is quantitatively accurate and the use of a 96-well plate-based filtration device allows rapid processing of a large number of samples (Terlecky and Dice, 1993).

8.1. Radiolabeling of CMA substrates

Purified CMA substrates are radiolabeled with [^{14}C]formaldehyde by reductive methylation. We routinely label glyceraldehyde-3-phosphate dehydrogenase (GAPDH; Sigma-Aldrich G5262) and ribonuclease A (RNase A: Rockland Immunochemicals, MB-113-0005) because both are commercially available as purified proteins.

1. Dissolve the protein in reaction buffer (10 mM MES, pH 7) to a final concentration of 3 mg/ml.
2. Add [^{14}C]formaldehyde (250 μCi) (PerkinElmer NEC039H001MC) and sodium cyanoborohydrate (Sigma-Aldrich, 156159) (final concentration of 1.8 mg/ml in reaction buffer) and incubate this reaction mixture in a final volume of 500 μl at 25 °C for 1 h.
3. Separate the radiolabeled protein from the unincorporated radioisotope by gel filtration through an appropriate Sephadex matrix (according to the size

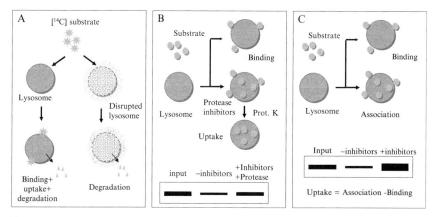

Figure 19.3 *In vitro* assays for the direct quantification of CMA. (A) Incubation of intact or broken lysosomes with radiolabeled CMA substrates allows quantification of the amount of protein processed into soluble amino acids (degradation). In intact lysosomes, substrates need to bind to the lysosomal membrane and translocate before they can be degraded. (B–C). Incubation of CMA substrates with intact lysosomes treated or not with protease inhibitors allows determination of the amount of substrate bound to the lysosomal membrane via immunoblot against the substrate after collecting the lysosomes by centrifugation. The amount of substrate translocated into the lysosomal lumen can be calculated in lysosomes treated with protease inhibitors after degradation of the substrate bound to the cytosolic side of the lysosomal membrane with an exogenous protease (B) or by subtracting the amount of bound substrate from the total amount of substrate associated with the protease-inhibited lysosomes (C).

of the protein). We routinely use minispin columns packed with the matrix (we prepare 1-ml columns with insulin syringes filled with the matrix, but there are also some commercially available alternatives (e.g., Pierce 89849) previously blocked with 20 mg/ml BSA (5 vol, for 30 min at 25 °C) (to prevent nonspecific binding) and equilibrated with the reaction buffer (10 vol). Spinning time is adjusted depending on the protein of interest and the characteristics of the minispin column (for most minispin columns and proteins in the 100-30 kDa range, the spin time varies from 1–5 min).

4. Collect the eluted radiolabeled protein in separate aliquots.
5. Measure the amount of radiolabeled protein and free radioisotope in each aliquot by determining the radioactivity associated with the acid-precipitable fraction (radiolabeled protein) and acid-soluble fraction (free radioisotope) using TCA precipitation (as described in the "Pulse and chase experiments" section).

8.2. Protein degradation with isolated lysosomes

1. Incubate freshly isolated intact lysosomes (25 μg protein in 10 μl final volume after dilution in proteolysis buffer (10 mM 3-(N-morpholino) propanesulfonic acid (MOPS), pH 7.3, 0.3 M sucrose, 5.4 μM cysteine,

1 mM DTT) with 10 µl of radiolabeled protein (260 nM, 2000 dpm/µl in the same buffer), 1 µl of the 6x energy-regenerating system (10 mM MgCl$_2$, 10 mM ATP, 2 mM phosphocreatine (Sigma-Aldrich, P-6502), 50g/ml creatinephosphokinase (Sigma-Aldrich, C-3755), and 10 µg/ml of GST-hsc70 or hsc70 purified from bovine brain or liver (as described previously (Welch and Feramisco, 1985)) in a final volume of 60 µl (adjusted with proteolysis buffer) for 30 min at 37 °C in a 0.22-µm 96-well filter plate (previously wet with sterile water for 10 min at room temperature and rinsed out).
2. Include a blank well containing all the reagents except for the lysosomes to account for the amount of protein spontaneously cleaved, and the possible contaminant amount of free radiolabeled amino acids present in the purified labeled protein fraction.
3. Stop the reaction by adding TCA (10% final concentration) and BSA to a final concentration of 0.5 mg/ml to favor protein precipitation.
4. After incubation at 4 °C for at least 30 min, collect the acid-soluble flow-through using the Millipore multiscreen vacuum system (as described in the "Pulse and chase experiments" section) and transfer to 5-ml vials, add scintillation liquid and count in a liquid scintillation counter.

Proteolysis is calculated as the amount of acid-precipitable radioactivity (protein) transformed into acid-soluble radioactivity (amino acids and small peptides) at the end of the incubation: [([dpm flow-through sample − dpm flow-through blank]/dpm pellet at time 0) × 100].

8.3. Protease protection assay

The previous assay measures proteolysis of substrates translocated into lysosomes by CMA and recapitulates binding, uptake, and translocation. However, individual CMA steps cannot be separately analyzed by this procedure. To dissect the two initial steps of CMA, a second *in vitro* assay with isolated lysosomes was developed in which association of substrate proteins to lysosomes is quantified by immunoblot (Aniento *et al.*, 1993). The original version of this assay was based on the protease-protection assays widely used for the study of translocation of proteins into different organelles (Fig. 19.3B). Briefly, incubation of substrates with intact lysosomes results in their translocation and rapid degradation in the lumen. Consequently, when pulled down, the only lysosome-associated substrate protein is that bound to the cytosolic side of the lysosomal membrane. However, if the substrate proteins are incubated with lysosomes previously treated with protease inhibitors, the translocated protein remains inside the lumen. To quantify the protein present in the lumen of the lysosomes, an exogenous protease is added. The protease degrades the substrate bound to the cytosolic side of the lysosomal membrane but cannot access the one present in the lumen. A recent variation of this experiment (to avoid problems associated with resistance to cleavage by the exogenous protease) has been proposed in which the amount of internalized protein is calculated by subtracting the amount of bound substrate from the amount of substrate that is associated with lysosomes treated with protease inhibitors (bound + internalized) (Fig. 19.3C) (Salvador *et al.*, 2000). Finally, binding and uptake can also be distinguished by modifying the incubation temperature. At temperatures below 10 °C binding occurs but substrates do not translocate, whereas at temperatures above this both binding and uptake via CMA are coupled (Aniento *et al.*, 1993). For all these assays, it is essential to include strict controls with lysosomes incubated alone (to account for any endogenous lysosomal protein recognized by the antibody against the substrate), lysosomes in which the membrane is disrupted with detergent (to demonstrate that the exogenous protease can degrade the substrate if access is allowed), and incubations in the presence of other CMA substrate and nonsubstrate proteins (to account for any translocation not mediated by CMA). To control for the amount of degradation of substrate due to lysed or leaky lysosomes, the same amount of lysosomes added per reaction should be centrifuged (25,000g for 5 min at 4 °C) and the supernatant fraction (that will contain the enzymes leaking from the lysosomes) should be incubated with the substrate protein under the same conditions.

1. Incubate freshly isolated intact lysosomes with a (x100) protease inhibitor cocktail (10 mM leupeptin, 10 mM AEBSF, 1 mM pepstatin and 100 mM EDTA) for 10 min on ice.
2. In 0.5-ml microcentrifuge tubes, add freshly isolated intact lysosomes (100 μg of protein) pretreated or not with protease inhibitor, along with 10–50 μg of CMA substrate (GAPDH, RNase A or any other protein of interest), 5 μl of (6x) energy regenerating system and 10 μg/ml GST-hsc70 or purified brain or liver bovine hsc70 in 30 μl final volume of incubation buffer (10 mM MOPS, pH 7.3, 0.3 M sucrose).
3. Incubate the samples for 20 min at 37 °C.
4. At the end of the incubation, cool down half of the tubes pretreated with protease inhibitors on ice (1 min) and add proteinase K (5 μl of a 1 mg/ml stock in 1mM CaCl$_2$, 50 mM Tris-HCl, pH 8). The proteinase K solution should be made fresh or kept frozen to prevent self-degradation.
5. Incubate the samples on ice for 10 min, add AEBSF (5 μl) and centrifuge all samples at 25,000g for 5 min at 4 °C.
6. Aspirate the supernatant fractions and wash the pellet fractions twice with 100 μl of incubation buffer to eliminate any protein bound nonspecifically to the lysosome surface.
7. Resuspend the final pellets in Laemmli buffer (Laemmli, 1970) with protease inhibitors, boil for 5 min at 95 °C and perform SDS-PAGE and immunoblot with antibodies specific for the substrate of choice.

Include in the gel a lane containing {1/10} of the amount of substrate added to the incubation to use it as reference in the calculations of the amount of substrate bound or translocated.

Calculations: Using densitometry of the immunoblotted membranes calculate substrate binding and uptake as follows:

(i) binding = the percentage of total added substrate associated with lysosomes untreated with protease inhibitors
(ii) association = the percentage of substrate recovered in lysosomes treated with protease inhibitors
(iii) uptake = either the difference between association and binding or the percentage of substrate associated with lysosomes treated with protease inhibitors after proteinase K treatment.

Although lysosomal internalization of proteins by other autophagic pathways (macroautophagy or microautophagy) cannot be reproduced *in vitro*, at least in the conditions used in this assay, as new proteins are considered as possible CMA substrate candidates, we cannot discard the possibility of the existence of other yet to be identified mechanisms for direct translocation of soluble proteins in lysosomes. Consequently, to confirm that the binding/uptake/degradation of putative CMA substrates assayed by the two methods described here is certainly occurring via CMA, it is advisable that one or both of the following assays are performed:

a. Competition assays with well-characterized CMA substrates: If the putative substrate is translocated into lysosomes by CMA, addition of equimolar concentrations of GAPDH or RNase A (two well-characterized CMA substrates) to any of the incubations indicated previously should decrease binding, uptake, and degradation of the tested substrate (as they compete for the same lysosomal machinery for degradation) (Cuervo *et al.*, 1994; Terlecky and Dice, 1993).
b. Blockage of the CMA receptor: Binding of CMA substrates to the cytosolic tail of LAMP-2A is required for their lysosomal translocation. Consequently, preincubation of the lysosomes added in the assays with the specific antibody against the cytosolic tail of LAMP-2A or supplementation of the incubation medium with a peptide of the same amino acid composition as the cytosolic tail of LAMP-2A should reduce CMA of the putative substrate (Cuervo and Dice, 1996).

9. Concluding Remarks

In conclusion, using the battery of assays described in this work it is possible to evaluate changes in CMA activity in cultured cells and different tissues from rodents under different physiological and pathological conditions.

The most accurate assays to measure possible changes in CMA activity in different samples are those reproducing *in vitro* the translocation of known CMA substrates in lysosomes, as the activity of other proteolytic pathways will not be measured in those assays. However, the large amount of cultured cells required for the isolation of lysosomes active for CMA and the training (to some extent) required to become proficient in these procedures (the characteristic instability of the lysosomal membrane upon isolation makes it necessary to perform all of these procedures rapidly in minimal amount of time) relegates often the *in vitro* uptake assays as a final confirmatory assay once evidence suggestive of changes in CMA have been gathered using several of the other assays.

Future efforts are oriented to the development of image-based reporters incorporated in cells or in animals (i.e., transgenic mouse models with the CMA reporter expressed in all tissues) that allow tracking changes in CMA through changes in the association of the reporter with lysosomes and/or its degradation in this cellular compartment.

ACKNOWLEDGMENTS

Work in our laboratory is supported by grants from the National Institute of Health AG021904, AG031782, and DK041918, and by a Glenn Award.

REFERENCES

Agarraberes, F., Terlecky, S., and Dice, J. (1997). An intralysosomal hsp70 is required for a selective pathway of lysosomal protein degradation. *J. Cell Biol.* **137,** 825–834.

Aniento, F., Emans, N., Griffiths, G., and Gruenberg, J. (1993). Uptake and degradation of glyceraldehyde-3- phosphate dehydrogenase by rat liver lysosomes. *J. Biol. Chem.* **268,** 10463–10470.

Auteri, J., Okada, A., Bochaki, V., and Dice, J. (1983). Regulation of intracellular protein degradation in IMR- 90 human diploid fibroblasts. *J. Cell Physiol.* **115,** 159–166.

Bandyopadhyay, U., Kaushik, S., Vartikovsky, L., and Cuervo, A. M. (2008). The chaperone-mediated autophagy receptor organizes in dynamic protein complexes at the lysosomal membrane. *Mol. Cell Biol.* **28,** 5747–5763.

Brown, C. R., Liu, J., Hung, G. C., Carter, D., Cui, D., and Chiang, H. L. (2003). The Vid vesicle to vacuole trafficking event requires components of the SNARE membrane fusion machinery. *J. Biol. Chem.* **278,** 25688–25699.

Carlsson, S. R., Roth, J., Piller, F., and Fukuda, M. (1988). Isolation and characterization of human lysosomal membrane glycoproteins, h-lamp-1 and h-lamp-2. *J. Biol. Chem.* **263,** 18911–18919.

Chiang, H., Terlecky, S., Plant, C., and Dice, J. (1989). A role for a 70 kDa heat shock protein in lysosomal degradation of intracellular protein. *Science* **246,** 382–385.

Cuervo, A. (2004a). Autophagy: in sickness and in health. *Trends Cell Biol.* **14,** 70–77.

Cuervo, A., and Dice, J. (1996). A receptor for the selective uptake and degradation of proteins by lysosomes. *Science* **273,** 501–503.

Cuervo, A., and Dice, J. (2000a). Regulation of lamp2a levels in the lysosomal membrane. *Traffic* **1,** 570–583.
Cuervo, A., and Dice, J. (2000b). Unique properties of lamp2a compared to other lamp2 isoforms. *J. Cell Sci.* **113,** 4441–4450.
Cuervo, A., Dice, J., and Knecht, E. (1997). A population of rat liver lysosomes responsible for the hsc73-mediated degradation of cytosolic proteins in lysosomes. *J. Biol. Chem.* **272,** 5606–5615.
Cuervo, A., Hildebrand, H., Bomhard, E., and Dice, J. (1999). Direct lysosomal uptake of α_2-microglobulin contributes to chemically induced nephropathy. *Kidney Int.* **55,** 529–545.
Cuervo, A., Hu, W., Lim, B., and Dice, J. (1998). IκB is a substrate for a selective pathway of lysosomal proteolysis. *Mol. Biol. Cell* **9,** 1995–2010.
Cuervo, A., Knecht, E., Terlecky, S., and Dice, J. (1995a). Activation of a selective pathway of lysosomal proteolysis in rat liver by prolonged starvation. *Am. J. Physiol.* **269,** C1200–C1208.
Cuervo, A., Palmer, A., Rivett, A., and Knecht, E. (1995b). Degradation of proteasomes by lysosomes in rat liver. *Eur. J. Biochem.* **227,** 792–800.
Cuervo, A., Terlecky, S., Dice, J., and Knecht, E. (1994). Selective binding and uptake of ribonuclease A and glyceraldehyde-3-phosphate dehydrogenase by isolated rat liver lysosomes. *J. Biol. Chem.* **269,** 26374–26380.
Cuervo, A. M. (2004b). Autophagy: Many pathways to the same end. *Mol. Cell Biochem.* **263,** 55–72.
Cuervo, A. M., and Dice, J. F. (2000c). Age-related decline in chaperone-mediated autophagy. *J. Biol. Chem.* **275,** 31505–31513.
Cuervo, A. M., Mann, L., Bonten, E., d'Azzo, A., and Dice, J. (2003). Cathepsin A regulates chaperone-mediated autophagy through cleavage of the lysosomal receptor. *EMBO J.* **22,** 12–19.
Cuervo, A. M., Stefanis, L., Fredenburg, R., Lansbury, P. T., and Sulzer, D. (2004c). Impaired degradation of mutant α-synuclein by chaperone-mediated autophagy. *Science* **305,** 1292–1295.
Dice, J. (1982). Altered degradation of proteins microinjected into senescent human fibroblasts. *J. Biol. Chem.* **257,** 14624–14627.
Dice, J. (1990). Peptide sequences that target cytosolic proteins for lysosomal proteolysis. *Trends Biochem. Sci.* **15,** 305–309.
Dice, J. (2007). Chaperone-mediated autophagy. *Autophagy* **3,** 295–299.
Eskelinen, E.-L., Illert, A. L., Tanaka, Y., Schwarzmann, G., Blanz, J., Von Figura, K., and Saftig, P. (2002). Role of LAMP-2 in lysosome biogenesis and autophagy. *Mol. Biol. Cell* **13,** 3355–3368.
Eskelinen, E.-L., Schmidt, C., Neu, S., Willenborg, M., Fuertes, G., Salvador, N., Tanaka, Y., Lullmann-Rauch, R., Hartmann, D., Heeren, J., von Figura, K., Knecht, E., and Saftig, P. (2004). Disturbed cholesterol traffic but normal proteolytic function in LAMP-1/LAMP-2 double-deficient fibroblasts. *Mol. Biol. Cell* **15,** 3132–3145.
Eskelinen, E.-L., Schmidt, C., Neu, S., Willenborg, M., Fuertes, G., Salvador, N., Tanaka, Y., Lullmann-Rauch, R., Hartmann, D., Heeren, J., von Figura, K., Knecht, E., and Saftig, P. (2005). Unifying nomenclature for the isoforms of the lysosomal membrane protein LAMP-2. *Traffic* **6,** 1058–1061.
Eskelinen, E.-L., Tanaka, Y., and Saftig, P. (2003). At the acidic edge: Emerging functions for lysosomal membrane proteins. *Trends Cell Biol.* **13,** 137–145.
Finn, P., Mesires, N., Vine, M., and Dice, J. F. (2005). Effects of small molecules on chaperone-mediated autophagy. *Autophagy* **1,** 141–145.

Finn, P. F., and Dice, J. F. (2005). Ketone bodies stimulate chaperone-mediated autophagy. *J. Biol. Chem.* **280,** 25864–25870.

Fuertes, G., Martin De Llano, J., Villarroya, A., Rivett, A., and Knecht, E. (2003). Changes in the proteolytic activities of proteasomes and lysosomes in human fibroblasts produced by serum withdrawal, amino-acid deprivation and confluent conditions. *Biochem. J.* **375,** 75–86.

Huynh, K., Eskelinen, E., Scott, C., Malevanets, A., Saftig, P., and Grinstein, S. (2007). LAMP proteins are required for fusion of lysosomes with phagosomes. *EMBO J.* **26,** 313–324.

Kaushik, S., Massey, A. C., and Cuervo, A. M. (2006). Lysosome membrane lipid microdomains: Novel regulators of chaperone-mediated autophagy. *EMBO J.* **25,** 3921–3933.

Kiffin, R., Christian, C., Knecht, E., and Cuervo, A. (2004). Activation of chaperone-mediated autophagy during oxidative stress. *Mol. Biol. Cell* **15,** 4829–4840.

Kiffin, R., Kaushik, S., Zeng, M., Bandyopadhyay, U., Zhang, C., Massey, A., Martinez-Vicente, M., and Cuervo, A. (2007). Altered dynamics of the lysosomal receptor for chaperone-mediated autophagy with age. *J. Cell Sci.* **120,** 782–791.

Klionsky, D., Cuervo, A., and Seglen, P. (2007). Methods for monitoring autophagy from yeast to human. *Autophagy* **3,** 181–206.

Konecki, D. S., Foetisch, K., Zimmer, K., Schlotter, M., and Lichter-Konecki, U. (1995). An alternatively spliced form of the human lysosome-associated membrane protein-2 gene is expressed in a tissue-specific manner. *Biochem. Biophys. Res. Comm.* **215,** 757–767.

Laemmli, U. (1970). Cleavage of structural proteins during the assembly of the head of the bacteriophage T4. *Nature* **227,** 680–685.

Levine, B., and Klionsky, D. J. (2004). Development by self-digestion: Molecular mechanisms and biological functions of autophagy. *Dev. Cell* **6,** 463–477.

Majeski, A., and Dice, J. (2004). Mechanisms of chaperone-mediated autophagy. *Int. J. Biochem. Cell Biol.* **36,** 2435–2444.

Martinez-Vicente, M., Martinez-Vicente, M., Talloczy, Z., Kaushik, S., Massey, A., Mazzulli, J., Mosharov, E., Hodara, R., Fredenburg, R., Wu, D., Follenzi, A., Dauer, W., Przedborski, S., Ischiropoulos, H., Lansbury, P., Sulzer, D., and Cuervo, A. (2008). Dopamine-modified α-synuclein blocks chaperone-mediated autophagy. *J. Clin. Invest.* **118,** 777–788.

Marzella, L., Ahlberg, J., and Glaumann, H. (1982). Isolation of autophagic vacuoles from rat liver: Morphological and biochemical characterization. *J. Cell Biol.* **93,** 144–154.

Massey, A., Kiffin, and Cuervo, A. (2004). Pathophysiology of chaperone-mediated autophagy. *Int. J. Biochem. Cell Biol.* **36,** 2420–2434.

Massey, A., Kiffin, R., and Cuervo, A. (2006a). Autophagic defects in aging: Looking for an "emergency exit"? *Cell Cycle* **5,** 1292–1296.

Massey, A., Zhang, C., and Cuervo, A. (2006b). Chaperone-mediated autophagy in aging and disease. *Curr. Top Dev. Biol.* **73,** 205–235.

Massey, A. C., Kaushik, S., Sovak, G., Kiffin, R., and Cuervo, A. M. (2006c). Consequences of the selective blockage of chaperone-mediated autophagy. *Proc. Nat. Acad. Sci. USA* **103,** 5905–5910.

Ohsumi, Y., Ishikawa, T., and Kato, K. (1983). A rapid and simplified method for the preparation of lysosomal membranes from rat liver. *J. Biochem.* **93,** 547–556.

Salvador, N., Aguado, C., Horst, M., and Knecht, E. (2000). Import of a cytosolic protein into lysosomes by chaperone-mediated autophagy depends on its folding state. *Journal of Biological Chemistry* **275,** 27447–27456.

Shintani, T., and Klionsky, D. (2004). Autophagy in health and disease: A double-edged sword. *Science* **306,** 990–995.

Sooparb, S., Price, S. R., Shaoguang, J., and Franch, H. A. (2004). Suppression of chaperone-mediated autophagy in the renal cortex during acute diabetes mellitus. *Kidney Int.* **65,** 2135–2144.

Storrie, B., and Madden, E. (1990). Isolation of subcellular organelles. *Methods Enzymol.* **182,** 203–225.

Tanaka, Y., Guhde, G., Suter, A., Eskelinen, E.-L., Hartmann, D., Lullmann-Rauch, R., Janssen, P., Blanz, J., von Figura, K., and Saftig, P. (2000). Accumulation of autophagic vacuoles and cardiomyopathy in Lamp-2-deficient mice. *Nature* **406,** 902–906.

Terlecky, S., Chiang, H.-L., Olson, T., and Dice, J. (1992). Protein and peptide binding and stimulation of *in vitro* lysosomal proteolysis by the 73-kDa heat shock cognate protein. *J. Biol. Chem.* **267,** 9202–9209.

Terlecky, S., and Dice, J. (1993). Polypeptide import and degradation by isolated lysosomes. *J. Biol. Chem.* **268,** 23490–23495.

Wattiaux, R., Wattiaux-De Coninck, S., Ronveaux-Dupal, M., and Dubois, F. (1978). Isolation of rat liver lysosomes by isopycnic centrifugation in a metrizamide gradient. *J. Cell Biol.* **78,** 349–368.

Welch, W., and Feramisco, J. (1985). Rapid purification of mammalian 70,000-dalton stress proteins: affinity of the proteins for nucleotides. *Mol. Cell Biol.* **5,** 1229–1237.

Wing, S., Chiang, H. L., Goldberg, A. L., and Dice, J. F. (1991). Proteins containing peptide sequences related to Lys-Phe-Glu-Arg-Gln are selectively depleted in liver and heart, but not skeletal muscle, of fasted rats. *Biochem. J.* **275,** 165–169.

Zhang, C., and Cuervo, A. M. (2008). Restoration of chaperone-mediated autophagy in aging liver improves cellular maintenance and hepatic function. *Nat. Med.* **14,** 959–965.

CHAPTER TWENTY

Methods to Monitor Autophagy of *Salmonella enterica* Serovar Typhimurium

Cheryl L. Birmingham[*] *and* John H. Brumell[*,†]

Contents

1. Introduction — 326
   1.1. The *Salmonella*-containing vacuole — 326
   1.2. Type III secretion systems — 328
   1.3. Intracellular bacterial populations — 329
2. Characteristics of Autophagy of *S.* Typhimurium — 330
   2.1. Markers — 330
   2.2. Kinetics — 332
   2.3. Differences from canonical autophagy — 332
   2.4. Advantages of *S.* Typhimurium autophagy as a model system — 333
   2.5. Relevance to disease: Role of Atg16L1 in *S.* Typhimurium autophagy and Crohn's disease — 333
3. *S.* Typhimurium Infection and Autophagy — 334
   3.1. Bacterial growth and infection techniques — 334
   3.2. Monitoring autophagy of bacteria by immunofluorescence — 335
4. Controls for Autophagy of Bacteria — 337
   4.1. Inhibiting bacterial protein synthesis — 337
   4.2. SPI-1 T3SS-deficient bacteria — 337
   4.3. Autophagy inhibition — 338
5. Concluding Remarks — 339
References — 339

Abstract

Autophagy is an important component of the mammalian innate immune system and is able to specifically target intracellular bacterial pathogens. *Salmonella enterica* serovar Typhimurium is an intracellular pathogen that causes gastroenteritis in humans. Autophagy has been shown to target

[*] Department of Molecular Genetics and Institute of Medical Science, University of Toronto, Toronto, Ontario, Canada
[†] Cell Biology Program, Hospital for Sick Children, Toronto, Ontario, Canada

S. Typhimurium during *in vitro* infection of mammalian cultured cells and protects the cytosol of these cells from bacterial colonization. Here we discuss autophagic sequestration of *S.* Typhimurium and how it can be used as a model system to study the mechanisms of autophagy.

1. INTRODUCTION

Salmonella enterica are facultative Gram-negative bacteria that infect a variety of hosts and result in different disease manifestations (Haraga *et al.*, 2008; Rabsch *et al.*, 2002). These food-borne bacteria are a serious public health issue in areas with poor hygiene and contaminated water resources. Infection by *Salmonella enterica* serovar Typhi, which is restricted to humans and higher primates, results in the disease typhoid fever (Kuhle and Hensel, 2004; Rabsch *et al.*, 2002). Upon oral infection, this bacterium is able to cross the epithelial cell layer of the small intestine, disseminate to other areas of the body, including the lymph nodes, liver, and spleen, and cause systemic infection. On the other hand, *Salmonella enterica* serovar Typhimurium (*S.* Typhimurium) is a nontyphoidal serovar that typically causes gastroenteritis in humans. In up to 10% of infections, these bacteria can also cause reactive arthritis (Maki-Ikola and Granfors, 1992). *S.* Typhimurium tends to colonize the human gastrointestinal tract but does not spread systemically, resulting in a self-limiting infection. However, colonization of susceptible mice results in systemic bacterial infection and a typhoidlike disease manifestation, allowing mouse infection by *S.* Typhimurium to be used as an animal model for typhoid fever (Kuhle and Hensel, 2004; Prost *et al.*, 2007).

S. Typhimurium is a vacuole-adapted intracellular pathogen that survives and replicates in both professional and nonprofessional phagocytic cells. Bacterial invasion of intestinal epithelial cells is required for the initiation of infection and the establishment of gastroenteritis (Hueck, 1998; Penheiter *et al.*, 1997; Tsolis *et al.*, 1999b). However, bacterial replication in host phagocytic cells is necessary for the establishment of systemic infection (Holden, 2002).

1.1. The *Salmonella*-containing vacuole

Upon entry into a host cell, *S.* Typhimurium resides and replicates in a membrane-bound compartment called the *Salmonella*-containing vacuole (SCV) (Brumell *et al.*, 2002a) (Fig. 20.1A). *S.* Typhimurium actively modifies this compartment into a niche permissive for bacterial replication. The SCV matures in a manner that parallels phagosome maturation to a point, but then diverges from this pathway to avoid acquiring the degradative properties of the phagolysosome (Brumell and Grinstein, 2004;

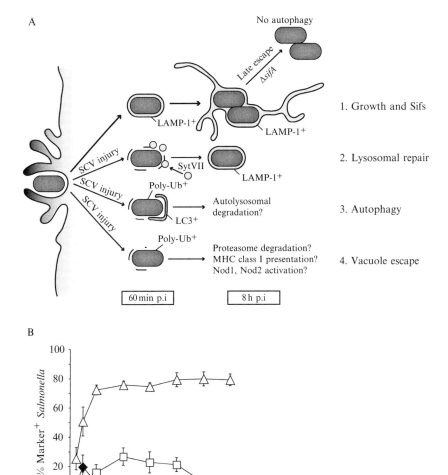

Figure 20.1 S. Typhimurium *in vitro* infection. (A) Model of intracellular bacterial populations during *in vitro* infection. After invasion, at least 4 different intracellular populations of S. Typhimurium are observed. **Population 1:** The majority of S. Typhimurium reside in SCVs that acquire LAMP-1. These bacteria direct SCV maturation to allow establishment of a niche permissive for growth. Sif formation is typically observed at 8 h p.i. and is associated with these bacteria. **Population 2:** S. Typhimurium can damage the SCV early after invasion via the SPI-1 T3SS. This allows release of calcium into the cytosol and triggers the recruitment of LAMP-1$^+$ lysosomes via the calcium sensor SytVII. The lysosomes fuse with the SCV and release their contents into the vacuole, possibly degrading the bacteria (Roy *et al.*, 2004). The bacteria are maintained in vacuoles but would not be predicted to grow. **Population 3:** SPI-1 T3SS-dependent damage to the SCV gives rise to a population of bacteria targeted by the autophagy system. The consequences of this event are still unknown, though interaction with the endocytic pathway and degradation of the bacteria in autolysosomes is probable.

Steele-Mortimer, 2008). The early SCV interacts with the endocytic system to acquire typical early endosomal markers, such as early endosomal antigen 1 and the transferrin receptor (Steele-Mortimer et al., 1999). The SCV then continues to mature, acquiring the late endocytic markers lysosome-associated membrane proteins (LAMPs) and Rab7, and acidifying to a pH of 4.0 to 5.5 (Garcia-del Portillo and Finlay, 1995; Meresse et al., 1999; Rathman et al., 1996). However, the bacteria actively inhibit complete maturation into a degradative compartment, as the SCV does not acquire the mannose-6-phosphate receptor or mature degradative lysosomal enzymes (Brumell et al., 2001; Garcia-del Portillo and Finlay, 1995; Mills and Finlay, 1998). In vitro, bacteria start replicating in the SCV 4–6 h after invasion. During this time, the bacteria actively induce the formation of dynamic tubular extensions of the SCV called *Salmonella*-induced filaments (Sifs). These structures are associated with rapidly replicating bacteria but are otherwise of unknown function (Birmingham et al., 2005; Brumell and Grinstein, 2004; Garcia-del Portillo F, 1993; Prost et al., 2007).

1.2. Type III secretion systems

The genome of S. Typhimurium possesses two pathogenicity islands, *Salmonella* pathogenicity island (SPI)-1 and SPI-2, that are indispensable for disease progression and encode type III secretion systems (T3SSs). T3SSs are needle-like structures that span bacterial and host membranes to inject bacterial effector proteins into host cells (Haraga et al., 2008; Schlumberger and Hardt, 2006). Importantly, these structures have been shown to damage host cell membranes (Blocker et al., 1999; Hakansson et al., 1996). In particular, the S. Typhimurium SPI-1 T3SS can damage the SCV early after invasion, causing the activation of a vacuole repair mechanism in which lysosomes are recruited to the damaged compartment (Roy et al., 2004).

Population 4: Damaged SCVs that are not repaired by lysosome fusion or targeted by the autophagy system eventually release S. Typhimurium into the cytosol of the host cell. These bacteria become heavily decorated with ubiquitinated proteins (Poly-Ub$^+$) (Birmingham et al., 2006; Perrin et al., 2004). The fate of these bacteria is cell type dependent: S. Typhimurium can grow in the cytosol of epithelial cells (Beuzón et al., 2002; Brumell et al., 2002b) and fibroblasts (Birmingham et al., 2006) but are killed by factors present in the cytosol of macrophages (Beuzón et al., 2000, 2002). **(B)** A graph of the markers acquired by different populations of intracellular S. Typhimurium over time. HeLa cells were transfected with GFP-LC3 and infected with S. Typhimurium. Cells were fixed at the indicated time points, and co-stained for bacteria and either ubiquitinated proteins or LAMP-1. The percentage of ubiquitinated protein$^+$ (□), LAMP-1$^+$ (△), or GFP-LC3$^+$ (◆) bacteria was enumerated by fluorescence microscopy. Error bars indicate +/- the standard deviation. Figures taken from Birmingham et al. (2006).

Bacterial proteins transferred into host cells by T3SSs can affect a variety of host cell processes, including host cytoskeleton remodeling and SCV trafficking (Haraga et al., 2008). The SPI-1 and SPI-2 T3SSs function during different times of the S. Typhimurium life cycle. SPI-1 is necessary for the invasion of nonprofessional phagocytes and the establishment of gastroenteritis in animal models (Hueck, 1998; Tsolis et al., 1999a; Wood et al., 1998). After the early stages of infection, the SPI-1 system is down-regulated and expression of the SPI-2 T3SS is induced, which is required for proper SCV trafficking and intracellular bacterial replication (Eriksson et al., 2003; Hensel et al., 1995; Ochman et al., 1996; Pfeifer et al., 1999; Valdivia and Falkow, 1996).

1.3. Intracellular bacterial populations

During *in vitro* infection of mammalian cultured cells, more than one population of intracellular S. Typhimurium is formed (Fig. 20.1A). Although the majority of internalized bacteria reside in SCVs, 10%–20% disrupt the vacuole and enter the cytosol (Perrin et al., 2004). In macrophages, SCV stability is essential for bacterial replication (Beuzón et al., 2002). However, cytosolic bacteria replicate more efficiently than those in SCVs in epithelial cells and fibroblasts (Beuzón et al., 2002; Brumell et al., 2002c). Interestingly, S. Typhimurium present in the cytosol associate with ubiquitinated proteins (Perrin et al., 2004), a phenotype that can be used as an indicator of bacterial exposure to the cytosolic environment. Whether a cytosolic population of S. Typhimurium exists *in vivo* remains to be determined.

During bacterial infection of nonprofessional phagocytic cells, the SPI-1 T3SS is required for escape of S. Typhimurium into the cytosol and association of bacteria with ubiquitinated proteins (Birmingham et al., 2006). A small population of intracellular bacteria label with ubiquitinated proteins while still retaining SCV markers, a phenotype that is lost during infection with SPI-1 T3SS-deficient bacteria (Birmingham et al., 2006). These results suggest that a population of S. Typhimurium damages the SCV membrane through expression of the SPI-1 T3SS, resulting in exposure of the bacteria to the cytosol through gaps in the SCV membrane. SPI-1 T3SS-induced SCV damage may be responsible for eventual SCV destabilization and release of bacteria into the cytosol.

Roy et al. (2004) have shown that lysosomes are recruited in a calcium-dependent manner to repair host cell membranes damaged by the S. Typhimurium SPI-1 T3SS in macrophages. As well, a population of S. Typhimurium is targeted by the autophagy system in epithelial cells and fibroblasts in a SPI-1 T3SS-dependent manner (Birmingham et al., 2006). Bacteria targeted by autophagy also associate with ubiquitinated proteins and retain SCV markers. The results suggest that bacteria in SCVs damaged by the

SPI-1 T3SS are targeted by autophagy, which may act as another mechanism to repair damaged intracellular compartments (Birmingham et al., 2006). However, the signal that drives autophagy to target damaged SCVs remains to be determined. Autophagy targeting of S. Typhimurium appears to protect the cell cytosol (which is permissive for bacterial replication in these cell types) from bacterial colonization (Birmingham et al., 2006).

2. Characteristics of Autophagy of S. Typhimurium

In vitro, approximately 20% of total intracellular S. Typhimurium are targeted by autophagy by 1 h post infection (p.i.) in nonprofessional phagocytic cells (including HeLa cells and mouse embryonic fibroblasts) (Birmingham et al., 2006). Autophagy-deficient ($atg5^{-/-}$) mouse embryonic fibroblasts are more permissive for intracellular growth by S. Typhimurium than wild-type fibroblasts because more bacteria escape into the cytosol (Birmingham et al., 2006). Therefore, autophagy appears to retain intracellular S. Typhimurium in vacuoles early after infection to protect the cytosol from bacterial colonization.

2.1. Markers

A standard autophagosome marker is microtubule-associated protein 1-light chain 3 (LC3) (Kabeya et al., 2000). This protein is covalently conjugated to phosphatidylethanolamine and is localized to autophagosomes during their formation, transport to, and fusion with lysosomes. LC3 is degraded inside the autolysosome along with the inner autophagosomal membrane (Tanida et al., 2005). Constructs expressing green fluorescent protein (GFP) or red fluorescent protein (RFP) fusions to LC3 localize to intracellular S. Typhimurium targeted by autophagy (Birmingham et al., 2006). LC3 usually forms a tight-fitting ring around the targeted bacterium but can also be visualized as a cup-shaped structure around one pole of the bacterium (presumably a forming autophagosome in the process of elongating and fusing to enclose the bacterium) (Fig. 20.2).

Bacteria targeted by autophagy colocalize with markers of the SCV yet are still exposed to the cytosol, indicative of the localization of these bacteria in damaged SCVs (Birmingham et al., 2006). Markers of the SCV that have been localized to bacteria targeted by autophagy include: class I major histocompatibility complex (MHC), CD44 and LAMP-1 (Birmingham et al., 2006). Class I MHC from the cell surface is present on the SCV from bacterial uptake to approximately 3 h p.i., at which time it is recycled back to the plasma membrane (Garcia-del Portillo et al., 1994; Smith et al.,

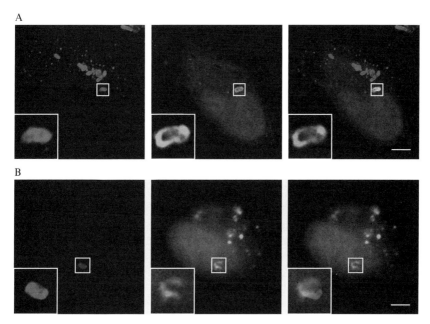

Figure 20.2 Immunofluorescence of S. Typhimurium targeted by autophagy. (A) HeLa cells were transiently transfected with GFP-LC3 (green) and infected with S. Typhimurium. At 1 h p. i., cells were fixed and stained for bacteria (blue). Shown is a confocal Z-slice of a tight-fitting LC3$^+$ autophagosome around a bacterium. Magnified images of the boxed areas are shown in the bottom left of the corresponding panels. Size bars = 5 μm. (B) HeLa cells were transiently transfected with RFP-LC3 (red), infected, fixed and stained as in A. Shown is an epifluorescence image of a cup-shaped LC3$^+$ structure around one pole of a bacterium (image taken by S. Shahnazari, Hospital for Sick Children, Toronto). This structure presumably represents a forming autophagosome elongating to form a complete structure around the targeted bacterium. (See Color Insert.)

2005). CD44 is present on the SCV until approximately 1 h p.i. and is similarly recycled (Garcia-del Portillo et al., 1994; Smith et al., 2005). LAMP-1 is a commonly used marker of the SCV during later stages of infection. This membrane protein is acquired by the SCV approximately 1 h p.i. and is maintained on this compartment throughout the bacterial life cycle. All of these SCV markers exhibit a patchy staining around bacteria targeted by autophagy (Birmingham et al., 2006). Furthermore, regions around bacteria lacking the SCV marker staining often contain GFP-LC3, suggesting that autophagy may serve to patch holes in the damaged SCV membrane (Birmingham et al., 2006) similar to the lysosomal repair system suggested by Roy et al. (2004).

S. Typhimurium labels strongly with ubiquitinated proteins when exposed to the mammalian cell cytosol (Perrin et al., 2004). Therefore, we use immunostaining for ubiquitinated proteins as a marker for bacterial exposure to the cytosol. We find that approximately 50% of LC3$^+$ bacteria also label with ubiquitinated proteins (Birmingham et al., 2006).

2.2. Kinetics

In HeLa cells, colocalization of intracellular S. Typhimurium with GFP-LC3 peaks at 1 h p.i. when approximately 20% of bacteria label with this autophagy marker (Birmingham et al., 2006) (Fig. 20.1B). GFP-LC3 colocalization with intracellular bacteria drops to a background level of <5% by 2 h p.i. This is consistent with autophagy targeting of S. Typhimurium in response to early SCV damage. On the other hand, LAMP-1 localizes to the SCV by 1 h p.i. (approximately 75%–80% of total intracellular bacteria) and remains on the SCV throughout infection (Beuzón et al., 2000, 2002; Birmingham et al., 2006; Brumell et al., 2002b) (Fig. 20.1B). The association of intracellular bacteria with ubiquitinated proteins reaches maximal levels (approximately 25% of intracellular bacteria) by 4 h p.i. and does not drop off to background levels (5%–10%) until approximately 10 h p.i. (Birmingham et al., 2006) (Fig. 20.1B). Therefore, it appears that autophagy does not target S. Typhimurium in the cytosol after 1 h p.i. Consistent with this, mutant bacteria that lose the SCV at later times in infection (6 to 8 h p.i.) are not targeted by autophagy (Birmingham et al., 2006).

In mouse embryonic fibroblasts, LC3 also colocalizes with approximately 20% of intracellular bacteria by 1 h p.i. However, this level of autophagy targeting is maintained for up to 4 h p.i., and does not drop down to background levels (3%–5%) until as late as 12 h p.i. (Birmingham et al., 2006). Therefore, there are cell-type differences in the autophagy response to S. Typhimurium.

2.3. Differences from canonical autophagy

Autophagy of bacteria probably occurs by a different mechanism than that of autophagy of the bulk cytosol (Yoshimori, 2006). Differences between these two types of autophagy include, first, the size of the autophagosome: Canonical autophagosomes (those degrading bulk cytosol) are typically 0.5–1.5 μm in diameter (Mizushima et al., 2002). S. Typhimurium are rod-shaped bacteria of approximately 2–3 μm in length, and therefore autophagosomes containing these bacteria must be of at least this size. Furthermore, autophagosomes surrounding S. Typhimurium are no longer circular but rather conform to the shape of the bacterium (Fig. 20.2). Second is the kinetics of autophagosome maturation: In a study by (Nakagawa et al. (2004) autophagosomes containing the normally extracellular pathogen Group A Streptococcus take approximately 2–3 h to mature into autolysosomes, whereas canonical autophagosomes mature within 10–15 min (Bampton et al., 2005; Nakagawa et al., 2004). Third, the specificity of autophagy targeting: Autophagy of bulk cytosol is considered to be a non-selective process. Autophagy of bacteria, on the other hand, is a specific process that requires targeting mechanisms. For example, cytosolic *Shigella flexneri* lacking expression of the *icsB* gene are targeted by autophagy via direct binding of Atg5 to a bacterial surface protein (Ogawa et al., 2005). Another signal for autophagy targeting is damage of phagosomes or phagosome-like structures, as occurs during autophagy of S. Typhimurium (Birmingham et al., 2006). Specific autophagy targeting of bacteria results in the formation of tight-fitting autophagosomes that are largely devoid of cytosol (Fig. 20.2). Importantly, the mechanisms determining the differences between autophagy of bacteria and canonical autophagy are not understood.

2.4. Advantages of S. Typhimurium autophagy as a model system

Autophagy of S. Typhimurium during *in vitro* infection is a very consistent and easily manipulated system that is ideal for studying autophagy of a specific target substrate. We routinely perform fixed-cell immunofluorescence or live cell imaging to observe autophagy of S. Typhimurium. Using this system, it is possible to observe and control when and where autophagy occurs. Furthermore, autophagy targeting of bacteria can be inhibited through the use of commercially available drugs, bacterial mutants or knockout cell-lines. Notably, S. Typhimurium can efficiently invade most cell types *in vitro*.

Of particular note, the most commonly used fluorescent autophagosome marker, GFP-LC3, has been shown to form nonspecific aggregates that can be mistaken for autophagosomes (Klionsky et al., 2007). Using autophagy of a large specific substrate (i.e., S. Typhimurium) allows for the automatic exclusion of these nonspecific $LC3^+$ artifacts.

2.5. Relevance to disease: Role of Atg16L1 in S. Typhimurium autophagy and Crohn's disease

Recently, the autophagy protein Atg16L1 was identified in several genetic screens for factors associated with inflammatory bowel disorders. In particular, this protein was linked to the incidence of Crohn's disease, which is the continuous inflammation of the gut caused by an inappropriate immune response to commensal enteric bacteria (WTCCC, 2007; Hampe et al., 2007; Rioux et al., 2007). Atg16L1 (Atg16 in yeast) is a conserved component of the core machinery required for autophagosome formation (Xie and Klionsky, 2007). This protein has also been shown to be involved in autophagy of S. Typhimurium during *in vitro* infection (Rioux et al., 2007).

Other genetic factors identified for Crohn's disease include: the NOD-like receptor nucleotide oligomerization domain 2 (NOD2) and the immunity-related GTPase IRGM (Hugot *et al.*, 2001; Massey and Parkes, 2007; Parkes *et al.*, 2007). Notably, IRGM has also been shown to be involved in autophagy of *Mycobacterium tuberculosis* in human cells (WTCCC, 2007; Parkes *et al.*, 2007; Singh *et al.*, 2006). Therefore, autophagy of bacteria may play a role in the pathogenesis of inflammatory bowel disorders.

3. *S.* Typhimurium Infection and Autophagy

S. Typhimurium actively invades nonprofessional phagocytes, including epithelial cells and fibroblasts, by injecting bacterial effector proteins into the host cell using the SPI-1 T3SS (Schlumberger and Hardt, 2006). *S.* Typhimurium can also infect macrophages in SPI-1-dependent and SPI-1-independent manners. We have characterized autophagy targeting of *S.* Typhimurium in nonprofessional phagocytic cell lines and therefore will restrict our discussion to these cell types. During *in vitro* infection, *S.* Typhimurium is grown under SPI-1 T3SS-inducing conditions to favor optimal invasion efficiency, as has been previously described (Steele-Mortimer *et al.*, 1999). This technique causes a population of bacteria to highly express the SPI-1 T3SS. For example, approximately 26% of *S.* Typhimurium grown under *in vitro* infection conditions express very high levels of the SPI-1 effector SipA (Schlumberger *et al.*, 2005). It is likely that this population of bacteria is the one that damages the SCV early after infection in a SPI-1-dependent manner and, subsequently, is targeted by autophagy.

3.1. Bacterial growth and infection techniques

To study autophagy targeting of *S.* Typhimurium, HeLa cells or fibroblasts are plated at 2.5×10^4 cells/well in 24-well tissue culture plates. The following day, the cells are transiently transfected with a fluorescently labeled autophagy marker, such as GFP-LC3. Alternatively, a cell line stably expressing a fluorescently labeled autophagy marker (i.e., GFP-LC3 (Mizushima *et al.*, 2004)) can be used, and plated at 5×10^4 cells/well. After approximately 16 h, the cells are infected with *S.* Typhimurium grown to late-log as follows:

1. Bacteria are inoculated in 2 mL of Luria-Bertani (LB) broth (10 g/L tryptone, 5 g/L yeast extract, 10 g/L NaCl, pH 7.0) containing antibiotics appropriate for selecting the bacterial population (i.e., for SL1344 wild-type *S.* Typhimurium, 50 μg/mL streptomycin is used).

The bacteria are then grown for approximately 16 h at 37 °C with shaking at approximately 250 rpm (results in an OD_{600} of approximately 1.7–2.0).

2. Bacteria are subcultured 1:33 in 10 mL of LB broth without antibiotics for 3 h at 37 °C with shaking (in a 125-mL flask to allow for good aeration). This results in an OD_{600} of approximately 1.2–1.5, corresponding to approximately 1×10^8 bacteria/mL.
3. 1 mL of bacteria is harvested by pelleting at 10,000×g for 2 min in a 1.5-mL microcentrifuge tube, then resuspended in 1 mL of sterile phosphate-buffered saline containing calcium and magnesium ions, pH 7.2 ($PBS^{+/+}$; 8 g/L NaCl, 1.15 g/L Na_2HPO_4, 0.2 g/L KCl, 0.2 g/L KH_2PO_4, 0.1 g/L $MgCl_2$ $6H_2O$, 0.1 g/L $CaCl_2$; from HyClone, Logan, Utah).
4. Bacteria are diluted 1:100 in sterile $PBS^{+/+}$.
5. 0.5 mL of the diluted bacteria is added to each well of the plated and transfected adherent cells.
6. Cells are incubated at 37 °C in 5% CO_2 for 10 min to allow bacterial invasion.
7. At 10 min p.i., extracellular bacteria are removed by extensive washing with $PBS^{+/+}$. First, the infection inoculum is aspirated from the wells carefully without scraping off or disturbing the adherent cells. Approximately 1 mL of $PBS^{+/+}$ is added to each well and aspirated off. This is repeated three times. Growth medium (Dulbecco's Modified Eagle's Medium (from HyClone, catalog number SH30243) supplemented with 10% fetal bovine serum) is added to the cells (0.5 mL/well), which are then placed back at 37 °C.
8. At 30 min p.i., the growth medium is aspirated off and new growth medium containing 100 μg/mL gentamicin (to kill any extracellular bacteria) is added to the cells (0.5 mL/well, placed back at 37 °C).
9. At 1 h p.i., the infected cells are analyzed for autophagy of bacteria. If later time points are required, add new growth medium containing 10 μg/mL gentamicin to the cells at 2 h p.i.

3.2. Monitoring autophagy of bacteria by immunofluorescence

We routinely analyze autophagy of S. Typhimurium at 1 h p.i. However, a full kinetic analysis should be done to determine the maximal level of autophagy targeting during different infection conditions or in different cell lines. For analysis of fixed samples by immunofluorescence, we generally use the following protocol:

1. Cells plated (5.0×10^4 cells/mL) on coverslips and infected as desired are fixed with 2.5% paraformaldehyde (PFA) in PBS, pH 7.2, for 10 min at 37 °C. Another common method is to fix samples with methanol (precooled at -20 °C) for 10 min at -20 °C.

2. The fixed cells are permeabilized and blocked in 0.2% saponin (diluted in water) and 10% normal goat serum (SS-PBS; 300 μL/well) for a minimum of 30 min.
3. SS-PBS is aspirated off and 12 μL of primary antibody (diluted in SS-PBS) is carefully layered on each coverslip. The coverslips are then incubated with the primary antibody for the desired length of time (usually 1 h) with monitoring every 10 min to ensure that the entire coverslip remains covered with antibody.
4. Primary antibody is washed off by adding approximately 1 mL of PBS to each well for 15 min. This is repeated 3 times.
5. Cells are repermeabilized in SS-PBS as in step 2 for 5 min.
6. Secondary antibody is layered on each coverslip as in step 3, except the incubation is done in the dark (i.e., in a drawer) for at least 30 min.
7. Cells are washed as in step 4.
8. Samples are mounted on slides using fluorescent mounting medium (Dako Fluorescent Mounting Medium, DakoCytomation, Mississauga, Ontario, Canada) and examined by microscopy.

Using this method, the percentage of bacteria targeted by autophagy under different conditions or the colocalization of proteins of interest with $LC3^+$ bacteria can be determined. Although we generally use fluorescently labeled LC3 to follow autophagosome formation by immunofluorescence, there are now commercially available antibodies to some autophagy markers.

Not all bacteria successfully invade cells, and some remain adhered to the outside of the host cell. Therefore, it is often necessary to distinguish between intra- and extracellular bacterial populations. Extracellular and total bacteria can be differentially stained as follows:

1. Fixed samples are blocked in 10% normal goat serum only (300 μL/well). This does not permeabilize the cells.
2. Samples are stained with primary antibodies to bacteria and an Alexa-Fluor 350 secondary antibody (as earlier; antibodies are diluted in 10% normal goat serum) to label only extracellular bacteria.
3. Samples are permeabilized with SS-PBS for 30 min (300 μL/well).
4. Samples are then stained with primary antibodies to bacteria and an AlexaFluor 568 secondary antibody (as earlier) to label total bacteria. At this stage, samples can also be stained with antibodies to other intracellular markers of interest.

In this way, intracellular bacteria are only visible in the red channel, whereas extracellular bacteria are visible in both the red and the blue channels. Note that if the preceding protocol is followed, the red and blue channels are reserved for staining bacteria. Other channels can be used for staining of other proteins, including green (AlexaFluor 488) and far red (AlexaFluor 647).

We have found that bacteria expressing fluorescent molecules, such as GFP or monomeric RFP (mRFP), are consistently targeted by autophagy to a greater degree than nonexpressing bacteria. For example, bacteria expressing mRFP or GFP colocalized with LC3 approximately 30%–35% of the time at 1 h p.i., respectively, compared to approximately 20% for nonexpressing bacteria (Birmingham et al., 2006). The reason for this increased targeting is currently unknown. However, bacteria expressing mRFP or GFP can be used to boost the amount of bacteria targeted by autophagy if required. These bacteria can be used for either analysis of fixed samples or live imaging. As well, bacteria can be labeled with amine-reactive succinimidyl esters conjugated to various fluorochromes (e.g., the AlexaFluor 647 carboxylic acid, succinimidyl ester available from Molecular Probes) for live imaging experiments in which more than one marker is followed.

4. Controls for Autophagy of Bacteria

Autophagy targeting of intracellular S. Typhimurium requires bacterial protein synthesis and SPI-1 T3SS-dependent damage to the SCV (Birmingham et al., 2006). Therefore, autophagy targeting of bacteria does not occur if either of these processes is inhibited.

4.1. Inhibiting bacterial protein synthesis

Chloramphenicol is an eukaryotic cell-permeant bacteriostatic antibiotic that inhibits bacterial protein synthesis (Brock, 1961; Gale and Folkes, 1953). This drug can be added to the infected cells by replacing the media with new media containing the antibiotic (200 μg/mL) at 10 min p.i. (after infection of the host cell has been established). Inhibition of bacterial protein synthesis in this manner blocks the association of ubiquitinated proteins with intracellular S. Typhimurium, and results in the delivery of more bacteria to LAMP-1$^+$ compartments (Birmingham et al., 2006). As well, chloramphenicol treatment reduces the percentage of LC3$^+$ bacteria at 1 h p.i. from approximately 20% without treatment to <5% with treatment (Birmingham et al., 2006).

4.2. SPI-1 T3SS-deficient bacteria

Bacteria lacking expression of the SPI-1 T3SS are not targeted by autophagy and are often delivered to the phagolysosome (Birmingham et al., 2006). S. Typhimurium harboring deletions in the invA gene are SPI-1 T3SSdeficient. *invA* encodes an essential component of the SPI-1 T3SS, and thus bacterial mutants of this gene are unable to form the translocation needle and deliver SPI-1 effectors into host cells (Collazo and Galan, 1997; Kubori et al., 2000; Sukhan et al., 2001). Consequently, SPI-1 T3SS-deficient bacteria are unable to invade nonprofessional phagocytic cells. To allow invasion of these cell types, *invA* mutant bacteria can be engineered to express the *Yersinia pseudotuberculosis* Invasin (*Inv*) protein (Steele-Mortimer et al., 2002). The Inv protein binds to β-integrins on the host cell surface, inducing bacterial uptake into phagosomes by a zipperlike mechanism and delivery of the bacteria to lysosomes (Pace et al., 1993; Steele-Mortimer et al., 2002). SPI-1 T3SS-deficient bacteria expressing *Yersinia* Inv ($\Delta invA$/ p *inv*) do not damage the SCV or enter the cytosol after invasion (Birmingham et al., 2006). These bacteria are not targeted by the host autophagy system and can be used as controls for autophagy targeting experiments (Birmingham et al., 2006).

Another method that allows $\Delta invA$ bacteria to enter host cells is coinfection of this strain with wild-type bacteria expressing mRFP (mRFP expression is necessary to differentiate between the two bacterial strains). Dramatic ruffling of the plasma membrane caused by wild-type bacterial invasion causes efficient uptake of the noninvasive $\Delta invA$ bacteria into the same host cell (Pace et al., 1993; Steele-Mortimer et al., 2002). After uptake, the wild-type and $\Delta invA$ bacteria enter distinct vacuoles and presumably undergo different intracellular trafficking routes. Contrary to wild-type bacteria, $\Delta invA$ bacteria do not replicate after coinfection (Steele-Mortimer et al., 2002). As both strains are taken up by the same invasion process, the only difference between the two bacterial populations early after invasion is the lack of the SPI-1 T3SS secretion apparatus by the $\Delta invA$ S. Typhimurium. In this coinfection model, $\Delta invA$ S. Typhimurium mutants are not targeted by autophagy, whereas the wild-type bacteria are targeted, even in the same cell (Birmingham et al., 2006).

4.3. Autophagy inhibition

The host autophagy system can be inhibited in a variety of ways. We have successfully blocked autophagy of S. Typhimurium using phosphatidylinositol-3-kinase (PI3K) inhibitors, siRNA to autophagy components, and knockout mouse embryonic fibroblast cell lines. Autophagy inhibition results in the colonization of the host cytosol by bacteria and enhanced bacterial replication (Birmingham et al., 2006).

We routinely use the following concentrations of PI3K inhibitors: 3-methyladenine, 10 mM; wortmannin, 100 nM; and LY294002, 100 μM. Unfortunately, these drugs are nonspecific and affect other processes in the host cell besides autophagy. Therefore, other methods of autophagy inhibition should be used in conjunction with PI3K inhibitors.

siRNA to essential autophagy genes can effectively inhibit autophagy of bacteria. For example, *atg12* siRNA treatment for 48 h results in a 2.5-fold reduction in LC3 colocalization with intracellular S. Typhimurium compared to treatment with control siRNA (unpublished observations).

Cell lines lacking the expression of essential autophagy genes are valuable assets in the autophagy field. We have used an $atg5^{-/-}$ mouse embryonic fibroblast cell line developed by Dr. N. Mizushima (Tokyo Medical and Dental University, Japan) to study autophagy of S. Typhimurium. Atg5 is essential for the early steps in autophagosome formation, and therefore cells lacking this protein are completely deficient in autophagy (Kuma *et al.*, 2004). As mentioned previously, S. Typhimurium infection of normal mouse embryonic fibroblasts results in approximately 20% of intracellular bacteria targeted by autophagy at 1 h p.i. However, in $atg5^{-/-}$ fibroblasts, <5% of intracellular bacteria colocalize with LC3 at 1 h p.i. and significantly more bacteria enter the cytosol (Birmingham *et al.*, 2006).

5. CONCLUDING REMARKS

Autophagy is proving to be an important component of mammalian innate immunity. This system protects the host cell cytosol from colonization by bacteria not adapted to life in this compartment (Birmingham *et al.*, 2006). Interestingly, bacteria with cytosol-adapted intracellular lifestyles have evolved mechanisms to avoid autophagy targeting (Birmingham, 2007; Ogawa *et al.*, 2005; Py *et al.*, 2007). The interactions between the host autophagy system and invading bacteria are just beginning to be understood, and many questions remain. Using *in vitro* infection of epithelial and fibroblast cell lines with S. Typhimurium, localized autophagy targeting of bacteria can be induced and controlled. Therefore, this is an ideal model system for investigating how autophagy targeting occurs, and what mechanisms are involved in autophagy induction and autophagosome formation.

REFERENCES

Bampton, E. T., Goemans, C. G., Niranjan, D., Mizushima, N., and Tolkovsky, A. M. (2005). The dynamics of autophagy visualized in live cells: From autophagosome formation to fusion with endo/lysosomes. *Autophagy* **1,** 23–36.

Beuzón, C. R., Méresse, S., Unsworth, K. E., Ruíz-Albert, J., Garvis, S., Waterman, S. R., Ryder, T. A., Boucrot, E., and Holden, D. W. (2000). Salmonella maintains the integrity of its intracellular vacuole through the action of SifA. *EMBO J.* **19,** 3235–3249.

Beuzón, C. R., Salcedo, S. P., and Holden, D. W. (2002). Growth and killing of a *Salmonella enterica* serovar Typhimurium sifA mutant strain in the cytosol of different host cell lines. *Microbiology* **148,** 2705–2715.

Birmingham, C. L., Canadien, V., Gouin, E., Troy, E. B., Yoshimori, T., Cossart, P., Higgins, D. E., and Brumell, J. H. (2007). *Listeria monocytogenes* evades killing by autophagy during colonization of host cells. *Autophagy* **3**, 442–451.

Birmingham, C. L., Jiang, X., Ohlson, M. B., Miller, S. I., and Brumell, J. H. (2005). *Salmonella*-induced filament formation is a dynamic phenotype induced by rapidly replicating *Salmonella enterica* serovar Typhimurium in epithelial cells. *Infect. Immun.* **73**, 1204–1208.

Birmingham, C. L., Smith, A. C., Bakowski, M. A., Yoshimori, T., and Brumell, J. H. (2006). Autophagy controls *Salmonella* infection in response to damage to the *Salmonella*-containing vacuole. *J. Biol. Chem.* **281**, 11374–11383.

Blocker, A., Gounon, P., Larquet, E., Niebuhr, K., Cabiaux, V., Parsot, C., and Sansonetti, P. (1999). The tripartite type III secreton of *Shigella flexneri* inserts IpaB and IpaC into host membranes. *J. Cell Biol.* **147**, 683–693.

Brock, T. D. (1961). Chloramphenicol. *Bacteriol. Rev.* **25**, 32–48.

Brumell, J. H., and Grinstein, S. (2004). *Salmonella* redirects phagosomal maturation. *Curr. Opin. Microbiol.* **7**, 78–84.

Brumell, J. H., Perrin, A. J., Goosney, D. L., and Finlay, B. B. (2002a). Microbial pathogenesis: New niches for *Salmonella*. *Curr. Biol.* **12**, R15–R17.

Brumell, J. H., Tang, P., Mills, S. D., and Finlay, B. B. (2001). Characterization of *Salmonella*-induced filaments (Sifs) reveals a delayed interaction between *Salmonella*-containing vacuoles and late endocytic compartments. *Traffic* **2**, 643–653.

Brumell, J. H., Tang, P., Zaharik, M. L., and Finlay, B. B. (2002b). Disruption of the *Salmonella*-containing vacuole leads to increased replication of *Salmonella enterica* serovar Typhimurium in the cytosol of epithelial cells. *Infect. Immun.* **70**, 3264–3270.

Collazo, C. M., and Galan, J. E. (1997). The invasion-associated type III system of *Salmonella typhimurium* directs the translocation of Sip proteins into the host cell. *Mol. Microbiol.* **24**, 747–756.

Eriksson, S., Lucchini, S., Thompson, A., Rhen, M., and Hinton, J. C. (2003). Unravelling the biology of macrophage infection by gene expression profiling of intracellular *Salmonella enterica*. *Mol. Microbiol.* **47**, 103–118.

Gale, E. F., and Folkes, J. P. (1953). The assimilation of amino acids by bacteria. 19. The inhibition of phenylalanine incorporation in *Staphylococcus aureus* by chloramphenicol and p-chlorophenylalanine. *Biochem. J.* **55**, 730–735.

Garcia-del Portillo, F., and Finlay, B. B. (1995). Targeting of *Salmonella typhimurium* to vesicles containing lysosomal membrane glycoproteins bypasses compartments with mannose 6-phosphate receptors. *J. Cell Biol.* **129**, 81–97.

Garcia-del Portillo, F., Pucciarelli, M. G., Jefferies, W. A., and Finlay, B. B. (1994). *Salmonella typhimurium* induces selective aggregation and internalization of host cell surface proteins during invasion of epithelial cells. *J. Cell Sci.* **107**(Pt 7), 2005–2020.

Garcia-del Portillo, F. Z. M., Leung, K. Y., and Finlay, B. B. (1993). *Salmonella* induces the formation of filamentous structures containing lysosomal membrane glycoproteins in epithelial cells. *Proc. Natl. Acad. Sci. USA* **90**, 10544–10548.

Hakansson, S., Schesser, K., Persson, C., Galyov, E. E., Rosqvist, R., Homble, F., and Wolf-Watz, H. (1996). The YopB protein of *Yersinia pseudotuberculosis* is essential for the translocation of Yop effector proteins across the target cell plasma membrane and displays a contact-dependent membrane disrupting activity. *EMBO J.* **15**, 5812–5823.

Hampe, J., Franke, A., Rosenstiel, P., Till, A., Teuber, M., Huse, K., Albrecht, M., Mayr, G., De La Vega, F. M., Briggs, J., Gunther, S., Prescott, N. J., et al. (2007). A genome-wide association scan of nonsynonymous SNPs identifies a susceptibility variant for Crohn disease in ATG16L1. *Nat. Genet.* **39**, 207–211.

Haraga, A., Ohlson, M. B., and Miller, S. I. (2008). *Salmonellae* interplay with host cells. *Nat. Rev. Microbiol.* **6**, 53–66.

Hensel, M., Shea, J. E., Gleeson, C., Jones, M. D., Dalton, E., and Holden, D. W. (1995). Simultaneous identification of bacterial virulence genes by negative selection. *Science* **269,** 400–403.

Holden, D. W. (2002). Trafficking of the *Salmonella* vacuole in macrophages. *Traffic* **3,** 161–169.

Hueck, C. J. (1998). Type III protein secretion systems in bacterial pathogens of animals and plants. *Microbiol. Mol. Biol. Rev.* **62,** 379–433.

Hugot, J. P., Chamaillard, M., Zouali, H., Lesage, S., Cezard, J. P., Belaiche, J., Almer, S., Tysk, C., O'Morain, C. A., Gassull, M., Binder, V., Finkel, Y., *et al.* (2001). Association of NOD2 leucine-rich repeat variants with susceptibility to Crohn's disease. *Nature* **411,** 599–603.

Kabeya, Y., Mizushima, N., Ueno, T., Yamamoto, A., Kirisako, T., Noda, T., Kominami, E., Ohsumi, Y., and Yoshimori, T. (2000). LC3, a mammalian homologue of yeast Apg8p, is localized in autophagosome membranes after processing. *EMBO J.* **19,** 5720–5728.

Klionsky, D. J., Abeliovich, H., Agostinis, P., Agrawal, D. K., Aliev, G., Askew, D. S., Baba, M., Baehrecke, E. H., Bahr, B. A., Ballabio, A., *et al.* (2007). Guidelines for the use and interpretation of assays for monitoring autophagy in higher eukaryotes. *Autophagy* **4,** 139–140.

Kubori, T., Sukhan, A., Aizawa, S. I., and Galan, J. E. (2000). Molecular characterization and assembly of the needle complex of the *Salmonella typhimurium* type III protein secretion system. *Proc. Natl. Acad. Sci. USA* **97,** 10225–10230.

Kuhle, V., and Hensel, M. (2004). Cellular microbiology of intracellular *Salmonella enterica*: functions of the type III secretion system encoded by *Salmonella* pathogenicity island 2. *Cell. Mol. Life Sci.* **61,** 2812–2826.

Kuma, A., Hatano, M., Matsui, M., Yamamoto, A., Nakaya, H., Yoshimori, T., Ohsumi, Y., Tokuhisa, T., and Mizushima, N. (2004). The role of autophagy during the early neonatal starvation period. *Nature* **432,** 1032–1036.

Maki-Ikola, O., and Granfors, K. (1992). *Salmonella*-triggered reactive arthritis. *Lancet* **339,** 1096–1098.

Massey, D. C., and Parkes, M. (2007). Genome-wide association scanning highlights two autophagy genes, ATG16L1 and IRGM, as being significantly associated with Crohn's disease. *Autophagy* **3,** 649–651.

Meresse, S., Steele-Mortimer, O., Finlay, B. B., and Gorvel, J. P. (1999). The rab7 GTPase controls the maturation of *Salmonella typhimurium*-containing vacuoles in HeLa cells. *EMBO J.* **18,** 4394–4403.

Mills, S. D., and Finlay, B. B. (1998). Isolation and characterization of *Salmonella typhimurium* and *Yersinia pseudotuberculosis*-containing phagosomes from infected mouse macrophages: *Y. pseudotuberculosis* traffics to terminal lysosomes where they are degraded. *Eur. J. Cell Biol.* **77,** 35–47.

Mizushima, N., Ohsumi, Y., and Yoshimori, T. (2002). Autophagosome formation in mammalian cells. *Cell Struct. Funct.* **27,** 421–429.

Mizushima, N., Yamamoto, A., Matsui, M., Yoshimori, T., and Ohsumi, Y. (2004). In vivo analysis of autophagy in response to nutrient starvation using transgenic mice expressing a fluorescent autophagosome marker. *Mol. Biol. Cell* **15,** 1101–1111.

Nakagawa, I., Amano, A., Mizushima, N., Yamamoto, A., Yamaguchi, H., Kamimoto, T., Nara, A., Funao, J., Nakata, M., Tsuda, K., Hamada, S., and Yoshimori, T. (2004). Autophagy defends cells against invading group A *Streptococcus*. *Science* **306,** 1037–1040.

Ochman, H., Soncini, F. C., Solomon, F., and Groisman, E. A. (1996). Identification of a pathogenicity island required for *Salmonella* survival in host cells. *Proc. Natl. Acad. Sci. USA* **93,** 7800–7804.

Ogawa, M., Yoshimori, T., Suzuki, T., Sagara, H., Mizushima, N., and Sasakawa, C. (2005). Escape of intracellular *Shigella* from autophagy. *Science* **307,** 727–731.

Pace, J., Hayman, M. J., and Galan, J. E. (1993). Signal transduction and invasion of epithelial cells by *S. typhimurium*. *Cell* **72,** 505–514.
Parkes, M., Barrett, J. C., Prescott, N. J., Tremelling, M., Anderson, C. A., Fisher, S. A., Roberts, R. G., Nimmo, E. R., Cummings, F. R., Soars, D., Drummond, H., Lees, C. W., *et al.* (2007). Sequence variants in the autophagy gene IRGM and multiple other replicating loci contribute to Crohn's disease susceptibility. *Nat. Genet.* **39,** 830–832.
Penheiter, K. L., Mathur, N., Giles, D., Fahlen, T., and Jones, B. D. (1997). Non-invasive Salmonella typhimurium mutants are avirulent because of an inability to enter and destroy M cells of ileal Peyer's patches. *Mol. Microbiol.* **24,** 697–709.
Perrin, A. J., Jiang, X., Birmingham, C. L., So, N. S., and Brumell, J. H. (2004). Recognition of bacteria in the cytosol of mammalian cells by the ubiquitin system. *Curr. Biol.* **14,** 806–811.
Pfeifer, C. G., Marcus, S. L., Steele-Mortimer, O., Knodler, L. A., and Finlay, B. B. (1999). *Salmonella typhimurium* virulence genes are induced upon bacterial invasion into phagocytic and nonphagocytic cells. *Infect. Immun.* **67,** 5690–5698.
Prost, L. R., Sanowar, S., and Miller, S. I. (2007). *Salmonella* sensing of anti-microbial mechanisms to promote survival within macrophages. *Immunol. Rev.* **219,** 55–65.
Py, B. F., Lipinski, M. M., and Yuan, J. (2007). Autophagy Limits *Listeria monocytogenes* Intracellular Growth in the Early Phase of Primary Infection. *Autophagy* **3,** 117–125.
Rabsch, W., Andrews, H. L., Kingsley, R. A., Prager, R., Tschape, H., Adams, L. G., and Baumler, A. J. (2002). *Salmonella enterica* serotype Typhimurium and its host-adapted variants. *Infect. Immun.* **70,** 2249–2255.
Rathman, M., Sjaastad, M. D., and Falkow, S. (1996). Acidification of phagosomes containing *Salmonella typhimurium* in murine macrophages. *Infect. Immun.* **64,** 2765–2773.
Rioux, J. D., Xavier, R. J., Taylor, K. D., Silverberg, M. S., Goyette, P., Huett, A., Green, T., Kuballa, P., Barmada, M. M., Datta, L. W., Shugart, Y. Y., Griffiths, A. M., *et al.* (2007). Genome-wide association study identifies new susceptibility loci for Crohn disease and implicates autophagy in disease pathogenesis. *Nat. Genet.* **39,** 596–604.
Roy, D., Liston, D. R., Idone, V. J., Di, A., Nelson, D. J., Pujol, C., Bliska, J. B., Chakrabarti, S., and Andrews, N. W. (2004). A process for controlling intracellular bacterial infections induced by membrane injury. *Science* **304,** 1515–1518.
Schlumberger, M. C., and Hardt, W.-D. (2006). *Salmonella* type III secretion effectors: pulling the host cell's strings. *Curr. Opin. Microbiol.* **9,** 46–54.
Schlumberger, M. C., Müller, A. J., Ehrbar, K., Winnen, B., Duss, I., Stecher, B., and Hardt, W.-D. (2005). Real-time imaging of type III secretion: *Salmonella* SipA injection into host cells. *Proc. Natl. Acad. Sci. USA* **102,** 12548–12553.
Singh, S. B., Davis, A. S., Taylor, G. A., and Deretic, V. (2006). Human IRGM Induces Autophagy to Eliminate Intracellular Mycobacteria. *Science* **313,** 1438–1441.
Smith, A. C., Cirulis, J. T., Casanova, J. E., Scidmore, M. A., and Brumell, J. H. (2005). Interaction of the *Salmonella*-containing vacuole with the endocytic recycling system. *J. Biol. Chem.* **280,** 24634–24641.
Steele-Mortimer, O. (2008). The *Salmonella*-containing vacuole-Moving with the times. *Curr. Opin. Microbiol.* **11,** 38–45.
Steele-Mortimer, O., Brumell, J. H., Knodler, L. A., Meresse, S., Lopez, A., and Finlay, B. B. (2002). The invasion-associated type III secretion system of *Salmonella enterica* serovar Typhimurium is necessary for intracellular proliferation and vacuole biogenesis in epithelial cells. *Cell Microbiol.* **4,** 43–54.
Steele-Mortimer, O., Meresse, S., Gorvel, J. P., Toh, B. H., and Finlay, B. B. (1999). Biogenesis of *Salmonella typhimurium*-containing vacuoles in epithelial cells involves interactions with the early endocytic pathway. *Cell Microbiol.* **1,** 33–49.

Sukhan, A., Kubori, T., Wilson, J., and Galan, J. E. (2001). Genetic analysis of assembly of the *Salmonella enterica* serovar Typhimurium type III secretion-associated needle complex. *J. Bacteriol.* **183,** 1159–1167.

Tanida, I., Minematsu-Ikeguchi, N., Ueno, T., and Kominami, E. (2005). Lysosomal turnover, but not a cellular level, of endogenous LC3 is a marker for autophagy. *Autophagy* **1,** 84–91.

Tsolis, R. M., Kingsley, R. A., Townsend, S. M., Ficht, T. A., Adams, L. G., and Baumler, A. J. (1999a). Of mice, calves, and men. Comparison of the mouse typhoid model with other *Salmonella* infections. *Adv. Exp. Med. Biol.* **473,** 261–274.

Tsolis, R. M., Kingsley, R. A., Townsend, S. M., Ficht, T. A., Adams, L. G., and Baumler, A. J. (1999b). Of mice, calves, and men. Comparison of the mouse typhoid model with other *Salmonella* infections. *Adv. Exp. Med. Biol.* **473,** 261–274.

Valdivia, R. H., and Falkow, S. (1996). Bacterial genetics by flow cytometry: rapid isolation of *Salmonella typhimurium* acid-inducible promoters by differential fluorescence induction. *Mol. Microbiol.* **22,** 367–378.

Wood, M. W., Jones, M. A., Watson, P. R., Hedges, S., Wallis, T. S., and Galyov, E. E. (1998). Identification of a pathogenicity island required for *Salmonella* enteropathogenicity. *Mol. Microbiol.* **29,** 883–891.

WTCCC (2007). Genome-wide association study of 14,000 cases of seven common diseases and 3,000 shared controls. *Nature* **447,** 661–678.

Xie, Z., and Klionsky, D. J. (2007). Autophagosome formation: core machinery and adaptations. *Nat. Cell Biol.* **9,** 1102–1109.

Yoshimori, T. (2006). Autophagy vs. Group A *Streptococcus*. *Autophagy* **2,** 154–155.

CHAPTER TWENTY-ONE

Monitoring Autophagy during *Mycobacterium tuberculosis* Infection

Marisa Ponpuak,* Monica A. Delgado,* Rasha A. Elmaoued,* *and* Vojo Deretic*

Contents

1. Introduction	346
2. Methods	347
2.1. Autophagic killing assay and the use of inhibitors	348
2.2. siRNA-mediated protein knockdown supplying the autophagic killing	351
2.3. LysoTracker colocalization analysis to monitor the acidification of the *M. tuberculosis* phagosome	354
2.4. Monitoring lysosomal protease delivery to the *M. tuberculosis* phagosome	356
Acknowledgments	359
References	359

Abstract

Tuberculosis is one of the world's most prevalent infectious diseases. The causative agent, *M. tuberculosis*, asymptomatically infects more than 30% of the world population and causes 8 million cases of active disease and 2 million deaths annually. Its pathogenic success stems from its ability to block phagolysosome biogenesis and subsequent destruction in the host macrophages. Recently, our laboratory has uncovered autophagy as a new means of overcoming this block and promoting the killing of mycobacteria. Here we describe the methods to study autophagy during *M. tuberculosis* infection of macrophages. The described assays can be used to investigate and identify factors important for autophagic elimination of mycobacteria that could potentially provide new therapeutic targets to defeat this disease.

* Department of Molecular Genetics and Microbiology, University of New Mexico, Health Sciences Center, Albuquerque, New Mexico, USA.

1. INTRODUCTION

Mycobacterium tuberculosis is the causative agent of tuberculosis, the leading global cause of infectious disease-associated morbidity and mortality in humans. Approximately 2 billion people or one-third of the world population are latently infected, and nearly 2 million deaths from 8 million cases of active tuberculosis yearly are associated with the disease (WHO, 2008). In most cases, the mycobacteria can be contained by the host immune system and infected individuals are in the majority of cases asymptomatic for long periods of time (Russell, 2001; Kaufmann *et al.*, 2006). However, nearly 10% of infected individuals will eventually develop active disease in their lifetime due to weakening of the immune system caused by many factors such as immunosuppression, malnutrition, or old age (Chan and Flynn, 2004). The typical 10%-lifetime risk increases substantially to an annual risk of 10% if an individual is coinfected with human immunodeficiency virus (HIV) (Corbett *et al.*, 2003). Antibiotics can be used to treat the disease. However, as mycobacteria are slow growing, the treatment requires several months to completely eliminate the microbe (Budha *et al.*, 2008). In addition, new strains that are resistant to all available drugs (dubbed XDR) have emerged (Ducati *et al.*, 2006). Currently, there are no new drugs available or vaccines effective in adults.

The success of *M. tuberculosis* as a human pathogen is contributed by its ability to survive and replicate in macrophages (Vergne *et al.*, 2004). After the mycobacteria are inhaled into the lung, the bacteria are phagocytosed into phagosomes by alveolar macrophages (Russell, 2007). To avoid destruction by the host, the mycobacteria interfere with macrophage membrane trafficking and block the fusion of the phagosome with lysosomes (Deretic *et al.*, 2004; Vergne *et al.*, 2004). This blocking event is characterized by the lack of complete acidification and acquisition of lysosomal components by the mycobacterial phagosome (Russell *et al.*, 2002) and is mediated by several mycobacterial secreted factors including specific lipids (Beatty *et al.*, 2000; Vergne *et al.*, 2003). This results in mycobacterial survival, disease latency, and the risk of reactivation later on in life (Russell *et al.*, 2002). Recently, however, induction of host autophagy was shown to overcome this block and eliminate mycobacteria in macrophages (Gutierrez *et al.*, 2004; Singh *et al.*, 2006; Delgado *et al.*, 2008).

Autophagy is an evolutionarily conserved intracellular process of cytoplasmic degradation. It begins with the sequestration of cytoplasmic contents into a double-membrane bound organelle called an autophagosome (Levine and Klionsky, 2004; Mizushima, 2004). Autophagosomes then fuse with lysosomes to acquire lysosomal hydrolases resulting in the degradation of its engulfed constituents (Klionsky *et al.*, 2007; Mizushima and Klionsky, 2007). Autophagy has been linked to many cellular functions. These include the maintenance of intracellular metabolism, the removal of defective organelles and protein aggregates, the elimination of pathogens such as bacteria, parasites, and viruses, and the processing of self and foreign proteins for antigen presentation (Talloczy *et al.*, 2002; Gutierrez *et al.*, 2004; Nakagawa *et al.*, 2004; Ogawa *et al.*, 2005; Schmid and Münz, 2005; Ling *et al.*, 2006; Rubinsztein, 2006; Sakai *et al.*, 2006; Singh *et al.*, 2006; Williams *et al.*, 2006; Bernales *et al.*, 2007; Devenish, 2007; Schmid *et al.*, 2007; Liang *et al.*, 2008). In the context of *M. tuberculosis* infection, induction of autophagy in macrophages by starvation or rapamycin treatment has been shown to overcome the phagolysosome biogenesis block resulting in increases in mycobacterial phagosome acidification and lysosomal component acquisition (Gutierrez *et al.*, 2004). These effects are inhibitable by known pharmacological inhibitors of autophagy such as 3-methyladenine (3-MA) and wortmannin (Gutierrez *et al.*, 2004) and Beclin 1and Atg5 siRNA knockdown (Harris *et al.*, 2007; Delgado *et al.*, 2008). The outcome is the reduction in *M. tuberculosis* survival in macrophages. A model by which autophagic induction leads to the fusion of mycobacterial phagosomes with autophagic organelles and hence the acquisition of lysosomal contents to overcome the phagolysosome biogenesis block and promote the killing of mycobacteria has been proposed (Harris, 2006). Moreover, a specific delivery of mycobactericidal agents, such as fragments of ubiquitin, to the *M. tuberculosis* phagosome has been shown following induction of autophagy (Alonso *et al.*, 2007). However, the factors and mechanisms involved remain largely uncharacterized. In this chapter, we describe assays to monitor autophagy during *M. tuberculosis* infection. The assays include the monitoring of mycobacterial phagosome acidification, mycobacterial phagosome acquisition of a lysosomal protease, and mycobacterial elimination on autophagic induction. In addition, methods for siRNA-mediated knockdown of proteins in macrophages are outlined. The combination of these assays are well suited and can be used to characterize important factors involved in autophagic elimination of mycobacteria which could potentially pave the way for pharmacological development to combat this noxious disease.

2. Methods

In the following text, we describe methods to monitor autophagy during *M. tuberculosis* infection in murine RAW264.7 macrophages. These methods can be adapted to other murine and human macrophage cell lines and primary cells (Harris, 2006).

2.1. Autophagic killing assay and the use of inhibitors

2.1.1. Materials

2.1.1.1. Growth and use of RAW264.7 cell line

1. Murine RAW264.7 macrophage cell line (ATCC, No. TIB-71).
2. Dulbecco's modified Eagle's medium (DMEM) (Gibco), 10% fetal bovine serum (FBS; Hyclone), and 4 mM L-glutamine (Gibco) (Full medium): Add 100 mL of FBS and 10 mL of 200 mM L-glutamine to 890 mL of DMEM. Filter through 0.2-μm filter. Store the medium at 4 °C.
3. 175-cm^2 tissue culture flasks (Greiner Bio-One).
4. Disposable cell scraper (Sarstedt).
5. Earle's balanced salt solution (EBSS) (Starvation medium) (Sigma).
6. 12-well tissue culture plates.

2.1.1.2. Growth and preparation of mycobacteria

1. Middlebrook albumin dextrose catalase (ADC) enrichment (BD Biosciences).
2. 20% Tween 80 (Sigma): Prepare 20% v/v stock at 4 °C and filter through 0.2-μm filter. Store at 4 °C.
3. Middlebrook 7H9 broth (Difco): Dissolve 4.7 g of powder in 900 mL of deionized water with 2 mL of glycerol (EMD Chemicals) and autoclave. Temper to room temperature. Add 100 mL of ADC and 2.5 mL of 20% Tween 80. Store at 4 °C.
4. Middlebrook 7H11 agar (Difco): Dissolve 21 g of powder in 900 mL deionized water with 5 mL of glycerol (EMD Chemicals) and autoclave. Temper to 50 °C. Add 100 mL of ADC and 2.5 mL of 20% Tween 80 and dispense into Petri dishes. Store at 4 °C.
5. Sterile, 150-mL tissue-culture roller bottles.
6. Autoclaved 7-mL Dounce glass homogenizer (Kontes Scientific Glassware).
7. Phosphate-buffered saline (PBS, 10X) (Gibco): Prepare 1X PBS by adding 100 mL into 900 mL of deionized water and filter through 0.2-μm filter. Store at room temperature.
8. PBS-Tween: Add 100 mL of 10X PBS and 2.5 mL of 20% Tween 80 into 897.5 mL of deionized water. Filter through 0.2-μm filter and store at 4 °C.
9. 96-well tissue culture plates.
10. Polyethylene bags (Fisher).

2.1.1.3. Inhibitor preparation

1. Rapamycin (FW = 914.19 g/mol) (LC Laboratories, R-5000): Prepare 50 mg/mL stock solution by dissolving the powder with 100% DMSO. Aliquot and store at −20 °C. For the experiment, prepare working

solution by diluting the stock solution 1:1000 in complete DMEM to yield a final concentration of 50 μg/mL.
2. 3-methyladenine (FW = 149.16 g/mol) (Sigma, M9281]): Prepare 100 mM stock solution by dissolving 14.916 mg of powder in 1 mL of deionized water. Heat to 60–80 °C. Prepare a working solution by adding 100 μL of 100 mM stock into 900 μL of starvation media, EBSS, to obtain a final concentration of 10 mM. Always prepare fresh solution for each experiment.
3. Wortmannin (FW 428.47 g/mol) (Sigma, W1628): Prepare 1 mM stock solution by dissolving 1 mg of powder in 2.064 mL 100% DMSO. From this solution, prepare 100 μM stock solution by performing a 1:10 dilution in 1X PBS. Prepare a working solution by diluting the 100 μM stock solution 1:1000 in EBSS to achieve a final concentration of 100 nM. Always prepare fresh solution for each experiment.

2.1.2. Methods
2.1.2.1. Macrophage cell preparation

1. Murine RAW264.7 macrophages are maintained in complete DMEM at 37 °C, 5% CO_2.
2. Subculture RAW264.7 cells to reach 70%–80% confluency by scraping off the cells into 12 mL complete DMEM. Pipet up and down to obtain a single cell suspension. Transfer 1.5 mL of cells into a 175-cm^2 tissue culture flask containing 25 mL of complete DMEM. Incubate for approximately 2 d at 37 °C, 5% CO_2.
3. One day before the experiment, scrape off the cells into 12 mL of complete DMEM and determine cell density using trypan blue staining and a hemocytometer.
4. Dilute cells to 3×10^5 cells/mL in complete DMEM and dispense 1 mL into each well of a 12-well plate. Plate 3 wells (triplicate) for each experimental condition. Incubate the plate at 37 °C, 5% CO_2 overnight.

2.1.2.2. Infection of macrophages

1. *Mycobacterium tuberculosis* var. *bovis* BCG (BCG) culture is maintained in Middlebrook 7H9 broth supplemented with 10% ADC, 0.05% Tween 80, and 0.2% glycerol at 37 °C on a roller.
2. Transfer 10 mL of a log-phase BCG culture into a 15-mL conical tube. Centrifuge to pellet mycobacteria at 2500 rpm for 5 min at room temperature.
3. Remove and discard the supernatant fraction and wash the cell pellet once with 10 mL of 1X PBS.
4. Spin at 2500 rpm for 5 min and remove and discard the supernatant fraction.

5. Resuspend the mycobacteria pellet in 6 mL of complete DMEM and transfer the suspension to a 7-mL Dounce homogenizer. Homogenize 35 times to generate a single cell suspension.
6. Measure OD_{600} of the 1:10 dilution of the homogenized culture.
7. Prepare mycobacterial inoculum in complete DMEM at the concentration of 3×10^6 BCG/mL (MOI = 10) using the following formula:(3×10^6); (total mL needed for infection)/($OD_{600} \times 10^9$) = mL homogenate needed.
8. Remove medium from cells and add 1 mL of 3×10^6 BCG/mL inoculum into each well of RAW264.7 macrophages.
9. Spin at 1200 rpm for 5 min at room temperature to settle mycobacteria onto macrophages. Incubate the plate for 1 h at 37 °C, 5% CO_2.

2.1.2.3. Autophagic induction and inhibitor treatment

1. For complete medium control (Full medium) and rapamycin treatment, remove the medium containing mycobacteria from the wells. Quickly wash each well three times with 2 mL of complete DMEM to remove noninternalized mycobacteria and add 1 mL of the respective media/treatment into each well (see Notes 1 and 2).
2. For starvation, 3-methyladenine, and wortmannin treatments, remove the medium containing mycobacteria from the wells. Quickly wash each well 3 times with 2 mL of EBSS and add 1 mL of the respective media/treatment into the well.
3. Incubate the plate for 2 h at 37 °C, 5% CO_2 (see Note 3).
4. For wortmannin treatment, at 1 h after incubation, remove the old media/treatment and add freshly prepared media/treatment to each well.

2.1.2.4. Recovery of mycobacteria and data analysis

1. After 2 h of incubation, wash each well twice with 1 mL of 1X PBS.
2. Add 0.5 mL of cold deionized water to each well and incubate the plate at 4 °C for 10–15 min to lyse the cells and release intracellular mycobacteria.
3. Prepare a dilution plate by adding 180 μL of PBS-Tween into each well of a 96-well plate.
4. Pipet the cell lysate up and down to mix, and transfer 20 μL into a well of the dilution plate containing 180 μL of PBS-Tween. This is the 10^{-1} dilution. Perform 10-fold serial dilutions to obtain 10^{-2}, 10^{-3}, and 10^{-4} dilutions. Repeat the process for each well. Since there are three wells (triplicate) for each treatment, there will be 12 dilutions total per each condition.
5. Inoculate 5 μL of each dilution onto Middlebrook 7H11 agar plates. To save plates, 12 dilutions can be inoculated onto different slots on the same plate.

6. Let the plate stand for 5 min to absorb the inoculum. Put plates in a plastic bag. Close the bag tightly (see Note 4). Incubate the plates at 37 °C for 14 d. Colonies should be visible at this time.
7. Count colonies from dilutions that yield good resolution (1–100 visible colonies).
8. Repeat the entire assay two more times to obtain data from three different experiments for analysis and plotting (Fig. 21.1A).

2.2. siRNA-mediated protein knockdown supplying the autophagic killing

2.2.1. Materials

2.2.1.1. Growth and use of RAW264.7 cell line

1. Murine RAW264.7 macrophage cell line (ATCC, No. TIB-71).
2. Dulbecco's modified Eagle's medium (DMEM) (Gibco), 10% fetal bovine serum (FBS; Hyclone), and 4 mM L-glutamine (Gibco) (Full medium): Add 100 mL of FBS and 10 mL of 200 mM L-glutamine to 890 mL of DMEM. Filter through 0.2-μm filter. Store the medium at 4 °C.
3. 175-cm^2 and 75-cm^2 tissue culture flasks (Greiner Bio-One).
4. Disposable cell scraper (Sarstedt).
5. 12-well tissue culture plates.

2.2.1.2. Transfection of macrophages with siRNAs

1. Amaxa Nucleofector Device and Nucleofector solution V (Amaxa).
2. siRNA duplexes (Dharmacon) designed to specifically target a gene of interest.

2.2.1.3. Immunoblot analysis of siRNA-mediated protein knockdown

1. BCA protein assay kit (Pierce).
2. Immunoblotting lysis buffer: 50 mM Tris-HCl, pH 7.4, 150 mM NaCl, 1% Nonidet-P40, 0.25% Na-deoxycholate, 1 mM EDTA, 1mM PMSF, 1 μg/mL aprotinin, 10 μg/mL leupeptin, 1 μg/mL pepstatin, 1 mM Na$_3$VO$_4$, and 1 mM NaF.
3. 12% precast BioRad minigel (BioRad).
4. MiniProtean 3 Western blot system (BioRad).
5. SuperSignal West Dura Extended Duration Substrate for western blot detection (Pierce).

2.2.2. Methods

2.2.2.1. Transfection of macrophages with siRNAs

1. Murine RAW264.7 macrophages are maintained in Dulbecco's modified Eagle's medium (DMEM) supplemented with 10% fetal bovine serum (FBS) and 4 mM L-glutamine at 37 °C, 5% CO$_2$.

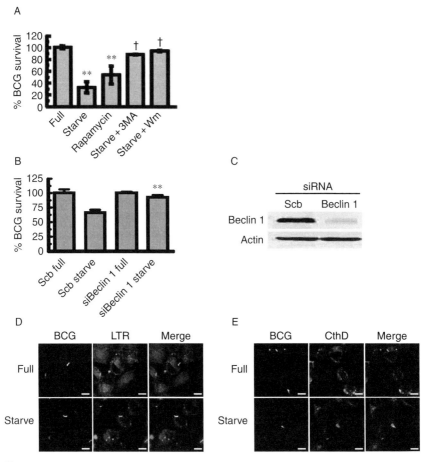

Figure 21.1 Autophagy increases *M. tuberculosis* phagosome maturation and eliminates mycobacteria in macrophages. A. Autophagy reduces mycobacterial survival. RAW264.7 macrophages were infected with BCG for 1 h and incubated with or without starvation medium in the presence or absence of rapamycin, 3-methyladenine (3MA), or wortmannin (Wm) for 2 h. Cells were lysed to determine mycobacterial viability. B. and C. Beclin 1 is important for autophagic killing of mycobacteria. RAW264.7 cells were transfected with siRNAs against Beclin 1or scramble control. Protein knockdown was allowed to proceed for 48 h. Mycobacterial viability was determined as in A. Immunoblot analysis was performed to validate Beclin 1knockdown level using Actin as a loading control. D. and E. Autophagic induction enhances acidification and acquisition of a lysosomal protease by the mycobacterial phagosome. RAW264.7 cells were transfected with siRNAs against proteins of interest. Cells were infected with BCG and stained with LysoTracker Red or antibodies against cathepsin D. Quantitative analysis of percent colocalization was performed. Confocal images of cells transfected with siRNAs scramble control subjected to starvation treatment are shown as examples. Data, means ± SEM from three different experiments, $\star\star p \leq 0.01$, $^{\dagger}p \geq 0.05$. Fig. 21.1A is modified from Gutierrez *et al.* (2004). Figs. 21.1B and 21.C are modified from Delgado *et al.* (2008). (See Color Insert.)

2. Subculture the cells at 1:3 dilution 24 h before transfection by scraping off the cells into 12 mL of complete DMEM. Pipette up and down to obtain single cell suspension. Transfer 4 mL of cells into a 175-cm^2 tissue culture flask containing 25 mL of complete DMEM. Incubate cells overnight at 37 °C, 5% CO_2.
3. Transfection of cells with siRNAs is performed using an Amaxa Nucleofector Device according to the manufacturer's protocol (see Note 5). Scrape cells into 12 mL of complete DMEM and aliquot 3 mL of cells (approximately 2×10^7 cells) into 15-mL conical tubes.
4. Pellet the cells at 700 rpm for 5 min at room temperature.
5. Completely remove DMEM from cells using a vacuum suction system.
6. Resuspend cells with 100 μL of Nucleofector Solution V containing 1.5 μg of siRNAs and transfer the cell suspension into a cuvette.
7. Nucleoporate the cells using program D-032.
8. Transfer cells into a 75-cm^2 tissue culture flask containing 12 mL of complete DMEM. Incubate cells for 5–6 h at 37 °C, 5% CO_2.
9. Remove the old medium and add fresh complete DMEM to cells.
10. At 24 h after transfection, scrape cells into 15-mL conical tubes and determine the cell density using trypan blue staining and a hemocytometer.
11. For the killing assay outlined in section 2, dilute cells to obtain a density of 3×10^5 cells/mL in complete DMEM and dispense 1 mL into each well of a 12-well plate. Plate 3 wells (triplicate) for each experimental condition. Incubate the plate at 37 °C, 5% CO_2 overnight. Plate the rest of the cells back into a 75-cm^2 tissue culture flask. Incubate the flasks overnight at 37 °C, 5% CO_2.
12. Perform killing assay (see section 2) at 48 h after transfection (Fig. 21.1B) (see Note 6).

2.2.2.2. Immunoblot analysis of siRNA-mediated protein knockdown

1. Collect cells from each knockdown condition at 48 h after transfection from the respective 75-cm^2 tissue culture flask by scraping off the cells into 15-mL conical tubes.
2. Pellet the cells at 700 rpm for 5 min at room temperature.
3. Wash the cells once with 1X PBS, pellet by centrifugation, and remove and discard the supernatant fraction.
4. Lyse the cells in 0.5 mL of cold lysis buffer on ice for 30 min. Transfer the lysate into 1.5-mL microcentrifuge tubes. Spin at 13,000 rpm for 15 min at 4 °C. Transfer the supernatant fraction containing the protein lysate into new 1.5-mL microcentrifuge tubes.
5. Determine the protein concentration using BCA protein assay according to the manufacturer's protocol.

6. Prepare and load 50 µg of total protein per each condition using the standard method of Laemmli and the MiniProtean 3 Western blot system according to the manufacturer's protocol.
7. Block blots with 5% nonfat dry milk in 1X PBS containing 0.1% Tween 20 for 1 h at room temperature.
8. Incubate blots with primary antibodies according to the manufacturer's protocol followed by the appropriate horseradish peroxidase-conjugated secondary antibodies.
9. Develop blots with SuperSignal West Dura Extended Duration Substrate according to the manufacturer's protocol and standard autoradiography (Fig. 21.1C).

2.3. LysoTracker colocalization analysis to monitor the acidification of the *M. tuberculosis* phagosome

2.3.1. Materials
2.3.1.1. Growth and use of RAW264.7 cell line

1. Murine RAW264.7 macrophage cell line (ATCC, No. TIB-71).
2. Dulbecco's modified Eagle's medium (DMEM) (Gibco), 10% fetal bovine serum (FBS; Hyclone), and 4 mM L-glutamine (Gibco) (Full medium): Add 100 mL of FBS and 10 mL of 200 mM L-glutamine to 890 mL of DMEM. Filter through 0.2-µm filter. Store the medium at 4 °C.
3. 175-cm^2 tissue culture flasks (Greiner Bio-One).
4. Disposable cell scraper (Sarstedt).
5. Earle's balanced salt solution (EBSS) (Starvation medium) (Sigma).
6. 12-well tissue culture plates.
7. Circular microscope cover glasses (18 mm in diameter) (Fisher).
8. LysoTracker Red DND-99 1 mM stock solution in DMSO (Molecular Probes).
9. Fixing solution: Dissolve 2 g of paraformaldehyde in 100 mL of 1X PBS. Heat to 50–60 °C inside a chemical hood until the powder is solubilized. Cool the solution to room temperature, aliquot, and store at −20 °C.
10. PermaFluor mounting medium (Thermo Scientific, 434990).

2.3.1.2. Growth and preparation of mycobacteria

1. Middlebrook albumin dextrose catalase (ADC) enrichment (BD Biosciences).
2. 20% Tween 80 (Sigma): Prepare 20% v/v stock at 4 °C and filter through 0.2-µm filter. Store at 4 °C.
3. Middlebrook 7H9 broth (Difco): Dissolve 4.7 g of powder in 900 mL deionized water with 2 mL of glycerol (EMD Chemicals, Inc.) and autoclave. Temper to room temperature. Add 100 mL of ADC and 2.5 mL of 20% Tween 80. Store at 4 °C.

4. Sterile, 150-mL tissue-culture roller bottles.
5. Autoclaved 7-mL Dounce glass homogenizer (Kontes Scientific Glassware).
6. Phosphate-buffered saline (PBS, 10X) (Gibco): Prepare 1X PBS by adding 100 mL in 900 mL of deionized water and filter through 0.2-μm filter and store at room temperature.
7. Alexa 488 carboxylic acid succinimidyl ester (Molecular Probes): Prepare 1 mg/mL stock solution by adding 1 mL of 100% DMSO to 1 mg of powder. Store at $-20\,^{\circ}$C.

2.3.2. Methods
2.3.2.1. Macrophage cell preparation

1. Dispense 1 mL of cells (3×10^5 cells/mL) into each well of a 12-well plate containing a sterile microscope cover glass. Plate 2 wells for each experimental condition. Incubate the plate at 37 $^{\circ}$C, 5% CO_2 overnight.
2. For siRNA-mediated protein knockdown and macrophage cell preparation, see section 2. Plate 3×10^5 cells/well in duplicate for each siRNA knockdown/experimental condition into a 12-well plate containing sterile microscope cover glasses. Incubate the plate at 37 $^{\circ}$C, 5% CO_2 overnight.

2.3.2.2. M .tuberculosis var. bovis BCG (BCG) labeling with Alexa 488

1. Transfer 10 mL of a log-phase BCG culture into a 15-mL conical tube. Centrifuge to pellet mycobacteria at 2500 rpm for 5 min at room temperature.
2. Remove and discard the supernatant fraction and wash once with 10 mL of 1X PBS.
3. Spin at 2500 rpm for 5 min and remove and discard the supernatant fraction.
4. Resuspend the pellet fraction in 1 mL of 1X PBS and transfer the cell suspension into a 1.5-mL microcentrifuge tube. Add 10 μL of Alexa 488 caboxylic acid succinimidyl ester stock solution to the tube to achieve a final concentration of 10 μg/mL. Wrap the tube with foil and incubate at room temperature for 45–60 min on a shaker.
5. Pellet mycobacteria at 8000 rpm for 3 min at room temperature. Remove the supernatant and wash twice with 1 mL of 1X PBS.
6. Resuspend the mycobacteria pellet in 6 mL of complete DMEM. Transfer to a 7-mL Dounce homogenizer and homogenize 35 times to generate single cell suspension.
7. Measure OD_{600} of the 1:10 dilution of the homogenized culture.
8. Prepare mycobacterial inoculum in complete DMEM at the concentration of 3×10^6 BCG/mL (MOI = 10) using the following formula:(3×10^6); (total mL needed for infection)/($OD_{600} \times 10^9$) = mL homogenate needed.
9. Add LysoTracker Red 1 mM stock solution to the inoculum to yield a final concentration of 0.25 μM.

2.3.2.3. Infection of macrophages and autophagic induction

1. At 2 h before infection, remove the old medium from the macrophages and stain the lysosomes with LysoTracker Red by adding 1 mL of complete DMEM containing 0.25 μM LysoTracker Red into each well. Incubate the plate at 37 °C, 5% CO_2.
2. After 2 h of incubation, remove the medium from the macrophages and add 1 mL of Alexa 488-labeled 3×10^6 BCG/mL inoculum into each well.
3. Spin at 1,200 rpm for 5 min to settle mycobacteria onto macrophages. Incubate the plate at 37 °C, 5% CO_2 for 15 min (pulse period).
4. Remove the inoculum and quickly wash each well 3 times with complete DMEM. Add 1 mL of complete DMEM containing 0.25 μM LysoTracker Red into each well and incubate the plate at 37 °C, 5% CO_2 for 60 min (chase period).
5. To induce autophagy, remove complete DMEM from the respective wells and quickly wash 3 times with 2 mL of starvation medium, EBSS. Add 1 mL of EBSS containing 0.25 μM LysoTracker Red into these wells and put the plate back into the incubator. Incubate for 2 h.

2.3.2.4. Immunofluorescence confocal microscopy

1. After 2 h of incubation, fix the cells with 1 mL of 2% paraformaldehyde/PBS for 10 min at room temperature.
2. Wash the cells 3 times for 5 min each with 1X PBS.
3. Mount coverslips onto microscope slides using PermaFluor mounting medium.
4. Collect 0.7-μm-thick optical sections using a 63x oil objective on a LSM META 510 (Carl Zeiss) and prepare images using Zeiss LSM Image Browser.
5. Count at least 50 phagosomes per each condition and calculate the percentage of LysoTracker Red–BCG colocalization.
6. Repeat the entire assay two more times to obtain data from three different experiments for analysis and plotting (Fig. 21.1D).

2.4. Monitoring lysosomal protease delivery to the *M. tuberculosis* phagosome

2.4.1. Materials
2.4.1.1. Growth and use of RAW264.7 cell line

1. Murine RAW264.7 macrophage cell line (ATCC, No. TIB-71).
2. Dulbecco's modified Eagle's medium (DMEM) (Gibco), 10% fetal bovine serum (FBS; Hyclone), and 4 mM L-glutamine (Gibco) (Full medium): Add 100 mL of FBS and 10 mL of 200 mM L-glutamine to 890 mL of DMEM. Filter through 0.2-μm filter. Store the medium at 4 °C.
3. 175-cm^2 tissue culture flasks (Greiner Bio-One).

4. Disposable cell scraper (Sarstedt).
5. Earle's balanced salt solution (EBSS) (Starvation medium) (Sigma).
6. 12-well tissue culture plates.
7. Circular microscope cover glasses (18 mm in diameter) (Fisher).
8. Fixing solution: Dissolve 2 g of paraformaldehyde in 100 mL of 1X PBS. Heat to 50–60 °C inside a chemical hood until the powder is solubilized. Cool the solution to room temperature, aliquot, and store at −20 °C.
9. Permeabilization solution: 0.1% (v/v) Triton-X 100 in 1X PBS.
10. Blocking solution: Dissolve 3 g of bovine serum albumin Fraction V (BSA) (Sigma) in 1X PBS containing 2% (v/v) of the appropriate serum matching the host of the secondary antibody.
11. Primary antibody: goat anticathepsin D antibody (R&D Systems).
12. Secondary antibody: Alexa 568 donkey antigoat IgG (Molecular Probes).
13. PermaFluor mounting medium (Thermo Scientific, [Cat. No. 434990]).

2.4.1.2. Growth and preparation of mycobacteria

1. Middlebrook albumin dextrose catalase (ADC) enrichment (BD Biosciences).
2. 20% Tween 80 (Sigma): Prepare 20% v/v stock at 4 °C and filter through 0.2-μm filter. Store at 4 °C.
3. Middlebrook 7H9 broth (Difco): Dissolve 4.7 g of powder in 900 mL deionized water with 2 mL of glycerol (EMD Chemicals) and autoclave. Temper to room temperature. Add 100 mL of ADC and 2.5 mL of 20% Tween 80. Store at 4 °C.
4. Sterile, 150-mL tissue-culture roller bottles.
5. Autoclaved 7-mL Dounce glass homogenizer (Kontes Scientific Glassware).
6. Phosphate-buffered saline (PBS, 10X) (Gibco): Prepare 1X PBS by adding 100 mL in 900 mL of deionized water. Filter through 0.2-μm filter and store at room temperature.
7. Alexa 488 carboxylic acid succinimidyl ester (Molecular Probes): Prepare 1 mg/mL stock solution by adding 1 mL of 100% DMSO to 1 mg of powder. Store at −20 °C.

2.4.2. Methods
2.4.2.1. Macrophage cell preparation

2.4.2.2. M. tuberculosis var. bovis BCG (BCG) labeling with Alexa 488 (see section 2.3.2.2., omit LysoTracker red in step 9)

2.4.2.3. Infection of macrophages and autophagic induction

1. Remove the old medium from macrophages and add 1 mL of Alexa 488-labeled 3×10^6 BCG/mL inoculum into each well.

2. Spin at 1200 rpm for 5 min to settle mycobacteria onto macrophages. Incubate the plate at 37 °C, 5% CO_2 for 15 min (pulse period).
3. Remove the inoculum and quickly wash each well 3 times with 1 mL of complete DMEM. Add 1 mL of complete DMEM into each well and incubate the plate at 37 °C, 5% CO_2 for 60 min (chase period).
4. To induce autophagy, remove complete DMEM from the respective wells and quickly wash 3 times with 2 mL of starvation medium, EBSS. Add 1 mL of EBSS into the wells. Incubate the plate for 4 h at 37 °C, 5% CO_2 (see Note 7).

2.4.2.4. Immunofluorescence confocal microscopy

1. Remove the media from each well and fix the cells with 1 mL of 2% paraformaldehyde/PBS for 10 min at room temperature.
2. Wash the cells 3 times for 5 min each with 1X PBS.
3. Permeabilize the cells with 1 mL of 0.1% Triton X-100/PBS for 3 min at room temperature.
4. Wash 3 times for 5 min each with 1X PBS.
5. Block the cells for 1 h at room temperature using the blocking solution.
6. Prepare primary antibody in blocking solution according to the manufacturer's protocol.
7. Invert cover slips onto 50-uL droplets of primary antibody/blocking solution dispensed onto Parafilm placed over a hard surface.
8. Cover samples, seal with Parafilm, and incubate overnight at 4 °C.
9. Transfer the cover slips back into the plate and wash the cover slips 3 times for 5 min each with 1X PBS.
10. Prepare appropriate secondary antibody in blocking solution at 1:500 dilution.
11. Add 1 mL into each well and incubate for 2 h at room temperature.
12. Wash 3 times for 5 min each with 1X PBS.
13. Mount coverslips onto microscope slides using PermaFluor mounting medium.
14. Collect 0.7-μm-thick optical sections using a 63x oil objective on a LSM META 510 (Carl Zeiss) and prepare images using Zeiss LSM Image Browser.
15. Count at least 50 phagosomes per each condition and calculate the percentage of cathepsin D-BCG colocalization.
16. Repeat the entire assay two more times to obtain data from three different experiments for analysis and plotting (Fig. 21.1E).

    Notes

    1. To prevent premature autophagic induction, it is necessary to wash the cells using complete DMEM instead of using 1X PBS.

2. To prevent cell detachment during the assay, it is important to wash the cells gently by adding the media along the side of the plastic well and aspirate off the media from the well using a vacuum suction system with a pipette tip attached to the end of a glass pipette to reduce the suction force.
3. The incubation time depends on the experimental conditions being tested. For autophagic killing of mycobacteria, 4 h incubation is routinely used for the assay.
4. Putting the plates in a plastic bag helps to prevent the plates from drying out and from contamination with other microorganisms.
5. Optimized transfection protocols for many cell types are available from the manufacturer's website (http://www.amaxa.com).
6. The incubation time to obtain the desirable knockdown level of a protein of interest needs to be predetermined for each individual protein.
7. The incubation time depends on the experimental conditions being tested. For analysis of autophagic-mediated acidification of *M. tuberculosis* phagosomes using LysoTracker Red staining, 2 h is found to be optimum. For analysis of autophagic-mediated delivery of a lysosomal protease into the *M. tuberculosis* phagosome using cathepsin D staining, 4 h is found to be best.

ACKNOWLEDGMENTS

We thank Sharon S. Master and Alexander S. Davis for sharing techniques and materials and Esteban A. Roberts for critical reading of the manuscript. This work is supported by NIH grants A1069345, A145148, and A142999.

REFERENCES

Alonso, S., Pethe, K., Russell, D. G., and Purdy, G. E. (2007). Lysosomal killing of Mycobacterium mediated by ubiquitin-derived peptides is enhanced by autophagy. *Proc. Natl. Acad. Sci. USA* **104**, 6031–6036.

Beatty, W. L., Rhoades, E. R., Ullrich, H. J., Chatterjee, D., Heuser, J. E., and Russell, D. G. (2000). Trafficking and release of mycobacterial lipids from infected macrophages. *Traffic* **1**, 235–247.

Bernales, S., Schuck, S., and Walter, P. (2007). ER-phagy: selective autophagy of the endoplasmic reticulum. *Autophagy* **3**, 285–287.

Budha, N. R., Lee, R. E., and Meibohm, B. (2008). Biopharmaceutics, pharmacokinetics and pharmacodynamics of antituberculosis drugs. *Curr. Med. Chem.* **15**, 809–825.

Chan, J., and Flynn, J. (2004). The immunological aspects of latency in tuberculosis. *Clin. Immunol.* **110**, 2–12.

Corbett, E. L., Watt, C. J., Walker, N., Maher, D., Williams, B. G., Raviglione, M. C., and Dye, C. (2003). The growing burden of tuberculosis: global trends and interactions with the HIV epidemic. *Arch. Intern. Med.* **163**, 1009–1021.

Delgado, M. A., Elmaoued, R. A., Davis, A. S., Kyei, G., and Deretic, V. (2008). Toll-like receptors control autophagy. *EMBO J.* **27,** 1110–1121.

Deretic, V., Vergne, I., Chua, J., Master, S., Singh, S. B., Fazio, J. A., and Kyei, G. (2004). Endosomal membrane traffic: Convergence point targeted by *Mycobacterium tuberculosis* and HIV. *Cell. Microbiol.* **6,** 999–1009.

Devenish, R. J. (2007). Mitophagy: Growing in intricacy. *Autophagy* **3,** 293–294.

Ducati, R. G., Ruffino-Netto, A., Basso, L. A., and Santos, D. S. (2006). The resumption of consumption–a review on tuberculosis. *Mem. Inst. Oswaldo Cruz* **101,** 697–714.

Gutierrez, M. G., Master, S. S., Singh, S. B., Taylor, G. A., Colombo, M. I., and Deretic, V. (2004). Autophagy is a defense mechanism inhibiting BCG and *Mycobacterium tuberculosis* survival in infected macrophages. *Cell* **119,** 753–766.

Harris, J., De Haro, S. A., Master, S. S., Keane, J., Roberts, E. A., Delgado, M., and Deretic, V. (2007). T helper 2 cytokines inhibit autophagic control of intracellular *Mycobacterium tuberculosis*. *Immunity* **27,** 505–517.

Harris, J., De Haro, S., and Deretic, V. (2006). Autophagy and *Mycobacterium tuberculosis*. In Autophagy in Immunity and Infection, (V. Deretic, ed.), WILEY-VCH, Weinheim. pp.129–138.

Kaufmann, S. H., Baumann, S., and Nasser Eddine, A. (2006). Exploiting immunology and molecular genetics for rational vaccine design against tuberculosis. *Int. J. Tuberc. Lung. Dis.* **10,** 1068–1079.

Klionsky, D. J., Cuervo, A. M., and Seglen, P. O. (2007). Methods for monitoring autophagy from yeast to human. *Autophagy* **3,** 181–206.

Levine, B., and Klionsky, D. J. (2004). Development by self-digestion: molecular mechanisms and biological functions of autophagy. *Dev. Cell.* **6,** 463–477.

Liang, C., E, X, and Jung, J. U. (2008). Downregulation of autophagy by herpesvirus Bcl-2 homologs. *Autophagy* **4,** 268–272.

Ling, Y. M., Shaw, M. H., Ayala, C., Coppens, I., Taylor, G. A., Ferguson, D. J., and Yap, G. S. (2006). Vacuolar and plasma membrane stripping and autophagic elimination of *Toxoplasma gondii* in primed effector macrophages. *J. Exp. Med.* **203,** 2063–2071.

Mizushima, N. (2004). Methods for monitoring autophagy. *Int. J. Biochem. Cell. Biol.* **36,** 2491–2502.

Mizushima, N., and Klionsky, D. J. (2007). Protein turnover via autophagy: implications for metabolism. *Annu. Rev. Nutr.* **27,** 19–40.

Nakagawa, I., Amano, A., Mizushima, N., Yamamoto, A., Yamaguchi, H., Kamimoto, T., Nara, A., Funao, J., Nakata, M., Tsuda, K., Hamada, S., and Yoshimori, T. (2004). Autophagy defends cells against invading group A *Streptococcus*. *Science* **306,** 1037–1040.

Ogawa, M., Yoshimori, T., Suzuki, T., Sagara, H., Mizushima, N., and Sasakawa, C. (2005). Escape of intracellular *Shigella* from autophagy. *Science* **307,** 727–731.

Rubinsztein, D. C. (2006). The roles of intracellular protein-degradation pathways in neurodegeneration. *Nature* **443,** 780–786.

Russell, D. G. (2001). *Mycobacterium tuberculosis*: here today, and here tomorrow. *Nat. Rev. Mol. Cell Biol.* **2,** 569–577.

Russell, D. G. (2007). Who puts the tubercle in tuberculosis? *Nat Rev Microbiol.* **5,** 39–47.

Russell, D. G., Mwandumba, H. C., and Rhoades, E. E. (2002). Mycobacterium and the coat of many lipids. *J. Cell Biol.* **158,** 421–426.

Sakai, Y., Oku, M., van der Klei, I. J., and Kiel, J. A. W. K. (2006). Pexophagy: autophagic degradation of peroxisomes. *Biochim. Biophys. Acta.* **1763,** 1767–1775.

Schmid, D., and Münz, C. (2005). Immune surveillance of intracellular pathogens via autophagy. *Cell. Death. Differ.* **12**(Suppl. 2), 1519–1527.

Schmid, D., Pypaert, M., and Münz, C. (2007). Antigen-loading compartments for major histocompatibility complex class II molecules continuously receive input from autophagosomes. *Immunity* **26,** 79–92.

Singh, S. B., Davis, A. S., Taylor, G. A., and Deretic, V. (2006). Human IRGM induces autophagy to eliminate intracellular mycobacteria. *Science* **313,** 1438–1441.

Tallóczy, Z., Jiang, W., Virgin, H. W. IV,, Leib, D. A., Scheuner, D., Kaufman, R. J., Eskelinen, E.-L., and Levine, B. (2002). Regulation of starvation- and virus-induced autophagy by the eIF2α kinase signaling pathway. *Proc. Natl. Acad. Sci. USA* **99,** 190–195.

Vergne, I., Chua, J., and Deretic, V. (2003). Tuberculosis toxin blocking phagosome maturation inhibits a novel Ca^{2+}/calmodulin-PI3K hVPS34 cascade. *J. Exp. Med.* **198,** 653–659.

Vergne, I., Chua, J., Singh, S. B., and Deretic, V. (2004). Cell biology of *mycobacterium tuberculosis* phagosome. *Annu. Rev. Cell Dev. Biol.* **20,** 367–394.

WHO (2008). Global tuberculosis control-surveillance, planning, and financing http://www.who.int/tb/publications/global_report/2008/summary/en/index.html.

Williams, A., Jahreiss, L., Sarkar, S., Saiki, S., Menzies, F. M., Ravikumar, B., and Rubinsztein, D. C. (2006). Aggregate-prone proteins are cleared from the cytosol by autophagy: Therapeutic implications. *Curr. Top. Dev. Biol.* **76,** 89–101.

CHAPTER TWENTY-TWO

Streptococcus-, *Shigella*-, and *Listeria*-Induced Autophagy

Michinaga Ogawa,[*] Ichiro Nakagawa,[†] Yuko Yoshikawa,[*] Torsten Hain,[‡] Trinad Chakraborty,[‡] *and* Chihiro Sasakawa[*,§]

Contents

1. Introduction	364
2. GAS Infection and Autophagy	365
2.1. Overview	365
2.2. Infection protocol for group A *Streptococci*	366
2.3. Analysis of the subcellular localization of GAS	367
2.4. Measurement of intracellular invasion and intracellular degradation rates	369
3. *Shigella* Infection and Autophagy	371
3.1. Overview	371
3.2. Infection protocol for *Shigella* in MDCK cells	372
3.3. Infection protocol for *Shigella* in BHK or MEF cells	373
3.4. Analysis of the intracellular spreading of *Shigella* using a plaque-forming assay	373
3.5. Assay for intracellular bacterial multiplication	374
3.6. Analysis of *Shigella*-containing autophagosomes using TEM	374
3.7. Visualization of *Shigella*-containing autophagosomes using GFP-LC3 and GFP-Atg5	375
4. *Listeria* Infection and Autophagy	376
4.1. Overview	376
4.2. Infection protocol for *Listeria monocytogenes* in epithelial cells	376
4.3. Infection protocol for *Listeria monocytogenes* in macrophages	377
4.4. Electron microscopy and immunoelectron microscopy	378
4.5. Intracellular bacterial multiplication assay	379

[*] Division of Bacterial Infection, Department of Microbiology and Immunology, Institute of Medical Science, University of Tokyo, Tokyo, Japan
[†] Division of Bacteriology, Department of Infectious Disease Control, International Research Center for Infectious Diseases, Institute of Medical Science, University of Tokyo, Tokyo, Japan
[‡] Institute of Medical Microbiology, Justus-Liebig University Gissen, Giessen, Germany
[§] Department of Infectious Disease Control, International Research Center for Infectious Diseases, Institute of Medical Science, University of Tokyo, Tokyo, Japan, and CREST, Japan Science and Technology Agency, Kawaguchi, Japan

Methods in Enzymology, Volume 452   © 2009 Elsevier Inc.
ISSN 0076-6879, DOI: 10.1016/S0076-6879(08)03622-7   All rights reserved.

5. Concluding Remarks 379
Acknowledgments 379
References 379

Abstract

Autophagy has recently been described as an intrinsic host defense system for recognizing and eliminating intracellular-invading bacterial pathogens. Some cytoplasmic-invading bacteria are trapped through the process of autophagy and are ultimately degraded within autolysosomal compartments. However, others exhibit highly evolved maneuvers for circumventing autophagic recognition and are capable of surviving and replicating within the cytoplasm. In this chapter, we describe bacterial infectious systems using group A *Streptococcus*, *Shigella flexneri*, and *Listeria monocytogenes* as examples of the interplay between bacteria and autophagy; in addition, methods for investigating bacterial activities related to the recognition of bacteria by the autophagic machinery or the escape of bacteria from autophagy are introduced.

1. INTRODUCTION

Many pathogenic bacteria are capable of invading various host cells by inducing endocytic or phagocytic events. With either of these invading mechanisms, if the bacterium is unable to escape from the vacuole enclosing the invading bacterium, to modify maturation of the vacuole, or to prevent the fusion of the vacuole with the lysosomal compartments, it will eventually be delivered to the lysosome and be degraded by the resident hydrolytic enzymes (Sinai and Joiner, 1997). Conversely, if the host cells are less capable of sensing pathogens or incapable of activating the autophagic pathway, some intracellular pathogens, such as *Brucella abortus*, *Coxiella brunetii*, and *Legionella pneumophila*, can block phagosome maturation and reside within the membrane compartments (Amer and Swanson, 2005; Dorn *et al.*, 2002; Gutierrez *et al.*, 2004; Gutierrez *et al.*, 2005; Sinai and Joiner, 1997). Others, such as *Burkholderia psuedomallei*, *Listeria monocytogenes*, *Mycobacterium marinum*, *Rickettsia* and *Shigella*, seem to escape from autophagy through their ability to disrupt the surrounding vacuole rapidly, replicate within the cytoplasm, and move by conducting actin polymerization at one pole of the bacterium, thus allowing the bacteria to disseminate into neighboring cells (Cossart and Sansonetti, 2004; Gouin *et al.*, 2005; Sinai and Joiner, 1997). In response to bacterial invasion, the host cells can sense bacteria through a variety of means, such as the toll-like receptors or the Nod-like receptor (NLR) family, thus leading to the activation of the innate immune system and the production/secretion of chemokines and cytokines. In addition, as described previously, host cells also sense bacteria through autophagy. Indeed, recent studies have highlighted autophagy as a critical intracellular surveillance system for the recognition of bacterial surface molecules (Birmingham et al., 2007; Dorn et al., 2002; Gutierrez et al., 2004; Gutierrez et al., 2005; Kirkegaard et al., 2004; Levine, 2005; Levine and Klionsky, 2004; Nakagawa et al., 2004; Ogawa and Sasakawa, 2006a; Ogawa and Sasakawa, 2006b; Ogawa et al., 2005; Py et al., 2007; Webster, 2006). For example, when group A *Streptococcus* (GAS) is internalized into the host cytoplasm, it is trapped by autophagy and undergoes lysosomal degradation. Importantly, some intracellular motile bacteria such as *Shigella* and *L. monocytogenes* (and perhaps all of the motile bacterial pathogens) possess systems that allow them to circumvent autophagic recognition, thus allowing them to multiply and disseminate into the surrounding host cells. These studies strongly suggest that a bacteria's ability to camouflage itself from autophagic recognition or to subvert autophagic activation or the maturation process is a prominent pathogenic feature that promotes the survival of intracellular bacteria.

In this chapter, we briefly describe the protocol for infecting epithelial cells with GAS, *Shigella*, and *L. monocytogenes* and for evaluating the autophagy induced by bacterial infection.

2. GAS INFECTION AND AUTOPHAGY

2.1. Overview

Streptococcus pyogenes (GAS) is the etiological agent for a diverse collection of human diseases ranging from self-limiting suppurative infections of the upper respiratory tract and skin to deeper and more serious invasive infections, such as toxic shock–like syndrome. Although GAS was not previously thought to be an intracellular pathogen, recent studies have revealed that it effectively attaches to and invades pharyngeal and skin epithelial cells. However, the fate of GAS after it has invaded the host cells is not clearly understood. Usually, most of the phagocytosed or invasive bacteria are eliminated through the phagocytic/endocytic degradation pathway. Therefore, the phagocytic/endocytic degradation pathway has been thought to be the only system to act against microbial pathogens that have entered the host cells. In contrast, macroautophagy, usually referred to simply as autophagy, is a physiologically important cellular process for the bulk degradation of organelles and cytosolic proteins, and this process is thought to degrade endogenous materials. However, intracellular GAS is selectively captured by an autophagosome-like compartment and is degraded through the fusion of this compartment with lysosomes in non-phagocytic cells. In autophagosome formation-deficient $atg5^{-/-}$ cells, such compartments are not formed and the number of intracellular GAS is markedly higher than that in control cells. In addition, the bacterial hemolytic toxin streptolysin O-dependent escape of GAS from endosomes is necessary for the induction of autophagy enclosing GAS (Nakagawa et al., 2004). In this section, we discuss some examples of microscopy and biochemical techniques for analyzing the localization and fate of GAS within host cells.

2.2. Infection protocol for group A *Streptococci*

2.2.1. Cell lines

Human cervical epithelial HeLa cells (CCL-2) and HeLa cells stably transfected with EGFP-LC3 (Kabeya et al., 2000) are cultured in a growth medium of Dulbecco's Modified Eagle Medium (DMEM; GIBCO BRL) supplemented with 50 μg/mL gentamicin, 10% FCS, 2 mM L-glutamine. $atg5^{+/+}$ and $atg5^{-/-}$ mouse embryonic fibroblasts (MEFs) were kindly provided by Dr. N. Mizushima (Tokyo Medical and Dental University) (Hara et al., 2006). For fluorescence imaging, cells are cultured on a cover glass (diameter, 12 mm and thickness, 0.17 mm; Matsunami Glass) or a glass-bottom culture dish (diameter, 35 mm). The glass surfaces are pretreated with poly-$_L$-lysine (500 μg/mL) or gelatin (0.2%–0.5% in PBS) to increase cellular attachment.

2.2.2. Bacterial strains

GAS is usually grown in Todd-Hewitt broth (THY broth; Difco) supplemented with 0.2% yeast extract (THY medium) at 37 °C. Before experimentation, the colony-forming unit (cfu; the number of colonies that form from a specific volume of culture at a particular OD_{600}) of the mid-log phase ($OD_{600} = 0.4$–0.8) should be determined using a serial dilution of the bacterial culture and plating on THY agar plates. In strain JRS4 (M6$^+$, F1$^+$), the average cfu was 8.0×10^8 /mL during the mid-log phase.

2.2.3. Infection of GAS

1. Bacteria are grown in 5 mL of THY broth grown at 37 °C until mid-log phase. A portion of the bacterial culture (1 mL) is transferred into a microcentrifuge tube, then harvested by centrifugation at 9000 rpm for 2 min. After washing twice with 1 mL of phosphate-buffered saline (PBS; pH 7.4), the cells are suspended in DMEM medium.
2. Cultured cells (4×10^4 HeLa cells or 1×10^5 $atg5^{+/+}$ or $atg5^{-/-}$ MEF cells in a 24-well culture plate, or 1×10^5 HeLa cells in a 60-mm culture dish) are infected with GAS at a 50:1 multiplicity of infection (MOI; 50 bacteria per 1 target cell), without antibiotics at 37 °C in 5% CO_2. The experimental infection is started (time zero of infection).
3. The cells are further incubated with GAS for 1 h to allow invasion, and the infected cells are thoroughly washed with PBS by aspiration. Then 10% FCS/DMEM containing antibiotics (100 μg/mL gentamicin and 100 U/mL penicillin G) is added to each well and incubated for approximately 1–4 h, and cells are recovered at 1-h intervals. Under this

Table 22.1 GFP fusion proteins used to analyze the intracellular localization of GAS

Protein	Localization	Reference
FYVE domain of EEA-1	Phagosome/early endosome	(Birkeland and Stenmark, 2004)
p40-PX domain	Phagosome/early endosome	(Birkeland and Stenmark, 2004)
Rab5	Phagosome/early endosome	(Simpson and Jones, 2005)
Rab7	Late endosome/ autophagosome	(Jager et al., 2004)
Lgp85	Lysosome	(Kuronita et al., 2002)
LC3 (Atg8)	Autophagosome	(Kabeya et al., 2000)

condition, typically 80% of the cells are infected with GAS, and at least 10 bacteria are found in each cell. It should be noted that GAS exists as a chainlike structure, so this GAS chain is the form found in the infected cells at 1 h infection.

2.3. Analysis of the subcellular localization of GAS

After adhering to the host cells via the extracellular matrix and integrins, GAS efficiently invades the cells by being engulfed by the plasma membrane of epithelial cells (Cue et al., 2000; Cywes and Wessels, 2001). Several studies have indicated that the intracellular bacteria is localized in membrane-bound compartments, but others have shown the organisms to be free in the cytoplasm (Medina et al., 2003; Nakagawa et al., 2001). To clarify the destination of GAS after internalization, several organelle-specific markers tagged with green fluorescent protein (GFP) or its derivatives followed by immunostaining with specific antibodies are usually used as an invaluable tool for investigating the distribution of the organisms using fluorescence or confocal microscopy (Table 22.1).

2.3.1. Transfection with fluorescent protein overexpression vectors

The fluorescent protein markers listed below are used for both live-cell imaging and fixed cells to trace the intracellular localization of GAS. The vectors for the expression of these marker proteins are transfected using an appropriate transfection reagent. However, a GFP-fusion protein may exert a dominant-negative effect or alter the phenotype of a cell simply by its abundance. Because in living cells the activities of some proteins are optimized, the subcellular distribution of the markers should be confirmed using another method, such as immunofluorescent staining.

2.3.2. Immunofluorescent staining of infected cells

Localization by immunofluorescence uses antibodies that are specific to the biological molecules of interest and that allow the visualization and optional quantification of the amount of specific labeling in terms of the number of distinct fluorescent objects and/or their intensities.

1. Cells are washed once with PBS then soaked at least 10 min in 4% (w/v) paraformaldehyde/PBS at 37 °C. Alternatively, cells may be fixed in 100% methanol at $-20\,°C$ for 3 min.
2. After washing in PBS, the cells are permeabilized with PBS containing 0.1% (w/v) Triton X-100 for 10 min.
3. Cells are washed with PBS containing 1% (w/v) sodium borohydride to reduce the free aldehyde groups; this step is crucial for reducing the specific background. If methanol fixation is used, Triton X-100 treatment can be omitted.
4. Cells are incubated with the appropriate concentration of primary antibody for 1 h at room temperature in a humidified chamber. If a 12-mm-diameter cover glass is being used, the cover glass is inverted onto a piece of Parafilm and a 30-μL aliquot of diluted antibody is placed on the cover glass.
5. After washing 3 times with PBS, the cells are incubated with the appropriate concentration of secondary antibody for 1 h at room temperature in a humidified chamber. (e.g., FITC, Texas-Red, Alexa dyes, or Cy5-conjugate).

2.3.3. Detection of intracellular GAS

As described previously, immunostaining with a GAS-specific antibody can be used to detect GAS (Nakagawa et al., 2001). Alternatively, infected cells can be stained with 0.2 μg/mL of propidium iodide (PI) at room temperature for 20 min or 0.2 μg/mL 4′,6-diamidino-2-phenylindole (DAPI) at room temperature for 15 min to stain the bacterial and cellular DNA. This method is simple and effective for analyzing the localization of bacteria when used in combination with fluorescent protein-expressing cells or immunostaining. Fluorescent protein-expressing GAS (Nakagawa et al., 2001) can also be used to analyze by live-cell imaging. Fluorochrome labeling, such as the bacterial surface labeling with fluorescein isothiocyanate (FITC) or DNA staining by DAPI, has been widely used for bacterial staining; however, it should be noted that direct labeling of the bacterial surface may change its binding properties, and DNA staining may affect the bacterial viability.

2.3.4. Measurement of GAS-containing compartments using morphometric analysis

To determine the percentage of cells with GAS-containing compartments using morphometric analysis, confocal microscopy images of GFP fusion protein-expressing cells are analyzed using Adobe Photoshop software to change the contrast. In brighter images of GFP-positive cells, the cell margin can be detected by the distribution of GFP diffusing throughout the cytoplasm; the area of the cytoplasm in each cell can then be measured using Scion Image software (Scion Corporation) or ImageJ software (Wayne Rasband; the Research Services Branch, National Institute of Mental Health, Bethesda, Maryland, USA). Darker images can also be used to measure the area of GFP-positive compartments. To determine the intracellular numbers of GAS, both PI-stained images and brighter images of GFP-stained cells are used. The percentage of (the number of bacteria within the GFP-positive compartments)/(the number of PI-positive intracellular bacteria) is calculated. At least 3 individual experiments should be performed, and at least 20 cell sections should be analyzed at each time point.

2.4. Measurement of intracellular invasion and intracellular degradation rates

Visualization using fluorescence microscopy is an excellent tool for qualitative analyses of the localization of GAS within host cells, and quantitative analyses of intracellular GAS are important for understanding the fate of this organism. The three methods discussed herein are useful for analyzing bacterial replication and degradation within the host cells.

2.4.1. Bacterial viability assay

1. HeLa and MEF cells are cultured in 24-well culture plates (4×10^4 cells for HeLa, 1×10^5 cells for MEF cells per well). The infection protocol is described previously.
2. After an appropriate incubation time, infected cells are lysed in sterile distilled water by vigorous pipetting and serial dilutions of the lysate are plated on THY agar plates. These agar plates are incubated at 37 °C for 24 h. The number of intracellular GAS can be determined by colony counting.

It should be noted that gentamicin and penicillin treatment are important for removing extracellular bacteria. To negate the possibility that GAS may have escaped outside of the host cells, tannic acid treatment is also effective. Tannic acid is a cell-impermeable fixative that prevents fusion during exocytosis but does not affect intracellular vesicle trafficking (Nakagawa et al., 2004).

1. At 1 h of infection, the remaining extracellular GAS is killed by the administration of antibiotics (100 U/mL penicillin G and 100 μg/mL gentamicin) as described in section 2.2.
2. 0.5% tannic acid is added to the medium twice for 5 min, both 15 min before the antibiotics treatment and at 2 h post-infection.

3. After approximately 1–4 h post-infecton, the number of intracellular bacteria is determined using a bacterial viability assay at 1-h intervals, as described earlier.

2.4.2. Measurement of intracellular replication of GAS

1. $atg5^{+/+}$ and $atg5^{-/-}$ MEF cells (1×10^6 cells in a 10-cm dish containing 10 mL 5% FCS/DMEM) are metabolically labeled with 1.85 MBq of [^{35}S]-Met+Cys for 48 h.
2. The cells are washed with PBS three times and infected with GAS as described earlier. In this condition, extracellular GAS are removed by antibiotic treatment, and therefore GAS acquire the majority of their nutrients within the host cells.
3. After an appropriate incubation time for infection, the cells are lysed with 2 mL of 1% (w/v) Triton X-100 in 150 mM NaCl and 50 mM Tris-HCl, pH 7.4, in the presence of protease inhibitors (Roche, Complete Proteinase Inhibitor Cocktail) (Triton lysis buffer) and the lysates are layered over discontinuous Percoll solutions (from 1.02 to 1.12 g/mL in 0.15 M NaCl, 0.02 g/mL steps). This Percoll gradient consists of 12 mL of total volume, and each of the Percoll solutions are 2 mL.
4. After centrifugation at 2000 rpm for 20 min, GAS can be separated from between the 1.08 g/mL and the 1.10 g/mL fractions.
5. The GAS-containing fraction is recovered and measured using a liquid scintillation counter.

2.4.3. Measurement of intracellular degradation of invasive GAS

1. GAS cultured in 5 mL of medium overnight is labeled with 1.85 MBq (0.05 mCi) of [^{35}S]-labeled methionine (Met)+cysteine (Cys) (GE) in 5 mL of THY medium for 5 h.
2. The labeled bacteria are washed with PBS three times and then used for infection as described previously.
3. After an appropriate incubation time for infection, the labeled GAS-infected cells are lysed in Triton lysis buffer. Note that it may be helpful to take multiple time points, including a time zero point, to measure the change in degradation over time. In addition, it may be useful to add excess unlabeled methionine and cysteine during the infection and degradation period to prevent reincorporation of released radioactive amino acids.
4. Cellular lysates are centrifuged at 15,000 rpm for 10 min at 4 °C to remove cellular debris.
5. The supernatant fraction is further separated by the addition of trichloroacetate (TCA; 10% final concentration) to remove long-chained proteins or polypeptides, and free-[^{35}S]-Met+Cys in the cytoplasm can be measured using a liquid scintillation counter.

3. *SHIGELLA* INFECTION AND AUTOPHAGY

3.1. Overview

3.1.1. *Shigella* infection of intestinal epithelium

Shigella are highly adapted human pathogens that cause bacillary dysentery, a disease that is manifested by severe bloody and mucous diarrhea. Following ingestion via a fecal-oral route, *Shigella* finally reach the colon and rectum, where they translocate through the epithelial barrier via the M (microfold) cells. Once they reach the M cells, *Shigella* invade the resident macrophages, and the infecting bacteria escape from the phagosomes into the cytoplasm, where they multiply and induce rapid cell death. *Shigella* released from the dead macrophages immediately enter the surrounding enterocytes through their basolateral surface by inducing a macropinocytic event.

When *Shigella* come into contact with epithelial cells, the type III secretion system (TTSS) is activated and delivers effectors into the host cells and surrounding bacterial space. The secreted effectors are capable of modulating various host functions engaged in remodelling the surface architecture of host cells, and thus escape from the host innate defense system.

As soon as a bacterium is surrounded by the membrane vacuole of infected epithelial cells, it disrupts the vacuole membrane and escapes into the cytoplasm. *Shigella* multiply in the cytoplasm, where they move by inducing local actin polymerization at one pole of the bacterium (Ogawa *et al.*, 2008; Ogawa and Sasakawa, 2006b).

3.1.2. Escape of *Shigella* from autophagy

IcsB, an effector secreted by intracellular *Shigella* via the TTSS, has been shown to play a pivotal role in the escape of *Shigella* from autophagy (Ogawa *et al.*, 2003; Ogawa *et al.*, 2005). IcsB is a 52-kDa protein composed of 494 amino acids encoded by the *icsB* gene, which is located upstream of the *ipaBCD* genes on the large 220-kb plasmid of *Shigella*. Our previous study shows that an *icsB*-deleted mutant of *S. flexneri* loses its ability to multiply within host cells. Indeed, the *icsB* mutant multiplies and moved normally for approximately 3 h in BHK cells, but intracellular multiplication eventually reaches a plateau at 4 h after invasion. When MDCK cells expressing GFP-LC3 (MDCK/pGFP-LC3) are infected with the *icsB* mutant or the wild type, around 40% of the intracellular *icsB* mutants and 8% of intracellular wild-type *Shigella* are associated with LC3 signals. When MDCK cells are treated with an autophagy inhibitor such as 3-methyladenine (an inhibitor of class III phosphatidylinositol 3-kinase), the proportion of LC3-positive *icsB* decreases markedly.

The inability of the *icsB* mutant to circumvent autophagy can also be demonstrated by using this mutant to infect *atg5*-knockout mouse embryonic fibroblasts ($atg5^{-/-}$ MEFs), which are defective in regard to autophagy. Although LC3-positive icsB mutant bacteria are detectable in normal in $atg5^{+/+}$MEF cells expressing GFP-LC3, hardly any bacterial signals are detected in $atg5^{-/-}$ MEFs expressing GFP-LC3 that are infected with the icsB mutant. As expected, intracellular multiplication of the icsB mutant is restored to the level of the wild type in $atg5^{-/-}$ MEFs.

When the icsB mutant is examined using thin-section electron microscopy (EM) after 3–4 h of infection, the icsB mutant is frequently enclosed by lamellar membranous structures. Immunogold EM with anti-GFP antibody has shown specific labeling for lamellar membrane around the icsB mutant in MDCK/pGFP-LC3 cells. Thus, the intracellular behavior of the icsB mutant strongly suggests that mutant Shigella that do not produce IcsB are readily recognized by autophagy and are destroyed within the lysosome.

3.2. Infection protocol for *Shigella* in MDCK cells

1. MDCK (Madin-Darby canine kidney) cells stably expressing GFP-LC3 (MDCK/pGFP-LC3) are seeded on collagen-coated glass coverslips (18 × 24 mm and thickness, 0.17 mm; Matsunami Glass) in 6-well plates and then cultured for 4 days to generate a confluent monolayer. Glass coverslips are coated by 9% Cellmatrix Type I-C (Nitta Gelatin) diluted in 1 mM HCl at room temperature for 1 h. MDCK/pGFP-LC3 were made by transfection of the pEGFP-LC3 vector into MDCK cells by Lipofectamin2000 (Invitrogen) followed by selection with 1000 μg/mL of G418 (Sigma) containing DMEM/FCS. Single MDCK/pGFP-LC3 colonies were isolated by the limited dilution method.

2. The cells are treated with modified Hanks's balanced salt solution (Cat. No. H2387, Sigma) supplemented with 10 μM EGTA (ethylene glycol-bis (b-aminoethylether)-N, N, N', N'-tetraacetic acid) at 37 °C for 1 h in 5% CO_2, which allows Shigella to enter from the basolateral surface. EGTA is a Ca^{2+}-chelater that can disrupt E-cadherin-to-E-cadherin binding, thereby exposing the basolateral surface of a polarized epithelial monolayer.

3. Red-colored colonies of Shigella plated on TSA (trypticase soy agar, Difco) supplemented with 0.01% Congo red, are inoculated into MH (Meuller Hinton, Difco) broth, and cultured overnight at 30 °C. 100 μL of overnight culture is inoculated into 5 ml of BHI (brain heart infusion, Difco) broth at 37 °C for 2.5 h, and bacteria are centrifuged for 3 min at 2,800×g and diluted 1:100 in Hanks's balanced salt solution supplemented with 10 μM EGTA.

4. 500 μl of Shigella suspension is added to each well and incubated at 37 °C for 2 h in 5% CO_2.

5. The medium is replaced with fresh medium supplemented with 100 μg/mL of gentamicin and 60 μg/mL of kanamycin.
6. The cell culture is continued at 37 °C for 2, 4, or 6 h in 5% CO_2.
7. After washing with PBS 3 times, the cells are fixed with 4% paraformaldehyde in PBS for 10 min at room temperature.
8. The number of bacteria associated with GFP-LC3 is determined by visual counting using fluorescence microscopy.

3.3. Infection protocol for *Shigella* in BHK or MEF cells

1. BHK (baby hamster kidney) or MEF cells stably expressing GFP-LC3 are seeded at 2×10^5/well on glass cover slips in a 6-well plate.
2. *Shigella* plated on TSA containing Congo red are inoculated into MH and cultured overnight at 30 °C. 100 μL of overnight culture is inoculated into 5 ml of BHI. Bacteria grown in BHI broth at 37 °C for 2 h are centrifuged for 3 min at $2,800 \times g$ and diluted 1:7.5 in MEM/FCS, and 1.5 ml of bacterial suspension is added to each well.
3. The cells are centrifuged for 10 min at $900 \times g$ at room temperature.
4. Plates are incubated for 30 min at 37 °C in 5% CO_2 to allow bacterial invasion.
5. After washing with PBS 3 times, the cells are incubated in a fresh medium supplemented with 100 μg/mL of gentamicin and 60 μg/mL of kanamycin for 1, 2, or 3 h at 37 °C in 5% CO_2.
6. After washing with PBS 3 times, the cells are fixed with 4% paraformaldehyde in PBS for 10 min at room temperature.
7. The number of bacteria associated with GFP-LC3 is determined by visual counting using fluorescence microscopy.

3.4. Analysis of the intracellular spreading of *Shigella* using a plaque-forming assay

The ability of motile bacteria to disseminate into adjacent epithelial cells can be easily evaluated using a plaque-forming assay. The following protocol has been developed to evaluate the ability of intracellular *Shigella* to spread into neighboring cells, as *Shigella* efficiently enters polarized epithelial monolayer cells from the basolateral surface (Ogawa *et al.*, 2003).

1. MDCK cells are seeded on 24-well plates and cultured for 5 days to a confluent monolayer.
2. The cells are treated with Hanks's balanced salt solution supplemented with 100 μM EGTA for 1 h at 37 °C in 5% CO_2.
3. As described in section 3.2.1, Bacteria grown in BHI broth at 37 °C for 2.5 h are centrifuged for 3 min at $2,800 \times g$ and suspended in Hanks' balanced salt solution supplemented with 100 μM EGTA.

4. Cells are infected with *Shigella* at MOI 10 and incubated at 37 °C for 2 h.
5. The medium is changed into fresh medium supplemented with 100 μg/mL of gentamicin and 60 μg/mL of kanamycin, and the cells are cultured at 37 °C for 1 or 2 days in 5% CO_2.
6. The diameter of the plaques can be monitored using a phase-contrast microscope, equipped with a CCD, and IPLab Spectrum software.

3.5. Assay for intracellular bacterial multiplication

1. BHK cells seeded at 6×10^4/well or MEF cells seeded at 1.8×10^4/well on 24-well plates are infected with *Shigella* strains at MOI 100 as described previously.
2. The plates are centrifuged for 10 min at $900\times g$ at room temperature.
3. The plates are incubated at 37 °C for 20 min in 5% CO_2 to allow bacterial invasion.
4. After washing the infected cells with PBS, fresh medium supplemented with 100 μg/mL of gentamicin is added to kill the extracellular bacteria, and the culture is continued at 37 °C in 5% CO_2 for the appropriate periods to allow intracellular bacterial growth.
5. At each time point, the cells are washed and lysed with PBS containing 0.5% Triton X-100. The lysates are diluted with PBS (1:10 to $1:10^4$) and plated onto LB-agar plates, and the number of intracellular bacteria is counted as cfu.

3.6. Analysis of *Shigella*-containing autophagosomes using TEM

1. MDCK cells seeded onto 35-mm dishes are infected with *Shigella* at 37 °C for 2 or 4 h, as described in 3.2.1.
2. Cells are washed with PBS 3 times, and the cells are fixed with 2% glutaraldehyde in PBS for 60 min at room temperature, post-fixed in 2% OsO_4, dehydrated with a graded ethanol series (50% → 70% → 80% → 90% → anhydrous 100%), and embedded in epoxy resin (see also the chapter by Anttila *et al.*, in this volume).
3. Ultrathin sections are double-stained with 1% uranyl acetate for 5 min and 2.5% lead citrate for 5 min.

For immunoelectron microscopy, the postembedding immunogold method is recommended:

1. MDCK/pGFP-LC3 cells seeded onto 35-mm dishes are infected with *Shigella* at 37 °C for 4 h, as described earlier.
2. Cells are fixed with 0.1% glutaraldehyde in 100 mM phosphate buffer, pH 7.4, containing 4% paraformaldehyde for 60 min at room temperature, and embedded in LR-White resin.

3. Ultrathin sections on an EM grid are treated with saturated sodium periodate solution for 10 min at room temperature.
4. After washing with PBS, the samples are treated with 5% BSA in PBS-T (PBS containing 0.02% Tween 20) for 30 min at room temperature.
5. The samples are stained with anti-GFP rabbit polyclonal antibody (MBL, 598) diluted with PBS at 37 °C for 1 h.
6. After washing with PBS-T for 5 min 3 times, the samples are stained with protein A conjugated with gold colloidal particles (10 nm diameter; EY laboratory, GP-01-10) diluted with PBS at 37 °C for 1 h.
7. After washing with PBS-T for 5 min 5 times, the samples are fixed with 2% glutaraldehyde solution for 10 min at room temperature.
8. The samples are washed with water and air-dried.
9. Before observation using TEM, the samples are double-stained with uranyl acetate and lead citrate as described earlier.

3.7. Visualization of *Shigella*-containing autophagosomes using GFP-LC3 and GFP-Atg5

To visualize *Shigella*-containing autophagosomes, MDCK/pGFP-LC3, BHK cells stably expressing GFP-LC3 or GFP-Atg5 (BHK/pGFP-LC3 or BHK/pGFP-Atg5) are infected with *Shigella* and fixed as described earlier. The immunostaining protocol is as follows (Ogawa et al., 2005).

1. Fixed samples are treated with 50 mM NH$_4$Cl in PBS for 10 min at room temperature.
2. The cells are treated with 0.2% Triton X-100 in PBS for 10 min at room temperature.
3. The cells are subsequently treated with 2% BSA in PBS for 30 min at room temperature.
4. The samples are stained with anti-*Shigella flexneri* 2a LPS rabbit polyclonal antibody diluted with TBS (50 mM Tris-HCl, pH 7.4, 150 mM NaCl) at 37 °C for 1 h.
5. After washing with TBS-T (TBS containing 0.05% Triton X-100) for 5 min three times, the samples are stained with TRITC-conjugated anti-rabbit IgG antibody (Sigma, T6778) diluted with TBS at 37 °C for 1 h.
6. After washing with TBS-T for 5 min 3 times, the samples are desalted by washing with distilled water 3 times and mounted with VECTA-SHIELD (Vector Laboratories).
7. The LC3 or Atg5 signals associated with bacteria are detected by visualizing the GFP signals using confocal laser scanning microscopy.

4. *Listeria* Infection and Autophagy

4.1. Overview

Once inside host cells, *L. monocytogenes* immediately lyse the surrounding vacuole membranes by secreting listeriolysin O and escape into the host cytoplasm, where they multiply and spread into adjacent cells by inducing actin polymerization at one pole of the bacterium. Rich *et al.* (2003) report that cytoplasmic *L. monocytogenes* metabolically arrested by chloramphenicol are frequently sequestered by autophagosomes. When *L. monocytogenes* in macrophages are exposed to chloramphenicol, the population of the bacteria residing within autophagosomes is approximately 20% by 3 h, and that eventually increases to 90% by 21 h. Intriguingly, this event is not affected by ActA, which is essential for actin-based bacterial motility, suggesting that bacterial movement does not contribute to escape from autophagy in macrophages. A later study using MEF cells reports that *L. monocytogenes* entrapped by autophagosomes at 1 hour after infection escape into the cytoplasm by secreting listeriolysin O (LLO), a bacterial toxin that disrupts phagocytic membranes (Py *et al.*, 2007). Another recent study using macrophages reports that bacterial phosphatidylinositol-specific- (PI-PLC) and phosphatidylcholine-specific- (PC-PLC) phospholipases are required (Birmingham *et al.*, 2007). These studies have indicated that *de novo Listeria* protein synthesis during infection is involved in evading autophagy.

4.2. Infection protocol for *Listeria monocytogenes* in epithelial cells

1. MDCK/pGFP-LC3 cells are seeded at 4×10^4 cells/mL on glass cover slips in 6-well plates and cultured for 4 days at 37 °C in 5% CO_2.
2. *Listeria* is cultured in BHI broth overnight at 30 °C to stationary phase prior to infection (Pistor *et al.*, 2000).
3. 5 μL of bacterial suspensions are diluted in 1.5 mL of fresh cell culture medium without antibiotics and added to each well in 6-well plates.
4. The plates are centrifuged at $700\times g$ for 10 min at room temperature.
5. Cells are incubated for 1 h at 37 °C in 5% CO_2 to allow invasion.
6. Cells are washed with PBS three times, and DMEM supplemented with 100 μg/mL of gentamicin is added to each well to kill any extracellular bacteria.
7. After incubation for 1–3 h at 37 °C in 5% CO_2, the cells are washed with PBS 3 times and fixed with 4% paraformaldehyde in PBS for 10 min at room temperature.
8. The fixed cells are immunostained as described earlier.

4.3. Infection protocol for *Listeria monocytogenes* in macrophages

4.3.1. Preparation of BMMs (bone marrow-derived macrophages) from mice

1. The femurs and tibias are dissected from GFP-LC3 transgenic mice (7–8 weeks after birth) (Mizushima *et al.*, 2004).
2. The bones are soaked in RPMI 1640 (Gibco)/10% FCS.
3. The muscles are separated from the bones.
4. The bones are soaked with 70% ethanol for 1 min.
5. The soaked bones are immersed with fresh RPMI 1640/10% FCS, and both ends of each bone are cut off.
6. The bone marrow is flushed from each bone using a syringe and a 26G needle and then incubated in 5 mL of fresh RPMI/10% FCS.
7. The bone marrow is combined, and the cells are scattered by pipetting up and down using a Pasteur pipette; a filtered cell suspension is created using 100-μm cell strainers (BD Falcon, 352360).
8. After centrifugation at $280 \times g$ at 4 °C for 5 min, the supernatant fraction is removed and discarded and the cells are suspended in 5 mL of ACS solution (20 mM Tris-HCl, pH7.4, 140 mM NH$_4$Cl) and incubated for 5 min at room temperature. This step causes the lysis of the red blood cells.
9. A 5-mL aliquot of RPMI/10% FCS is added, the sample is centrifuged at $280 \times g$ at 4 °C for 5 min, and the supernatant fraction is removed and discarded.
10. The cells are suspended with 20 mL of conditioned medium (RPMI/10% FCS + 30% L929 sup. +Streptomycin/Penicillin (50 μg/ml and 50 units/ml respectively). L929 sup. is prepared from the culture supernatant of L929 cells. L929 cells are cultured in MEM/10% FCS/non-essential amino acids solution/sodium pyruvate/streptomycin/penicillin at 37 °C for 5 days at $5 \times 10^5 / 20$ mL.
11. The cells are seeded at 4×10^6 in 100-mm culture dishes and cultured for 5 days at 37 °C.
12. The cells are washed with cold PBS and incubated for 20 min at 4 °C to detach the cells.
13. The cells are detached by pipetting and seeded in 6-well plates at 1×10^6/mL $\times$ 2 mL.

4.3.2. Listeria infection protocol in BMM and macrophages

1. BMMs and J774 cells suspended in RPMI1640 supplemented with 10% FCS are seeded at 1×10^5 cells/well in 6-well plates.
2. The cells are infected with *Listeria* at MOI 10.

3. The plates are centrifuged at 700×g for 10 min at room temperature to synchronize the stage of infection.
4. After centrifugation, the plates are incubated at 37 °C for 30 min to allow invasion.
5. The cells are washed with PBS 3 times, fresh RPMI1640 medium containing gentamicin (100 µg/mL) is added to each well, and the cells are incubated for 60 min at 37 °C. After 1≈4 h of incubation, the cells are washed with PBS 3 times and fixed with 4% paraformaldehyde in PBS.

The fixed cells are immunostained, as described earlier.

4.4. Electron microscopy and immunoelectron microscopy

1. MDCK cells seeded in 35-mm dishes are infected with *Listeria* for 2 or 4 h, as described earlier.
2. Cells are fixed with 2% glutaraldehyde in PBS for 1 h at room tempertature, post-fixed in 2% OsO_4, and embedded in epoxy resin.
3. Ultrathin sections are double-stained with uranyl acetate and lead citrate as described previously.
4. The number of bacteria associated with autophagosomes is determined by visual counting under transmission electron microscopy. Approximately 30 independent bacteria are examined in triplicate for each experimental condition.

For immunoelectron microscopy, the postembedding immunogold method is recommended.

1. Infected cells are fixed with 0.1% glutaraldehyde and 4% paraformaldehyde in PBS, and embedded in LR-White resin.
2. Ultrathin sections on an EM grid are treated with 5% BSA in PBS-T for 30 min at room temperature.
3. The samples are stained with rabbit- or mouse-derived primary antibody diluted with PBS at 37 °C for 1 h.
4. After washing with PBS-T for 5 min 3 times, the samples are stained with gold colloidal particles-conjugated antimouse IgG (10 nm diameter; EY Laboratory, GAF-011-10) or gold colloidal particles-conjugated antirabbit IgG (5 nm diameter; EY Laboratory, GAF-012-10) diluted with PBS at 37 °C for 1 h.
5. After washing with PBS-T for 5 min 5 times, the samples are fixed with 2% glutaraldehyde solution for 10 min at room temperature.
6. The samples are washed with water and dried.
7. Before observation by TEM, samples are double-stained with uranyl acetate and lead citrate as described previously.

4.5. Intracellular bacterial multiplication assay

1. MDCK or MEF cells seeded at 2×10^4/well or 5×10^4/well, respectively on 24-well plates are infected with *Listeria* strains at MOI 100.
2. The plates are centrifuged for 10 min at $700 \times g$ at room temperature.
3. The plates are incubated at 37 °C for 60 min to allow bacterial invasion.
4. After washing the infected cells with PBS 3 times, fresh medium supplemented with gentamicin (100 µg/mL) is added to kill extracellular bacteria and the culture is continued at 37 °C for the appropriate periods to allow intracellular bacterial growth.
5. At each time point, the cells are washed with PBS 3 times and lysed with PBS containing 0.5% Triton X-100. The lysates are diluted with PBS and plated onto BHI-agar plates; the number of intracellular bacteria is counted as cfu.

5. Concluding Remarks

As described in this chapter, pathogenic bacteria possess highly evolved mechanisms to circumvent the innate host response and survive in hostile intracellular conditions. Although our understanding of the mechanisms of autophagy activation in response to bacterial infection is still limited, there is no doubt that the bacterial defense system against autophagy is a critical pathogenic feature of invasive and intracellular parasitic bacteria that allows them to successfully colonize infected cells. Identification of the bacterial mechanisms that override autophagy as well as the bacterial components targeted by autophagy will provide clues and tools for investigating the mechanisms of the onset of autophagy and its activation process in mammalian cells.

ACKNOWLEDGMENTS

This work was supported by a Grant-in-Aid for the Scientific Research on Priority Areas, a Grant-in-aid for Scientific Research (C) and a Grant-in-aid for Young Scientists (B) from the Ministry of Education, Culture, Sports, Science and Technology of Japan (MEXT) and the Special Coordination Funds for Promoting Science from Japan Science and Technology Agency (JSTA).

REFERENCES

Amer, A. O., and Swanson, M. S. (2005). Autophagy is an immediate macrophage response to *Legionella pneumophila*. *Cell Microbiol.* **7,** 765–778.

Birkeland, H. C., and Stenmark, H. (2004). Protein targeting to endosomes and phagosomes via FYVE and PX domains. *Curr. Top. Microbiol. Immunol.* **282,** 89–115.

Birmingham, C. L., Canadien, V., Gouin, E., Troy, E. B., Yoshimori, T., Cossart, P., Higgins, D. E., and Brumell, J. H. (2007). *Listeria monocytogenes* evades killing by autophagy during colonization of host cells. *Autophagy* **3,** 442–451.

Cossart, P., and Sansonetti, P. J. (2004). Bacterial invasion: The paradigms of enteroinvasive pathogens. *Science* **304,** 242–248.

Cue, D., Southern, S. O., Southern, P. J., Prabhakar, J., Lorelli, W., Smallheer, J. M., Mousa, S. A., and Cleary, P. P. (2000). A nonpeptide integrin antagonist can inhibit epithelial cell ingestion of *Streptococcus pyogenes* by blocking formation of integrin alpha 5beta 1-fibronectin-M1 protein complexes. *Proc. Natl. Acad. Sci. USA* **97,** 2858–2863.

Cywes, C., and Wessels, M. R. (2001). Group A *Streptococcus* tissue invasion by CD44-mediated cell signalling. *Nature* **414,** 648–652.

Dorn, B. R., Dunn, W. A. Jr., and Progulske-Fox, A. (2002). Bacterial interactions with the autophagic pathway. *Cell Microbiol.* **4,** 1–10.

Gouin, E., Welch, M. D., and Cossart, P. (2005). Actin-based motility of intracellular pathogens. *Curr. Opin. Microbiol.* **8,** 35–45.

Gutierrez, M. G., Master, S. S., Singh, S. B., Taylor, G. A., Colombo, M. I., and Deretic, V. (2004). Autophagy is a defense mechanism inhibiting BCG and *Mycobacterium tuberculosis* survival in infected macrophages. *Cell* **119,** 753–766.

Gutierrez, M. G., Vazquez, C. L., Munafo, D. B., Zoppino, F. C., Beron, W., Rabinovitch, M., and Colombo, M. I. (2005). Autophagy induction favours the generation and maturation of the *Coxiella*-replicative vacuoles. *Cell Microbiol.* **7,** 981–993.

Hara, T., Nakamura, K., Matsui, M., Yamamoto, A., Nakahara, Y., Suzuki-Migishima, R., Yokoyama, M., Mishima, K., Saito, I., Okano, H., and Mizushima, N. (2006). Suppression of basal autophagy in neural cells causes neurodegenerative disease in mice. *Nature* **441,** 885–889.

Jager, S., Bucci, C., Tanida, I., Ueno, T., Kominami, E., Saftig, P., and Eskelinen, E. L. (2004). Role for Rab7 in maturation of late autophagic vacuoles. *J. Cell. Sci.* **117,** 4837–4848.

Kabeya, Y., Mizushima, N., Ueno, T., Yamamoto, A., Kirisako, T., Noda, T., Kominami, E., Ohsumi, Y., and Yoshimori, T. (2000). LC3, a mammalian homologue of yeast Apg8p, is localized in autophagosome membranes after processing. *EMBO J.* **19,** 5720–5728.

Kirkegaard, K., Taylor, M. P., and Jackson, W. T. (2004). Cellular autophagy: Surrender, avoidance and subversion by microorganisms. *Nat. Rev. Microbiol.* **2,** 301–314.

Kuronita, T., Eskelinen, E.-L., Fujita, H., Saftig, P., Himeno, M., and Tanaka, Y. (2002). A role for the lysosomal membrane protein LGP85 in the biogenesis and maintenance of endosomal and lysosomal morphology. *J. Cell Sci.* **115,** 4117–4131.

Levine, B. (2005). Eating oneself and uninvited guests: Autophagy-related pathways in cellular defense. *Cell* **120,** 159–162.

Levine, B., and Klionsky, D. J. (2004). Development by self-digestion: Molecular mechanisms and biological functions of autophagy. *Dev. Cell* **6,** 463–477.

Medina, E., Rohde, M., and Chhatwal, G. S. (2003). Intracellular survival of *Streptococcus pyogenes* in polymorphonuclear cells results in increased bacterial virulence. *Infect. Immun.* **71,** 5376–5380.

Mizushima, N., Yamamoto, A., Matsui, M., Yoshimori, T., and Ohsumi, Y. (2004). In vivo analysis of autophagy in response to nutrient starvation using transgenic mice expressing a fluorescent autophagosome marker. *Mol. Biol. Cell* **15,** 1101–1111.

Nakagawa, I., Amano, A., Mizushima, N., Yamamoto, A., Yamaguchi, H., Kamimoto, T., Nara, A., Funao, J., Nakata, M., Tsuda, K., Hamada, S., and Yoshimori, T. (2004). Autophagy defends cells against invading group A *Streptococcus*. *Science* **306,** 1037–1040.

Nakagawa, I., Nakata, M., Kawabata, S., and Hamada, S. (2001). Cytochrome c-mediated caspase-9 activation triggers apoptosis in *Streptococcus pyogenes*-infected epithelial cells. *Cell Microbiol.* **3,** 395–405.

Ogawa, M., Handa, Y., Ashida, H., Suzuki, M., and Sasakawa, C. (2008). The versatility of *Shigella* effectors. *Nat. Rev. Microbiol.* **6,** 11–16.

Ogawa, M., and Sasakawa, C. (2006a). Bacterial evasion of the autophagic defense system. *Curr. Opin. Microbiol.* **9,** 62–68.

Ogawa, M., and Sasakawa, C. (2006b). Intracellular survival of *Shigella. Cell Microbiol.* **8,** 177–184.

Ogawa, M., Suzuki, T., Tatsuno, I., Abe, H., and Sasakawa, C. (2003). IcsB, secreted via the type III secretion system, is chaperoned by IpgA and required at the post-invasion stage of *Shigella* pathogenicity. *Mol. Microbiol.* **48,** 913–931.

Ogawa, M., Yoshimori, T., Suzuki, T., Sagara, H., Mizushima, N., and Sasakawa, C. (2005). Escape of intracellular *Shigella* from autophagy. *Science* **307,** 727–731.

Pistor, S., Grobe, L., Sechi, A. S., Domann, E., Gerstel, B., Machesky, L. M., Chakraborty, T., and Wehland, J. (2000). Mutations of arginine residues within the 146-KKRRK-150 motif of the ActA protein of *Listeria monocytogenes* abolish intracellular motility by interfering with the recruitment of the Arp2/3 complex. *J. Cell Sci.* **113,** 3277–3287.

Py, B. F., Lipinski, M. M., and Yuan, J. (2007). Autophagy limits *Listeria monocytogenes* intracellular growth in the early phase of primary infection. *Autophagy* **3,** 117–125.

Rich, K. A., Burkett, C., and Webster, P. (2003). Cytoplasmic bacteria can be targets for autophagy. *Cell Microbiol.* **5,** 455–468.

Simpson, J. C., and Jones, A. T. (2005). Early endocytic Rabs: functional prediction to functional characterization. *Biochem. Soc. Symp.* 99–108.

Sinai, A. P., and Joiner, K. A. (1997). Safe haven: The cell biology of nonfusogenic pathogen vacuoles. *Annu. Rev. Microbiol.* **51,** 415–462.

Webster, P. (2006). Cytoplasmic bacteria and the autophagic pathway. *Autophagy* **2,** 159–161.

CHAPTER TWENTY-THREE

Kinetic Analysis of Autophagosome Formation and Turnover in Primary Mouse Macrophages

Michele S. Swanson, Brenda G. Byrne, *and* Jean-Francois Dubuisson

Contents

1. Introduction	384
2. Monitoring Flux Through the Autophagy Pathway	386
3. Assessing the Impact of Infection on the Autophagy Pathway	388
4. LC3 localization in Primary Mouse Macrophages	390
5. Isolation of Bone Marrow-Derived Macrophages	393
5.1. Prepare media	393
5.2. Setup	394
5.3. Dissection to obtain both femurs and tibias	394
5.4. Collect bone marrow cells	395
5.5. Replate macrophages for experiments	395
5.6. Cell densities	396
5.7. Quantification of autophagosome dynamics by immunofluorescence microscopy	396
5.8. Controls	397
5.9. Hank's buffer	398
5.10. Western analysis of lipidated LC3	399
6. Conclusion	400
References	400

Abstract

Macrophages enlist autophagy to combat infection by a variety of bacteria, viruses, and parasites. In response to this selective pressure, some pathogenic microbes have acquired strategies to evade or tolerate autophagy. Accordingly, infected cells may accumulate numerous autophagic vacuoles/autophagosomes when microbial products either stimulate their formation or inhibit

Department of Microbiology and Immunology, University of Michigan Medical School, Ann Arbor, Michigan, USA their maturation. To distinguish between the two mechanisms, we describe methods to assess the impact of infection on the kinetics and amplitude of autophagosome formation and maturation within mouse macrophages by microscopy or Western analysis using antibodies specific for endogenous or recombinant LC3 protein.

1. INTRODUCTION

In addition to its roles in morphogenesis, cellular differentiation, and tissue remodeling, autophagy combats infection (Kirkegaard *et al.*, 2004; Swanson, 2006; Levine and Deretic, 2007). Its contribution to immunity became apparent when, after activation by cytokines, mouse endothelial cells engulfed and digested cytosolic *Rickettsia conorii* within autophagic vacuoles (Walker *et al.*, 1997). Likewise, activated macrophages also degrade vacuolar *M. tuberculosis* and *T. gondii* by autophagy (Gutierrez *et al.*, 2004; Andrade *et al.*, 2006).

The innate immune system has also co-opted this ancient membrane-trafficking pathway. When certain extracellular microbe-associated molecular patterns (MAMPs) are detected by their cognate toll-like Receptors (TLRs; Kawai and Akira, 2006), macrophages induce autophagy (Xu *et al.*, 2007; Sanjuan *et al.*, 2007; Delgado *et al.*, 2008). Autophagy also sequesters and degrades a variety of bacteria that invade the cytoplasm, including some Group A *Streptococcus, Francisella tularensis, Salmonella enterica,* and *Listeria monocytogenes,* which escape into the cytosol (Py *et al.*, 2007; Checroun *et al.*, 2006; Birmingham *et al.*, 2006; Birmingham *et al.*, 2008; Rich *et al.*, 2003). Several pathogens that persist within vacuoles also encounter the autophagy pathway when their specialized secretion systems deliver virulence factors into the cytoplasm. For example, *S. enterica* induces autophagy when its type III secretion system perforates the phagosomal membrane (Birmingham and Brumell, 2006), whereas *L. pneumophila, B. abortus,* and likely *C. burnetii* each need type IV secretion for sustained interactions between their phagosomes and the autophagic machinery (Swanson and Isberg, 1995; Celli *et al.*, 2003; Amer *et al.*, 2005; Romano *et al.*, 2006). In each case, mutant bacteria that fail to breach their phagosomal membranes do not activate the autophagy machinery. Accordingly, autophagy also appears to be activated by a host surveillance system that monitors the cytosol. Indeed, both human and mouse studies implicate nucleotide oligomerization domain-like receptors (NLRs; Wilmanski *et al.*, 2008) as regulators of autophagy. For example, the amount of the mouse NLR protein Naip5 correlates not only with robust proinflammatory death but also with the speed of autophagosome maturation and delivery of *L. pneumophila* to lysosomes (Amer *et al.*, 2005). The mouse macrophage response to *S. flexneri* also reveals a regulatory link between pyroptosis and autophagy (Suzuki *et al.*, 2007).

Thus, host cells may elevate autophagy as a barrier to infection when pathogens either invade the cytosol or deposit in the cytoplasm virulence factors tainted with MAMPs (Swanson and Molofsky, 2005; Swanson, 2006).

Genetic studies of a human inflammatory bowel disease also point to a role for NLR proteins as critical regulators of autophagy that modulate the inflammatory response to microbial products. Humans with Crohn's disease suffer chronic inflammation, apparently due to a dysregulated response to the gastrointestinal tract flora. For some patients, the disease is conferred by a mutant NLR protein, NOD2 (Hugot et al., 2001; Ogura et al., 2001); others encode a variant of the autophagy gene Atg16L1 (Parkes et al., 2007; Rioux et al., 2007). Accordingly, it is urgent that methods be developed to allow researchers to exploit both mouse and bacterial genetics to identify the host and microbial factors that equip macrophages to recruit the autophagy pathway as an arm of the acquired and innate immune systems.

The observation that extracellular or intracellular microbes or MAMPs can trigger accumulation of large numbers of autophagosomes in macrophages can be interpreted in two different ways: Infection either stimulates autophagosome formation or inhibits autophagosome maturation (Fig. 23.1). To

1) Low rate of constitutive autophagy

2) Stimulated autophagosome formation
Microbe

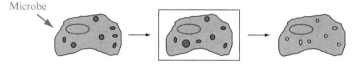

3) Inhibited autophagosome maturation
Microbe

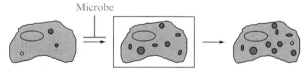

Figure 23.1 The number of autophagosomes in a cell is determined by the rates of both formation and maturation. When the constitutive rate of autophagy (1, box) is stimulated, the number of LC3-GFP$^+$ vacuoles in cells increases (2, box). When the rate of autophagosome maturation is slower than normal, the number of vacuoles also increases (3, box). Whether a bacterial or host factor stimulates or inhibits the autophagy machinery can be deduced from time-course experiments performed in the absence or presence of reagents known either to stimulate autophagosome formation or to inhibit their progression to autolysosomes.

distinguish between the possibilities, it is necessary to measure flux through the pathway (Klionsky et al., 2008; Mizushima and Yoshimori, 2007; Tanida et al., 2005). However, these experiments present technical challenges: in primary macrophages, autophagosomes are short lived, and reagents to label these organelles are limited. To facilitate kinetic and quantitative studies of the impact of infection on the autophagy pathway of mouse macrophages, we have developed a sensitive and specific assay that exploits LC3 to analyze the maturation of a large, synchronous population of preformed autophagosomes.

2. Monitoring Flux Through the Autophagy Pathway

To induce autophagosome formation, primary macrophages derived from the bone marrow of mice are incubated in Hank's buffer, which lacks amino acids. After incubation for 5–55 min, the macrophages are fixed, autophagosomes are stained fluorescently, and then the number of autophagosomes in each cell is quantified. Under these conditions, within 25 min the majority of C57B1/6 mouse macrophages contain multiple large autophagosomes (Fig. 23.2), organelles easily identified by their decoration with LC3, a mammalian homolog of the yeast autophagy-related (Atg) protein Atg8 (Yamamoto et al., 2001; Kabeya et al., 2000) and the most stable and specific component of autophagosomes identified to date (Klionsky et al., 2008).

In macrophages, autophagosomes rapidly merge with lysosomes, wherein both the contents and the LC3 protein are degraded. To determine how quickly the LC3 autophagosomal marker is extinguished within lysosomes in primary mouse macrophages, cells are subjected to starvation for 30 min and then transferred to rich medium. Within 30 min, the population of LC3 autophagosomes is diminished to baseline levels (Fig. 23.3).

To verify that fusion of autophagosomes with lysosomes accounts for the extinction of the GFP-LC3 signal, vacuole maturation is inhibited pharmacologically (Tanida et al., 2005; Mizushima and Yoshimori, 2007). After stimulating autophagy by starving macrophages for 40 min, the pH of the endosomal compartment is neutralized by incubating the cells for another 20 min with the proton pump inhibitor bafilomycin A_1 (Yamamoto et al., 1998). Indeed, when progression to lysosomes is blocked, the fraction of cells that contain >5 GFP-LC3$^+$ vacuoles increases from 19 to 59% (Fig. 23.4).

To verify results obtained by microscopy, the kinetics of activation and turnover of LC3 is assessed by Western analysis (Mizushima and Yoshimori, 2007; Tanida et al., 2005). When activated, LC3 is covalently linked directly to phosphatidylethanolamine, a modification that speeds migration of the enzyme during SDS-PAGE (Tanida et al., 2005). Accordingly, the kinetics and amplitude of LC3 activation can be assessed by Western analysis of lysates

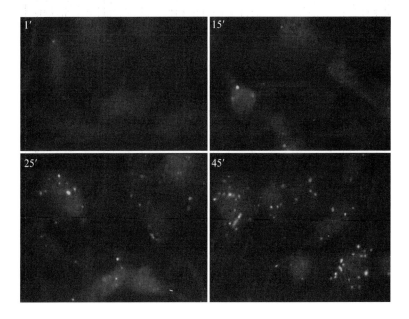

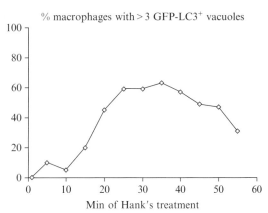

Figure 23.2 Primary mouse macrophages respond rapidly to autophagy stimulants. Macrophages derived from the bone marrow of GFP-LC3 transgenic C57BL/6 mice were incubated with Hank's buffer for the periods shown, fixed with PLP-sucrose, stained with GFP-specific primary antibody (1:250 dilution; Roche, #11814460001) and Oregon Green-conjugated secondary antibody (1:1000 dilution; Invitrogen, #O-6383), and viewed by fluorescence microscopy (A), and then the fraction of macrophages containing >3 GFP-LC3$^+$ vacuoles was scored (B).

obtained from macrophages subjected to a pulse of starvation. Control macrophages contain primarily nonlipidated LC3, whereas starved cells also contain the lipidated protein (LC3-II), the form that is preferentially recognized by an

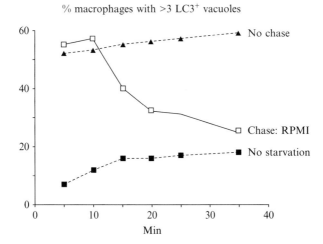

Figure 23.3 Autophagosomes mature quickly in primary mouse macrophages. To generate autophagosomes, C57B1/6 GFP-LC3 transgenic macrophages were subjected to amino acid starvation by incubating for 30 min with Hank's buffer. To measure the rate of autophagosome maturation, cells were then fed (chase: RPMI). To determine the maximal and minimal autophagy response, macrophages were continuously starved (no chase), or fed (no starvation), respectively. At the times shown, the fraction of macrophages containing >3 GFP-LC3$^+$ vacuoles was quantified by microscopy.

antibody available from Novus Biologicals (Fig. 23.5). If the rapid loss of GFP-LC3 antigenicity detected by fluorescence microscopy reflects robust fusion with lysosomes, expect to observe by Western analysis a progression from nonlipidated LC3 to lipidated LC3 (LC3-II), to a decline of lipidated protein (Mizushima and Yoshimori, 2007). In addition, expect treatment of cells with pharmacological inhibitors of either the proton pump or lysosomal proteases to stabilize the lipidated LC3 (LC3-II; Mizushima and Yoshimori, 2007; see also the chapter by Kimura *et al.*, in this volume).

3. Assessing the Impact of Infection on the Autophagy Pathway

To test whether a particular microbe stimulates flux through the autophagy pathway, starved macrophages are transferred to rich medium that contains or lacks the microbe of interest, in our case *L. pneumophila*. To gauge the specificity and sensitivity of these assays, we assess in parallel the impact of virulent and mutant bacteria on autophagosome maturation. For example, macrophages are subjected to starvation by incubating with

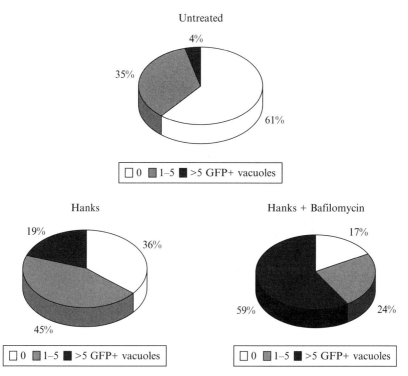

Figure 23.4 The GFP-LC3 signal persists when macrophage proton pumps are inhibited with bafilomycin A$_1$, which retards the progression of autophagosomes to autolysosomes. To stimulate autophagosome formation, macrophages derived from the bone marrow of C57BL/6 mice were incubated in Hank's buffer for 40 min. To measure the impact of vacuole acidification on autophagosome maturation, the macrophages were incubated for an additional 20 min in the presence or absence of bafilomycin A$_1$. The fraction of macrophages that contained 0 (white), 1–5 (gray), or >5 (black) autophagic vacuoles/autophagosomes was quantified by fluorescence microscopy after amplifying the GFP-LC3 signal with GFP-specific antibody (Roche).

Hank's buffer for 20 min, and then the cells are incubated for 5–40 min with either wild-type *L. pneumophila* or *dotA* mutant bacteria, which are defective for type IV secretion. Even within 5 min of infection, it is evident that *L. pneumophila* stimulates autophagosome maturation, as the number of cells that contain GFP-LC3$^+$ vacuoles declines at a rate markedly faster than that observed for uninfected macrophages (data not shown). After 40 min of infection with virulent bacteria, macrophages are devoid of autophagosomes, as judged by their loss of GFP-LC3 antigenicity (Fig. 23.6). Moreover, rapid turnover of the autophagosomes is triggered by either the activity or a substrate(s) of the type IV secretion apparatus, as macrophages retain numerous autophagosomes, similar to uninfected macrophages, when they harbor motile bacteria that are defective in type IV secretion

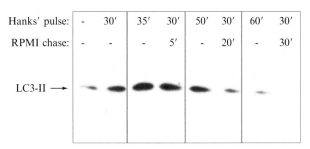

Figure 23.5 Kinetic analysis of lipidation and turnover of endogenous LC3 protein by mouse macrophages using a commercial antibody. To induce a synchronous pulse of autophagy, C57Bl/6 mouse macrophages were incubated in rich medium (RPMI) or amino acid-free Hank's buffer (Hank's) for the periods shown before transfer to rich medium for the times indicated. Lysates of 10^6 cells were separated by SDS-PAGE on a 4%–20% gradient acrylamide gel, transferred to membrane, probed with rabbit polyclonal anti-LC3 (Novus Biologicals, #NB100-2331) and an HRP-conjugated secondary antibody (Santa Cruz Biotechnology, sc-2030), and then developed to visualize the activated LC3-II protein. Size markers indicate that LC3-II is approximately 14 kDa, and a nonspecific band of approximately 50 kDa indicates that each lane contains a similar quantity of protein (not shown).

(*dotA* mutants). Thus, in response to a microbe equipped to pierce its phagosome and deliver microbial products to the cytosol, macrophages derived from the bone marrow of C57BL/6 mice stimulate rapid autophagosome maturation.

4. LC3 LOCALIZATION IN PRIMARY MOUSE MACROPHAGES

In primary mouse macrophages, LC3 can be localized by one of four methods. Mizushima and colleagues have generated and characterized a transgenic line of C57BL/6 mice that encode recombinant GFP-LC3 (Mizushima *et al.*, 2004; see also the chapter by N. Mizushima in this volume). By using macrophages derived from the bone marrow of this mouse strain, investigators are assured that every cell expresses similar amounts of GFP-LC3 while avoiding artifactual LC3 aggregation due to overexpression (Klionsky *et al.*, 2008). With our equipment, it is beneficial to amplify the signal using a GFP-specific primary antibody (Roche, #11814460001) and a fluorophore-conjugated secondary antibody (Invitrogen, Oregon Green #O-6383).

As an alternative to monitoring recombinant transgenic GFP-LC3, a polyclonal antibody that is specific to LC3 and suitable for detecting the endogenous protein in mouse macrophages by immunofluorescence

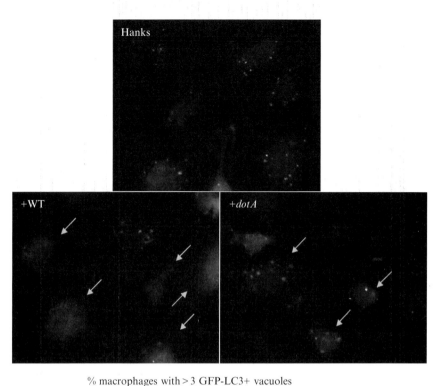

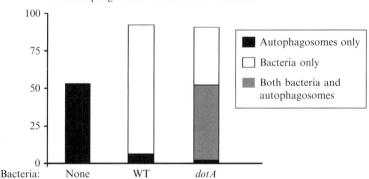

Figure 23.6 Type IV secretion-competent *Legionella* stimulate autophagosome maturation. GFP-LC3 transgenic macrophages were treated for 20 min with Hank's buffer, cultured for 40 min with wild-type or mutant bacteria as shown, fixed with PLP-sucrose, stained with GFP-specific primary (1:250 dilution; Roche) and Oregon Green-conjugated secondary antibody (1:1000 dilution; Molecular Probes), and viewed by fluorescence microscopy (A). Arrows indicate infected cells. The fraction of macrophages that contained >3 GFP-LC3$^+$ vacuoles (black), bacteria only (white), or both (gray) was scored (B).

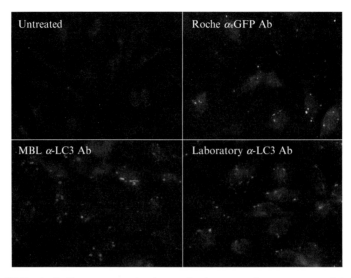

Figure 23.7 A similar response is observed when autophagosomes are labeled via endogenous LC3 or the GFP-LC3 transgene. Macrophages derived from the bone marrow of commercial (A, C, D) or transgenic GFP-LC3 (B) C57Bl/6 mice were incubated for 25 min in rich medium (A, RPMI) or Hank's buffer (B-D), then autophagosomes were visualized using antibody specific to GFP (1:250 dilution, Roche; B) or an LC3 antibody prepared commercially (1:150 dilution, MBL; C), or by a laboratory protocol (Taylor and Kirkegaard, 2007; D).

microscopy is currently available from MBL International Corporation (#M152-3; Fig. 23.7). To analyze the impact of particular mouse mutations on macrophage autophagy, this is currently the best option.

The Kirkegaard laboratory has also published a protocol to generate polyclonal antibodies against recombinant human LC3 that cross-reacts with the mouse protein (Taylor and Kirkegaard, 2007). Antibody quality is evaluated by comparing sera obtained from multiple rabbits before and after immunization using Western analysis of macrophage lysates as the initial screen, and localization of LC3 in fed and starved macrophages as the secondary screen. Note that some antibody preparations are suitable for Western analysis but not immunofluorescence microscopy. Using antibody prepared by this protocol, endogenous LC3 protein is readily detected on autophagosomes formed when primary C57Bl/6 macrophages are subjected to amino acid starvation (Fig. 23.7.)

Another general strategy is to transduce mammalian cells with recombinant LC3 protein. For this purpose, a newly developed specific probe, tandem-tagged fluorescent LC3 (tfILC3; Kimura *et al.*, 2007; see also the chapter by Kimura *et al.*, in this volume) offers an important advantage: the ability to monitor progression of autophagosomes to autolysosomes. Recombinant mRFP-GFP-LC3 protein exploits the observation that

GFP is degraded and therefore quenched in acidic lysosomes, whereas RFP is relatively stable (Mizushima and Yoshimori, 2007). Therefore, in mammalian cells, immature autophagosomes are visible in both the red and green channels, but acidic autolysosomes fluoresce only red. The primary drawbacks of using recombinant proteins to analyze autophagy are the time and expense required to introduce the transgene either into the cells or the animals of each genotype of interest. In transfection experiments, cell-to-cell variability in expression and artifactual patterns due to aggregation of the recombinant protein can also be problematic (Klionsky *et al.*, 2008).

5. Isolation of Bone Marrow-Derived Macrophages

This method is adapted from Celada *et al.* (1984) and typically yields after 1 week in culture 8×10^7 macrophages per mouse.

5.1. Prepare media

L-cell supernatant: L-cells are a mouse fibroblast cell line that secretes Macrophage Colony Stimulating Factor (M-CSF). Culture L-cells in RPMI + 10% FBS-heat inactivated + pen/strep. A subconfluent culture of L-cells can be used to seed flasks; alternatively, thaw a fresh aliquot of cells stored in liquid nitrogen for each supernatant preparation. The shelf life at $-70\,^{\circ}\text{C}$ is at least 6 months; at $-20\,^{\circ}\text{C}$, 2–3 months; at $4\,^{\circ}\text{C}$, only 2–3 weeks.

1. Transfer from liquid nitrogen storage an aliquot of L-cells to a 37 °C bath, then resuspend the thawed cells in 9 ml of 37 °C medium.
2. Collect the cells by centrifugation at $250 \times g$ for 10 min.
3. Aspirate the supernatant fraction and resuspend the cell pellet in 6 ml of medium.
4. Incubate in a 25-cm² flask overnight at 37 °C and 5% CO_2.
5. Wash away debris with warm medium and replace with 6 ml of fresh medium.
6. When confluent, trypsinize the monolayer: Remove the medium, add 5 mL of 0.05% Trypsin-EDTA (Gibco #25300) and let stand at RT for 5 min. Tap the flask and pipette up and down gently a few times to release any remaining attached cells. Transfer the cell suspension to a 15-mL conical tube and collect by centrifugation at $250 \times g$ for 10 min at 4 °C. Remove the supernatant fraction and resuspend the cell pellet in 8 mL of RPMI containing 10% FBS. Count the cells using a hemacytometer.
7. Replate 3×10^5 cells in 60 ml of medium per 150-cm² flask.
8. Incubate until there is a confluent monolayer (approximately 5 days).

9. Collect the supernatant: Pour the conditioned medium through a bottle-top 0.22-μ filter into sterile bottles.
10. Freeze aliquots at $-20\ °C$, then store at $-70\ °C$.

Bone marrow macrophage culture medium: For each mouse, prepare 100 ml in RPMI.

30% L-cell supernatant
1% Penicillin-Streptomycin (pen/strep; Gibco, #15140)
20% FBS, heat-inactivated (Gibco, #10437-028; To heat inactivate complement proteins, thaw the bottle of FBS overnight at 4 °C, heat to 56 °C for 30 min, and store aliquots at $-20\ °C$)

5.2. Setup

1. Chill to 4 °C:
   - 10-ml syringe equipped with 25-gauge needle and filled with 10 ml of RPMI
   - 2 15-ml tubes
   - Centrifuge
2. Have ready sterile:
   - Forceps
   - Scissors
   - Scalpel
   - 100-mm Petri plates

5.3. Dissection to obtain both femurs and tibias

1. Sacrifice 1 mouse.
2. Pin the mouse spread-eagle with the abdomen facing up on the dissecting tray.
3. Sterilize the mouse abdomen with 70% ethanol.
4. Using forceps, lift the skin at the hip joint away from the animal, then snip with scissors.
5. Peel the skin back to expose the quadriceps muscle from the hip to the knee joint.
6. Using the forceps and scalpel, remove each leg above the hip.
7. Using the forceps to grasp the quadriceps muscle, detach to expose each femur and tibia.
8. Using the forceps, detach each femur by cutting with the scalpel above the proximal ball joint and below the distal ball joint. Repeat for the second femur.
9. Using the scalpel, scrape the connective tissue from the femurs and tibias.
10. Wash the femurs and tibias with 70% ethanol.

5.4. Collect bone marrow cells

1. Move the femurs and tibias, the sterilized scissors and the forceps into a sterile hood.
2. Cut both ball joints from one femur.
3. Insert the needle into the hollow of the femur and pulse with 2.5 ml of medium, flushing the marrow cells into the tube on ice. Repeat from the other end of the femur.
4. Repeat with the other femur and both tibias.
5. Release cells from the marrow by gentle pipetting with a 10-ml pipette and by inverting the closed tube several times.
6. Allow large flocculent material to settle to the bottom of the tube by standing on ice for 5 min.
7. Place $3–4 \times 10^6$ cells in approximately 15 ml of bone marrow macrophage medium in each 100-mm Petri dish. Typically, if the cell harvest went smoothly, we divide the cells evenly among 5 dishes. Do not use tissue-culture grade plastic, as the tightly adherent mature macrophages will be difficult to dislodge. Instead, we use deep Petri dishes, 100×25 mm (Nunc, Lab-Tek #4031).
8. On day 3, feed the cells by adding approximately 5 ml of bone marrow macrophage medium.

5.5. Replate macrophages for experiments

On day 6 or 7, replate adherent cells to tissue culture wells or coverslips. Because primary macrophages readily adhere to tubes and pipettes, best yields are obtained when reagents and tissue culture supplies are ice cold.

1. Aspirate the supernatant and add approximately 10 ml of ice-cold PBS that lacks divalent cations (Gibco, #10010) to inhibit adherence mediated by integrin receptors.
2. Let dish stand on ice for approximately 5 min to promote cell detachment.
3. Using gentle pipetting, flush the monolayer from the dish.
4. Transfer the suspension to an ice-cold 50-ml conical tube.
5. Wash the dish with an additional 5–10 ml of ice-cold PBS.
6. Collect the cells by centrifugation at $400 \times g$ for 10 min at 4 °C, and discard the supernatant fraction by decanting or aspiration.
7. Resuspend the cells in 10 ml of 10% FBS in RPMI (Gibco, #11875) and count using a hemacytometer, expecting $12–18 \times 10^6$ macrophages per dish.
8. Adjust to the desired concentration in 10% FBS in RPMI (see subsequently) and plate for experiments. For studies of the impact of bacteria on macrophage biology, omit antibiotics.

Cells attach and spread within a 30-min incubation at 37 °C and 5% CO_2. In our experience, cells can be used after approximately 90 min or within the next 4–5 days (the period varies with mouse genotype). With time, the cells grow larger and more thinly spread.

5.6. Cell densities

- 2.5×10^5 macrophages per 0.5 ml per well of a 24-well plate will form a confluent monolayer for bacterial growth curve studies.
- 10^6 macrophages per 35 mm well of a 6-well plate for Western analysis.
- 7.5×10^4 to 10^5 macrophages per glass coverslip (#1 thickness, Fisher Scientific, #12-545-80) for microscopy.

5.7. Quantification of autophagosome dynamics by immunofluorescence microscopy

1. Culture 7.5×10^4 to 10^5 macrophages per 0.5 ml of RPMI + 10% FBS per glass coverslip (#1 thickness, Fisher Scientific, #12-545-80) in 24-well plates.
2. To induce a pulse of autophagy, aspirate the medium and replace with 0.5 ml of Hank's buffer (see subsequently) prewarmed to 37 °C and incubate at 37 °C for the desired period. (For bone marrow–derived macrophages from C57B1/6 mice, the maximum number of autophagosomes is observed after approximately 25 min of incubation with Hank's buffer; Fig. 23.2.)
3. Option A: To allow the population of autophagosomes to mature, aspirate Hank's buffer and replace with RPMI/FBS prewarmed to 37 °C. (For bone marrow–derived macrophages from C57B1/6 mice, the number of autophagosomes returns to baseline after approximately 20 min incubation with RPMI/FBS; Fig. 23.3.)
   Option B: To investigate the impact of microbes or microbial products on autophagy, replace Hank's buffer with RPMI/FBS that contains or lacks the microbe or product of interest at an optimal concentration, determined empirically, and then incubate for the desired period(s), also determined empirically (e.g., Fig. 23.6).
4. Replace Hank's buffer or RPMI/FBS with freshly prepared modified PLP-sucrose (see subsequently) prewarmed to 37 °C for 30 min at room temperature.
5. Wash 3 times with PBS.
6. Methanol extract to permeabilize the membranes: Using a fine-tip forceps, gently swish each coverslip for 5 s in -20 °C methanol held on ice.
7. Rinse 3 times with PBS.

8. Block 5 min at RT with PBS + 2% heat-inactivated goat serum (PBS/GS; Gibco #16210-064). To heat-inactivate complement components in the serum, heat in a water bath to 56 °C for 30 min; store aliquots at −20 °C.
9. Incubate overnight at 4 °C with the appropriate dilution of primary antibody in PBS/GS. To minimize the volume of antibody needed, prepare an incubation chamber: invert a 24-well tissue culture plate, stretch Parafilm across the solid bottom surface, and place 25–50 µL drops of diluted antibody onto the Parafilm, using the well outlines as a guide. Using a fine-tip forceps, remove the coverslip from the well and invert onto a drop of antibody; repeat for all the samples. Cover the coverslips using the top of the 24-well culture plate. At the end of the overnight incubation, transfer each coverslip, right side up, back into the original 24-well plate for washes.
10. Wash 3 times for 5 min with PBS/GS.
11. Incubate 1 h at 37 °C in a humidified incubator with 25–200 µl of the appropriate dilution of secondary antibody in PBS/GS.
12. Wash 3 times for 5 min with PBS.
13. Remove each coverslip from its well, gently wipe the PBS off the back of the coverslip, then place the coverslip cell-side-down onto a tiny drop of mounting medium (Prolong Gold with DAPI, Invitrogen/Molecular Probes, #P369935) on a plain glass microscope slide (25 × 75 × 1 mm). Aspirate excess mounting medium that seeps out from the coverslip. Let dry overnight at RT protected from light.
14. Add a tiny drop of immersion oil and observe fluorescence using the appropriate filter sets.

5.8. Controls

1. To gauge the background level of autophagy, fix one untreated sample.
2. To gauge nonspecific staining by the secondary antibody, omit the primary antibody from one sample.
3. To determine the appropriate dilution of the primary antibody, analyze a series of dilutions (e.g., 1:50, 1:100, 1:250) using a constant secondary antibody concentration. Choose the dilution that yields the optimal signal:noise ratio.
4. To determine the appropriate dilution of secondary antibody, analyze a series of dilutions (e.g., 1:100, 1:200, 1:400) using a constant primary antibody concentration and a control lacking primary antibody. Choose the dilution that yields the optimal signal:noise ratio.
5. As a guard against physical damage or loss during handling, use duplicate coverslips for each sample.
6. To determine the specificity of the response to the microbial product of interest, analyze the appropriate controls in parallel (e.g., isogenic

mutant that lacks product, mutant protein that lacks activity, formalin-killed microbes, denatured wild-type protein, or the like).

5.9. Hank's buffer

To induce autophagy, macrophages are transferred to Hank's Balanced Salt Solution (Gibco, #14025), which lacks amino acids.

Note: Be certain the buffer contains calcium, which is essential for a macrophage autophagy response.

5.9.1. Periodate-lysine-paraformaldehyde fixative (PLP) + 5% sucrose

This fixative has been modified from McLean and Nakane (1974) to include sucrose as an osmotic support for fragile membrane networks, including the endosomal compartment and the endoplasmic reticulum. The final solution contains 10 mM $NaIO_4$, 75 mM lysine, 37.5 mM $NaPO_4$, 2% paraformaldehyde, and 4.5% sucrose.

40 mls Stock A

20 ml 0.2 M Lysine
15.5 ml 0.1 M Na_2HPO_4
4.5 ml 40% sucrose
Store at 4 °C.

8% Paraformaldehyde

1. Weigh out 1–2 g of paraformaldehyde (toxic: wear gloves, use mask, weigh in chemical fume hood).
2. Add water to bring up to 8%.
3. Heat with stirring in a fume hood to approximately 70 °C.
4. Add a drop of 1 N NaOH; the solution should clear.
5. When cool enough to handle, filter to remove particles.
6. Store at 4 °C and protected from light by wrapping in foil.
7. For best results, use on the day of preparation. Alternatively, store up to approximately 1 month at −20 °C as 2.5-ml aliquots in 15-ml amber disposable tubes; thaw and prewarm to 37 °C until the solution clears before use.

5.9.2. 10 ml PLP-sucrose fixative

Dissolve 21.4 mg of sodium periodate (toxic: wear gloves) in 7.5 ml of Stock A and 2.5 ml of 8% paraformaldehyde. For best results (i.e., preservation of endoplasmic reticulum or endosomal network) make fresh and warm to 37 °C before use.

5.10. Western analysis of lipidated LC3

In our experience, the limit of detection for endogenous LC3 protein per lane requires a protein extract from 10^6 bone marrow–derived mouse macrophages. The LC3 protein is also labile: even during storage of boiled protein samples overnight at $-20\,°C$, the signal deteriorates.

1. Culture 10^6 macrophages per well of a 6-well tissue culture plate (e.g., Costar #3516).
2. To assess activation of autophagy, aspirate the culture medium, replace with Hank's buffer (Gibco, #14025), and incubate in a humidified incubator at 37 °C and 5% CO_2 for the desired periods. (For C57Bl/6 macrophages, we observe peak LC3-II approximately 35 min after exposure to Hanks's buffer; Fig. 23.5.)
3. To assess maturation of autophagosomes, treat macrophages with a pulse of Hank's buffer for the period determined empirically to generate maximal levels of the LC3-II protein, then replace Hanks's buffer with prewarmed RPMI + 10% FBS for the desired periods. (For C57B1/6 macrophages, we observe that LC3-II has diminished to baseline levels approximately 20 min after a return to rich medium; Fig. 23.5.)
4. Rinse monolayers twice with 4 °C PBS, then add 1 ml of cold PBS/well.
5. Dislodge the cells using a Disposable Cell Scraper (Fisher, #08-773-2) and transfer the cell suspension to an ice-cold microcentrifuge tube.
6. Repeat for each sample, keeping all of the cell suspensions on ice.
7. Collect the cells by centrifugation for 10 min at 5000 rpm in a microcentrifuge.
8. Aspirate the supernatant fraction and resuspend the cell pellet in 15 μl of ice-cold PBS.
9. Lyse the cells by addition of 15 μl of 2X Laemmli sample buffer and boil in a water bath for 5 min.
10. Clarify the sample by centrifugation for 5 min at 10,000 rpm in a microcentrifuge.
11. Load 30 μl of each sample (equivalent to 10^6 macrophages) per lane of either a 4%–20% linear gradient SDS-polyacrylamide gel (Fig. 23.5) or a 15% SDS-polyacrylamide gel, and separate by electrophoresis using a constant 70 V.
12. Using a semidry blotting system, transfer to PVDF membranes (Bio-Rad) by applying 15 V for 45 min.
13. Block the membranes by incubating overnight at 4 °C in 5% dry milk and 0.1% Tween 20 in Tris-buffered saline (TBS, Bio-Rad, #170-6435).
14. To detect endogenous protein, incubate with LC3-specific rabbit antibody (1:500 dilution; Novus Biological, #NB 100-2331) for 1 h at RT with agitation on a platform shaker. Alternatively, to detect

recombinant GFP-LC3 protein, incubate with GFP-specific mouse antibody (Roche, #11814460001; 1:1000 dilution).
15. Wash with TBS + 0.1% Tween 20, 5 times, 5 min per wash, at RT.
16. Incubate with HRP-conjugated anti-Rabbit IgG secondary antibody (1:3000 dilution; Santa Cruz Biotechnology, sc-2030). Alternatively, to detect recombinant GFP-LC3 protein, incubate with peroxidase-conjugated mouse IgG-specific secondary antibody (Sigma, A9917, 1:4,000) for 1 h at RT with agitation on a platform shaker.
17. Wash with TBS + 0.1% Tween 20, 5 times, 5 min per wash, at RT with agitation on a platform shaker.
18. Detect specific protein using the enhanced chemiluminescence protocol described by the manufacturer (Pierce; Supersignal West Pico).

6. Conclusion

Our autophagy assay offers several technical advantages. By first subjecting macrophages to amino acid starvation, we generate a large and synchronous population of autophagosomes in the absence of confounding MAMPs, ubiquitous microbial products likely to affect the pathway (Delgado et al., 2008; Sanjuan et al., 2007; Xu et al., 2007). By restricting our infection and assay periods to <1 h, we limit the impact of developmental changes by either the macrophages or the microbes. Rapid effects of bacterial factors on turnover of GFP-LC3$^+$ vacuoles can be quantified within single cells by microscopy. Alternatively, autophagosome formation and flux can be assessed by Western analysis. Thus, using the protocols provided here and reagents now available, it is possible to analyze the impact of microbes or their products on the amplitude and kinetics of autophagosome formation and maturation within primary macrophages derived from the bone marrow of wild-type or mutant mice.

REFERENCES

Amer, A. O., Byrne, B. G., and Swanson, M. S. (2005). Macrophages rapidly transfer pathogens from lipid raft vacuoles to autophagosomes. *Autophagy* **1**, 53–58.

Andrade, R. M., Wessendarp, M., Gubbels, M. J., Striepen, B., and Subauste, C. S. (2006). CD40 induces macrophage anti-*Toxoplasma gondii* activity by triggering autophagy-dependent fusion of pathogen-containing vacuoles and lysosomes. *J. Clin. Invest.* **116**, 2366–2377.

Birmingham, C. L., and Brumell, J. H. (2006). Autophagy recognizes intracellular *Salmonella enterica* serovar *Typhimurium* in damaged vacuoles. *Autophagy* **2**, 156–158.

Birmingham, C. L., Smith, A. C., Bakowski, M. A., Yoshimori, T., and Brumell, J. H. (2006). Autophagy controls *Salmonella* infection in response to damage to the *Salmonella*-containing vacuole. *J. Biol. Chem.* **281,** 11374–11383.

Birmingham, C. L., Canadien, V., Kaniuk, N. A., Steinberg, B. E., Higgins, D. E., and Brumell, J. H. (2008). Listeriolysin O allows *Listeria monocytogenes* replication in macrophage vacuoles. *Nature* **451,** 350–354.

Celada, A., Gray, P. W., Rinderknecht, E., and Schreiber, R. D. (1984). Evidence for a gamma-interferon receptor that regulates macrophage tumoricidal activity. *J. Exp. Med.* **160,** 55–74.

Celli, J., de Chastellier, C., Franchini, D.-M., Pizarro-Cerda, J., Moreno, E., and Gorvel, J.-P. (2003). *Brucella* evades macrophage killing via VirB-dependent sustained interactions with the endoplasmic reticulum. *J. Exp. Med.* **198,** 545–556.

Checroun, C., Wehrly, T. D., Fischer, E. R., Hayes, S. F., and Celli, J. (2006). Autophagy-mediated reentry of *Francisella tularensis* into the endocytic compartment after cytoplasmic replication. *Proc. Natl. Acad. Sci. USA* **103,** 14578–14583.

Delgado, M. A., Elmaoued, R. A., Davis, A. S., Kyei, G., and Deretic, V. (2008). Toll-like receptors control autophagy. *EMBO J.* **27,** 1110–1121.

Gutierrez, M. G., Master, S. S., Singh, S. B., Taylor, G. A., Colombo, M. I., and Deretic, V. (2004). Autophagy is a defense mechanism inhibiting BCG and *Mycobacterium tuberculosis* survival in infected macrophages. *Cell* **119,** 753–766.

Hugot, J. P., Chamaillard, M., Zouali, H., Lesage, S., Cezard, J. P., Belaiche, J., et al. (2001). Association of NOD2 leucine-rich repeat variants with susceptibility to Crohn's disease. *Nature* **411,** 599–603.

Kabeya, Y., Mizushima, N., Ueno, T., Yamamoto, A., Kirisako, T., Noda, T., et al. (2000). LC3, a mammalian homologue of yeast Apg8p, is localized in autophagosome membranes after processing. *EMBO J.* **19,** 5720–5728.

Kimura, S., Noda, T., and Yoshimori, T. (2007). Dissection of the autophagosome maturation process by a novel reporter protein, tandem fluorescent-tagged LC3. *Autophagy* **3,** 452–460.

Kirkegaard, K., Taylor, M. P., and Jackson, W. T. (2004). Cellular autophagy: Surrender, avoidance and subversion by microorganisms. *Nat. Rev. Microbiol.* **2,** 301–314.

Klionsky, D. J., Abeliovich, H., Agostinis, P., Agrawal, D. K., Aliev, G., Askew, D. S., et al. (2008). Guidelines for the use and interpretation of assays for monitoring autophagy in higher eukaryotes. *Autophagy* **4,** 151–175.

Levine, B., and Deretic, V. (2007). Unveiling the roles of autophagy in innate and adaptive immunity. *Nat. Rev. Immunol.* **7,** 767–777.

McLean, I. W., and Nakane, P. K. (1974). Periodate-lysine-paraformaldehyde fixative. A new fixative for immunoelectron microscopy. *J. Histochem. Cytochem.* **22,** 1077–1083.

Mizushima, N., and Yoshimori, T. (2007). How to interpret LC3 immunoblotting. *Autophagy* **3,** 542–545.

Mizushima, N., Yamamoto, A., Matsui, M., Yoshimori, T., and Ohsumi, Y. (2004). In vivo analysis of autophagy in response to nutrient starvation using transgenic mice expressing a fluorescent autophagosome marker. *Mol. Biol. Cell* **15,** 1101–1111.

Ogura, Y., Bonen, D. K., Inohara, N., Nicolae, D. L., Chen, F. F., Ramos, R., et al. (2001). A frameshift mutation in NOD2 associated with susceptibility to Crohn's disease. *Nature* **411,** 603–606.

Parkes, M., Barrett, J. C., Prescott, N. J., Tremelling, M., Anderson, C. A., Fisher, S. A., et al. (2007). Sequence variants in the autophagy gene IRGM and multiple other replicating loci contribute to Crohn's disease susceptibility. *Nat. Genet.* **39,** 830–832.

Py, B. F., Lipinski, M. M., and Yuan, J. (2007). Autophagy limits *Listeria monocytogenes* intracellular growth in the early phase of primary infection. *Autophagy* **3,** 117–125.

Rich, K. A., Burkett, C., and Webster, P. (2003). Cytoplasmic bacteria can be targets for autophagy. *Cell. Microbiol.* **5,** 455–468.

Rioux, J. D., Xavier, R. J., Taylor, K. D., Silverberg, M. S., Goyette, P., Huett, A., et al. (2007). Genome-wide association study identifies new susceptibility loci for Crohn's disease and implicates autophagy in disease pathogenesis. *Nat. Genet.* **39**, 596–604.

Romano, P. S., Gutierrez, M. G., Beron, W., Rabinovitch, M., and Colombo, M. I. (2006). The autophagic pathway is actively modulated by phase II *Coxiella burnetii* to efficiently replicate in the host cell. *Cell. Microbiol.* **9**, 891–909.

Sanjuan, M. A., Dillon, C. P., Tait, S. W., Moshiach, S., Dorsey, F., Connell, S., et al. (2007). Toll-like receptor signalling in macrophages links the autophagy pathway to phagocytosis. *Nature* **450**, 1253–1257.

Suzuki, T., Franchi, L., Toma, C., Ashida, H., Ogawa, M., Yoshikawa, Y., et al. (2007). Differential regulation of caspase-1 activation, pyroptosis, and autophagy via Ipaf and ASC in *Shigella*-infected macrophages. *PLoS Pathog.* **3**, e111.

Swanson, M. S. (2006). Autophagy: Eating for good health. *J. Immunol.* **177**, 4945–4951.

Swanson, M. S., and Isberg, R. R. (1995). Association of *Legionella pneumophila* with the macrophage endoplasmic reticulum. *Infect. Immun.* **63**, 3609–3620.

Swanson, M. S., and Molofsky, A. B. (2005). Autophagy and inflammatory cell death, partners of innate immunity. *Autophagy* **1**, 174–176.

Tanida, I., Minematsu-Ikeguchi, N., Ueno, T., and Kominami, E. (2005). Lysosomal turnover, but not a cellular level, of endogenous LC3 is a marker for autophagy. *Autophagy* **1**, 84–91.

Taylor, M. P., and Kirkegaard, K. (2007). Modification of cellular autophagy protein LC3 by poliovirus. *J. Virol.* **81**, 12543–12553.

Walker, D. H., Popov, V. L., Crocquet-Valdes, P. A., Welsh, C. J., and Feng, H. M. (1997). Cytokine-induced, nitric oxide-dependent, intracellular antirickettsial activity of mouse endothelial cells. *Lab. Invest.* **76**, 129–138.

Wilmanski, J. M., Petnicki-Ocwieja, T., and Kobayashi, K. S. (2008). NLR proteins: Integral members of innate immunity and mediators of inflammatory diseases. *J. Leukoc. Biol.* **83**, 13–30.

Xu, Y., Jagannath, C., Liu, X. D., Sharafkhaneh, A., Kolodziejska, K. E., and Eissa, N. T. (2007). Toll-like receptor 4 is a sensor for autophagy associated with innate immunity. *Immunity* **27**, 135–144.

Yamamoto, A., Tagawa, Y., Yoshimori, T., Moriyama, Y., Masaki, R., and Tashiro, Y. (1998). Bafilomycin A1 prevents maturation of autophagic vacuoles by inhibiting fusion between autophagosomes and lysosomes in rat hepatoma cell line, H-4-II-E cells. *Cell Struct. Funct.* **23**, 33–42.

Yamamoto, A., Hatano, M., Kobayashi, Y., Kabeya, Y., Suzuki, K., Tokuhisa, T., et al. (2001). LC3, a mammalian homologue of yeast Apg8p, is localized in autophagosome membranes after processing. *J. Cell Biol.* **152**, 657–668.

CHAPTER TWENTY-FOUR

MONITORING MACROAUTOPHAGY BY MAJOR HISTOCOMPATIBILITY COMPLEX CLASS II PRESENTATION OF TARGETED ANTIGENS

Monique Gannagé *and* Christian Münz

Contents

1. Introduction 404
2. Antigen Presentation on MHC Class II Molecules After Macroautophagy 404
3. Substrates of Endogenous Antigen Processing for MHC Class II Presentation via Macroautophagy 405
4. Methods to Measure Antigen Processing for MHC Class II Presentation After Macroautophagy 407
   4.1. Expression of monitoring constructs in suitable target cells 407
   4.2. Analysis of targeted antigen colocalization with MHC class II containing compartments (MIIC) 411
   4.3. T cell assays to monitor macroautophagy substrate presentation on MHC class II molecules 416
5. Concluding Remarks 419
Acknowledgments 419
References 419

Abstract

Major histocompatibility complex (MHC) class I and II molecules can both present cytosolic and nuclear antigens to $CD8^+$ and $CD4^+$ T cells, respectively. However, MHC class I displays proteasomal, whereas MHC class II molecules display lysosomal, degradation products. One pathway by which intracellular antigens gain access to lysosomal degradation is macroautophagy. Therefore, MHC class II presentation of antigens that are targeted to autophagosomes can be used to investigate regulation events of the macroautophagy pathway. We fuse antigens to Atg8/LC3 for targeting to autophagosomes, because this

Viral Immunobiology, Institute of Experimental Immunology, University Hospital of Zürich, Zürich, Switzerland

ubiquitin-like protein is selectively coupled to autophagosome membranes, and the portion that is coupled to the inner autophagosome membrane is degraded with this membrane in lysosomes. The localization of these fusion antigens in MHC class II loading compartments can be visualized by immunofluorescence and electron microscopy, and used as a measure of autophagic amphisome generation. In addition, MHC class II presentation of autophagosome-targeted antigens can be monitored by CD4$^+$ T cell recognition and indicates completion of macroautophagy. Together these immunological assays are well suited to investigate autophagic flux and analyze experimental conditions and physiological perturbations for their influence on macroautophagy.

1. Introduction

Adaptive immune responses rely on peptide presentation by major histocompatibility complex (MHC) molecules to T cells, an immunological phenomenon called MHC restriction (Bevan, 2004; Bryant and Ploegh, 2004; Chapman, 2006; Doherty and Zinkernagel, 1975). The peptide ligands for MHC class I and II molecules that stimulate CD8$^+$ and CD4$^+$ T cells, respectively, are generated by largely separate proteolytic machineries. MHC class I peptide ligands are thought to be primarily the products of proteasomal, and MHC class II ligands primarily the product of lysosomal degradation. Substrates for lysosomal proteolysis can reach this vesicular compartment by different routes. Apart from endocytic pathways, autophagy delivers cytoplasmic constituents to lysosomes (Mizushima and Klionsky, 2007). For this purpose, several distinct pathways are used, which include macroautophagy, microautophagy, and chaperone-mediated autophagy. While the contribution of the latter two pathways to constitutive autophagy in higher eukaryotes is still debated, macroautophagy is thought to deliver cell organelles, protein aggregates, and intracellular pathogens for lysosomal degradation (Mizushima *et al.*, 2008; Schmid and Münz, 2007). In this chapter, we will describe assays that monitor MHC class II presentation of macroautophagic substrates as a means to analyze macroautophagic flux in antigen presenting cells.

2. Antigen Presentation on MHC Class II Molecules After Macroautophagy

MHC class II molecules are loaded with their peptide ligands in late endosomal vesicles, which have been named MHC class II containing compartments (MIIC) and are often multivesicular or multilamellar bodies (Rocha and Neefjes, 2008). MHC class II molecules reach these vesicles in association with the invariant chain (Ii), a chaperone molecule that prevents premature peptide binding during transport and guides MHC class II molecules via its cytosolic YLL sorting signal to the MIIC. Ii is then degraded by lysosomal hydrolysis in parallel with the antigens, and its last peptide remnant, the class II-associated Ii chain peptide (CLIP), is exchanged for high affinity peptides under the influence of another chaperone called HLA-DM. The stable MHC class II molecules in a complex with peptide are then transported to the cell surface for immune surveillance by $CD4^+$ T cells.

Endocytosis of extracellular antigens has been described as the main pathway by which antigens reach the MIIC in phagocytes such as dendritic cells (DCs), and these cells are thought to initiate or prime $CD4^+$ T cells in response to infections (Steinman and Banchereau, 2007). However, in addition autophagy delivers intracellular material for MHC class II loading, and both macroautophagy and chaperone-mediated autophagy have been suggested to contribute to antigen processing of cytosolic and nuclear antigens (Schmid et al., 2007; Zhou et al., 2005). Especially in tissues that express MHC class II molecules on their surfaces but show little endocytic potential this pathway might contribute to immune surveillance by $CD4^+$ T cells as well as tolerance induction in this T cell compartment. Along these lines, the analysis of MHC class II presentation after macroautophagy in epithelial cells of the thymus (Anderson et al., 2002), secondary lymphoid tissues (Lee et al., 2007), and sites of inflammation (Heller et al., 2006) should be of special interest. Autophagosomes seem to fuse frequently with MIICs in epithelial, B, and dendritic cells, as characterized by a more than 50% overlap between specific autophagosome markers and components of the MHC class II loading machinery. In addition, antigen targeting to autophagosomes enhances MHC class II presentation up to 20-fold (Schmid et al., 2007). Similarly, overexpression of components of the machinery for chaperone-mediated autophagy increase MHC class II presentation of select autoantigens (Zhou et al., 2005). Therefore, autophagic pathways seem to be alternative routes by which intracellular antigens in addition to extracellular endocytosed antigens can gain access to MHC class II presentation.

3. Substrates of Endogenous Antigen Processing for MHC Class II Presentation via Macroautophagy

Even though up to 20% of peptide ligands eluted from affinity purified MHC class II molecules are derived from cytosolic and nuclear source proteins (Chicz et al., 1993; Dengjel et al., 2005; Dongre et al., 2001; Rammensee et al., 1999), only a handful of proteins have been demonstrated to be both autophagy substrates and presented by MHC class II (Schmid and

Münz, 2007). Among these, peptide fragments of the essential autophagy protein Atg8 have been isolated. This protein is selectively coupled to autophagosomal membranes in an ubiquitin-like reaction and partially degraded with these vesicles after fusion with lysosomes (Ohsumi, 2001). Of the three Atg8 homologs in mammals, microtubule-associated protein 1 light chain 3 (MAP1LC3), gamma-aminobutyric-acid-type-A-receptor-associated protein (GABARAP), and Golgi-associated ATPase enhancer of 16 kDa (GATE16), two of them give rise to peptide ligands that have been eluted from MHC class II molecules (MAP1LC3B$_{93-109}$, MAP1LC3B$_{93-110}$, and GABARAP$_{29-45}$) (Dengjel et al., 2005; Suri et al., 2008). Interestingly, the GABARAP-derived peptide is even recognized by CD4$^+$ T cells from pancreatic lymph nodes of prediabetic NOD mice (Suri et al., 2008). In addition, some classical autophagy substrates are frequently found in natural MHC class II ligand isolates. For example, glyceraldehyde-3-phosphate dehydrogenase (GAPDH) is a known substrate of chaperone-mediated autophagy (Anieto et al., 1993) and purifies with autophagosomes (Fengsrud et al., 2000). MHC class II bound peptides of GAPDH have been isolated from four different MHC class II alleles, making it one of the most consistent source proteins for MHC class II ligands identified to date (Rammensee et al., 1999).

Other intracellular proteins, whose MHC class II presentation is enhanced upon macroautophagy induction via starvation, are primarily found among cytosolic and nuclear antigens, whereas this treatment does not influence MHC class II presentation of membrane and secreted proteins (Dengjel et al., 2005). Peptide ligands, whose MHC class II presentation is most dramatically enhanced upon macroautophagy stimulation, include elongation factor 1-alpha 1, heat shock 70kDa protein 1 and RAD23 homolog B (Dengjel et al., 2005). Because RAD23 blocks its degradation by the proteasome in *cis* via its UBA2 domain (Heessen et al., 2005), it is tempting to speculate that this redirects the protein towards macroautophagy. A similar *cis*-inhibition of proteasomal processing has been described for the nuclear antigen 1 (EBNA1) of the Epstein Barr virus (EBV) (Lee et al., 2004; Levitskaya et al., 1995, 1997). EBNA1's *cis*-inhibition of proteasomal degradation is mediated by its glycine-alanine repeat (Hoyt et al., 2006; Levitskaya et al., 1997). Whereas this inhibits turnover by the proteasome, full-length EBNA1 is processed by lysosomes after macroautophagy, and this pathway contributes to its presentation on MHC class II molecules to CD4$^+$ T cells (Paludan et al., 2005). In addition to EBNA1, CD4$^+$ T cell recognition of three other antigens has been reported. Pharmacological inhibition of macroautophagy inhibits CD4$^+$ T cell recognition of complement C5 after transfection into mouse macrophage and B cell lines (Brazil et al., 1997). In human cells, MHC class II presentation of the Mucin 1 (MUC1) tumor antigen and the bacterial neomycin phosphotransferase II (NeoR) by dendritic cells and EBV transformed B cells as well as renal cell carcinoma cell lines, respectively, is sensitive to pharmacological inhibition of macroautophagy (Dorfel et al., 2005; Nimmerjahn et al., 2003). Therefore, several antigens are presented on MHC class II molecules after autophagic antigen presentation, and it will be exciting to unravel the contribution of this pathway to immune control in vivo.

4. Methods to Measure Antigen Processing for MHC Class II Presentation After Macroautophagy

In the following we describe protocols to monitor autophagic flux by MHC class II presentation of antigens targeted to autophagosomes in human epithelial cell lines, EBV transformed B cell lines and primary human monocyte-derived dendritic cells. These are cell types relevant to our particular studies, but we suggest that our methods can be used in different species and different tissues that express MHC class II molecules and use lysosomal antigen processing for MHC class II presentation to $CD4^+$ T cells. Similarly, we describe here the use of one particular model antigen, the influenza matrix protein 1 (MP1) for these targeting studies, but we believe that other $CD4^+$ T cell antigens can also be used after confirming that the proposed modifications efficiently target these proteins to autophagosomes. Therefore, the described protocols should be useful to monitor autophagic flux in MHC class II-positive tissues in general.

4.1. Expression of monitoring constructs in suitable target cells

4.1.1. Transfection with GFP-LC3 and antigen-LC3 in suitable target cells

4.1.1.1. Materials

1. Cell lines: HaCat keratinocyte cell line, a gift of Rajiv Khanna, Brisbane, Australia. MDAMC human breast carcinoma cell line, a gift of Irene Joab, Paris, France. HEK293T/DR4 transfected human embryonic kidney epithelial cell line, a gift of Rong-Fu Wang, Houston, TX. Cell lines were chosen based on their ability to up-regulate MHC class II expression upon IFN-γ treatment (Schmid et al., 2007) and their HLA haplotype. HaCat is HLA-DRB1★15-, -DRB1★04-, -DQBI★03- and -DQB1★03-positive, and therefore can be used in MHC class II presentation assays with MP1-specific $CD4^+$ T cell clones from a healthy lab donor with the following MHC haplotype: HLA-A★0201, -A★6801, -B★4402, -B★0702, -C★0501, -C★0702, -DRB1★1501, -DRB1★0401, -DRB5★01, -DRB4★01, -DQBI★0602 and -DQB1★0301. MDAMC is HLA-A★02-positive and therefore can be used in MHC class I presentation assays with HLA-A2-restricted MP1-specific $CD8^+$ T cell

clones derived from the same donor. In addition, HLA-DR4 transfected HEK293T/DR4 cells are also positive for HLA-A2 and HLA-DR4 and can be alternatively used for presentation assays to the MP1-specific CD4$^+$ and CD8$^+$ T cells. If other cell lines have to be used, determine MHC class II expression/up-regulation upon IFN-γ treatment and compare their HLA haplotype with that of T cell donors.

2. Cell culture medium: Dulbecco's Modified Eagle's Medium (DMEM, Gibco) with 10% fetal calf serum (FCS, Sigma), 2 mM glutamine, 110 μg/ml sodium pyruvate and 2 μg/ml gentamicin (Gibco).
3. Sterile Dulbecco's Phosphate Buffered Saline (DPBS, Gibco) with 0.5 mM ethylenediamine tetraacetic acid (EDTA, Sigma). To prepare, make 0.5 M EDTA stock, pH 8.0 in H_2O, filter through a 0.2-μm filter to sterilize and dilute 1:1000 in sterile PBS.
4. Solution of 0.05% trypsin/0.53 mM EDTA (Gibco).
5. Recombinant human Interferon-γ (IFN-γ, ProSpec-Tany TechnoGene LTD, Israel). Reconstitute in sterile H_2O + 0.1% human serum albumin (HSA, Sigma) to prepare a 2×10^6 U/ml stock and freeze in aliquots at $-20\,°C$.
6. Mammalian expression vector coding for LC3 fusion protein of influenza A matrix protein 1 (MP1, genebank entry X08088), as described previously (Schmid et al., 2007). The construct was designed by cloning the human Atg8/LC3 cDNA sequence (genebank entry NM_022818) into the mammalian expression vector pEGFP-C2 (BD Biosciences) and subsequently replacing in some constructs the EGFP coding sequence with the MP1 antigen coding sequence (without a stop codon at the 3' end). As controls, mammalian expression vectors coding for MP1 or MP1-LC3($G_{120}A$) (Schmid et al., 2007) should be used in parallel with MP1-LC3. These constructs were designed the same way, except that the MP1 construct contains a stop codon at the 3' end of the MP1 sequence and the MP1-LC3($G_{120}A$) construct contains a point mutation at nucleotide 358 of the Atg8/LC3 sequence (A instead of G). For best transfection results, maxipreps of the different DNA plasmids should be prepared and DNA eluted in sterile DPBS. Determine DNA concentration and purity by the OD_{260}/OD_{280} reading in a spectrophotometer. The OD_{260}/OD_{280} ratio should be >1.8. Store DNA in aliquots at $-20\,°C$.
7. Cell culture medium for transfection (see subsequent sections) without gentamicin. For transfection with lipofectamine 2000, cell culture medium should not contain any antibiotics. Therefore, omit gentamicin.
8. Lipofectamine 2000 transfection reagent (Invitrogen).
9. OptiMEM-I medium (Gibco).

4.1.1.2. Methods

1. HaCat, MDAMC, and HEK293T/DR4 cell lines are maintained in DMEM + 10% FCS medium in 75-cm^2 tissue culture flasks or 100-mm plates until they approach confluence.

2. To split cells, wash the monolayer once with 5 ml of DPBS/0.5 mM EDTA and incubate with 2 ml of trypsin/EDTA solution at 37 °C for 2–3 min (MDAMC) or 10–15 min (HaCat) to detach cells. To set up new maintenance cultures, replate 1/20 th of the cells and add fresh culture medium. These cultures will approach confluence after 2–3 days.
3. To induce expression of MHC class II machinery, cells have to be cultured with 200 U/ml IFN-γ for 36 h (for HaCat) or 48 h (for MDAMC). At least 50% of the cells then will express MHC class II on their cell surface.
4. To set cells up for transfection, detach cells as described previously and plate onto a 6-well tissue culture plate in antibiotic-free culture medium at a density of 2×10^5 cells/well. For transfection with lipofectamine 2000, the cell culture medium should not contain any antibiotics. Therefore, omit gentamicin.
5. The next day, cultures should be about 70%–80% confluent. For each well to be transfected, dilute 2.5 μg of plasmid DNA in 250 μl of OptiMEM-I medium and mix by pipetting up and down. In a separate tube, dilute 7.5 μl of lipofectamine 2000 in 250 μl of OptiMEM-I medium and mix gently by pipetting up and down. Incubate both tubes for 5 min at room temperature.
6. Combine both solutions, mix gently by pipetting up and down, and incubate 20 min at room temperature to allow formation of DNA-lipofectamine complexes.
7. Add complexes in a dropwise manner to the culture medium of cells in the 6-well plate and incubate at 37 °C for 4–6 h.
8. After 4–6 h, replace the complex-containing medium with fresh culture medium and culture cells for 18–20 h at 37 °C. Removal of complex-containing medium 4–6 h after transfection was found to improve viability of MDAMC and HaCat cell lines.
9. 24 h post-transfection, cells are ready for further analysis in the assays described subsequently. Alternatively, stable transfectants of the above listed cell lines can be easily established by recombinant lentiviral infection as outlined in the following chapter.

4.1.2. Expression of GFP-LC3 and antigen-LC3 by recombinant lentiviral infection of dendritic cells

4.1.2.1. Materials

1. For lentiviral constructs, the EGFP-LC3, MP1-LC3 or MP1Stop-LC3 sequences were subcloned into the lentiviral vector pHR-SIN-CSGWΔNotI (a gift from Jeremy Luban, Geneva, Switzerland).
2. For production of lentiviral particles, lentiviral vectors are co-transfected with the helper plasmids pCMVΔR8.91 and pMDG into 293T cells by calcium phosphate transfection and cultured in DMEM medium with 10% FCS (Sigma), 2 mM glutamine, 110 μg/ml sodium pyruvate, and 2 μg/ml gentamicin (Gibco).

3. Culture supernatants containing recombinant viral particles are harvested on day 1, 2, and 3 after transfection, filtered through a 0.45 µm filter and frozen at −80 °C.

4.1.2.2. Methods

1. PBMCs are isolated by density gradient centrifugation on Ficoll-Paque Plus (GE Healthcare) from human peripheral blood.
2. $CD14^+$ monocytes/macrophages are isolated by positive magnetic cell separation (MACS) using anti-CD14 MicroBeads (Miltenyi Biotec).
3. To generate monocyte-derived DCs, 3×10^6 $CD14^+$ cells are plated into each well of a 6-well plate in 3 ml of Roswell Park Memorial Institute-1640 medium (RPMI 1640, Gibco) with 1% single-donor plasma 2 mM glutamine, 110 µg/ml sodium pyruvate, and 2 µg/ml gentamicin plus recombinant human IL-4 (rhIL-4, 500 U/ml, Peprotech) and rhGMCSF (1000 U/ml, Leukine from Berlex). rhIL-4 and rhGMCSF are added again on day 2 and 4 and floating immature DCs are collected on day 5.
4. For lentiviral transduction of DCs, $CD14^+$ monocytes are infected with lentivirus at a MOI of 10 on day 1 after isolation, in the presence of 8 µg/ml polybrene (Sigma), rhIL-4 and rhGM-CSF.
5. For maturation, immature DCs are transferred to new plates on day 5 and half of the medium is replaced with fresh medium containing proinflammatory cytokines [IL-1β (10 ng/ml), IL-6 (1000 U/ml), TNF-α (10 ng/ml) from R&D Systems, and PGE_2 (1 µg/ml, Sigma)] or 200 ng/ml lipopolysaccharide (LPS, Sigma).

4.1.3. Electroporation of GFP-LC3 and antigen-LC3 into EBV transformed B cell lines (LCLs)

4.1.3.1. Materials

1. PBMCs are isolated by density gradient centrifugation on Ficoll-Paque Plus (GE Healthcare) from human peripheral blood.
2. For the generation of LCLs, add 1 ml of B95-8 cell line (Miller *et al.*, 1975) supernatant to $1-2 \times 10^6$ PBMCs in RPMI-1640 medium with 20% FCS (Sigma), 2 mM glutamine, 110 µg/ml sodium pyruvate, 2 µg/ml gentamicin (Gibco), and 1 µg/ml cyclosporine A (Sigma).
3. Established LCLs are then cultured in RPMI-1640 medium with 10% FCS (Sigma), 2 mM glutamine, 110 µg/ml sodium pyruvate, and 2 µg/ml gentamicin (Gibco).

4.1.3.2. Methods

1. Resuspend $1-2 \times 10^6$ LCLs in 300 µl of Opti-MEM (Gibco).
2. Transfer into 0.2-cm electroporation cuvette (BioRad).

3. Add 10–20 μg plasmid DNA and electroporate immediately with 300 V and 150 μF in Bio-Rad Gene Pulser with Capacitance Extender.
4. Immediately transfer back into RPMI-1640 medium with 10% FCS (Sigma), 2 mM glutamine, 110 μg/ml sodium pyruvate, and 2 μg/ml gentamicin (Gibco).
5. Monitor transfection efficiency with GFP.

4.2. Analysis of targeted antigen colocalization with MHC class II containing compartments (MIIC)

4.2.1. Immunofluorescence microscopy to characterize MIIC/autophagosome fusion

4.2.1.1. Materials

1. Circular 1.5-mm coverslips (Fisher).
2. 70% (v/v) ethanol and sterile DPBS.
3. Recombinant human IFN-γ (see section 4.1.1.5.).
4. Chloroquine (CQ, Sigma). Prepare a 20 mM stock in ddH$_2$O and sterilize through a 0.2-μm filter. Freeze in aliquots at −20 °C. Use at a 1:400 dilution.
5. Phosphate buffered saline (PBS): Prepare 10x stock (1.37 M NaCl, 27 mM KCl, 100 mM Na$_2$HPO$_4$, 18 mM KH$_2$PO$_4$, adjust pH to 7.4 with HCl) and autoclave before storage at room temperature. Prepare 1x PBS by dilution of 1 part with 9 parts ddH$_2$O.
6. 4% paraformaldehyde (PFA): Prepare a 36.5% (w/v) solution in PBS by heating PFA (Sigma) in PBS in a covered jar until PFA dissolves (carefully, as PFA is toxic and vapors should not be inhaled). Let cool to room temperature for immediate use or store in aliquots at −20 °C. Thaw a fresh aliquot and dilute to 4% (w/v) PFA in PBS for each experiment.
7. Permeabilization solution: 0.1% (v/v) Triton-X 100 in PBS.
8. Blocking buffer: PBS + 1% bovine serum albumin (BSA, Sigma) + 5% normal donkey serum (NDS) + 0.1% saponin (Calbiochem). Normal donkey serum is used as blocking reagent because secondary antibodies are derived from donkey. If secondary antibodies come from a different species, use 5% normal serum from that species as blocking reagent.
9. Wash buffer: PBS + 0.1% saponin.
10. Primary antibodies: Rabbit polyclonal MP1-specific antiserum (use at 1:2000) and mouse monoclonal HLA-DR/DP/DQ-specific hybridoma IVA12 (ATCC). Grow hybridoma in RPMI-1640 with 10% FBS, 2 mM glutamine, and 2 μg/ml gentamicin, and harvest the supernatant by spinning cells down at 300g for 10 min. Filter supernatant through a 0.2-μm filter and store at 4 °C. Use at a 1:10 dilution.
11. Secondary antibodies: Donkey antirabbit IgG-Alexa 488 (Invitrogen-Molecular Probes, use at 1:500) and donkey antimouse IgG-Rhodamine-Red-X (Jackson ImmunoResearch, use at 1:300). For choice of

fluorophores, consider which lasers and filter sets are available for your confocal microscope. For colocalization analysis it is important that emission spectra of the green and red fluorochromes do not overlap. Also consider brightness and photostability of fluorophores.

12. DAPI nucleic acid stain: Prepare a 5 mg/ml stock of 4,6-diamidino-2-phenylindole (DAPI, Invitrogen-Molecular Probes) in ddH$_2$O and store in aliquots at $-20\,°C$. Dilute 1:10,000 in PBS to prepare working solution and store at 4 $°C$ wrapped in aluminium foil.
13. Mounting medium: Prolong Gold Antifade Reagent (Invitrogen-Molecular Probes).

4.2.1.2. Methods

1. To induce expression of MHC class II machinery, cells have to be cultured with 200 U/ml IFN-γ for 36 h (for HaCat) or 48 h (for MDAMC). At least 50% of cells then will express MHC class II on their cell surface.
2. Place round 1.5-mm microscopy coverslips into 24-well tissue culture plate. Use 8 extra coverslips for control stainings. For correct interpretation of results, the following control stainings should be included:
    a. Replace primary and secondary antibodies with blocking buffer. The stainings should be completely negative.
    b. Replace primary antibody with blocking buffer, but use secondary antibodies. Background from secondary antibodies should be low. If background turns out to be too high, titer down concentration of secondary antibodies.
    c. Use untransfected cells and cells that were not treated with IFN-γ. Stainings should be negative.
    d. Do single-labeling of cells and check signal in the wrong channel (red channel for Alexa488 labeling and green channel for Rhodamine-Red X-labeling). There should be no bleed-through into the wrong channel.
    e. Do single-labeling with wrong secondary antibodies. There should be no cross-reactivity with the wrong species.
3. Sterilize coverslips by washing once with 70% ethanol and twice with sterile DPBS. Remove any traces of ethanol by completely aspirating off ethanol and wash solutions with vacuum suction flask.
4. Trypsinize cells transfected with different MP1 or GFP-LC3 constructs and plate onto sterilized coverslips in cell culture medium, at a density of 5×10^4 cells/well. Plate 2 wells of each sample, so that cells can be analyzed with and without chloroquine treatment. Chloroquine inhibits lysosomal acidification and thus prevents degradation of lysosomal substrates, including LC3 fusion proteins (Schmid et al., 2007).

Without inhibition of lysosomal proteases, GFP-LC3 and MP1-LC3 are rapidly degraded and their detection in lysosomal compartments is more difficult. However, when cells are treated with chloroquine, the fusion protein accumulates in lysosomal compartments/MIICs and now can be visualized much more readily (see Fig.24.1; Schmid et al., 2007). Therefore, it is recommended that one set of cells be treated with chloroquine for 6 h prior to the staining.

5. Treat cells with 200 U/ml IFN-γ for 36 h (HaCat cells) or 48 h (MDAMC cells) to induce expression of MHC class II molecules.
6. During the last 6 h of the culture, treat one set of cells with 50 μM chloroquine (CQ), to prevent degradation of GFP-LC3 and MP1-LC3 by lysosomal proteases. Leave the other set of cells untreated.
7. Wash cells once in PBS (0.5 ml/well) and fix in 4% paraformaldehyde (PFA, 200 μl/well) for 15 min at room temperature. From the fixation step onward, cells can be handled outside sterile hood on a laboratory bench. Use a vacuum suction flask to change solutions, exchange plastic tip of suction device when handling different solutions (e.g., antibodies). All incubation steps are done at room temperature, unless noted otherwise.
8. Wash cells once in PBS (0.5 ml/well) and permeabilize in 0.1% Triton X-100 (200 μl/well) for 5 min.
9. Wash cells once in PBS (0.5 ml/well) and add blocking buffer (200 μl/well) for 30 min.
10. Dilute primary antibodies in blocking buffer and add to cells (200 μl/well) for 60 min at room temperature or for longer periods (up to overnight) at 4 °C.
11. Wash cells 3 times in wash buffer (0.5 ml/well), incubate for 5 min each time.
12. Dilute secondary antibodies in blocking buffer and add to cells (200 μl/well) for 45 min.
13. Suck off secondary antibody solutions and add DAPI nucleic acid stain for 20–30 s (200 μl/well). Afterward, immediately wash cells as in step 9.
14. Wash cells once in PBS and mount coverslips by inverting them onto a drop of mounting medium on a microscope slide, up to 4 coverslips per slide. Carefully press down on coverslip, suck off excess mounting medium, and let dry at room temperature. Prolong Gold antifade mounting medium (Invitrogen-Molecular Probes) should be allowed to dry at room temperature overnight. During this time, the mounting medium will gel and its refractive index increases. Sealing of coverslips with nail polish is not necessary for Prolong Gold, but is recommended for other, water-based mounting media.
15. Afterward, slides can be stored in the dark at 4 °C for several months.
16. Analyze slides with a confocal laser-scanning microscope, using a high N.A. oil immersion lens (e.g., 63x/1.4 N.A.). Excitation at 405 nm induces DAPI fluorescence (blue emission), excitation at 488 nm induces

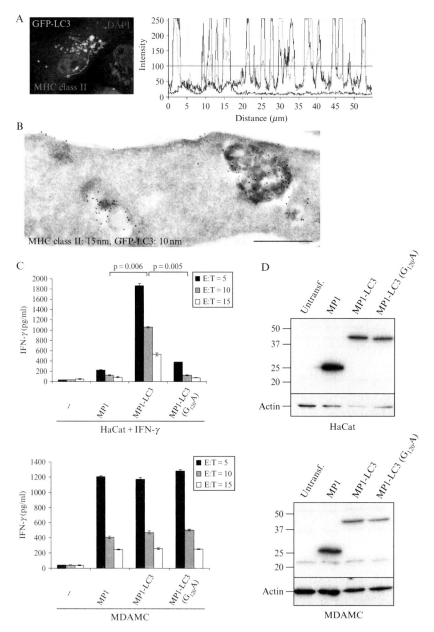

Figure 24.1 LC3 fusion proteins localize to MHC class II-containing compartments and are more efficiently presented on MHC class II molecules to CD4[+] T cells. (A) Example of a colocalization analysis of GFP-LC3 (green) and MHC class II molecules (red) in MDAMC breast carcinoma cells. Analysis with the LSM510 profile tool reveals double-positive vesicles along the path outlined in red in the picture on the left. The histogram on the right shows that 12 of 18 MHC class II-positive vesicles

Alexa 488 fluorescence (green emission) and excitation at 543 nm induces Rhodamine-Red-X fluorescence (red emission). Software can be used to overlay the different fluorescence channels and to quantify colocalization. If a confocal microscope is not available, alternatively slides can be analyzed with a conventional wide-field fluorescence microscope with a motorized z-stage. To remove out-of-focus light and accurately analyze colocalization of fluorochromes, z-stacks subsequently have to be deconvolved using deconvolution software.

17. An experiment with GFP-LC3 expressing MDAMC cells is shown as an example in Fig. 24.1.

4.2.2. Immunogold EM for GFP-LC3 colocalization with MHC class II molecules

4.2.2.1. Materials

1. Polyclonal rabbit anti-HLA-DR antiserum (C6861; a kind gift of Dr. Peter Cresswell, New Haven, CT) was used at a 1:50 dilution.
2. Polyclonal rabbit anti-GFP antiserum (Molecular Probes) was used at a 1:50 dilution.
3. 10- or 15-nm gold-protein A was purchased from the Department of Cell Biology, University of Utrecht, Netherlands.

4.2.2.2. Methods

1. MDAMC cells stably transfected with GFP-LC3 are fixed for 1 h at RT with 4% paraformaldehyde (PFA, Electron Microscopy Sciences) in 0.25 M HEPES, pH 7.4, followed by overnight fixation at 4 °C in 8% PFA/HEPES.
2. Cells are embedded in 5% gelatin in PBS, small pieces of gelatin pellets were infiltrated overnight at 4 °C with 2.3 M sucrose in PBS, mounted onto cryospecimen pins and frozen in liquid nitrogen.

---

contain significant amounts of GFP-LC3 (fluorescence intensity >100). (B) Immunoelectron microscopy, labeling MHC class II molecules with 15-nm gold particles and GFP of GFP-LC3 with 10-nm gold particles, reveals colocalization of GFP-LC3 and MHC class II molecules on intravesicular membranes of late multivesicular endosomes. In addition, some MHC class II staining is found on the cell surface. (C) The influenza matrix protein 1 (MP1)-specific $CD4^+$ T cell clone 11.46 or the MP1 specific $CD8^+$ T cell clone 9.2 (top and bottom, respectively) were stimulated at the indicated effector to target (E:T) ratios with HLA matched epithelial target cells expressing either wild-type MP1, MP1 fused to LC3 (MP1-LC3) or the fusion protein mutated at the C-terminal amino acid of LC3 (MP1-LC3 ($G_{120}A$)). T cell reactivity was measured the next day by IFN-γ secretion into the supernatant, as assessed by cytokine-specific ELISA. Error bars indicate standard deviations and p values are given for paired, one-tailed student T test statistics. (D) Expression levels of the different MP1 constructs were analyzed for the used target cells by Western blot analysis with anti-MP1 antiserum. Actin blots demonstrate the levels of protein loading for the analyzed SDS Page gels. Fig. 24.1B and parts of Fig. 24.1C have been published previously (Schmid *et al.*, 2007). (See Color Insert.)

3. Ultrathin sections (80 nm) are cut using a Leica ultracut ultramicrotome with an FCS cryoattachment at $-108\,°C$ and collected on Formvar- and carbon-coated nickel grids using a 1:1 mixture of 2% methyl cellulose (25 centipoises; Sigma) and 2.3 M sucrose in PBS.
4. Sections are quenched with 0.1 M NH$_4$Cl in PBS, and blocked in 1% fish skin gelatin (FSG, Sigma) in PBS for 30 min each.
5. The sections are then subsequently labeled with rabbit anti-HLA-DR antiserum and 10- or 15-nm protein A–gold overnight.
   After fixation in 1% glutaraldehyde and quenching with 0.1 M NH$_4$Cl for 10 min each, the same labeling procedure is repeated for the rabbit anti-GFP antibody.
7. After final fixation in 1% glutaraldehyde for 30 min, grids are washed 8 times in HPLC-grade water.
8. Sections are infiltrated for 10 min on ice with a mixture of 1.8% methylcellulose and 0.5% uranyl acetate (Electron Microscopy Sciences).
9. Sections are then washed 3 times in 0.5% uranyl acetate/1.8% methylcellulose and air-dried.
10. Samples are analyzed in a Tecnai 12 Biotwin (FEI) microscope and pictures are taken using Kodak 4489 film or an equivalent.

4.3. T cell assays to monitor macroautophagy substrate presentation on MHC class II molecules

4.3.1. T cell coculture with macroautophagy substrate expressing target cells

4.3.1.1. Materials

1. MP1-specific CD4$^+$ T cell clones, generated as described before (Fonteneau et al., 2001), and cultured in RPMI-1640 with 8% pooled human serum (PHS, Mediatech), 450 U/ml recombinant human IL-2 (Chiron), 2 mM glutamine, 2 μg/ml gentamicin in round bottom 96-well plates. Briefly, for the generation of MP1-specific CD4$^+$ T cell clones, CD14$^-$ cells isolated from a healthy lab donor were stimulated with autologous mature DCs that were electroporated with 10 μg of in vitro–transcribed MP1-RNA at a DC:CD14$^-$ ratio of 1:40. On day 8, the stimulation was repeated and 10 U/ml IL-2 were added to enhance T cell survival. On day 21, the surviving cells were cloned by limiting dilution at 10, 1, or 0.3 cells/well and expanded in RPMI-1640 + 8% PHS + 150 U/ml rhIL-2 (Chiron) + 1μg/ml PHA-L (Sigma). 10^5 irradiated PBMCs/well and 10^4 irradiated LCLs/well were added as feeder cells. On day 40, expanded cells were tested in split-well IFNγ ELISPOT assays for recognition of an MP1 peptide mix. T cell clones can be frozen and stored for several years in liquid nitrogen (about 10^6 cells/aliquot). Frozen cultures can be reexpanded as described previously.

2. Recombinant human Interferon-γ (IFN-γ) (see previous sections).
3. RPMI-1640.
4. Coculture medium: RPMI-1640 with 5% PHS, 2 mM glutamine, 2 μg/ml gentamicin.
5. Positive control stimulus: Specific MP1 peptide (1 mM stock in 10% DMSO, for T cell stimulation dilute 1:1000 in coculture medium) or phytohemagglutinin (PHA-L, Sigma, 1 mg/ml stock, for T cell stimulation dilute 1:1000 in coculture medium).

4.3.1.2. Methods

1. For the MHC class II presentation assay, use HaCat cells transfected with different MP1 constructs in a 6-well format, as described in section 4.1.1. Treat cells with 200 U/ml IFN-γ for 24 h to initiate expression of the MHC class II machinery.
2. Remove any traces of IFN-γ from cells by washing cell monolayers 3 times in RPMI-1640 medium. Trypsinize cells to prepare a cell suspension, wash once in coculture medium and count with hemacytometer. Prepare cell suspensions in coculture medium at three different cell concentrations (2×10^5, 10^5, and 6.67×10^4 cells/ml). The optimal number of target cells and T cell clones may vary, depending on the T cell clone and the type of target cell. Therefore it is recommended to try a range of different effector and target cell numbers, ranging from $10^4 - 2 \times 10^5$ T cells/well and $10^3 - 10^5$ target cells/well.
3. Collect MP1-specific CD4$^+$ T cell clones from 96-well culture plates, wash once in coculture medium and count. Adjust cell concentration to 2×10^6 cells/ml.
4. Set up cocultures of T cells and target cells in doublets (2 wells/condition) in a 96-well round-bottomed plate. Per well, plate 50 μl of T cell suspension (10^5 cells/well) and 100 μl of the different target cell suspensions (either 2×10^4, 10^4 or 6.67×10^3 cells/well). This will result in effector to target (E:T) ratios of 5, 10, and 15. As a positive control, stimulate T cell clones with specific MP1 peptide (1 μM) or PHA-L (1 μg/ml). As negative control, stimulate T cell clones with coculture medium only.
5. Culture cells overnight (18–24 h) at 37 °C.

4.3.2. Analyzing T cell reactivity by ELISA assay for IFN-γ from supernatants of T cell/target cell cocultures
4.3.2.1. Materials

1. High protein binding 96-well ELISA plate (e.g., Maxisorp from Nunc).
2. ELISA kit for Human Interferon-γ (IFNγ) (Mabtech). Prepare a 10 μg/ml stock of human recombinant IFN-γ provided with kit and freeze in aliquots at -20 °C. Use a freshly thawed aliquot for each experiment.

3. Coculture medium: RPMI-1640 with 5% PHS, 2 mM glutamine, 2 μg/ml gentamicin.
4. Blocking buffer: PBS + 1% bovine serum albumin (BSA, Sigma).
5. Wash buffer: PBS + 0.05% Tween-20.
6. Incubation buffer: PBS + 0.1% BSA + 0.05% Tween-20.
7. TMB peroxidase substrate solution (Sigma) and stop solution (1 N sulfuric acid).

4.3.2.2. Methods

1. One day prior to ELISA, coat high-protein-binding ELISA plate with primary anti-IFN-γ antibody (1-D1K, included in IFN-γ ELISA kit), diluted 1:500 in PBS, 100 μl/well. Incubate plate overnight at 4 °C.
2. The next day, wash plate 2 times with PBS (200 μl/well) and block with blocking buffer (200 μl/well) for 1 h at room temperature.
3. Thaw an aliquot of IFN-γ standard (10 μg/ml) and prepare serial dilutions in coculture medium (prepare 4000; 2000; 1000; 500; 250; 125; 62.5; 31.25, 0 pg/ml standards, at least 300 μl each).
4. To make sure that IFN-γ secreted by T cells is homogenously distributed in culture supernatants, mix supernatants by pipetting up and down with a multichannel pipette and pellet cells by centrifugation of plates at 300g for 5 min.
5. With a multichannel pipette, carefully remove 120 μl of supernatant from each well and transfer to a new 96-well plate.
6. Wash ELISA plate 4 times with wash buffer.
7. Add supernatants or IFN-γ standards (100 μl/well) and incubate for 2 h at room temperature. Freeze remaining 20 μl of supernatant at −20 °C, in case ELISA has to be repeated on diluted supernatants. In case IFN-γ levels in supernatants exceed the linear range of the ELISA (approximately 20–2000 pg/ml), dilute frozen supernatants 1:10 in coculture medium and repeat ELISA.
8. Wash as in step 6 and add 100 μl/well of secondary antibody (7-B6-1-biotin, provided in ELISA kit), diluted 1:1000 in incubation buffer. Incubate 1 h at room temperature.
9. Wash as in step 6 and add 100 μl/well of Streptavidin-HRP (provided in ELISA kit), diluted 1:1000 in incubation buffer. Incubate 1 h at room temperature.
10. Wash as in step 6 and add 100 μl/well of TMB peroxidase substrate. Incubate until blue reaction product has sufficiently developed, then stop reaction by adding 100 μl/well of Stop solution.
11. Measure optical density at 450 nm (OD_{450}) in an ELISA plate reader and convert OD_{450} values into IFN-γ concentration in pg/ml. In case IFN-γ levels in supernatants exceed the linear range of the ELISA (approximately 20–1,000 pg/ml), dilute frozen supernatants 1:10 in coculture medium and repeat ELISA. An example of the results produced is shown in Fig. 24.1.

5. Concluding Remarks

T cells are very sensitive components of the human immune system that detect peptide products of proteasomal and lysosomal degradation at nanomolar and even picomolar concentrations after presentation by MHC products (Gotch et al., 1987). Indeed, for some T cell specificities as little as one peptide/MHC complex has been suggested to be sufficient for T cell activation (Crotzer et al., 2000). $CD4^+$ T cells monitor antigen presentation on MHC class II molecules after lysosomal antigen processing, and can therefore be used to monitor autophagy, when they recognize natural MHC class II peptide ligands derived from autophagic substrates. Similarly, MHC class II containing compartments (MIIC) are one subgroup of late endosomes, and autophagosome fusion with MIICs can be used as an example of amphisome generation during macroautophagy. Therefore, T cells, very specific tools of the immune system, can be harnessed to measure autophagic flux, and such analysis might contribute to a better understanding of regulatory mechanisms influencing this important catabolic pathway.

ACKNOWLEDGMENTS

We thank Rajiv Khanna and Irene Joab for the gift of cell lines, and Ari Helenius and Peter Cresswell for the gift of antisera.

Our work is supported by the Arnold and Mabel Beckman Foundation, the Alexandrine and Alexander Sinsheimer Foundation, the Burroughs Wellcome Fund, the Dana Foundation's Neuroimmunology program, the National Cancer Institute (R01CA108609 and R01CA101741), the National Institute of Allergy and Infectious Diseases (RFP-NIH-NIAID-DAIDS-BAA-06-19), the Foundation for the National Institutes of Health (Grand Challenges in Global Health), the Starr Foundation (to C.M.), and an Institutional Clinical and Translational Science Award (to the Rockefeller University Hospital).

REFERENCES

Anderson, M. S., Venanzi, E. S., Klein, L., Chen, Z., Berzins, S. P., Turley, S. J., von Boehmer, H., Bronson, R., Dierich, A., Benoist, C., and Mathis, D. (2002). Projection of an immunological self shadow within the thymus by the aire protein. *Science* **298,** 1395–1401.

Aniento, F., Roche, E., Cuervo, A. M., and Knecht, E. (1993). Uptake and degradation of glyceraldehyde-3-phosphate dehydrogenase by rat liver lysosomes. *J. Biol. Chem.* **268,** 10463–10470.

Bevan, M. J. (2004). Helping the $CD8^+$ T-cell response. *Nat. Rev. Immunol.* **4,** 595–602.

Brazil, M. I., Weiss, S., and Stockinger, B. (1997). Excessive degradation of intracellular protein in macrophages prevents presentation in the context of major histocompatibility complex class II molecules. *Eur. J. Immunol.* **27,** 1506–1514.

Bryant, P., and Ploegh, H. (2004). Class II MHC peptide loading by the professionals. *Curr. Opin. Immunol.* **16,** 96–102.

Chapman, H. A. (2006). Endosomal proteases in antigen presentation. *Curr. Opin. Immunol.* **18,** 78–84.

Chicz, R. M., Urban, R. G., Gorga, J. C., Vignali, D. A., Lane, W. S., and Strominger, J. L. (1993). Specificity and promiscuity among naturally processed peptides bound to HLA-DR alleles. *J. Exp. Med.* **178,** 27–47.

Crotzer, V. L., Christian, R. E., Brooks, J. M., Shabanowitz, J., Settlage, R. E., Marto, J. A., White, F. M., Rickinson, A. B., Hunt, D. F., and Engelhard, V. H. (2000). Immunodominance among EBV-derived epitopes restricted by HLA-B27 does not correlate with epitope abundance in EBV-transformed B-lymphoblastoid cell lines. *J. Immunol.* **164,** 6120–6129.

Dengjel, J., Schoor, O., Fischer, R., Reich, M., Kraus, M., Müller, M., Kreymborg, K., Altenberend, F., Brandenburg, J., Kalbacher, H., Brock, R., Driessen, C., et al. (2005). Autophagy promotes MHC class II presentation of peptides from intracellular source proteins. *Proc. Natl. Acad. Sci. USA* **102,** 7922–7927.

Doherty, P. C., and Zinkernagel, R. M. (1975). H-2 compatibility is required for T-cell-mediated lysis of target cells infected with lymphocytic choriomeningitis virus. *J. Exp. Med.* **141,** 502–507.

Dongre, A. R., Kovats, S., deRoos, P., McCormack, A. L., Nakagawa, T., Paharkova-Vatchkova, V., Eng, J., Caldwell, H., Yates, J. R. 3rd., and Rudensky, A. Y. (2001). In vivo MHC class II presentation of cytosolic proteins revealed by rapid automated tandem mass spectrometry and functional analyses. *Eur. J. Immunol.* **31,** 1485–1494.

Dorfel, D., Appel, S., Grunebach, F., Weck, M. M., Müller, M. R., Heine, A., and Brossart, P. (2005). Processing and presentation of HLA class I and II epitopes by dendritic cells after transfection with in vitro transcribed MUC1 RNA. *Blood* **105,** 3199–3205.

Fengsrud, M., Raiborg, C., Berg, T. O., Stromhaug, P. E., Ueno, T., Erichsen, E. S., and Seglen, P. O. (2000). Autophagosome-associated variant isoforms of cytosolic enzymes. *Biochem. J.* **352**(Pt 3), 773–781.

Fonteneau, J. F., Larsson, M., Somersan, S., Sanders, C., Münz, C., Kwok, W. W., Bhardwaj, N., and Jotereau, F. (2001). Generation of high quantities of viral and tumor-specific human $CD4^+$ and $CD8^+$ T-cell clones using peptide pulsed mature dendritic cells. *J. Immunol. Methods* **258,** 111–126.

Gotch, F., Rothbard, J., Howland, K., Townsend, A., and McMichael, A. (1987). Cytotoxic T lymphocytes recognize a fragment of influenza virus matrix protein in association with HLA-A2. *Nature* **326,** 881–882.

Heessen, S., Masucci, M. G., and Dantuma, N. P. (2005). The UBA2 domain functions as an intrinsic stabilization signal that protects Rad23 from proteasomal degradation. *Mol. Cell* **18,** 225–235.

Heller, K. N., Gurer, C., and Münz, C. (2006). Virus-specific $CD4^+$ T cells: Ready for direct attack. *J. Exp. Med.* **203,** 805–808.

Hoyt, M. A., Zich, J., Takeuchi, J., Zhang, M., Govaerts, C., and Coffino, P. (2006). Glycine-alanine repeats impair proper substrate unfolding by the proteasome. *EMBO J.* **25,** 1720–1729.

Lee, J. W., Epardaud, M., Sun, J., Becker, J. E., Cheng, A. C., Yonekura, A. R., Heath, J. K., and Turley, S. J. (2007). Peripheral antigen display by lymph node stroma promotes T cell tolerance to intestinal self. *Nat. Immunol.* **8,** 181–190.

Lee, S. P., Brooks, J. M., Al-Jarrah, H., Thomas, W. A., Haigh, T. A., Taylor, G. S., Humme, S., Schepers, A., Hammerschmidt, W., Yates, J. L., Rickinson, A. B., and Blake, N. W. (2004). CD8 T cell recognition of endogenously expressed Epstein-Barr virus nuclear antigen 1. *J. Exp. Med.* **199,** 1409–1420.

Levitskaya, J., Coram, M., Levitsky, V., Imreh, S., Steigerwald-Mullen, P. M., Klein, G., Kurilla, M. G., and Masucci, M. G. (1995). Inhibition of antigen processing by the internal repeat region of the Epstein-Barr virus nuclear antigen-1. *Nature* **375,** 685–688.

Levitskaya, J., Sharipo, A., Leonchiks, A., Ciechanover, A., and Masucci, M. G. (1997). Inhibition of ubiquitin/proteasome-dependent protein degradation by the Gly-Ala repeat domain of the Epstein-Barr virus nuclear antigen 1. *Proc. Natl. Acad. Sci. USA* **94**, 12616–12621.

Miller, G., Robinson, J., Heston, L., and Lipman, M. (1975). Differences between laboratory strains of Epstein-Barr virus based on immortalization, abortive infection and interference. *IARC Sci. Publ.* **11**, 395–408.

Mizushima, N., and Klionsky, D. J. (2007). Protein turnover via autophagy: Implications for metabolism. *Annu. Rev. Nutr.* **27**, 19–40.

Mizushima, N., Levine, B., Cuervo, A. M., and Klionsky, D. J. (2008). Autophagy fights disease through cellular self-digestion. *Nature* **451**, 1069–1075.

Nimmerjahn, F., Milosevic, S., Behrends, U., Jaffee, E. M., Pardoll, D. M., Bornkamm, G. W., and Mautner, J. (2003). Major histocompatibility complex class II-restricted presentation of a cytosolic antigen by autophagy. *Eur. J. Immunol.* **33**, 1250–1259.

Ohsumi, Y. (2001). Molecular dissection of autophagy: Two ubiquitin-like systems. *Nat. Rev. Mol. Cell Biol.* **2**, 211–216.

Paludan, C., Schmid, D., Landthaler, M., Vockerodt, M., Kube, D., Tuschl, T., and Münz, C. (2005). Endogenous MHC class II processing of a viral nuclear antigen after autophagy. *Science* **307**, 593–596.

Rammensee, H., Bachmann, J., Emmerich, N. P., Bachor, O. A., and Stevanovic, S. (1999). SYFPEITHI: Database for MHC ligands and peptide motifs. *Immunogenetics* **50**, 213–219.

Rocha, N., and Neefjes, J. (2008). MHC class II molecules on the move for successful antigen presentation. *EMBO J.* **27**, 1–5.

Schmid, D., and Münz, C. (2007). Innate and adaptive immunity through autophagy. *Immunity* **26**, 11–21.

Schmid, D., Pypaert, M., and Münz, C. (2007). MHC class II antigen loading compartments continuously receive input from autophagosomes. *Immunity* **26**, 79–92.

Steinman, R. M., and Banchereau, J. (2007). Taking dendritic cells into medicine. *Nature* **449**, 419–426.

Suri, A., Walters, J. J., Rohrs, H. W., Gross, M. L., and Unanue, E. R. (2008). First signature of islet β-cell-derived naturally processed peptides selected by diabetogenic class II MHC molecules. *J. Immunol.* **180**, 3849–3856.

Zhou, D., Li, P., Lott, J. M., Hislop, A., Canaday, D. H., Brutkiewicz, R. R., and Blum, J. S. (2005). Lamp-2a facilitates MHC class II presentation of cytoplasmic antigens. *Immunity* **22**, 571–581.

CHAPTER TWENTY-FIVE

DETACHMENT-INDUCED AUTOPHAGY IN THREE-DIMENSIONAL EPITHELIAL CELL CULTURES

Jayanta Debnath

Contents

1. Introduction 424
2. Monitoring and Manipulating Autophagy In Mammalian Epithelial Cells 425
   2.1. Retroviral Vectors Encoding GFP-LC3 425
   2.2. Tandem Fluorescent Protein LC3 Chimeras 425
   2.3. Retroviral and lentiviral transduction in epithelial cells 426
3. Autophagy During Lumen Formation In 3D Epithelial Cultures 428
   3.1. Growth of MCF10A acini using overlay 3D culture and analysis of GFP-LC3 428
   3.2. Measuring lumenal cell death upon autophagy inhibition in 3D epithelial cultures 431
   3.3. Immunofluorescence for cleaved caspase-3 to detect lumenal apoptosis 432
   3.4. Ethidium bromide staining for analysis of lumenal cell death in 3D culture assays 433
4. Measuring Autophagy In Traditional Anoikis Assays 434
   4.1. Substratum detachment assay and immunoblotting for LC3 lipidation 434
   4.2. Visualizing GFP-LC3 in detached cells 436
5. Incubation of Cells With Integrin Function-Blocking Antibodies 437
6. Conclusions 437
Acknowledgments 438
References 438

Abstract

Integrin-mediated cell adhesion to extracellular matrix (ECM) is critical for normal epithelial cell survival; cells deprived of ECM contact rapidly undergo apoptotic cell death, termed anoikis. Recent work demonstrates that ECM

Department of Pathology, University of California, San Francisco, San Francisco, California, USA detachment also robustly induces autophagy, which protects epithelial cells from the stresses of matrix detachment, allowing them to survive provided they can reattach in a timely manner. This chapter details the methods used to measure and manipulate detachment-induced autophagy during lumen formation in three-dimensional *in vitro* glandular epithelial cultures as well as in traditional substratum detachment assays employed to study anoikis.

1. Introduction

Three dimensional (3D) organotypic culture systems have been increasingly used as powerful cell-based models to study epithelial cells within a tissue-relevant context and to rapidly identify genes and deconstruct signaling pathways that regulate glandular morphogenesis and epithelial cancer development. Using 3D culture conditions, epithelial cells proliferate to eventually form growth-arrested spherical acini characterized by polarized cells surrounding a hollow lumen (Hebner *et al.*, 2008). Studies of 3D lumen formation have also provided unique insight into the regulation of autophagy by integrin-mediated cell-matrix adhesion pathways. Although cells occupying the lumen undergo apoptosis due to extracellular matrix (ECM) deprivation (termed anoikis), autophagic vesicles have also been observed in these central cells prior to their clearance (Debnath *et al.*, 2002). Excessive self-eating was originally proposed to promote autophagic (type 2) cell death in the 3D lumen; however, recent work now demonstrates that inhibiting autophagy during 3D morphogenesis elicits both increased lumenal apoptosis and accelerated lumen formation (Fung *et al.*, 2008; Karantza-Wadsworth *et al.*, 2007). Similarly, autophagy delays the onset of anoikis in traditional substratum detachment assays (Fung *et al.*, 2008). These data are consistent with the hypothesis that autophagy mitigates the stresses of ECM detachment in cells occupying the lumen.

Because oncogenes protect cells from anoikis and confer long-term survival in the glandular lumen, two important topics for future investigation include determining how autophagy is regulated in ECM detached cells and delineating how autophagy contributes to the survival of cancer cells independently of cell-matrix contact (Lock and Debnath, 2008). To facilitate these studies, this chapter provides a collection of protocols to monitor and manipulate autophagy using *in vitro* three-dimensional culture systems as well as to measure the effects of autophagy inhibition on ECM detachment-induced cell death, via apoptosis or necrosis, in the lumen. Finally, as a complementary approach, methods to investigate autophagy during anoikis using substratum detachment assays are detailed.

2. Monitoring and Manipulating Autophagy in Mammalian Epithelial Cells

Numerous reagents have become available to monitor autophagy as well as to inhibit autophagy via RNAi mediated silencing of mammalian autophagy-related (ATG) genes. In this section, we focus on technical considerations unique to epithelial cell culture systems when creating the cell-based reagents necessary to monitor and manipulate mammalian autophagy. Most importantly, successful gene or small hairpin RNA delivery by standard transfection methods can prove difficult in both primary epithelial cells and several epithelial cell lines; fortunately, the retroviral or lentiviral transduction methods overviewed below largely overcome these technical barriers.

2.1. Retroviral Vectors Encoding GFP-LC3

To monitor autophagosome formation, my lab has created retroviral vectors encoding green fluorescent protein fused to the mammalian Atg8 ortholog, microtubule-associated protein 1 light chain 3 (GFP-LC3). During autophagy, LC3 is modified with the lipid phospatidylethanolamine (PE) through a ubiquitin-like conjugation process, upon which it specifically relocates to autophagosome precursors; the relocation of GFP-LC3 to easily visualized puncta has emerged as an important technique to monitor autophagosome formation (Kabeya et al., 2000). We have successfully transduced a number of epithelial cell types with retrovirus encoding GFP-LC3 to generate stable pools; these include epithelial cell lines, such as MCF10A human mammary cells, MCF7 carcinoma cells, Madin Darby Canine Kidney (MDCK) cells as well as primary human mammary epithelial cells (Fung et al., 2008). Importantly, although transient GFP-LC3 transfection is often limited by overexpression artifacts (Kuma et al., 2007), we have not observed such problems upon generating stable pools in epithelial cells ectopically expressing GFP-LC3 fusions (Fung et al., 2008).

2.2. Tandem Fluorescent Protein LC3 Chimeras

Similar to other GFP-LC3 ectopic expression strategies, an important limitation of this assay is that the acidic environment of the lysosome potently quenches the green fluorescent signal; as a result, punctate GFP-LC3 cannot be used to measure autophagosome maturation into autolysosomes (autophagic flux) unless cells are treated with lysosomal inhibitors (Klionsky et al., 2008). Alternatively, new experimental approaches using a tandem mCherry-GFP-tagged LC3 may now be used to assay autophagic flux (Kimura *et al.*, 2007; Pankiv *et al.*, 2007). Whereas the GFP signal is sensitive to the acidic and proteolytic conditions of the lysosome, mCherry (a variant of red fluorescent protein) is stable. Therefore, the punctate colocalization of mCherry and GFP fluorescence delineates an early autophagosome that has not fused with a lysosome; in contrast, an mCherry signal without GFP corresponds to an autolysosome (Pankiv *et al.*, 2007). An advantage of this technique is that it enables simultaneous quantification of autophagy induction and maturation to autolysosomes (flux) without any chemical treatments. We have created a retroviral construct encoding tandem mCherry-GFP-LC3 and generated MCF10A cells that stably express this fusion (Fig. 25.1).

2.3. Retroviral and lentiviral transduction in epithelial cells

For both ectopic gene expression and stable RNA interference, we have found that retroviral vectors using the long terminal repeat regions of Moloney Leukemia Virus to be very effective for stable gene expression (e.g., pBABE) (Morgenstern and Land, 1990). However, a technical constraint of retroviruses is that the viral long terminal repeat (LTR) regions used to promote ectopic gene expression are driven by upstream mitogenic inputs and cell cycle status. These become an important concern for studies of autophagy, because the starvation conditions commonly used to experimentally induce autophagy, notably incubation in saline buffer, theoretically will elicit reduced gene or shRNA expression and hence confound the results. Thus, the use of lentiviral strategies may prove to be a more useful and robust approach for stable gene and shRNA expression during studies of autophagy in epithelial cells (Salmon and Trono, 2007).

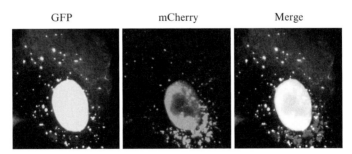

Figure 25.1 Stable expression of tandem mCherry-GFP-LC3 in mammary epithelial cells. MCF10A mammary cells were infected with a retrovirus encoding mCherry-GFP-LC3 to generate stable pools. Upon 2 h of starvation with Hanks's saline buffer, both double mCherry, GFP-positive LC3 puncta, which correspond to early autophagosomes, as well as single mCherry-positive LC3 puncta, which correspond to mature autolysosomes, are observed within cells. (See Color Insert.)

Finally, for cells of human origin, it is necessary to use a virus that either possesses an amphotropic envelope (i.e., a virus that can infect multiple species) or is pseudotyped with the envelope of vesicular stomatitis virus (VSV). For either retro- or lentivirus production, numerous producer lines exist, many of which are commercially available. Notably, these lines are usually derived from HEK293 cells, making them easily transfectable using calcium phosphate or lipid-based methods. Virus-containing supernatant can be collected and used to directly infect epithelial cells using the protocol herein. When establishing stable pools, it is desirable to infect with a multiplicity of infection (MOI) of three, which ensures that each host cell integrates a single copy of the retrovirus.

1. On the night prior to infection, plate early passage epithelial cells at 100,000 cells per well on a 6-well tissue culture dish. Cells should be cultured in their standard growth medium. If titer is being determined, several wells should be made to test several dilutions of the viral supernatant obtained from a producer line. In addition, if a drug selection scheme will be utilized, extra plates (that will not be infected) should be prepared to serve as positive controls for drug-induced killing. Cells will be 15%–20% confluent on the day of infection.
2. On the day of infection, virus-containing supernatant from the producer line is collected and filtered through a 0.45-cm filter to remove cell debris; alternatively, the collected medium can be spun at 1500 rpm for 10 min in a tissue-culture centrifuge to pellet the debris. Virus can be stored as aliquots at $-80\,°C$ until use; if a frozen aliquot is used for infection, it should be thawed quickly in a $37\,°C$ water bath. For VSV-pseudotyped viruses, aliquots can be subject to at least 2 cycles of repeated freezing and thawing. Although the retroviruses produced by most standard producer lines are replication-incompetent, strict BSL2 safety guidelines should be employed when working with these reagents.
3. Varying dilutions of the virus-containing supernatant are combined with the standard growth medium for a total volume of 4 ml. A range of dilutions from 1:4 to 1:4000 is a good starting point; once an appropriate titer is determined to obtain a MOI of 3 to 5, only that dilution needs to be used for subsequent infections.
4. Add polybrene (Sigma, H9268) to the virus incubation cocktail from step 3 at a final concentration of 8 $\mu g/ml$. A 1000x stock solution (8 mg/ml) of polybrene is prepared by dissolving the solid in sterile H_2O and sterile filtering this solution with a 0.2-μm filter. Aliquots are stored at $-20\,°C$ and can be repeatedly frozen and thawed.
5. Aspirate the medium from the cells and add the virus incubation cocktail. Return to a 5% CO_2 humidified incubator at $37\,°C$.
6. Five h post-infection, feed the cells with an additional 6 ml of medium.
7. Replace the medium on the following day (18 h post-infection).

8. Many retroviral and lentiviral vectors contain drug-resistance cassettes (e.g., puromycin, neomycin, or hygromycin) to select for resistant pools carrying the gene or shRNA of interest. For such vectors, begin drug selection at 36–48 h post-infection. Concentrations for commonly used drugs include: 0.5 μg/ml puromycin, 250–300 μg/ml active G418, and 50 μg/ml hygromycin. However, the optimal concentrations are highly dependent on both cell type and the retroviral vector being used. Drug selection should be continued until the uninfected drug-treated controls have completely died; for puromycin or hygromycin, this will occur by 4–5 days of treatment and for G418, after 9–10 days of treatment. If cells become confluent, they should be split (1:3 passage) to maintain the efficacy of the selection agent.
9. Following selection, the cells should be passed once to allow recovery from drug treatment. Thereafter, the stable pools can be utilized for the autophagy assays discussed below.
10. For retroviral- or lentiviral-driven small hairpin RNAs targeting ATG genes, we have found that stable knockdown is achieved within 1 week of culture. However, the efficacy of ATG knockdown is lost upon extended propagation of cells. As a result, it is recommended that freshly infected shRNA-expressing cells be used for assays. In 3D culture assays, robust ATG knockdown is maintained for at least 20 days.

3. Autophagy During Lumen Formation In 3D Epithelial Cultures

To form acini with a hollow lumen, epithelial cells are often cultured in a matrix of laminin-rich reconstituted basement membrane ECM derived from Engelbreth-Holm-Swarm tumor (EHS, commercially available as Matrigel); the exogenously provided laminin in the EHS-derived matrix is primarily responsible for driving lumen formation when epithelial cells are cultured three-dimensionally (Gudjonsson *et al.*, 2002). Although a variety of different protocols can be used to grow epithelial cells in 3D culture, the most tractable method, which is described herein, uses a thin gel bed (approximately 1 mm thick) of Matrigel upon which epithelial cells are seeded as single cells and overlaid with culture medium containing diluted Matrigel (Debnath *et al.*, 2003a).

3.1. Growth of MCF10A acini using overlay 3D culture and analysis of GFP-LC3

1. Thaw Growth Factor-Reduced Matrigel (BD Biosciences, 354230) at 4 °C. This usually takes several hours to overnight depending on the size of the aliquot. Matrigel remains as a viscous liquid on ice but rapidly solidifies at room temperature.

2. Add 30–40 µl of Matrigel to each well of an 8-well glass cover slip chamber slide (Nunc, 154409) and spread evenly in the well using the P-200 tip of a pipette. Coverslip slides are optimal because they facilitate visualization of punctate GFP-LC3 on an inverted confocal microscope. For standard immunofluorescent staining (see subsequently), glass chamber slides can be used (Nunc, 154534). While spreading, it is important not to overspread because this results in a high meniscus on the border. Place the slides in a cell-culture incubator to allow the basement membrane to solidify for at least 15 min.
3. While the Matrigel is solidifying, trypsinize a confluent plate of MCF10A cells expressing GFP-LC3 and resuspend in 2.0 ml Resuspension Medium (DMEM/F12 with 20% horse serum, 100 U/ml Penicillin G, and 100 µg/ml streptomycin). Use an additional 3 ml of this medium to rinse the plate and combine in a 15-ml conical tube. Details on the care and passage of MCF10A cells are available in the literature (Debnath et al., 2003a).
4. Spin the cells at 900 rpm in a tissue culture centrifuge for 3–5 min.
5. Resuspend the cells in 8 ml of assay medium (DMEM/F12 with 2% horse serum, penicillin/streptomycin, 10 µg/ml insulin (Sigma, I882), 0.5 µg/ml hydrocortisone (Sigma, H0888), 100 ng/ml cholera toxin (Sigma, C8052)) lacking epidermal growth factor (EGF). Details on the preparation of this medium as well as other medias to culture MCF10A cells are available in the literature (Debnath et al., 2003a).
6. Count cells using a hemocytometer and aliquot 200,000 cells to a fresh tube. Pellet in a tissue culture centrifuge at 900 rpm for 5 min.
7. Resuspend the 200,000 cells in 8 ml of assay medium for a stock of 25,000 cells/ml.
8. Prepare a stock of assay medium plus 4% Matrigel. Also add EGF (Peprotech, 100-15) or other growth factors, pharmacological agents, and so on, to this medium at 2x the final concentration that is desired. For standard assays in our laboratory, this corresponds to an EGF concentration of 10 ng/ml. It is necessary to make 200 µl of this stock for each assay that will be performed; furthermore it is recommended that enough stock solution for n + 2 assays be prepared to account for pipetting errors.
9. Mix the cells from part 6 (25,000 cells/ml) with the Matrigel-containing medium from part 7 in a 1:1 ratio.
10. Plate 400 µl of this mixture per well on top of the solidified Matrigel in each well of the chamber slide from part 1. This corresponds to a final overlay solution of 5000 cells/well in medium containing 2% Matrigel and 5 ng/ml EGF.
11. Allow the cells to grow in a 5% CO_2 humidified incubator at $37°C$. The cells should be refed with assay medium containing 2% Matrigel and 5 ng/ml EGF every 4 days. (The day that the assay is set up

corresponds to day 0; thus, feed on days 4, 8, 12, 16, etc.). The cells should form clusters by day 5–6 of 3D cultures and subsequently start forming hollow lumen.

12. Without treatment with lysosomal inhibitors, punctate GFP-LC3 can be observed at low levels throughout individual acini (Fig. 25.2A). Between 4 and 6 h prior to analysis, replace medium with EGF-containing assay medium of 10 μg/ml E64d and 10 μg/ml pepstatin A. Upon treatment, two distinct patterns of GFP-LC3 emerge. In developing acini that have not undergone lumenal clearance, the highest levels of punctate GFP-LC3 are found in cells occupying the lumenal space (Fig. 25.2B, top row). Interestingly, in acini possessing a hollow lumen, apically oriented GFP-LC3 clusters are observed in

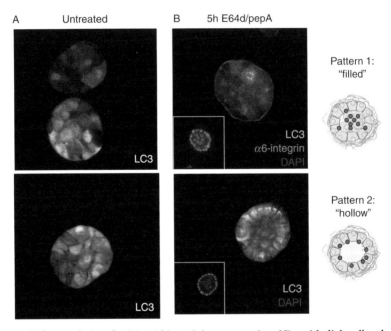

Figure 25.2 Analysis of LC3 within acini grown using 3D epithelial cell culture. (A) In day 8 MCF10A acini stably expressing GFP-LC3, puncta are observed throughout the structures. (B) MCF10A acini were treated with the lysosomal cathepsin inhibitors E64d and pepstatin A (pepA; 10 μg/ml each) to delineate sites of LC3 turnover in the lysosome. In these structures, two patterns emerge. First, in acini that have not cleared their lumen (top panel), punctate LC3 (green) is predominantly observed within central cells that lack contact with the underlying basement membrane ECM, which is delineated by stabilized α6 integrin receptor subunit located at sites of direct cell-matrix contact (red). Second, in acini in which the lumen has cleared (bottom panel), apically oriented clusters of GFP-LC3 are observed in the cells that contact the underlying basement membrane. (See Color Insert.)

cells that remain attached to the underlying basement membrane (Fig. 25.2B, bottom row).
13. At the time of analysis, cells can either be fixed with freshly prepared 2% paraformaldehyde or alternatively visualized without fixation using spinning disc confocal microscopy.

3.2. Measuring lumenal cell death upon autophagy inhibition in 3D epithelial cultures

In recent years, numerous oncogenes and growth factor signaling pathways have been demonstrated to aberrantly promote cell survival in the lumenal space of 3D structures; the resultant phenotypes recapitulate the lumenal filling observed in premalignant cancers (Debnath and Brugge, 2005). Because inhibiting apoptosis is not sufficient for long-term survival in the lumen, significant interest exists in identifying additional processes that facilitate the survival of oncogenic epithelial cells lacking ECM contact (Debnath et al., 2002; Mailleux et al., 2007). Autophagy may be one such mechanism for tumor cell survival in the lumen (Debnath, 2008). To test this hypothesis, additional studies assaying how autophagy inhibition affects lumenal apoptosis and 3D lumenal filling in oncogene-expressing acini must be carried out.

Two primary approaches to inhibit autophagy in 3D culture exist. First, ATGs can be stably depleted from cells prior to 3D culture assays using lentiviral or retroviral driven shRNAs; such reagents can be delivered to cells using the aforementioned transduction protocol. For experiments using RNAi-mediated ATG knockdown to inhibit autophagy, the knockdown of multiple ATGs must be analyzed to rule out phenotypic effects associated with an individual ATG. Second, autophagy can be inhibited by treating 3D cultures with chemical inhibitors of autophagy, such as 3-methyladenine (3-MA) or lysosomal function, such as hydroxychloroquine (HCQ) or bafilomycin A_1 (BafA). Unlike shRNA-mediated knockdown, the use of chemical inhibitors allows one to easily assay how autophagy inhibition has an acute impact on lumenal cell death at specific time points and stages during 3D morphogenesis. However, as none of the commonly used agents specifically inhibit autophagy, chemical inhibition studies are best viewed as a complementary approach, rather than as an alternative, to ATG knockdown when inhibiting autophagy in 3D culture (Klionsky et al., 2008).

Regardless of the method used to inhibit autophagy, techniques to precisely quantify programmed cell death in the lumenal space are imperative. Detailed protocols for two common methods to measure cell death in 3D culture are delineated in the following sections. The first technique uses immunostaining for the cleavage product of activated caspase-3; this method is highly specific for the involvement of an apoptotic process (type I cell death) because it detects activation of an executioner caspase

(Danial and Korsmeyer, 2004). The second protocol employs *in situ* staining with ethidium bromide (EtBr), a DNA intercalating dye that readily labels dying cells whose plasma membrane integrity has been compromised; accordingly, this technique is a simple and rapid assay to examine and quantify cell death resulting from either apoptosis or necrosis.

3.3. Immunofluorescence for cleaved caspase-3 to detect lumenal apoptosis

1. Fix acini with 2% paraformaldehyde (2% PFA in PBS, pH 7.4, freshly prepared) for 20 min at room temperature. For 8-well chamber slides, 500 μl volumes are appropriate for the fixation, permeabilization and all subsequent washing steps in this protocol.
2. Permeabilize with PBS containing 0.5% Triton X-100 for 10 min at 4 °C.
3. Rinse 2 times with PBS/Glycine (130 mM NaCl; 7 mM Na$_2$HPO$_4$; 3.5 mM NaH$_2$PO$_4$; 100 mM glycine, pH 7.4), 10–15 min per wash at room temperature.
4. Primary Block: Incubate with 200 μl/well of IF Buffer (130 mM NaCl; 7 mM Na$_2$HPO$_4$; 3.5 mM NaH$_2$PO$_4$; 7.7 mM NaN$_3$; 0.1% bovine serum albumin; 0.2% Triton X-100; 0.05% Tween-20, pH 7.4) + 10% goat serum (Invitrogen, 10000C) for 1–1.5 h at room temperature.
5. *Optional* secondary block: Aspirate the primary block and incubate with 100 μl/well of secondary block (IF Buffer + 10% goat serum + 20 μg/ml of goat antimouse F(ab')$_2$ fragment (Jackson ImmunoResearch, 015-000-007) for 30–40 min at room temperature.
6. *Primary antibody:* Dilute primary antibodies in block solution (see step 4) and incubate overnight (15–18 h) at 4 °C. Rabbit polyclonal antibody against cleaved caspase-3 (Cell Signaling Technologies, 9661) to mark apoptotic cells is used at 1:200 dilution and mouse monoclonal antibody against laminin 5 (Millipore, MAB19562) to mark the basement membrane is used at 1:500 dilution. Occasionally, overnight incubation at 4 °C elicits liquefaction of the basement membrane and extensive lifting of the acini during subsequent washing steps. This varies with each lot of Matrigel used for the morphogenesis assay, and unfortunately, cannot easily be predicted. If this problem arises, it is advisable to perform the primary antibody incubations overnight at room temperature rather than 4 °C.
7. Rinse 3 times (20 min each) with IF buffer at room temperature with gentle rocking.
8. *Secondary antibody:* Incubate with fluorescent conjugated secondary antibodies in IF buffer + 10% goat serum for 40–50 min at room temperature. We recommend Alexa conjugated, highly cross-absorbed secondary antibodies from Molecular Probes/Invitrogen used at 1:200 dilution; in our experience, the secondary reagents exhibit low levels of

background and minimal cross-reactivity between species, making them useful for double immunostaining procedures.
9. Rinse 3 times (20 min each) with IF buffer at room temperature with gentle rocking.
10. To counterstain nuclei, incubate with PBS containing 5 μM TOPRO-3 (Molecular Probes/Invitrogen, T3605) and/or 0.5 ng/ml 4′,6-diamidino-2-phenylindole (DAPI, Sigma, 9542) for 15 min at room temperature.
11. Rinse once with PBS for 5 min at room temperature.
12. Mount with freshly prepared Prolong Gold Antifade Reagent (Molecular Probes/Invitrogen, P36934) and allow to dry overnight at room temperature. Once dry, slides can be stored at 4 °C for up to 1 week or at −20 °C for up to 2 months.
13. When imaged using confocal microscopy, the highest levels of apoptosis are observed in the lumens of wild-type acini between day 8 and day 12 in 3D culture. To measure lumenal apoptosis, the number of cells occupying the lumen positive for cleaved caspase-3, is quantified at various time points during 3D culture. Cells occupying the lumen are defined as those lacking direct contact with the basement membrane, as delineated by immunostaining with laminin-5 (Debnath *et al.*, 2003b). Because the exact onset of lumenal apoptosis and lumen formation in 3D culture assays can vary among experiments, wild-type or empty vector-control acini should be directly compared with regard to experimental variables, such as ATG knockdown, for each and every experiment.
14. This protocol is generally suitable for a wide range of immunofluorescence studies; antibodies commonly used in the analysis of 3D cultures have been cataloged in the literature (Debnath *et al.*, 2003a).

3.4. Ethidium bromide staining for analysis of lumenal cell death in 3D culture assays

1. Prior to the experiment, prepare mixture of 1 μM ethidium bromide (EtBr, Sigma, E7637) in PBS. If desired, Hoeschst 33342 (2.5–5 μg/ml, Sigma, B2261) can be added to the staining solution to counterstain nuclei. Warm to 37 °C.
2. Remove the growth medium from the 3D cultures to be analyzed and carefully wash each well with 500 μl of PBS. Then, add the EtBr-containing solution to the 3D culture and incubate for 15–30 min at 37 °C in a tissue-culture incubator.
3. Analyze using indirect immunofluorescence on a microscope equipped with a mercury lamp. EtBr-positive cells are readily detected on the rhodamine channel available on standard filter sets. If a high autofluorescence background interferes with proper microscopy, the staining solution can be replaced with PBS.

4. For a quantitative measure of cell death within a specific culture, the culture is stained with EtBr for exactly 15 min and the percentage of acini containing one or more EtBr-positive cells within a culture is counted; at least 200 structures should be counted for such quantification. Alternatively, individual structures can be analyzed by confocal microscopy to enumerate the EtBr-positive cells present in the lumen.

4. Measuring Autophagy In Traditional Anoikis Assays

Traditionally, anoikis is assayed by culturing epithelial cells in low attachment (suspension) conditions, upon which they undergo apoptosis within hours to days (Frisch, 1999). Such assays have also provided direct evidence that the lack of extracellular matrix contact promotes autophagy in multiple cell types (Fung et al., 2008). Thus, suspension assays represent an important parallel approach to 3D cultures to interrogate how autophagy contributes to adhesion-independent survival. They are also experimentally tractable for delineating the intracellular signals that control detachment-induced autophagy.

4.1. Substratum detachment assay and immunoblotting for LC3 lipidation

1. Dissolve 6 mg/ml poly (2-hydroxyethyl methacrylate) (Poly-HEMA, Sigma, P3932) in 95% ethanol, which will require 1–2 days to solubilize at room temperature with stirring. Do not heat.
2. Coat tissue-culture plates with the 6 mg/ml poly-HEMA solution. Although any plate can be coated for suspension assays, we typically use 6-well dishes and coat each individual well with 1.5 ml of the poly-HEMA solution.
3. Incubate tissue-culture dishes at 37 °C in an oven for several days until dry.
4. Prior to using these dishes for suspension experiments, treat the coated plates with ultraviolet (UV) light to sterilize; this is most easily performed using the UV lamp in most standard biosafety cabinets. Rinse coated plates several times with sterile water, followed by PBS.
5. Detach cells to be assayed in their usual manner. Take care to generate a single cell suspension. Count cells and resuspend in its standard culture medium at a concentration of 200,000 per ml.
6. Suspend 400,000 cells (2 ml of the cell solution prepared in step 4) onto the poly-HEMA coated wells. Importantly, for each cell type or condition being analyzed, cells should be replated onto noncoated

tissue culture dishes to serve as attached controls. Return all cells to a 5% CO_2 humidified incubator at 37 °C.

7. Starting at 6–12h of suspension, monitor cells using phase microscopy to ensure they have not reattached to the poly-HEMA coated plates; if cells are found to reattach, the experiment must be repeated using plates more evenly coated with poly-HEMA.
8. To assay detachment-induced autophagy via immunoblotting for PE-modified LC3 (LC3-II), assays should be performed both in the presence and absence of lysosomal inhibitors to measure LC3-II turnover (autophagic flux) in the lysosome. To do so, two parallel suspension cultures should be initially prepared for each time point as well as the attached controls. At 2–4h prior to harvest of a time point, lysosomal cathepsin inhibitors, E64d and pepstatin A (E + P), are added directly to the culture medium of one of these two parallel cultures at 10 μg/ml each. Alternatively, cultures can be treated with 10 nM BafA for 30–60 min to inhibit lysosomal acidification.
9. Harvest each well containing the suspended cells into a 15-ml conical tube. Wash wells with sterile PBS to collect residual cells and add these washes to the harvested cells in the 15-ml tube.
10. Centrifuge the cells at 1200 rpm in a tissue culture centrifuge for 3–5 min. Carefully aspirate the medium, wash the cell pellet with sterile PBS, and transfer to a microcentrifuge tube. Repellet at 6000 rpm for 2 min, carefully aspirate the PBS, and lyse the pellet on ice (4 °C) using RIPA buffer (1% Triton X-100, 1% sodium deoxycholate, 0.1% SDS, 25 mM Tris, pH 7.6, 150 mM NaCl, 10 mM NaF, 10mM β-glycerophosphate (Sigma, G6251), 1 mM Na_3VO_3 (Sigma, S6508), 10nM calyculin A (Sigma, C5552) plus protease inhibitor cocktail (Sigma, P8340). The attached controls can be washed and lysed directly on the plate.
11. Lyse on ice (4 °C) for 20–30 min. Then freeze lysates at −80 °C in the microcentrifuge tube. For earlier time points, lysates can be stored at −80°C until the final time point is obtained; however, the final time point should be frozen and stored at −80°C for at least 3–5 h prior to the clarification of all lysates for immunoblot analysis. A single freeze-thaw will significantly improve protein yield in the lysates prepared from suspension cultures without adversely affecting LC3-II immunoblotting. Thaw and clarify all lysates by centrifugation at 15,000 rpm for 15–30 min at 4 °C and thereafter use standard techniques for Western blotting to detect PE-modified LC3-II. We use a rabbit polyclonal antibody generated in our laboratory against the N-terminus of LC3 at 1:500 to 1:1000 for immunoblotting (Fung et al., 2008) and α-tubulin (Sigma, T9026) at 1:10,000 as a loading control (Fig 25.3A).
12. Multiple time points should be analyzed following suspension. The onset of detachment-induced autophagy is dependent on cell type; typically, it is observed as early as 6–12 h and continues thereafter. The onset of

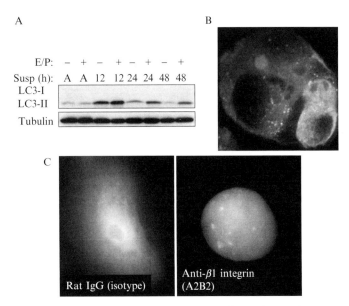

Figure 25.3 Analysis of LC3 during epithelial cell anoikis. (A) Primary human mammary epithelial cells were grown attached (A) or suspended over poly-HEMA-coated plates for the indicated times in full growth medium. Cells were lysed and subjected to immunoblotting with anti-LC3 and antitubulin antibodies. When indicated by E/P +, E64d and pepstatin A (10μg/ml each) were added to the cultures 4 h prior to lysis to measure LC3-II turnover. (B) MCF10A mammary epithelial cells expressing GFP-LC3 cells were grown detached on poly-HEMA-coated plates in complete growth medium for 48 h and analyzed using confocal microscopy. Punctate GFP-LC3 is detected throughout cells within suspended cell clusters. (C) MCF10A cells expressing GFP-LC3 were incubated with an isotype control (rat IgG) or a function-blocking antibody directed against the β1-integrin receptor subunit (A2B2) for 24 h. Upon blocking β1-integrin function, cells partially detach from the underlying substratum and exhibit a rounded phenotype, as well as increases in punctate GFP-LC3.

apoptosis (anoikis) also depends on cell type; in both primary human mammary epithelial cells and nontumorigenic cell lines, such as MCF10A cells, apoptosis starts at approximately 24–36 h of suspension, peaking at 48–72 h (Fung et al., 2008; Reginato et al., 2003).

4.2. Visualizing GFP-LC3 in detached cells

One can also monitor autophagosome formation during anoikis by visualizing punctate GFP-LC3 in suspended cells:

1. Place GFP-LC3-expressing cells into suspension as described for steps 1–6 in the substratum detachment protocol.

2. At the desired time point, harvest cells into a 15-ml conical tube and fix with 2% paraformaldehyde for 5–10 min at room temperature. Pellet the cells using a tissue culture centrifuge at 1200 rpm for 5–7 min, aspirate away the fixative, and resuspend pellet in PBS (1 ml).
3. Transfer to a microcentrifuge tube. Wash cell pellets twice with PBS (1 ml). Remove residual PBS and resuspend pellet in a small volume (less that 50 μl per cell pellet) of Immu-mount glycerol mounting medium (Thermo-Shandon, 9990402) Tease the resuspended pellet onto a glass slide and mount with a coverslip. Allow coverslip to dry completely and image using a confocal microscope. Puncta are visible throughout the cell clusters harvested from suspension cultures (Fig. 25.3B).

5. Incubation of Cells With Integrin Function-Blocking Antibodies

As an alternative to culturing cells on low attachment plates, treating epithelial cells with function-blocking antibodies directed against specific integrin receptor subunits can also be used to induce autophagy. This strategy allows one to test if signals downstream of specific integrin receptors are responsible for detachment-induced autophagy in a particular cell type. For example, blocking $\beta 1$ integrin function promotes autophagosome formation (LC3-II formation) in human mammary epithelial cells, suggesting that the $\beta 1$ integrin subunit conveys an important signal to suppress autophagy in attached conditions (Fung *et al.*, 2008). Furthermore, this technique offers technical advantages for imaging and enumerating punctate LC3 in detached cells, which can be difficult due to cell clustering in more traditional suspension assays. In contrast, when cells are treated with function-blocking antibodies against specific integrin subunits, they usually remain as single rounded cells and stay partially attached to the coverslip, which makes them better suited for quantifying punctate GFP-LC3 (Figure 25.3C).

6. Conclusions

The methods described in this chapter have been successfully used to demonstrate that autophagy is robustly induced during extracellular matrix detachment and that this fundamental intracellular self-digestion process promotes epithelial cell survival during anoikis (Fung *et al.*, 2008). These initial findings resemble other situations where autophagy facilitates mammalian cell survival during stress, such as nutrient deprivation, growth factor withdrawal, endoplasmic reticulum (ER) stress, and metabolic stress (Boya *et al.*, 2005; Degenhardt *et al.*, 2006; Kouroku *et al.*, 2007; Lum *et al.*, 2005; Ogata *et al.*, 2006). Given that ECM engagement of integrin adhesion receptors is essential for proper epithelial cell function, future studies of detachment-induced autophagy in both normal and cancerous cells will provide exciting new insights into this previously unrecognized aspect of cell adhesion receptor biology.

ACKNOWLEDGMENTS

Grant support to J.D. includes a NIH KO8 Award (CA098419), Culpeper Scholar Award (Partnership for Cures), an AACR/Genentech BioOncology Career Award, and an HHMI Early Career Award.

REFERENCES

Boya, P., Gonzalez-Polo, R. A., Casares, N., Perfettini, J. L., Dessen, P., Larochette, N., Metivier, D., Meley, D., Souquere, S., Yoshimori, T., Pierron, G., Codogno, P., *et al.* (2005). Inhibition of macroautophagy triggers apoptosis. *Mol. Cell. Biol.* **25,** 1025–1040.

Danial, N. N., and Korsmeyer, S. J. (2004). Cell death: Critical control points. *Cell* **116,** 205–219.

Debnath, J. (2008). Detachment-induced autophagy during anoikis and lumen formation in epithelial acini. *Autophagy* **4,** 351–353.

Debnath, J., and Brugge, J. S. (2005). Modelling glandular epithelial cancers in three-dimensional cultures. *Nat. Rev. Cancer* **5,** 675–688.

Debnath, J., Mills, K. R., Collins, N. L., Reginato, M. J., Muthuswamy, S. K., and Brugge, J. S. (2002). The role of apoptosis in creating and maintaining luminal space within normal and oncogene-expressing mammary acini. *Cell* **111,** 29–40.

Debnath, J., Muthuswamy, S. K., and Brugge, J. S. (2003a). Morphogenesis and oncogenesis of MCF-10A mammary epithelial acini grown in three-dimensional basement membrane cultures. *Methods* **30,** 256–268.

Debnath, J., Walker, S. J., and Brugge, J. S. (2003b). Akt activation disrupts mammary acinar architecture and enhances proliferation in an mTOR-dependent manner. *J. Cell Biol.* **163,** 315–326.

Degenhardt, K., Mathew, R., Beaudoin, B., Bray, K., Anderson, D., Chen, G., Mukherjee, C., Shi, Y., Gelinas, C., Fan, Y., Nelson, D. A., Jin, S., *et al.* (2006). Autophagy promotes tumor cell survival and restricts necrosis, inflammation, and tumorigenesis. *Cancer Cell* **10,** 51–64.

Frisch, S. M. (1999). Methods for studying anoikis. *Methods Mol. Biol.* **129,** 251–256.

Fung, C., Lock, R., Gao, S., Salas, E., and Debnath, J. (2008). Induction of autophagy during extracellular matrix detachment promotes cell survival. *Mol. Biol. Cell* **19,** 797–806.

Gudjonsson, T., Ronnov-Jessen, L., Villadsen, R., Rank, F., Bissell, M. J., and Petersen, O. W. (2002). Normal and tumor-derived myoepithelial cells differ in their ability to interact with luminal breast epithelial cells for polarity and basement membrane deposition. *J. Cell Sci.* **115,** 39–50.

Hebner, C., Weaver, V. M., and Debnath, J. (2008). Modeling morphogenesis and oncogenesis in three-dimensional breast epithelial cultures. *Annu. Rev. Pathol.* **3,** 313–339.

Kabeya, Y., Mizushima, N., Ueno, T., Yamamoto, A., Kirisako, T., Noda, T., Kominami, E., Ohsumi, Y., and Yoshimori, T. (2000). LC3, a mammalian homologue of yeast Apg8p, is localized in autophagosome membranes after processing. *EMBO J.* **19**, 5720–5728.

Karantza-Wadsworth, V., Patel, S., Kravchuk, O., Chen, G., Mathew, R., Jin, S., and White, E. (2007). Autophagy mitigates metabolic stress and genome damage in mammary tumorigenesis. *Genes Dev.* **21**, 1621–1635.

Kimura, S., Noda, T., and Yoshimori, T. (2007). Dissection of the autophagosome maturation process by a novel reporter protein, tandem fluorescent-tagged LC3. *Autophagy* **3**, 452–460.

Klionsky, D. J., Abeliovich, H., Agostinis, P., Agrawal, D. K., Aliev, G., Askew, D. S., Baba, M., Baehrecke, E. H., Bahr, B. A., Ballabio, A., Bamber, B. A., Bassham, D. C., *et al.* (2008). Guidelines for the use and interpretation of assays for monitoring autophagy in higher eukaryotes. *Autophagy* **4**, 151–175.

Kouroku, Y., Fujita, E., Tanida, I., Ueno, T., Isoai, A., Kumagai, H., Ogawa, S., Kaufman, R. J., Kominami, E., and Momoi, T. (2007). ER stress (PERK/eIF2alpha phosphorylation) mediates the polyglutamine-induced LC3 conversion, an essential step for autophagy formation. *Cell Death Differ.* **14**, 230–239.

Kuma, A., Matsui, M., and Mizushima, N. (2007). LC3, an autophagosome marker, can be incorporated into protein aggregates independent of autophagy: Caution in the interpretation of LC3 localization. *Autophagy* **3**, 323–328.

Lock, R., and Debnath, J. (2008). Extracellular matrix regulation of autophagy. *Curr. Opin. Cell Biol.* **20**, 583–588.

Lum, J. J., Bauer, D. E., Kong, M., Harris, M. H., Li, C., Lindsten, T., and Thompson, C. B. (2005). Growth factor regulation of autophagy and cell survival in the absence of apoptosis. *Cell* **120**, 237–248.

Mailleux, A. A., Overholtzer, M., Schmelzle, T., Bouillet, P., Strasser, A., and Brugge, J. S. (2007). BIM regulates apoptosis during mammary ductal morphogenesis, and its absence reveals alternative cell death mechanisms. *Dev. Cell* **12**, 221–234.

Morgenstern, J. P., and Land, H. (1990). Advanced mammalian gene transfer: High titre retroviral vectors with multiple drug selection markers and a complementary helper-free packaging cell line. *Nucleic Acids. Res.* **18**, 3587–3596.

Ogata, M., Hino, S., Saito, A., Morikawa, K., Kondo, S., Kanemoto, S., Murakami, T., Taniguchi, M., Tanii, I., Yoshinaga, K., Shiosaka, S., Hammarback, J. A., *et al.* (2006). Autophagy is activated for cell survival after endoplasmic reticulum stress. *Mol. Cell Biol.* **26**, 9220–9231.

Pankiv, S., Clausen, T. H., Lamark, T., Brech, A., Bruun, J. A., Outzen, H., Øvervatn, A., Bjørkøy, G., and Johansen, T. (2007). p62/SQSTM1 binds directly to Atg8/LC3 to facilitate degradation of ubiquitinated protein aggregates by autophagy. *J. Biol. Chem.* **282**, 24131–24145.

Reginato, M. J., Mills, K. R., Paulus, J. K., Lynch, D. K., Sgroi, D. C., Debnath, J., Muthuswamy, S. K., and Brugge, J. S. (2003). Integrins and EGFR coordinately regulate the pro-apoptotic protein Bim to prevent anoikis. *Nat. Cell. Biol.* **5**, 733–740.

Salmon, P., and Trono, D. (2007). Production and titration of lentiviral vectors. *Curr. Protoc. Hum. Genet.* **12**(Unit 12), 10.

CHAPTER TWENTY-SIX

Methods for Inducing and Monitoring Liver Autophagy Relative to Aging and Antiaging Caloric Restriction in Rats

Alessio Donati,* Gabriella Cavallini,* and Ettore Bergamini*

Contents

1. Overview	442
2. Models of Caloric Restriction	442
2.1. Animals	443
2.2. Daily dietary restriction paradigm (DR)	443
2.3. Timed meal feeding paradigm	443
3. Methods to Induce and Monitor Liver Autophagy	444
3.1. How to maximize autophagic proteolysis by the *in vivo* administration of antilipolytic drugs	445
3.2. Autophagic protein degradation in perfused liver	446
4. A Minimally Invasive Procedure for the Rapid Evaluation of the Induced Intensification of Autophagy	448
5. Autophagic Protein Degradation in Isolated Liver Cells	449
6. Concluding Remarks	453
References	454

Abstract

The functioning of macroautophagy declines with increasing age in the liver of *ad libitum* fed animals, whereas it is preserved in rats submitted to antiaging caloric restriction. In this perspective, monitoring autophagy during aging may provide a useful biomarker of aging. Here we describe a procedure for the quantification of the *ex vivo* functioning of autophagy by the use of single-pass liver perfusion to measure the rate of degradation of prelabeled long-lived proteins. The maximum rate of autophagy can be measured after the pharmacological suppression of the supply of free fatty acids during fasting, which intensifies the activation of autophagy by a physiological mechanism.

* Centro di Ricerca Interdipartimentale di Biologia e Patologia dell'Invecchiamento, Università di Pisa, Pisa, Italy

The effects of treatment on the plasma level of branched chain amino acids may be used as a minimally invasive indicator of the intensification of autophagy. The effects of aging on autophagic proteolysis and on amino acid and hormone control can also be assessed by measuring the rate of the 3-methyladenine-sensitive valine released from isolated liver cells incubated *in vitro*.

1. Overview

Aging is characterized by a progressive accumulation of damaged macromolecules, organelles, and cytomembranes, which may account for the age-associated malfunctioning of many biological processes. Inefficiency and failure of maintenance, repair, and turnover pathways may be the main cause of damage accumulation during aging (Rattan and Clark, 2005).

Macroautophagy is the degradation/recycling system ubiquitous in eukaryotic cells that contributes to the turnover and rejuvenation of cellular components (long-lived proteins, cytomembranes and organelles) and generates nutrients during fasting, under the control of amino acids and pancreatic hormones (e.g., Donati, 2006; Meijer and Codogno, 2006).

Recent evidence shows that a decline of macroautophagy is involved in aging (Bergamini, 2006; Bergamini *et al.*, 2003; Cuervo *et al.*, 2005; Terman *et al.*, 2007), whereas the chronic stimulation of the function by food deprivation may mediate the antiaging action of caloric restriction (CR) (Bergamini *et al.*, 2003; Jia and Levine, 2007; Wohlgemuth *et al.*, 2007). In this perspective, we describe how rat liver autophagy may be assayed to monitor biological aging in *ad libitum* fed rodents, and the benefits from caloric restriction and antiaging interventions.

2. Models of Caloric Restriction

Diet restriction (better if initiated at young age) is the only nutritional intervention that consistently extends the median and maximum life span and health span of animals (Masoro, 2005), counteracts age-related changes in tissue function (e.g., Payne, Dodd and Leeuwenburgh, 2003), and delays the age of onset and/or the rate of progression of most age-associated diseases (Masoro, 2005; Roth *et al.*, 2007).

The antiaging effects of dietary restriction can be induced by two different approaches: daily reduction of food consumption (e.g., 40%) or intermittent feeding (or every-other-day *ad libitum* feeding); the effects of treatments on health and life span are similar, but the effects on metabolism may be different (Bergamini *et al.*, 1990).

2.1. Animals

Rodents are the most widely used animal models for aging research, because they have a relatively short life span and small size compared to other mammals, and their upkeep requires a relatively small amount of resources. In addition, a large body of literature is available on their biological, behavioral, and gerontological characteristics. The most used rat strains in gerontological studies are Sprague-Dawley, Wistar, Long-Evans, Brown Norway, and Fisher 344. In our studies we use Sprague-Dawley albino male rats born and housed in the vivarium of the department and fed with a standard pellet chow diet (Harlan autoclavable Teklad diet, Harlan-Italy, containing: 12% water, 18.4% crude protein, 5.5% crude fiber, 5.6% crude ash; caloric content: 5.6 Kcal/g). Food is provided *ad libitum* until 2 months of age (1 month postweaning). At that time, rats are randomly assigned to dietary treatments. Rats in all groups have free access to water. The animals are kept in a 12 h light/12 h dark cycle (lights on at 6 a.m. and off at 6 p.m.). Temperature is 20–22 °C. Body weight, body conditions, caloric intake, and survival are assessed every month until age 24 months.

2.2. Daily dietary restriction paradigm (DR)

Food consumption by the ad libitum fed animals (AL, 4–5 animals per cage) is measured by weighing the given food and the amount of food left by the end of the timing period (usually 1 week). The DR animals are fed 60% of the AL food intake, and the amount is adjusted every month. In this paradigm, the animals are caged individually in order to feed every animal the same amount of food. An interesting effect is that the DR paradigm tends to transform the restricted rats from nibblers (as AL animals) to meal eaters, with daily long periods of fasting (see Fig. 26.1).

2.3. Timed meal feeding paradigm

Twenty Sprague-Dawley male rats are fed on an intermittent feeding regimen (they are fed *ad libitum* every other day and then fasted the day after). This paradigm allows for the multiple caging of the animals and therefore is much less expensive to perform. In addition, variance in body weight is similar to that found in the *ad libitum* fed population. The major disadvantage is that the degree of restriction is not totally under the control of the investigator. In the Sprague-Dawley rats, reduction of food intake may attain 30% and is associated with a 20% reduction of body weight.

The caloric intake, the body weight and the percent of surviving Sprague-Dawley male rats submitted to the different diets are shown in Table 26.1.

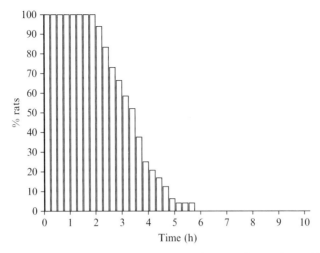

Figure 26.1 Time required for food consumption by rats submitted to 40% daily dietary restriction. On the ordinate; percentage rats still engaged in eating food.

Table 26.1 Body weight, caloric intake and percentage survival of *ad libitum*-fed and food-restricted male Sprague-Dawley rats (20 animals in each group)

Body weight (g)	AL	DR	EOD
2 mo	248 ± 9.6	248 ± 9.6	248 ± 9.6
12 mo	549 ± 12.7	388 ± 15.0	445 ± 17.6
24 mo	561 ± 22.2	365 ± 19.1	456 ± 29.5
Caloric intake (Kcal/d)	AL	DR	EOD
2 mo	72.2 ± 1.25	72.2 ± 1.25	72.2 ± 1.25
12 mo	77.6 ± 1.48	47	52.6 ± 2.89
24 mo	79.2 ± 1.87	48	56.6 ± 0.86
Percent survival	AL	DR	EOD
12 mo	97%	98%	98%
24 mo	55%	76%	82%

Notes: AL, *ad libitum* fed; EOD, fed intermittently (every other day); DR, 40% daily dietary restriction.

3. Methods to Induce and Monitor Liver Autophagy

Most of the studies on autophagy have been carried out in mammalian liver, presumably because of the quantitative importance of this organ during starvation (Blommaart *et al.*, 1997). In gerontological studies it is generally accepted that liver autophagy can be measured as the rate of autophagic proteolysis, which is responsible for the degradation of long-lived (resident) protein (Mortimore and Poso, 1987; Mortimore *et al.*, 1983; Schworer *et al.*, 1981). Usually, the rate of valine release *ex vivo* (from the single-pass perfused liver preparation) or *in vitro* (from isolated liver cells) is used to monitor proteolysis. (Valine is not appreciably metabolized in the liver and does not inhibit autophagy) (Mortimore and Poso, 1987). Valine release is measured in the presence of cycloheximide, to inhibit the reincorporation of valine into protein (Schworer *et al.*, 1981) (also see the chapter by Bauvy *et al.*, in this volume).

In the *ex vivo* assay, rats are injected intraperitoneally with ^{14}C-labeled valine 24 h prior to the liver perfusion (by this time the major fraction of the short-lived proteins has already been degraded). The rate of autophagic proteolysis is measured as the amount of labeled liver protein accounting for the ^{14}C-valine released in the perfusate (Bergamini *et al.*, 1993; Del Roso *et al.*, 2003). After the intensification of autophagy by an antilipolytic drug (see "How to maximize autophagic proteolysis by the *in vivo* administration of antilipolytic drugs"), the release of radioactivity from the perfused liver increases dramatically and data show that the increase always parallels changes in the release of valine (the specific activity of valine in the perfusate is unchanged) (Bergamini *et al.*, 1993). It may be mentioned that the levels of valine in liver blood increase with the same temporal pattern (Bergamini and Kovacs, 1990).

In the *in vitro* assay, the rate of proteolysis is assessed by the linear release of free valine (in a 10-min period) from the *in vitro* incubated isolated liver cells (Donati *et al.*, 2001; Venerando *et al.*, 1994) and the values of autophagic proteolysis can be obtained by subtracting the amount of valine released from cells incubated in the presence of the inhibitor of autophagy, 3-methyladenine (5m*M*) (Seglen and Gordon, 1982). It was shown that this parameter correlates linearly with changes in the amount of the sequestered protein in autophagic vacuoles (Donati *et al.*, 2001).

3.1. How to maximize autophagic proteolysis by the *in vivo* administration of antilipolytic drugs

It might be convenient to evaluate autophagic proteolysis in terms of percentage of maximum function, to evaluate change in maximum reserve capacity with increasing age. Maximum function of autophagic proteolysis can be observed after the intraperitoneal administration of an antilipolytic agent to fasting rats, which causes a sudden decrease in the free fatty acids supply to liver, leading to the intensification of autophagy (Bergamini *et al.*, 1993). A suitable antilipolytic agent is 3,5 dimethylpyrazole (Sigma Aldrich, D182001), and a useful (maximally active) dosage is 12 mg/kg body weight in 0.2 ml of saline. This tool provides a convenient (i.e., a safe, highly reproducible and timeable) physiological model to investigate maximum *in vivo* levels of autophagic function.

Fasted animals injected intraperitoneally with 3,5-dimethylpyrazole (DMP) show a sudden decrease in insulin and glucose plasma levels together with an increase in glucagon and corticosteroids: these changes precede the increase of liver autophagic vacuoles and autophagic proteolysis (Bergamini and Kovacs, 1990; Bergamini *et al.*, 1993; Pollera *et al.*, 1990). Treatment also increases the expression of LC3, a well-known marker of autophagosomes (Donati *et al.*, 2008a).

3.2. Autophagic protein degradation in perfused liver

To label long-lived (resident) liver proteins, rats are injected with ^{14}C-uniformly labeled valine (6 μCi/100 g body weight; 288.5 μCi/mole) intraperitoneally, 24 h prior to sacrifice. Food can be withdrawn 16–18 h before experimentation.

3.2.1. Liver perfusion

Livers are perfused *in situ* as described by Mortimore and Mondon (1970). Surgery is performed under pentobarbital anesthesia (50 mg/kg of body weight, given by an intraperiteoneal injection 5 min prior to surgery).

Briefly, we use the following protocol (see subsequently for additional details):

1. The anesthesized rat is placed into a thermostated cage (37 °C), the abdomen is opened and the liver is perfused with reconstituted blood (8 ml/min) by cannulating the exposed portal vein. The inferior caval vein is cannulated and the perfusate is collected in the test tube. Surgery can be performed in less than 3 min.
2. After a 7-min washout, two adjoining 1.5-min fractions of the outflow of the liver (approximately 12 ml each) are taken and centrifuged.
3. The supernatant fraction is used for the measurement of $^{14}C_6$-valine and of amino acid release.
4. By the end of the perfusion, samples of liver tissue are taken, proteins are purified, weighed, dissolved, counted in a scintillation counter and their specific activity is assessed (see "Determination of protein degradation and analytical procedure," steps 3–11).

3.2.2. Perfusion medium

To achieve an optimal oxygenation of liver tissue we use a perfusion medium containing erythrocytes (derived from outdated human blood stores or from bovine blood). Albumin is added as an oncotic agent.

1. The day before the experiment, 2000 ml of whole blood are centrifuged ($900 \times g$) for 10 min at 0 °C and the plasma (supernatant fraction) discarded.
2. Erythrocytes are washed twice in saline (0.18 M NaCl) and stored overnight at 2–4 °C.
3. A 10% (w/v) solution of bovine serum albumin in distilled water is dialyzed overnight in the cold (4–5 °C) against 10 volumes of glass-distilled water.
4. The morning after, the dialyzed albumin is filtered through 3-μm and then 0.2-μm Millipore filters under vacuum.
5. Erythrocytes are resuspended in Krebs Ringer bicarbonate (139 mM NaCl, 5.8 mM KCl, 2.9 mM CaCl$_2$, 1.4 mM KH$_2$PO$_4$, 1.4 mM, MgSO$_4$, 29 mM NaHCO$_3$) pH 7.4, 3% albumin, 10 mM glucose and 18 μM cycloheximide. Hematocrit is adjusted to 27% (v/v).

3.2.3. Determination of protein degradation and analytical procedure

1. The liver is perfused as described earlier. The collected perfusate is centrifuged to precipitate erythrocytes ($900 \times g$ for 10 min).
2. Ice-cold 25% TCA is added to the plasma (0.44 ml to 1.75 ml of plasma), stored on ice for 15 min and centrifuged ($1000 \times g$ for 15 min at 0 °C). Radioactivity in the 1.8-ml supernatant fraction is counted in a liquid scintillation counter to an error less than 3%.
3. Approximately 0.5 g of liver tissue is taken from the perfused liver and homogenized in ice cold 5% TCA (1:4 w/v).
4. Proteins are precipitated by centrifugation ($1000 \times g$ for 15 min at 0 °C), and washed 3 times with ice-cold 5% TCA.
5. The nucleic acids are removed by hydrolysis in 5% TCA at 90 °C for 15 min, and the hydrolysate is centrifuged at $1000 \times g$ for 15 min.
6. The supernatant fraction is discarded and the lipid is extracted with 2 ml of ethanol/diethylether (2:1, v/v) at 65 °C for 5 min.
7. The samples are centrifuged at $1000 \times g$ for 15 min and the supernatant fraction discarded; the extraction procedure is repeated 2 more times.
8. The protein pellet is washed 2 times with 2 ml of diethylether, dried at 60° for 1 h and weighed.
9. Then, proteins are dissolved in 2 ml of 1 N KOH at 100 °C for 30 min in sealed tubes, neutralized by the addition of 1 ml of 2 N perchloric acid (PCA), and centrifuged ($1000 \times g$ for 15 min).
10. The supernatant fraction is transferred to new tubes; the salt pellet is washed 3 times with 200 μl of ultrapure water and the supernatant fractions are combined.
11. Finally, the radioactivity in the 3-ml aliquot of hydrolyzed protein is counted in a Beckman LS TD 500 liquid scintillation counter to an

error less than 1% and liver protein degradation (mg/h) is computed from the radioactivity released in the perfusate (cpm/h per whole liver) divided by the specific activity of liver proteins (cpm/mg protein).

4. A Minimally Invasive Procedure for the Rapid Evaluation of the Induced Intensification of Autophagy

The intensification of autophagy by the administration of antilipolytic drugs to overnight fasted rats causes an increase in liver proteolysis and a release of the produced branched-chain amino acids into the blood (Bergamini et al., 1993; Donati et al., 2004, 2008). Therefore, after the induction of macroautophagy, unbalanced changes in the concentration of plasma amino acids can be detected with a decrease in gluconeogenic amino acids as glutamine and alanine, and an important increase in branched chain amino acids (valine, leucine and isoleucine). Thus, the rate of autophagic proteolysis in liver cells can be monitored by analyzing the changes in the plasma levels of valine (Fig. 26.2).

1. Blood samples (approximately 300 μl) are collected from the tail into test tubes containing 10 μl of 0.25 M EDTA and centrifuged at 2000$\times g$ at 4°C for 10 min.
2. Plasma (30 μl) is diluted 1:20 with physiological solution, deproteinized with 60% PCA (66 μl, 6% final concentration) and kept on ice for 30min.
3. The samples are centrifuged at 10,000$\times g$ at 4 °C for 10 min.
4. An internal standard (80 μl of 100 μM Norvaline, Sigma–Aldrich, N7627) is added to 400 μl of the supernatant fraction and the acidic sample is neutralized by the addition of 70 μl of 3 M KOH and 50 μl of 0.5 M Na$_2$CO$_3$, pH 9.5, kept on ice for 30 min and the pH adjusted (between 9.68 and 9.80) by an appropriate addition of small amounts of 3 M KOH and 6% PCA.
5. The samples are centrifuged at 10,000$\times g$ at 4 °C for 10 min.
6. For derivatization, dansyl chloride (200 μl; 5.56 mM in acetonitrile) is added to 400 μl of the supernatant fraction (Tapuhi et al., 1981); the mixture is shaken and allowed to stand at room temperature in the dark for 45 min. The reaction is terminated by the addition of 26 μl of 2% methylamine; then samples are filtered through a 0.45-μm syringe filter into HPLC autosampler vials.
7. Separation of dansylated amino acids is carried out on a 4.6 × 250 mm Bio-Sil ODS-5S column (particle size, 5 μm) in a HPLC system (e.g., a Beckman model equipped with 32 Karat software) (Fig. 26.3).

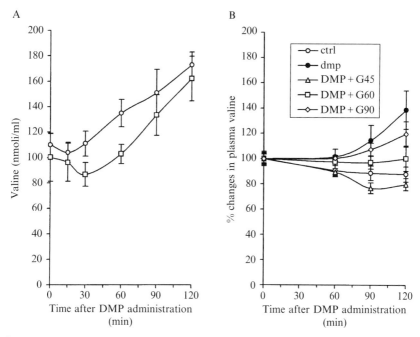

Figure 26.2 (A) Effects of DMP injection (12 mg/Kg body weight) to fasted rats on the plasma levels of valine in liver (○) and peripheral blood (□); (B) Changes in valine plasma levels after the stimulation of autophagic proteolysis by the injection of DMP and the inhibition of the effect of DMP by the injection of glucose (2 g/Kg body weight). Glucose was given to enhance insulin and decrease glucagon plasma levels and reverse the endocrine effects of DMP before [45 min after DMP: G45] or after [60 min (G60) and 90 min (G90) after DMP] the start of the proteolytic effect of the DMP administration.

The mobile phase gradient program and solvent compositions are listed in Table 26.2. Before use, all solvents are filtered through 0.22-μm membrane filter and sonicated for 10 min. The injection volume is 20 μl. Amino acids are determined by measuring the fluorescence of the dansylated derivatives (340 nm excitation, 525 emission).

5. Autophagic Protein Degradation in Isolated Liver Cells

Liver parenchymal cells are isolated by the collagenase (type IV collagenase from *Clostridium Hystoliticum* can be obtained from Sigma-Aldrich) method of Seglen (1976). Perfusion buffers are maintained at 37 °C by a water-jacketed coiled tube. Briefly, we use the following protocol:

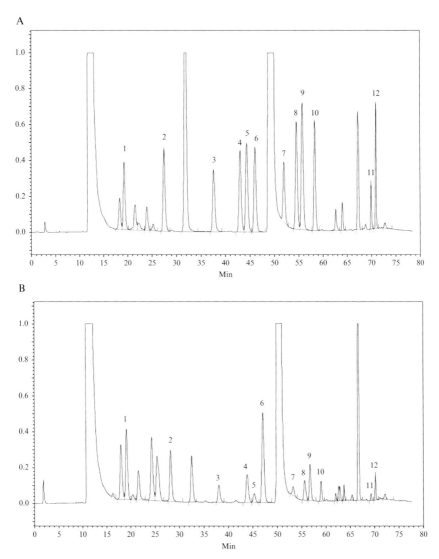

Figure 26.3 Representative HPLC separation of a standard amino acid mixture (each at a final concentration 8.5 μM, (A) and of a plasma sample (B) according to the procedure described in the text. In both cases 8.5 μM norvaline was added as internal standard. (1 = Gln, 2 = Ala, 3 = Pro, 4 = Val, 5 = Met, 6 = Nval, 7 = Trp, 8 = Ile, 9 = Leu, 10 = Phe, 11 = His, 12 = Tyr).

1. Overnight (16–18 h) fasted rats are anaesthetized by an intraperitoneal injection of pentobarbital (50 mg/kg of body weight).
2. The abdomen is opened; the portal vein is exposed and cannulated: the inferior caval vein is cannulated to collect the perfusate.

Table 26.2 Gradient program for the separation of a physiological amino acid mixture by the HPLC procedure

Time	Solvent A %	Solvent B %	Gradient duration (linear, min)
Starting condition (min)	68	32	
0	65.5	34.5	10
10	65	35	2
12	59	41	15
27	50	50	23
50	45	55	6
56	20	80	13
76	68	32	1
80 End			

Solvent composition. A: 50 mM sodium acetate, pH 5.6. B: 100% methanol

3. The liver is perfused first with PB buffer (0.14 M NaCl, 6.7 mM KCl, 10 mM HEPES, 5.5 mM NaOH, pH 7.4) 40 ml/min for 10 min; then, the collagenase-containing buffer (67 mM NaCl, 6,7 mM KCl, 4.8 mM CaCl$_2$, 0.1 M HEPES, 66 mM NaOH, 0.5 mg/ml collagenase, pH 7.6) is recirculated (20 ml/min). The time required for optimum digestion may vary between 7 and 10 min (check by the visual inspection of liver swelling).
4. The liver is removed and transferred to a square (10×10 cm) Petri dish containing 70 ml of suspension buffer (67 mM NaCl, 5.4 mM KCl, 1.22 mM CaCl$_2$, 0.64 mM MgCl$_2$, 1.1 mM KH$_2$PO$_4$, 0.7 mM, Na$_2$SO$_4$, 30 mM HEPES, 30 mM TES, 36 mM Tricine, 52.5 mM NaOH, pH 7.6).
5. The cells are liberated from the connective-vascular tissue by careful raking with a stainless steel comb. Then the cell suspension is filtered through a coarse nylon filter and incubated with gentle shaking (we use a PBI orbital shaker, set at a speed of 20 rpm) at 37 °C for 30 min.
6. Liver cells are washed two times with 70 ml of ice-cold washing buffer (0.14 M NaCl, 6.7 mM KCl, 1.22 mM CaCl$_2$, 10 mM HEPES, 5.5 mM NaOH, pH 7.4) and then resuspended with Krebs-Ringer bicarbonate buffer (0.12 M NaCl, 4.8 mM KCl, 2.5 mM CaCl$_2$, 1.2 mM KH$_2$PO$_4$, 1.46 mM, MgSO$_4$, 0.48 % albumin, pH 7.2–7.4). Cell viability is tested by trypan blue exclusion and should always be better than 90%.
7. Hepatocytes are suspended (12 mg of wet cell weight/ml) in Krebs-Ringer bicarbonate buffer and 3 ml of cell suspension are incubated. (4-rows of water-jacketed 10-ml conical flasks enclosed within a Lucite box attached to a Dubnoff apparatus with shaking may be used).

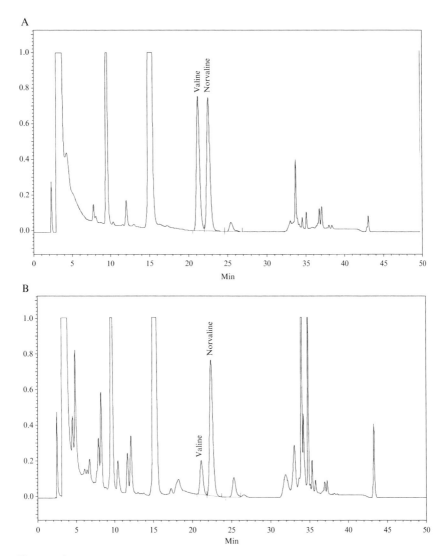

Figure 26.4 Chromatograms show separation of a standard (8.5 μM) pure valine sample (A) and of the valine released from isolated liver cells in 200 μl of incubation medium (B) (see text for procedure). In both cases 8.5 μM norvaline was added as internal standard.

Flask temperature is maintained at 37 °C by means of a constant-temperature recirculating water bath; optimal gas exchange and pH 7.4 are achieved by continuous flow (4 l/min) of humidified 95% O_2–5% CO_2 through the box.

8. To explore amino acid regulation, mixtures of plasma amino acids (without valine) may be added as fraction-multiples of a standard

Table 26.3 Gradient program for the HPLC assay of valine

Time	Solvent A%	Solvent B%	Gradient duration (linear, min)
Starting condition (min)	60	40	
0	50	50	10
30	0	100	1
40	60	40	0.5
50 End			

Solvent composition. A: 88% ultrapure water (Milli-Q), 12% acetonitrile, 0.03% glacial acetic acid, 0.0035% methylamine. B: 100% methanol.

reference mixture reflecting the physiological amino acid concentration in the portal plasma of fasted rats (Venerando et al., 1994). To explore hormone regulation, hormones may be added within the range of their physiological concentration (see Donati et al., 2008b).

9. After 30 min, cycloheximide (10 μM final concentration) is added and 0.5-ml samples of the hepatocyte suspension are taken by time 37 and 47 min after the start of the incubation.
10. Samples are deproteinized by the addition of 56 μl of ice-cold 60% PCA (6% final concentration), kept on ice for 30 min, and centrifuged at 10,000×g for 10 min. The supernatant fraction can be stored at $-20\,^{\circ}\text{C}$ until analysis.
11. An aliquot of the sample is diluted 1:1 with ultrapure water and then amino acids are derivatized as described using the previously mentioned procedure. HPLC assay for the quantification of dansylated valine is carried out on a 4.6X 250-mm Bio-Sil ODS-5S column (particle size, 5 μm) in a HPLC system (Fig. 26.4). The mobile phase gradient program and solvent compositions are listed in Table 26.3. Before use, all solvents are sonicated for 10 min. The injection volume of each sample is 20 μl.

The rate of proteolysis is assessed by the linear release of free valine in the 10-min period and is reported as nmoles of valine/min per gram of wet weight cells. Values are corrected by the subtraction of the valine released in the presence of an inhibitor of lysosomal proteolysis (5 mM 3-methyladenine).

6. Concluding Remarks

Many experimental procedures are available to evaluate autophagy and have been described elsewhere in this volume. It is well known that the process of aging may affect most factors involved in the function and regulation of autophagy (e.g., hormone action, signal transduction, acidification, endowment of lysosomes with hydrolytic enzymes). However, the key factors in aging and antiaging dietary intervention are related to the degradation and turnover of protein and cell organelles. Thus, the measurement of autophagic protein degradation appears to be the most useful method to monitor the function of autophagy in gerontological research.

REFERENCES

Bergamini, E. (2006). Autophagy: A cell repair mechanism that retards ageing and age-associated diseases and can be intensified pharmacologically. *Mol. Aspects Med.* **27,** 403–410.
Bergamini, E., Cavallini, G., Del Roso, A., De Tata, V., Fierabracci, V., Masiello, P., Masini, M., Novelli, M., and Simonetti, I. (1990). Different circadian variations of plasma glucose and insulin concentrations in rats submitted to 60% food restriction or intermittent feeding. *In* "Protein Metabolism in Aging" (H. L. Segal, M. Rothstein, and E. Bergamini, eds.), pp. 295–300. Wiley-Liss, New York.
Bergamini, E., Cavallini, G., Donati, A., and Gori, Z. (2003). The anti-ageing effects of caloric restriction may involve stimulation of macroautophagy and lysosomal degradation, and can be intensified pharmacologically. *Biomed. Pharmacother.* **57,** 203–208.
Bergamini, E., Del Roso, A., Fierabracci, V., Gori, Z., Masiello, P., Masini, M., and Pollera, M. (1993). A new method for the investigation of endocrine-regulated autophagy and protein degradation in rat liver. *Exp. Mol. Pathol.* **59,** 13–26.
Bergamini, E., and Kovacs, J. (1990). Exploring the age-related changes in the hormone-regulated protein breakdown by the use of a physiologic model of stimulation of liver autophagy. *In* "Protein Metabolism in Aging" (H. L. Segal, M. Rothstein, and E. Bergamini, eds.), pp. 361–370. Wiley-Liss, New York.
Cuervo, A. M., Bergamini, E., Brunk, U. T., Dröge, W., Ffrench, M., and Terman, A. (2005). Autophagy and aging: the importance of maintaining "clean" cells. *Autophagy* **1,** 131–140.
Donati, A. (2006). The involvement of macroautophagy in aging and anti-aging interventions. *Mol. Aspects Med.* **27,** 455–470.
Donati, A., Cavallini, G., Carresi, C., Gori, Z., Patentini, I., and Bergamini, E. (2004). Anti-aging effects of anti-lipolytic drugs. *Exp. Gerontol.* **39,** 1061–1067.
Donati, A., Taddei, M., Cavallini, G., and Bergamini, E. (2006). Stimulation of macroautophagy can rescue older cells from 8-OHdG mtDNA accumulation: A safe and easy way to meet goals in the SENS agenda. *Rejuvenation Res.* **9,** 408–412.
Donati, A., Ventruti, A., Cavallini, G., Masini, M., Vittorini, S., Chantret, I., Codogno, P., and Bergamini, E. (2008a). *In vivo* effect of an antilipolytic drug (3,5′-dimethylpyrazole) on autophagic proteolysis and autophagy-related gene expression in rat liver. *Biochem. Biophys. Res. Commun.* **366,** 786–792.
Donati, A., Recchia, G., Cavallini, G., and Bergamini, E. (2008b). Effect of aging and anti-aging caloric restriction on the endocrine regulation of rat liver autophagy. *J. Gerontol. A. Biol. Sci. Med. Sci.* **63,** 550–555.
Jia, K., and Levine, B. (2007). Autophagy is required for dietary restriction-mediated life span extension in C. elegans. *Autophagy* **3,** 597–599.
Masoro, E. J. (2005). Overview of caloric restriction and ageing. *Mech. Ageing Dev.* **126,** 913–922.

Meijer, A. J., and Codogno, P. (2006). Signalling and autophagy regulation in health, aging and disease. *Mol. Aspects Med.* **27,** 411–425.

Mortimore, G. E., and Pösö, A. R. (1987). Intracellular protein catabolism and its control during nutrient deprivation and supply. *Annu. Rev. Nutr.* **7,** 539–564.

Mortimore, G. E., Hutson, N. J., and Surmacz, C. A. (1983). Quantitative correlation between proteolysis and macro- and microautophagy in mouse hepatocytes during starvation and refeeding. *Proc. Natl. Acad. Sci. USA* **80,** 2179–2183.

Payne, A. M., Dodd, S. L., and Leeuwenburgh, C. (2003). Life-long calorie restriction in Fischer 344 rats attenuates age-related loss in skeletal muscle-specific force and reduces extracellular space. *J. Appl. Physiol.* **95,** 2554–2562.

Pollera, M., Masini, M., Del Roso, A., and Bergamini, E. (1990). Changes in the temporal pattern of stimulation of liver autophagy in rats during growth and aging. *In* "Protein Metabolism in Aging" (H. L. Segal, M. Rothstein, and E. Bergamini, eds.), pp. 169–173. Wiley-Liss, New York.

Rattan, S. I., and Clark, B. F. (2005). Understanding and modulating ageing. *IUBMB Life* **57,** 297–304.

Roth, G. S., Ingram, D. K., and Joseph, J. A. (2007). Nutritional interventions in aging and age-associated diseases. *Ann. N.Y. Acad. Sci.* **1114,** 369–371.

Schworer, C. M., Shiffer, K. A., and Mortimore, G. E. (1981). Quantitative relationship between autophagy and proteolysis during graded amino acid deprivation in perfused rat liver. *J. Biol. Chem.* **256,** 7652–7658.

Seglen, P. O. (1976). Preparation of isolated rat liver cells. *Methods Cell. Biol.* **13,** 29–83.

Seglen, P. O., and Gordon, P. B. (1982). 3-Methyladenine: Specific inhibitor of autophagic/lysosomal protein degradation in isolated rat hepatocytes. *Proc. Natl. Acad. Sci. USA* **79,** 1889–1892.

Tapuhi, Y., Schmidt, D. E., Lindner, W., and Karger, B. L. (1981). Dansylation of amino acids for high-performance liquid chromatography analysis. *Anal. Biochem.* **115,** 123–129.

Terman, A., Gustafsson, B., and Brunk, U. T. (2007). Autophagy, organelles and ageing. *J. Pathol.* **211,** 134–143.

Venerando, R., Miotto, G., Kadowaki, M., Siliprandi, N., and Mortimore, G. E. (1994). Multiphasic control of proteolysis by leucine and alanine in the isolated rat hepatocyte. *Am. J. Physiol.* **266,** C455–C461.

Wohlgemuth, S. E., Julian, D., Akin, D. E., Fried, J., Toscano, K., Leeuwenburgh, C., and Dunn, W. A. Jr. (2007). Autophagy in the heart and liver during normal aging and calorie restriction. *Rejuvenation Res.* **10,** 281–292.

CHAPTER TWENTY-SEVEN

Physiological Autophagy in the Syrian Hamster Harderian Gland

Ignacio Vega-Naredo* *and* Ana Coto-Montes*

Contents

1. Introduction	458
2. Methods to Detect Autophagy by Morphological Techniques	459
2.1. Optical microscopy: Tissue dissection, fixation, and sectioning	460
2.2. LC3 immunohistochemistry	462
2.3. Electron microscopy: Tissue dissection, fixation, and sectioning	463
2.4. LC3 immunocytochemistry: Preembedding method	464
3. Methods to Detect Autophagy by Biochemical and Molecular Biology Techniques	465
3.1. Procedure for animal tissue homogenization	465
4. Measurement of Cell Viability: Ratio of Cathepsins B:D	466
4.1. Cathepsin B activity	466
4.2. Cathepsin D activity	466
5. Analysis of the Lysosomal Pathway	467
5.1. Study of cathepsin D processing	467
5.2. Study of LAMP-2 expression	469
5.3. Study of cytoskeletal proteins	470
6. Western Blot Analysis of Autophagy-Related Proteins	471
6.1. LC3 immunoblotting	471
6.2. Immunoprecipitation of endogenous Beclin 1-Bcl-2 complexes	471
7. Methods to Detect Chaperone-Mediated Autophagy	472
8. Concluding Remarks	473
Acknowledgments	474
References	474

Abstract

The Syrian hamster Harderian gland (HG) displays a huge porphyrins metabolism with sexual dimorphism. Even in male Syrian hamsters with much lower porphyrins concentration than female HG, this activity is higher than in the liver.

* Departamento de Morfología y Biología Celular, Facultad de Medicina, Universidad de Oviedo, Oviedo, Spain

The damage derived from constant porphyrin production, displayed by reactive oxygen species, forces the gland to develop mechanisms that allow it to continue with its normal physiology. The survival strategy of the Harderian gland is mainly based on autophagic processes that are considered as a constant renovation system. Our results show different autophagy mechanisms in Syrian hamster HG, macroautophagy and other lysosomal-like processes such as chaperone-mediated autophagy, depending on sex and probably related to oxidative stress status. This chapter describes the methods used by us to characterize the autophagic processes that are being physiologically developed by this organ under normal conditions.

1. Introduction

The Syrian hamster Harderian gland (HG) is a retro-orbital gland consisting widely of tubule-alveolar units that, in sections, appear as round, oval, or roughly polygonal profiles smooth in outline, separated by connective strands containing blood vessels, nerve fibers, and mast cells. Besides lubricating the eye, their precise functions have not been determined, but some other potential functions, including the production of pheromones, a role in temperature regulation and the synthesis of indolamines, have been proposed (Tolivia et al., 1996). This gland exhibits morphological sex differences in porphyrin production, mast cell numbers, and secretory cell types that synthesize porphyrins and lipid droplets (Hoffman, 1971). Thus, the glands from females contain large quantities of porphyrin deposits, numerous mast cells, and a single cell type (type I), which is characterized by minute lipid droplets, whereas those from male hamsters do not accumulate intralumenal porphyrins and present two secretory cell types, type I (similar but not identical to the female type) and type II cells, which are filled with large lipid vacuoles and have a swollen appearance (type I and II) (Coto-Montes et al., 2001a). Also, there are distinct ultrastructural differences in the Harderian gland of both sexes (Lopez et al., 1992). Related to this, female glands present with high-frequency clear cells that seems to be damaged cells (Antolin et al., 1996) showing less electron density than intact cells. Many of these sexual differences are controlled by androgens, as shown by castration of males that typically leads to conversion of the Harderian glands to the female phenotype (Coto-Montes et al., 1994). Furthermore, the exposure of hamsters of both sexes to extreme short photoperiods induces the conversion of the glands to the opposite type; this phenomenon is prevented by pinealectomy. Therefore, the morphology of the Harderian gland is regulated by the light-pineal-gonadal axis (Coto-Montes et al., 1994).

In this organ, we discovered a huge oxidative stress due to the extremely high porphyrin production, a thousand times higher than in other porphyrinogenic organs, such as the liver. Accordingly, we consider the HG as a physiological model to study oxidative stress (Coto-Montes et al., 2001a). The porphyrin production, higher in female than in male HGs, follows circadian rhythms of activity (Coto-Montes et al., 2001b) and is counteracted by antioxidants such as melatonin (Tomas-Zapico et al., 2002a; Tomas-Zapico et al., 2002b). However, this physiological oxidative stress is so high that the gland needs additional survival mechanisms for fighting against it. These mechanisms are autophagy and invasive processes (Tomas-Zapico et al., 2005). Therefore, autophagy is considered to be a constant renovation system, which allows the organism to sustain normal gland activity.

The chief characteristics of the Syrian hamster Harderian gland are the sex differences in porphyrin synthesis and gland structure, together with hormonal control of both synthesis and structure in both sexes. Indeed, it is argued that the gland may be a useful model for clinical conditions such as acute intermittent porphyria (Spike et al., 1988), oxidative stress, cell repair, cancer, and most likely aging.

In the past decade, several articles have described abnormal structures as clear cells in the HG of both sexes (Antolin et al., 1994; Tolivia et al., 1996). However, the attempts to associate these characteristics with any type of known physiological processes were unsuccessful. The situation of physiological oxidative stress, which the HG suffers, requires massive cellular elimination. However, the possibility that this cellular death is due to apoptotic processes is not consistent (Antolin et al., 1996), indicating that the cellular turnover has to be performed at least in part by another mechanism. Thus, our group has shown that in the Syrian hamster HG the cell's suicide program involve the autophagic-lysosomal compartment. This chapter describes the methods followed by our laboratory to describe autophagy or lysosome-dependent cell death processes in the Syrian hamster HG, and the utilities for using this organ as an experimental model to study the relationship between autophagy and other related events such as oxidative stress and invasive processes.

2. Methods to Detect Autophagy by Morphological Techniques

Describing or characterizing autophagy in organs has additional difficulties because it is not possible to use the techniques performed for cell cultures.

The morphological detection of autophagic structures by electron microscopy is an important and necessary approach. Extensive autophagy, as occurs in the Syrian hamster HG, can result in obvious changes in cell morphology and the elimination of sufficient amounts of cytoplasm to cause a loss of electron density, thus receiving the adjective *clear* (Tomas-Zapico *et al.*, 2005). Another important approach is immunohistochemical staining, using this method to show the presence or absence of autophagic markers and their localization. Moreover, this is the easiest method for detecting differences in the autophagy machinery among the different cellular types from the same tissue (Klionsky *et al.*, 2008).

The morphological characteristics of autophagy have been described over the past four decades and the observation of this autophagic morphology has been considered an irrefutable proof of the presence of autophagy in a tissue. Accordingly, the use of these techniques is indispensable for demonstrating autophagy in a tissue in a physiological state. Even so, these techniques are also one of the most problematic and prone to misinterpretations.

Studies by electron microscopy can reveal autophagic characteristics such as the presence of cytoplasm filled with swollen Golgi complexes, or rich in mitochondria, some of them with clear cristae, and with a wide variety of vacuoles in the cells, and a profusion of late autophagosomes with unrecognizable content. Furthermore, the endoplasmic reticulum is usually observed surrounding mitochondria, as seen in the male and female Harderian gland (Tomas-Zapico *et al.*, 2005). In addition, electron microscopy is a valid method for studying changes in various autophagic structures: phagophores, autophagosomes, amphisomes, autolysosomes, and/or multi-lamellar whorls. However, methods that include quantitative studies are inoperative when an entire organ is the object of study.

2.1. Optical microscopy: Tissue dissection, fixation, and sectioning

The following protocol describes the technique routinely used to perform a morphological study, using tissue from at least three different animals. Thus, this protocol is common for the most tissues of animals, but some aspects should be considered depending on the tissue. For instance, the penetration rate of the fixative solution in mm/hour is variable. Therefore, here we describe the steps and periods optimized for the Syrian hamster Harderian gland.

1. To extract the Harderian gland and be able to study its morphology, the animals have to be decapitated and the tissue rapidly dissected from the retro-orbital part of the eye (Fig. 27.1).
2. Divide each gland into 2 or 3 4-mm thick pieces, and immerse them in a fixative medium containing 4% paraformaldehyde in 0.1 M phosphate buffer, pH 7.4, for 15 min.

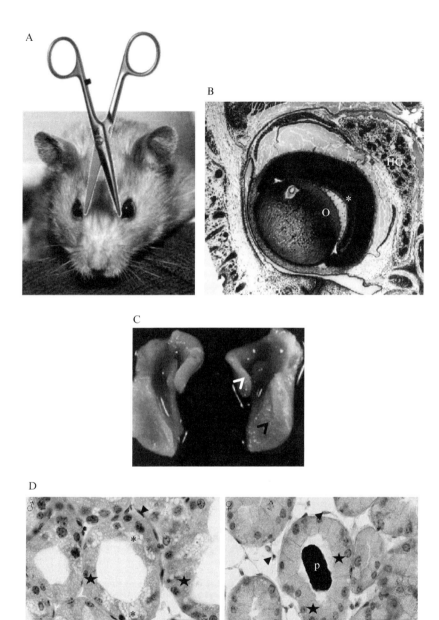

Figure 27.1 Surgical procedure for Harderian gland dissection. (A) Fracture the interorbital space using scissors to extract the gland. (B) Optic micrograph of a section from Syrian hamster embryo showing the localization of the Harderian gland (HG) close to the retina (asterisk), in the retro-orbital region. (C) Macroscopic picture of Harderian glands from Syrian hamster fixed in paraformaldehyde, showing the small (white arrowhead) and large lobes (black arrowhead). (D) Optic micrograph of the Harderian gland from male (♂) and female (♀) hamsters showing their classical acini with type I (asterisks) and II (stars) secretory cells in males and only type I (stars) cells in females. Note the presence of a porphyrin accretion (p) in the lumen of the female acinus and the myoepithelial cells surrounding the acini in both sexes. Bar: 50 μm.

3. Remove the fixative, and postfix in the same fixative liquid for at least 2 days.
4. Remove the water from the tissues by gradual dehydration in a series of alcohols (30, 50, 70, 96, and 100%), twice for 10 min in each percentage. Immerse the pieces in 100% isopropyl alcohol for 30 min and embed the specimens in paraffin by standard methods (see the chapter by Anttila et al., in this volume).
5. Cut the paraffin-embedded tissue into sections of 7–10 μm with a microtome, float them on a warm water bath, and pick them up on a glass Superfrost Plus slide to enhance the tissue adhesion.

2.2. LC3 immunohistochemistry

Microtubule-associated protein 1 light chain 3, LC3, is an autophagosomal ortholog of yeast Atg8. LC3-modification is essential for the autophagic process, as the protein LC3-II is localized to phagophores and autophagosomes and is considered an autophagosomal marker. LC3 immunohistochemistry can give indications as to whether autophagic phenomena are occurring and in which type of cells. The immunopositive reaction has to be observed in the cytoplasm at the level of membranes surrounding cytoplasmic portions (Tomas-Zapico et al., 2005). Here we describe the peroxidase-antiperoxidase (PAP) immunohistochemical method to show LC3 immunostaining using an antigoat LC3 antibody:

1. Deparaffinize the slides obtained by the preceding protocol at room temperature by immersing them in pure xylene twice for 10 min and hydrate through a succession of alcohols (100, 96, 70, 50, and 30%), twice for 10 min in each percentage, ending with distilled water.
2. Rinse the sections 3 times for 10 min in 0.01 M PBS containing 0.1% Triton X-100 and 0.25% bovine serum albumin (PBS-TB).
3. Incubate the sections 30 min in a solution (3% H_2O_2 in methanol) to inhibit the endogenous peroxidase.
4. Wash the slides 3 times with PBS-TB for 5 min.
5. Block with rabbit normal serum (1:30 in PBS-TB) for 30 min.
6. Incubate the sections with goat anti-LC3 from Santa Cruz Biotechnology, Inc (sc-16756, 1:100 in PBS-TB) for 24 h at 4 °C in a humid chamber.
7. Wash 3 times with PBS-TB for 5 min.
8. Incubate in peroxidase conjugated rabbit antigoat IgG (Sigma, A5420, 1:50 in PBS-TB) for at least 1 h at room temperature.
9. Wash 3 times with PBS-TB for 10 min.
10. Incubate with peroxidase antiperoxidase complexes (Sigma, P1901, 1:50 in PBS-TB) for 1 h at room temperature.

11. Wash the sections 2 times with PBS-TB for 10 min, and 10 min in 50 mM Tris-HCl, pH 7.6 buffer.
12. Reveal staining by an incubation in 0.05% 3,3′-diaminobenzidine-HCl (Sigma, D5637) with 0.005% H_2O_2 in 50 mM Tris-HCl for 5 to 15 minutes and stop the reaction with tap water by washing.
13. Dehydrate the sections, place a drop of Eukitt mounting medium onto the slide (O. Kindler GmbH) and cover it with the cover slip.

2.3. Electron microscopy: Tissue dissection, fixation, and sectioning

1. Prepare the fixative mixture. The standard fixative contains 1.25% paraformaldehyde, 2.5% glutaraldehyde in 0.1 M phosphate buffer, pH 7.4, but we have optimized a specific fixative for the Harderian gland: 4% paraformaldehyde, 1.25% glutaraldehyde in 0.1 M phosphate buffer, pH 7.4.
2. Perfuse the animals under deep anesthesia by intraventricular puncture with the fixative mixture and, afterward, dissect the tissue.
3. Cut the dissected tissue into small pieces and immerse into the fixative mixture overnight at 4 °C.
4. Wash 3 times with 0.1 M phosphate buffer, pH 7.4, for 10 min.
5. Divide the specimens in 1-mm^2 blocks.
6. Postfix in 1% OsO4 for 2 h.
7. Dehydrate in graded acetone maintaining the specimens in a microcentrifuge tube and changing the medium through the different acetones:
   i. 50% acetone, 3 times for 5 min.
   ii. 0.5% uranyl acetate in 70% acetone, 30 min.
   iii. 80% acetone, 3 times for 5 min.
   iv. 90% acetone, 3 times for 5 min.
   v. 100% acetone, 3 times for 5 min.
   vi. Anhydrous acetone, 3 times for 5 min.
8. Infiltrate the samples sequentially in epon:acetone mixtures 1:3, 1:1, and 3:1, 3 times for 15 min for each mixture.
9. Embed the pieces in Epon resin until it polymerizes.
10. Obtain semithin sections (1 cm) using an ultramicrotome and stain with 0.5% toluidine blue until obtaining the ideal staining.
11. The slides of these semithin sections should be studied using a light microscope to select the area of interest and then samples should be recut from the original Epon block.
12. Collect ultrathin (700-nm thick) sections on copper grids for electron microscopy.
13. Contrast with 2% uranyl acetate, 2% lead citrate in water.
14. Examine with a transmission electron microscope.

2.4. LC3 immunocytochemistry: Preembedding method

LC3 immunoreactivity under electron microscopy shows immunoreactive membranes, which delimit unrecognizable material together with a punctate staining. The punctate staining in addition to a diffuse pattern are also present in the cytoplasm (Fig. 27.2).

1. Cut the dissected tissue into small pieces and immerse in a fixative mixture containing 4% paraformaldehyde in PBS.
2. Cut 80-μm thick sections with a Vibratome Series 1000 (Technical Products International). Collect the sections floating on a PBS bath and pick them up with a common fine brush.
3. Submerge the sections in the first line of a 96-well microtiter plate filled with PBS. Maintain in a shaker rotating smoothly at 60 rpm for 15 min.
4. Pass the sections to the following line of the plate that contains 4% normal serum in PBS and incubate for 30 min at room temperature.
5. Incubate, using the following lines of the plate, with the anti-LC3 antibody (diluted 1:500 in PBS with 0.25% BSA) for 48 h at 4 °C. Cover control sections with an equal quantity of PBS.
6. Rinse 3 times for 10 min in PBS containing 0.25% BSA, using the next lines of the plate.

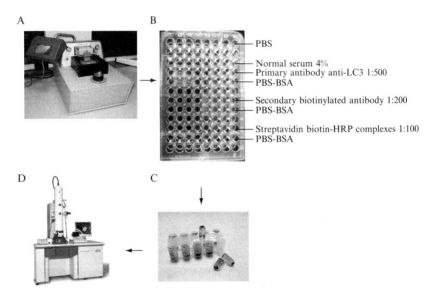

Figure 27.2 Essential steps to perform *preembedding* immunocytochemistry procedure. (A) Vibratome; (B) 96-well microtiter plate; (C) Epon-embedded tissues and (D) transmission electron microscope.

7. Incubate overnight with VECTASTAIN biotinylated antigoat IgG (Vector labs, PK-6105) following the manufacturer's instructions, in the corresponding line of the microtiter plate at 4 °C.
8. Rinse 3 times for 10 min in PBS with 0.25% BSA.
9. Incubate overnight with VECTASTAIN ABC Reagent (Avidin and Biotinylated horseradish peroxidase macromolecular complexes) (Vector labs, PK-6105) following the manufacturer's instructions.
10. Rinse in PBS containing 0.25% BSA.
11. Incubate with 0.05% 3,3′-diaminobenzidine-HCl containing 0.005% H_2O_2 in 50 mM Tris-HCl, pH 7.4, for 15 min.
12. Complete the fixation of the sections with 1.25% glutaraldehyde in PBS for 30 min.
13. After dehydration in graded acetone by standard methods, embed the pieces in Epon as described previously.
14. Obtain ultrathin sections using an ultramicrotome.
15. Contrast the sections with 2% uranyl acetate and 2% lead citrate in water.
16. Examine with a transmission electron microscope.

3. Methods to Detect Autophagy by Biochemical and Molecular Biology Techniques

Some biochemical assays may be used to at least provide indirect correlative data relating to autophagy, in particular when examining the role of autophagy in cell death. For instance, lysosomal enzymes such as cathepsins D and B are translocated from lysosomal compartments to the cytosol during programmed cell death, with cathepsin D acting as a death mediator even though its death-inducing activity is usually suppressed by cathepsin B (Isahara et al., 1999). Therefore, cellular viability is related to high cathepsin B and low cathepsin D activities. Accordingly, comparing to the same tissue under control conditions, or to a different tissue in the same organism, alterations in the cathepsin B:cathepsin D activity ratio indicate the onset of autophagic cell death, as it reflects particular aspects of the autophagy-lysosomal pathway (Klionsky et al., 2008; Tomas-Zapico et al., 2005).

There are other procedures that are also useful for characterizing physiological autophagic processes. We provide details for all of these procedures that are followed in our laboratory.

3.1. Procedure for animal tissue homogenization

1. Sacrifice the hamsters by decapitation and dissect the Harderian glands, freeze them in liquid nitrogen, and store them at −80 °C until the tissues are homogenized.

2. Homogenize 100 mg of tissue with a polytron homogenizer at 4 °C in 1 ml of lysis buffer (20 mM HEPES, pH 7.4, 2 mM EDTA, 1% Nonidet P-40, 5 mM dithiothreitol, and 0.25% Na-deoxycholate) with protease inhibitors (1 mM Na$_3$VO$_4$ and 1 μg/ml aprotinin).
3. Centrifuge the tissue homogenates at 1000g for 10 min at 4 °C to remove insoluble material. Collect the supernatant fractions and centrifuge again using the same conditions.

4. Measurement of Cell Viability: Ratio of Cathepsins B:D

4.1. Cathepsin B activity

The cysteine-proteinase cathepsin B (EC 3.4.22.1) is normally assayed fluorimetrically according to Barrett (1980) with minor modifications developed by our laboratory, using Z-Arg-Arg-aminomethylcoumarin as a specific substrate.

1. Prepare a standard curve with aminomethylcoumarin solutions of 0, 0.01, 0.025, 0.05, 0.075, 0.1, 0.25, 0.5 μM in incubation buffer (100 mM sodium acetate, pH 5.5, containing 1 mM EDTA, 5 mM dithiothreitol, and 0.1% Brij-35).
2. Dilute 40 μl of tissue homogenates (see preceding) with 300 μl of incubation buffer.
3. Add 50 μl of the solution into a 96-well fluorescence microtiter plate.
4. Start the reaction by adding 20 μl of substrate solution (40 μl Z-Arg-Arg-aminomethylcoumarin in incubation buffer) and incubate at 37 °C for 20 min.
5. Stop the reaction by adding 150 μl of stop buffer (33 mM sodium acetate, pH 4.3, and 33 mM sodium chloroacetate).
6. Measure using a fluorimeter with an excitation wavelength of 360 nm and an emission wavelength of 460 nm.
7. Results are expressed as enzymatic milliunits/mg protein. One enzymatic unit is the amount of enzyme necessary to release 1 μmol aminomethylcoumarin per minute.

4.2. Cathepsin D activity

The aspartate-proteinase cathepsin D (EC 3.4.23.5) is assayed spectrophotometrically according to Takahashi and Tang (1981), with minor modifications (Schreurs et al., 1995), using hemoglobin as substrate.

1. Mix 200 µl of tissue homogenates (see preceding) with 500 µl of substrate solution (3% hemoglobin in 200 mM acetic acid) and incubate at 37 °C for 30 min.
2. Stop the reaction by adding 500 µl of 15% trichloroacetic acid and keep the samples at 4 °C for 30 min.
3. Centrifuge the samples at 12,000g for 5 min.
4. Measure using a spectrophotometer the optical densities of the supernatant fractions at 280 nm. The cathepsin D activities are expressed as enzymatic units/mg protein. One enzymatic unit is the amount of enzyme that increases the optical density by 0.01 absorbance units.

5. Analysis of the Lysosomal Pathway

Traffic from the Golgi to lysosomes has to be analyzed to be able to know whether there is any abnormality in the endosomal system that carries out this process (Fig. 27.3).

The Syrian hamster HG presents alterations in the autophagic-lysosomal pathway, especially in female glands, which show a block in trafficking of procathepsin D from the *trans*-Golgi network, and changes in LAMP-2 and in cytoskeleton dynamics. The knowledge of these alterations could help to discover which form of autophagy is developed by an organ.

5.1. Study of cathepsin D processing

When analyzing cathepsin D, it is necessary to use western blot and activity assays. Activity measurements alone can be misleading because procathepsin D is also active. The levels of mature cathepsin D in a tissue that is undergoing autophagy are usually lower than expected. Procathepsin D is matured in lysosomes, and extensive vacuolization resulting from autophagy interferes with trafficking of the enzyme through the endosome, increasing procathepsin D levels (Zeng *et al.*, 2006). Therefore, indirect measures of autophagy may be a higher ratio of procathepsin D (52 and/or 46 kDa) to mature cathepsin D (33 kDa).

The procedure to show the kinetics of cathepin D processing includes an analysis of its expression. Thus, steady-state levels of intracellular cathepsin D are measured in whole-cell lysates by SDS-PAGE and immunoblot analysis.

1. Measure the protein content of tissue homogenates by a standard procedure such as the Bradford method (1976).
2. Place samples containing 50 µg of protein in load buffer (0.5 M Tris-HCl, pH 6.8, 10% glycerol, 2% SDS, 5% β-mercaptoethanol, 0.05% bromophenol blue) and denature by boiling at 95–100 °C for 5 min.

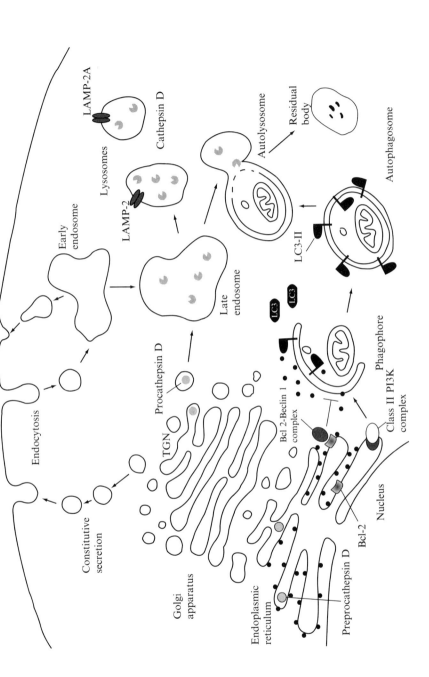

Figure 27.3 The autophagic-lysosomal pathway. Autophagy is a catabolic trafficking pathway for bulk destruction of long-lived proteins and organelles via regulated lysosomal degradation. The diagram shows the process and the convergence of the endocytic and autophagic pathways. The steps and structures of these pathways have some markers such as LAMP-2 (lysosome-associated membrane protein type 2), cathepsin D processing, Beclin 1, Bcl-2 and LC3 that could be used to characterize the autophagic process.

3. Fractionate the samples by 12% acrylamide/bisacrylamide SDS-PAGE (running buffer: 25 mM Tris, 192 mM glycine, and 0.1% SDS; do not adjust pH) and transfer (transfer buffer: 25 mM Tris, 190 mM glycine, 20% methanol; do not adjust pH) to PVDF membrane at 350 mA for 80 min.
4. Block the membrane with 5% skim milk in phosphate-buffered saline containing 0.05% Tween-20 (PBS-T) overnight at 4 °C.
5. Incubate the membrane with the primary antibody goat anticathepsin D from Santa Cruz Biotechnology (sc-6486, 1:1,000) for 4 h at 4 °C.
6. Wash three times with PBS-T for 5 min at room temperature.
7. Incubate the membrane with the corresponding horseradish peroxidase-conjugated secondary antibody (1:5000) for 2 h at 4 °C.
8. Wash three times with PBS-T for 20 min at room temperature.
9. Detect the immunoconjugates using the western blotting luminol reagent (sc-2048; Santa Cruz Biotechnology) according to the manufacturer's protocol and develop the film as usual.

5.2. Study of LAMP-2 expression

To complete the study of the autophagic-lysosomal pathway, the lysosome membrane-associated protein 2 (LAMP-2) acquires a special relevance. LAMP-2 is a late endosomal and lysosomal marker. This endosomal-lysosomal pathway plays a dynamic role in cellular processes such as macroautophagy, chaperone-mediated autophagy and receptor trafficking (Eskelinen *et al.*, 2002). In several LAMP-2-deficient tissues, including muscle, heart, pancreas, and liver, an accumulation of autophagic vacuoles is observed. In particular, in LAMP-2-deficient cells the half-life of early and late autophagic vacuoles is prolonged and endosomal/lysosomal constituents are delivered to autophagic vacuoles that accumulate (Eskelinen *et al.*, 2002). Likewise, cell death processes in cells lacking LAMP-2, adopt intermediate characteristics of type 1 and type 2 programmed cell death with caspase activation, chromatin condensation, and extensive autophagic vacuolation (Gonzalez-Polo *et al.*, 2005). Therefore, an analysis of LAMP-2 expression is necessary to evaluate if the accumulation of autophagic vacuoles is occurring as a result of the absence of this protein. The analysis can be performed by Western blotting, using the protocol described for cathepsin D with special attention to the following points:

1. Denature the samples by boiling at 95–100 °C for 5 min.
2. Fractionate the samples by 12% acrylamide/bisacrylamide SDS-PAGE and transfer to PVDF membrane at 400 mA for 90 min.
3. Incubate with the primary antibody anti-LAMP-2 (Santa Cruz Biotech., sc-8100, 1:500) overnight at 4 °C.
4. Incubate with the corresponding horseradish peroxidase-conjugated secondary antibody (1:2500) for 2 h at 4 °C.

5.3. Study of cytoskeletal proteins

It is well known that the cytoskeleton (microfilaments, intermediate filaments, and microtubules) is a dynamic structure that maintains cell shape, enables cellular motion, and plays important roles in intracellular transport. Due to this, the cytoskeleton exhibits different fates during autophagic and apoptotic cell death, specially the actin filaments and the intermediate filaments such as keratins. In apoptosis, the cell's preparatory as well as executional steps include depolymerization or cleavage of actin, cytokeratins, laminins, and other cytoskeletal proteins. In contrast, during autophagic death, the cytoskeleton is redistributed but largely preserved (Bursch, 2001; Tomas-Zapico et al., 2005).

Cytokeratin pattern should be analyzed using a broad-range anticytokeratin antibody, which recognizes various cytokeratins that include type II neutral-to-basic cytokeratin subfamily and type I acidic subfamily. The analysis can be performed by Western blotting, following the protocol described for cathepsin D with these conditions:

1. Fractionate the samples by 12% acrylamide/bisacrylamide SDS-PAGE and transfer to PVDF membrane at 350 mA for 80 min.
2. Incubate with the primary antibody mouse anticytokeratin (Sigma-Aldrich Co. C9687, 1:4000) for 4 h at 4 °C.
3. Incubate with the corresponding horseradish peroxidase-conjugated secondary antibody (1:2500) for 2 h at 4 °C.

Actin dynamics are important for trafficking from the *trans*-Golgi network to lysosomes (Carreno et al., 2004). Furthermore, it was described in yeast that the actin cytoskeleton is required for selective autophagy but is dispensable for the nonselective processes (Reggiori et al., 2005). Thus, the analysis of these proteins is useful for the study of autophagic processes, especially if the autophagy-lysosomal pathway is altered as occur in the Syrian hamster HG (Vega-Naredo I., Caballero B., Huidobro C., Coto-Montes A. *in preparation for publication*).

Actin filaments can be analyzed using different actin antibodies recognizing the different actin isoforms, for instance using antibodies from Santa Cruz Biotech. (sc-1615), which recognizes a broad range of actin isoforms and from Sigma (A5441), which recognizes only the β-actin isoform. The analysis can be performed by western blotting, following the protocol described for cathepsin D with these conditions:

1. Fractionate the samples by 12% acrylamide/bisacrylamide SDS-PAGE and transfer to PVDF membrane at 350 mA for 80 min.
2. Incubate with the primary antibody from Santa Cruz Biotech at 1:1000–2000 for 4 h at 4 °C, and for 30 min at 1:5000–10,000 in the case of the antibody from Sigma.
3. Incubate with the corresponding horseradish peroxidase-conjugated secondary antibody (1:5000–10,000) for 2 h at 4 °C.

6. Western Blot Analysis of Autophagy-Related Proteins

Actually, there is an important set of antibodies that has been developed against autophagy-related proteins that are useful to characterize an autophagic phenomenon. We describe LC3 expression analysis as being the main and the most essential autophagy marker, but this could be accompanied by others, such as Beclin 1, Vps34, and Atg 12.

6.1. LC3 immunoblotting

Analysis of LC3 is one of the most useful methods, and this protein is suggested as an ideal marker for autophagy. LC3 is first converted into LC3-I (18 kDa), and part of this modified protein is subsequently converted into a phagophore- and autophagosome-associating form, LC3-II (16 kDa). Therefore, the presence of LC3-II is the only reliable marker of autophagosomes. We use a rabbit anti-LC3 antibody from MBL (PD014) to carry out the following protocol (also see the chapter by Kimura *et al.*, in this volume):

1. Boil samples with load buffer for 3 min at 95–100 °C and load 100 mg of sample per lane in a 1-mm-thick SDS-PAGE 16% polyacrylamide gel.
2. Blot the protein to a PVDF membrane for 1 h in a semidry transfer system.
3. Soak the membrane in 10% skim milk in PBS-T for 1 h at room temperature.
4. Incubate the membrane with primary antibody diluted 1:1000 in PBS-T overnight at 4 °C.
5. Wash 3 times with PBS-T for 10 min at room temperature.
6. Incubate the membrane with the HRP-conjugated anti-rabbit IgG (1:10,000) for 2 h at room temperature.
7. Wash 3 times with PBS-T for 10 min at room temperature.
8. Develop as usual.

6.2. Immunoprecipitation of endogenous Beclin 1-Bcl-2 complexes

Beclin 1, a class III phosphatidylinositol 3-kinase-interacting protein, is a key regulator of autophagy formation since the Beclin 1-Vps34 complex is involved in autophagosome formation at an early stage. This step is regulated by Bcl-2. Bcl-2 not only functions as an antiapoptotic protein but also as an anti-autophagy protein via its inhibitory interaction with Beclin 1 preventing Beclin 1-Vps34 complex formation (Furuya *et al.*, 2005). To determine if Bcl-2 is a normal endogenous binding partner for Beclin 1, Beclin 1 can be immunoprecipitated from homogenates and the Beclin 1-Bcl-2 interaction probed by immunoblot analysis with an antibody against Bcl-2.

1. Incubate a volume of homogenated tissue containing 750 μg of protein with 5 μl goat anti-Beclin 1 (sc-10086, Santa Cruz Biotechnology) for 1 h at 4 °C, followed by an incubation with 20 μl protein G PLUS-Agarose (sc-2002, Santa Cruz) overnight at 4 °C, maintaining both incubations under orbital shaking conditions.
2. Centrifuge 1 minute at 16,000g and discard the supernatant fraction.
3. Add 1 ml of IP buffer I (50 mM Tris-HCl, pH 7.4, 150 mM NaCl, 1% Na-deoxycholate, 1% NP-40, 1 mM Na$_3$VO$_4$) to the pellet fraction.
4. Mix the sample on an oscillatory shaker for 20 min at 4 °C.
5. Centrifuge for 1 min at 16,000g and discard the supernatant fraction.
6. Repeat with IP buffer II (50 mM Tris-HCl, pH 7.4, 75 mM NaCl, 0.1% Na-deoxycholate, 0.1% NP-40, 1 mM Na$_3$VO$_4$).
7. Repeat with IP buffer III (50 mM Tris-HCl, pH 7.4, 0.05% Na-deoxycholate, 0.05% NP-40, 1 mM Na$_3$VO$_4$).
8. Centrifuge 1 min at 16,000g and discard the supernatant fraction.
9. Add 20 μl of load buffer and boil for 5 min.
10. Centrifuge 1 min at 16,000g.
11. Transfer the supernatant fraction to a new tube, subject it to SDS-PAGE and immunoblot analysis as previously described using primary antibodies against Beclin 1 (BD Bioscience, 1:1000) and Bcl-2 (Santa Cruz, 1:1000).

7. Methods to Detect Chaperone-Mediated Autophagy

LAMP-2 is a single gene that undergoes alternative splicing rendering three different mRNA species that encode the LAMP-2A, LAMP-2B, and LAMP-2C variants of this protein. Although it is possible that all the isoforms share common functions, isoform-specific functions have also been described. LAMP-2A is the only isoform shown to participate in CMA, and therefore is used as a marker of CMA activity (Cuervo and Dice, 1996) (see also the chapter by Kaushik and Cuervo in this volume). Our first approach to evaluate if this type of degradative process is occurring in an organ is a retrotranscription polymerase chain reaction (RT-PCR) assay:

1. Purify total RNA from tissue extracts by a single extraction with TRI Reagent (T9424, Sigma) following the manufacturer's recommendations and store at −80 °C until use.
2. Synthesize LAMP-2A cDNA using the Titan One Tube RT-PCR System (Roche Corporation, Indianapolis, IN, USA): Add 25 μl of purified total RNA (0.1 μg/μl) to 25 μl of a RT-PCR mixture containing 0.2 pmol of primers according to the manufacturer's instructions.

The amplification protocol includes a retrotranscription at 48 °C for 30 min, a denaturation at 94 °C for 2 min, followed by 30 cycles at 94 °C for 10 s, 55 °C for 30 s and 68 °C for 1 min, and a final cycle at 68 °C for 10 min.

3. Analyze the PCR products using 2% agarose-Tris-borate-EDTA gel electrophoresis. To prepare the gel, mix the powdered agarose with electrophoresis buffer at 2%. The gel is then melted by heating in a microwave oven. Add ethidium bromide (0.01%) at this stage to allow visualization of the separated fragments of DNA. Wait until the gel has cooled to approximately 60 °C and pour it into the gel-casting tray as the manufacturer indicates. Place the combs to form the wells immediately after pouring the gel. Allow it to cool at room temperature. Remove the combs and place the casting tray with the gel into the electrophoresis chamber. Fill the chamber with the electrophoresis buffer.
4. Mix DNA samples with loading buffer (1:10) from Novagen and transfer slowly and smoothly the DNA samples into the wells of the gel using a micropipette.
5. Run the gel at 80–100 mA for 45–60 min.
6. Finally, place the gel on a transilluminator (UV light of wavelength 254 nm), visualize the DNA, and photograph if possible for record keeping.

The primer pairs (LAMP-2A, 5′-GCAGTGCAGATGAAGACAAC-3′, 5′-AGTATGATGGCGCTTGAGAC-3′) are designed for amplification of a 120-bp fragment of the *Mus musculus* lysosomal-associated membrane protein 2 (LAMP-2), transcript variant 1, mRNA (NM 001017959). To avoid cross-contamination in PCR assays, rigid precautions are taken according to published recommendations (Kitchin *et al.*, 1990). Mainly:

1. Procedures of purification of nucleic acids are carried out in a room physically separated from that for performing PCR reactions.
2. Positive and negative controls (negative sample and water) are included in all PCR assays.
3. Purification of nucleic acids and PCR reactions are repeated by different people working in the laboratory.

The methods described in this chapter make it possible to describe autophagic processes occurring in a tissue *in vivo*, in physiological or experimental conditions, as these techniques allow a quantitative or semiquantitative comparison between samples. Furthermore, it is possible to discriminate between different types of autophagy, such as macroautophagy and CMA.

8. Concluding Remarks

Reactive oxygen species (ROS) are a group of highly reactive molecular forms of oxygen containing unpaired electrons. These ROS are continuously produced as a byproduct of the mitochondrial respiratory chain in normal, healthy cells. Due to their high reactivity, ROS can oxidize cell constituents such as lipids, proteins, and DNA, thus damaging cell structures and compromising vital functions. Because of these potentially lethal effects, cells maintain ROS at a tolerable level by means of antioxidants such as the redox system, superoxide dismutase, and catalase (Yu et al., 2006) or by means of endogenous antioxidant molecules such as melatonin (Coto-Montes and Hardeland, 1999). However, this display of defensive mechanisms is often not enough to completely avoid cellular injury, and a second front of defense, aimed at the repair and removal of damaged components, is required.

Syrian hamster Harderian glands provide a very interesting model of oxidative-stress-induced cell injury. This injury can be finely tuned by modulating up or down the intensity and the duration of the exposure to light, by changing the wavelength or by using antioxidants or pro-oxidant agents. Thus, this model could be very helpful to explore the functioning of autophagy as a cell repair mechanism, as this process shows important oxidative stress-related differences between sexes. Thus, extensive macroautophagy is the predominant form in female glands that support the highest oxidative pressure. Therefore, we believe that this gland is a very useful model in the study of the relationship between autophagy and oxidative stress signaling.

ACKNOWLEDGMENTS

Thanks to Dr. María Josefa Rodríguez-Colunga, Covadonga Huidobro Fernández and Dr. José Antonio Boga Riveiro for their suggestions and collaboration. This work was partially performed with grants FISS-06-RD06/0013/0011 from the Instituto de Salud Carlos III (Ministerio de Sanidad y Consumo), INIA-07-RTA2007-00087-C02-02 from INIA (Ministerio de Ciencia e Innovación) and FEDER Fund (European Union). I V-N is a FPU predoctoral fellow from Ministerio de Ciencia e Innovación, Spain; A C-M is a contractual professor from I3 Program application awared by Gobierno del Principado de Asturias, Spain.

REFERENCES

Antolin, I., Rodriguez, C., Uria, H., Sainz, R. M., Mayo, J. C., Kotler, M. L., Rodriguez-Colunga, M. J., Tolivia, D., and Menendez-Pelaez, A. (1996). Castration increases cell damage induced by porphyrins in the Harderian gland of male Syrian hamster. Necrosis and not apoptosis mediates the subsequent cell death. *J. Struct. Biol.* **116,** 377–389.

Antolin, I., Uria, H., Tolivia, D., Rodriguez-Colunga, M. J., Rodriguez, C., Kotler, M. L., and Menendez-Pelaez, A. (1994). Porphyrin accumulation in the harderian glands of female Syrian hamster results in mitochondrial damage and cell death. *Anat. Rec.* **239,** 349–359.

Barrett, A. J. (1980). Fluorimetric assays for cathepsin B and cathepsin H with methylcoumarylamide substrates. *Biochem. J.* **187,** 909–912.

Bradford, M. M. (1976). A rapid and sensitive method for the quantitation of microgram quantities of protein utilizing the principle of protein-dye binding. *Anal. Biochem.* **72,** 248–254.

Bursch, W. (2001). The autophagosomal-lysosomal compartment in programmed cell death. *Cell Death. Differ.* **8,** 569–581.

Carreno, S., Engqvist-Goldstein, A. E., Zhang, C. X., McDonald, K. L., and Drubin, D. G. (2004). Actin dynamics coupled to clathrin-coated vesicle formation at the trans-Golgi network. *J. Cell Biol.* **165,** 781–788.

Coto-Montes, A., Boga, J. A., Tomas-Zapico, C., Rodriguez-Colunga, M. J., Martinez-Fraga, J., Tolivia-Cadrecha, D., Menendez, G., Hardeland, R., and Tolivia, D. (2001a). Physiological oxidative stress model: Syrian hamster Harderian gland-sex differences in antioxidant enzymes. *Free Radic Biol. Med.* **30,** 785–792.

Coto-Montes, A., Boga, J. A., Tomas-Zapico, C., Rodriguez-Colunga, M. J., Martinez-Fraga, J., Tolivia-Cadrecha, D., Menendez, G., Hardeland, R., and Tolivia, D. (2001b). Porphyric enzymes in hamster Harderian gland, a model of damage by porphyrins and their precursors. A chronobiological study on the role of sex differences. *Chem. Biol. Interact.* **134,** 135–149.

Coto-Montes, A., and Hardeland, R. (1999). Antioxidative effects of melatonin in Drosophila melanogaster: antagonization of damage induced by the inhibition of catalase. *J. Pineal Res.* **27,** 154–158.

Coto-Montes, A. M., Rodriguez-Colunga, M. J., Uria, H., Antolin, I., Tolivia, D., Buzzell, G. R., and Menendez-Pelaez, A. (1994). Photoperiod and the pineal gland regulate the male phenotype of the Harderian glands of male Syrian hamsters after androgen withdrawal. *J. Pineal Res.* **17,** 48–54.

Cuervo, A. M., and Dice, J.F (1996). A receptor for the selective uptake and degradation of proteins by lysosomes. *Science.* **273,** 501–503.

Eskelinen, E.-L., Illert, A. L., Tanaka, Y., Schwarzmann, G., Blanz, J., Von Figura, K., and Saftig, P (2002). Role of LAMP-2 in lysosome biogenesis and autophagy. *Mol. Biol. Cell.* **13,** 3355–3368.

Furuya, N., Yu, J., Byfield, M., Pattingre, S., and Levine, B. (2005). The evolutionarily conserved domain of Beclin 1 is required for Vps34 binding, autophagy and tumor suppressor function. *Autophagy.* **1,** 46–52.

Gonzalez-Polo, R. A., Boya, P., Pauleau, A. L., Jalil, A., Larochette, N., Souquere, S., Eskelinen, E.-L., Pierron, G., Saftig, P., and Kroemer, G. (2005). The apoptosis/autophagy paradox: autophagic vacuolization before apoptotic death. *J. Cell. Sci.* **118,** 3091–3102.

Hoffman, R. A. (1971). Influence of some endocrine glands, hormones and blinding on the histology and porphyrins of the Harderian glands of golden hamsters. *Am. J. Anat.* **132,** 463–478.

Isahara, K., Ohsawa, Y., Kanamori, S., Shibata, M., Waguri, S., Sato, N., Gotow, T., Watanabe, T., Momoi, T., Urase, K., Kominami, E., and Uchiyama, Y. (1999). Regulation of a novel pathway for cell death by lysosomal aspartic and cysteine proteinases. *Neuroscience* **91,** 233–249.

Kitchin, P. A., Szotyori, Z., Fromholc, C., and Almond, N. (1990). Avoidance of PCR false positives [corrected]. *Nature* **344,** 201.

Klionsky, D. J., Abeliovich, H., Agostinis, P., Agrawal, D. K., Aliev, G., Askew, D. S., Baba, M., Baehrecke, E. H., Bahr, B. A., Ballabio, A., *et al.* (2008). Guidelines for the use and interpretation of assays for monitoring autophagy in higher eukaryotes. *Autophagy* **4,** 151–175.

Lopez, J. M., Tolivia, J., and Alvarez-Uria, M. (1992). Postnatal development of the harderian gland in the Syrian golden hamster (Mesocricetus auratus): a light and electron microscopic study. *Anat. Rec.* **233,** 597–616.

Reggiori, F., Monastyrska, I., Shintani, T., and Klionsky, D. J. (2005). The actin cytoskeleton is required for selective types of autophagy, but not nonspecific autophagy, in the yeast *Saccharomyces cerevisiae*. *Mol. Biol. Cell* **16**, 5843–5856.

Schreurs, F. J., van der Heide, D., Leenstra, F. R., and de Wit, W. (1995). Endogenous proteolytic enzymes in chicken muscles. Differences among strains with different growth rates and protein efficiencies. *Poult. Sci.* **74**, 523–537.

Spike, R. C., Payne, A. P., and Moore, M. R. (1988). The effects of age on the structure and porphyrin synthesis of the harderian gland of the female golden hamster. *J. Anat.* **160**, 157–166.

Takahashi, T., and Tang, J., Cathepsin D from porcine and bovine spleen. *In:* L. Loran, (Ed.), Proteolitic enzymes, Part C. Methods in Enzymology. Academic Press, New York, 1981, pp. 567–589

Tolivia, D., Uria, H., Mayo, J. C., Antolin, I., Rodriguez-Colunga, M. J., and Menendez-Pelaez, A. (1996). Invasive processes in the normal Harderian gland of Syrian hamster. *Microsc. Res. Tech.* **34**, 55–64.

Tomas-Zapico, C., Caballero, B., Sierra, V., Vega-Naredo, I., Alvarez-Garcia, O., Tolivia, D., Rodriguez-Colunga, M. J., and Coto-Montes, A. (2005). Survival mechanisms in a physiological oxidative stress model. *FASEB J.* **19**, 2066–2068.

Tomas-Zapico, C., Coto-Montes, A., Martinez-Fraga, J., Rodriguez-Colunga, M. J., Hardeland, R., and Tolivia, D. (2002a). Effects of δ-aminolevulinic acid and melatonin in the harderian gland of female Syrian hamsters. *Free Radic. Biol. Med.* **32**, 1197–1204.

Tomas-Zapico, C., Martinez-Fraga, J., Rodriguez-Colunga, M. J., Tolivia, D., Hardeland, R., and Coto-Montes, A. (2002b). Melatonin protects against delta-aminolevulinic acid-induced oxidative damage in male Syrian hamster Harderian glands. *Int. J. Biochem. Cell Biol.* **34**, 544–553.

Yu, L., Wan, F., Dutta, S., Welsh, S., Liu, Z., Freundt, E., Baehrecke, E. H., and Lenardo, M. (2006). Autophagic programmed cell death by selective catalase degradation. *Proc. Natl. Acad. Sci. USA* **103**, 4952–4957.

Zeng, X., Overmeyer, J. H., and Maltese, W. A. (2006). Functional specificity of the mammalian Beclin-Vps34 PI 3-kinase complex in macroautophagy versus endocytosis and lysosomal enzyme trafficking. *J. Cell Sci.* **119**, 259–270.

Author Index

A

Abe, H., 371, 373
Abeliovich, H., 51, 53, 57, 58, 79, 162, 194, 202, 211, 229, 278, 279, 333, 386, 390, 393, 425, 431, 460, 465
Abraham, R. T., 166, 169
Adams, L. G., 326, 329
Adam-Vizi, V., 120
Adibhatla, R. M., 120
Aebersold, R., 169, 170
Agarraberes, F., 298, 299, 308, 309, 311
Agostinis, P., 51, 53, 57, 58, 79, 162, 194, 211, 278, 279, 333, 386, 390, 393, 425, 431, 460, 465
Agrawal, A., 279, 287
Agrawal, D. K., 51, 53, 57, 58, 79, 162, 194, 211, 278, 279, 333, 386, 390, 393, 425, 431, 460, 465
Aguzzi, A., 182
Aizawa, S. I., 337
Akaishi, R., 53, 202
Akaji, K., 289
Akerboom, T. P., 72
Akin, D. E., 442
Aktories, K., 67
Alafuzoff, I., 182
Albrecht, M., 333
Al-Hajj, A., 288
Ali, S. M., 166, 167, 170, 173, 174, 175
Aliev, G., 51, 53, 57, 58, 79, 162, 194, 278, 279, 333, 386, 390, 393, 425, 431, 460, 465
Al-Jarrah, H., 406
Allen, T. D., 229
Allfrey, V. G., 67
Almer, S., 333
Almond, N., 473
Alonso, S., 347
Alroy, J., 228, 229
Alt, F. W., 182
Altenberend, F., 404, 406
Alvarez-Garcia, O., 459, 460, 462, 465, 470
Alvarez-Uria, M., 458
Amano, A., 332, 347, 365, 366, 369
Ambrosio, S., 120, 127
Amelotti, M., 120, 127
Amer, A. O., 364, 384
Ames, B. N., 121
Amiji, M., 286, 289

Anderson, C. A., 333, 334, 385
Anderson, D., 437
Anderson, M. S., 405
Anderson, O. R., 15
Andrade, R. M., 384
Andrews, H. L., 326
Andrews, N. W., 327, 328, 329, 331
Angermüller, S., 147, 156
Aniento, F., 318, 319, 406
Antolin, I., 458, 459
Antúnez de Mayolo, A., 279
Appel, S., 407
Appelles, A., 249
Araki, K., 16
Arnold, R. S., 120
Arstila, A. U., 144
Asahi, M., 182
Asato, K., 236
Ashida, H., 371, 384
Askew, D. S., 51, 53, 57, 58, 79, 162, 194, 278, 279, 333, 386, 390, 393, 425, 431, 460, 465
Asmyhr, T., 79
Atkinson, E. N., 290
Auteri, J., 303
Avruch, J., 166, 167, 169, 170, 173, 174
Ayala, C., 347

B

Baba, H., 16
Baba, M., 51, 53, 57, 58, 79, 162, 194, 239, 278, 279, 333, 425, 431, 460, 465
Baba, Y., 286
Bachmann, J., 405, 406
Bachor, O. A., 405, 406
Backer, J. M., 152
Bader, A., 279, 281, 285, 287, 290
Baehrecke, E. H., 51, 53, 57, 58, 65, 79, 120, 126, 162, 194, 278, 279, 333, 425, 431, 460, 465, 474
Baek, J. H., 241
Bagley, A. F., 166, 174
Bahr, B. A., 51, 53, 57, 58, 79, 162, 194, 278, 279, 333, 425, 431, 460, 465
Baier, J., 32
Baird, G. S., 134
Bakalova, R., 286
Bakeeva, L. E., 241
Bakowski, M. A., 328, 329, 330, 331, 332, 333, 337, 338, 339, 384

Ballabio, A., 51, 53, 57, 58, 79, 162, 182, 194, 278, 279, 333, 425, 431, 460, 465
Baltes, J., 87, 88
Bamber, B. A., 194
Bampton, E. T., 91, 92, 133, 332
Banchereau, J., 405
Bandhakavi, S., 167
Bandyopadhyay, U., 120, 299
Bannenberg, G. L., 121
Bao, G., 285
Barbry, P., 286
Barker, P. E., 289
Barmada, M. M., 333
Bar-Peled, L., 167
Barrachina, M., 120, 127
Barrett, A. J., 466
Barrett, J. C., 333, 334, 385
Barrow, R. K., 169
Barth, H., 248, 249
Bassham, D. C., 120, 194
Bassnett, S., 230
Basso, L. A., 346
Bauer, D. E., 437
Baumann, S., 346
Baumler, A. J., 326, 329
Bauserman, R. G., 230
Bauvy, C., 47, 49, 120, 127, 132, 203, 305, 445
Bawendi, M. G., 279
Bazer, F. W., 54
Bazzi, H. S., 287
Bearer, E. L., 286, 287
Beatty, W. L., 346
Beaudoin, B., 437
Becker, J. E., 405
Becker, L. B., 121
Bedard, K., 120
Behrends, U., 65, 407
Belaiche, J., 333, 385
Bellaiche, Y., 286
Benian, G. M., 120
Benoist, C., 405
Berciaud, S., 286
Berg, C., 254
Berg, T., 66
Berg, T. O., 65, 66, 73, 74, 406
Bergamini, E., 441, 442, 445, 446, 448, 453
Berger, Z., 58
Bergmeyer, H. U., 77
Bernales, S., 347
Berndt, E., 77
Beron, W., 88, 364, 365, 384
Berry, M. N., 52
Berzins, S. P., 405
Besancon, F., 120, 127
Bessoule, J. J., 120
Betts, M. R., 290
Beug, H., 230, 237
Beuzón, C. R., 328, 329, 332

Bevan, A., 248, 249
Bevan, M. J., 404
Bhagat, G., 16
Bhakdi, S. C., 67
Bhardwaj, N., 416
Bhatia, S., 229
Bhattacharya, J., 123
Bialik, S., 120
Biederbick, A., 86, 87, 90
Binder, V., 333
Birkeland, H. C., 367
Birmingham, C. L., 325, 328, 329, 330, 331, 332, 333, 337, 338, 339, 365, 376, 384
Bissell, M. J., 428
Bjørkøy, G., 181, 182, 185, 187, 192, 193, 274, 426
Blab, G. A., 286
Blake, N. W., 406
Blanz, J., 152, 162, 469
Blenis, J., 167, 173, 174
Bliska, J. D., 327, 328, 329, 331
Blocker, A., 328
Blommaart, E. F., 50, 53, 132
Blum, J. S., 405
Boga, J. A., 458, 459
Bohley, P., 65, 66, 70, 71, 74, 75, 76, 77, 79
Boland, B., 132
Bolton, J., 239
Bondurant, M. C., 229, 230, 236
Bonen, D. K., 385
Bonenfant, D., 166
Bonneau, S., 286
Boon, L., 53
Borchert, G. L., 287
Bornkamm, G. W., 65, 407
Borutaite, V., 50
Bos, J. L., 167
Bosch-Marce, M., 241
Botelho, R. J., 249, 258
Botti, J., 49, 203
Bouchier-Hayes, L., 286
Boucrot, E., 328, 329, 332
Bouillet, P., 431
Boulme, F., 230, 237
Bouzigues, C., 286
Boya, P., 89, 437, 469
Boyd, J. M., 239
Bozek, G., 239
Brachmann, S. M., 152, 162
Bradford, M. M., 467
Brandenburg, J., 404, 406
Bray, K., 437
Brazil, M. I., 406
Brech, A., 15, 66, 181, 182, 185, 187, 192, 193, 426
Briggs, J., 333
Brismar, H., 286
Brock, R., 404, 406

Brock, T. D., 337
Bronson, R., 405
Brooks, D. M., 230
Brooks, J. M., 406, 419
Brossart, P., 407
Brown, C. R., 300
Brown, E. B., 279
Brown, M., 167
Browne, C., 232
Bruchez, M. P., 290
Brugge, J. S., 424, 428, 429, 431, 433, 436
Brumell, J. H., 325, 326, 328, 329, 330, 331, 332, 333, 337, 338, 339, 365, 376, 384
Brunk, U. T., 229, 442
Brunn, G. J., 169
Bruns, N. E., 182
Bruska, J. S., 205
Brust, M., 286
Brutkiewicz, R. R., 405
Bruun, J. A., 182, 185, 193, 426
Bryant, N. J., 258
Bryant, P., 404
Bucci, C., 88, 145, 152, 162, 367
Budha, N. R., 346
Bulavina, L., 279, 281, 285, 287, 290
Burden, S. J., 57, 167
Burds, A. A., 167
Burkett, C., 376, 384
Burley, S. K., 170
Burnett, P. E., 169
Bursch, W., 470
Buzzell, G. R., 458
Byfield, M., 471
Byrne, B. G., 383, 384

C

Caballero, B., 459, 460, 462, 465, 470
Cabiaux, V., 328
Cadinanos, J., 182
Cai, X., 289
Caldwell, H., 405
Callisen, T. H., 286
Camougrand, N., 120, 229
Campbell, R. E., 134, 193
Canaday, D. H., 405
Canadien, V., 339, 365, 376, 384
Cantley, L. C., 167
Cao, L., 182
Cappello, G., 286
Carlsson, S. R., 311
Caro, L. H., 53, 56
Carpentier, S., 49, 203
Carpi, A., 200
Carr, S. A., 167, 174
Carreno, S., 470
Carresi, C., 448

Casanova, J. E., 330
Casares, N., 89, 437
Cathcart, R., 121
Cattoretti, G., 16
Cavallini, G., 441, 442, 446, 448, 453
Cederbaum, A. I., 128
Celada, A., 393
Celli, J., 384
Cezard, J. P., 333, 385
Chadwick, L. H., 115
Chait, B. T., 182
Chaki, S., 286, 289
Chakrabarti, S., 327, 328, 329, 331
Chakraborty, T., 363, 376
Chamaillard, M., 333, 385
Chan, E. Y. W., 14, 263, 274
Chan, H. W., 67
Chan, J., 346
Chan, P., 289
Chantret, I., 446, 448
Chapman, H. A., 404
Chasis, J. A., 230
Chatterjee, D., 346
Chattopadhyay, P. K., 290
Chaudry, Q., 290
Checroun, C., 384
Chen, C. A., 100
Chen, F. F., 385
Chen, G., 239, 424, 437
Chen, J.-J., 232
Chen, J. M., 103, 114
Chen, J. Y., 285, 286
Chen, L., 286, 287
Chen, M., 229, 239
Chen, R. J., 127
Chen, X., 205
Chen, Y., 120, 126, 127, 229, 236
Chen, Z., 405
Cheng, A. C., 405
Cheng, C. M., 279
Cheng, G., 120
Cheng, T. J., 127
Cheng, Z., 290
Cheong, H., 239
Chernyak, B. V., 241
Chhatwal, G. S., 367
Chiang, H., 298, 311
Chiba, T., 7, 20, 238, 240
Chicz, R. M., 405
Chikae, M., 286
Chin, G., 232
Chinnadurai, G., 239
Chinopoulos, C., 120
Choquet, D., 286
Choy, K. L., 289
Christian, C., 120
Christian, R. E., 419

Chu, S., 286, 287
Chua, J., 346
Chung, A. B., 120
Chung, L. W., 290
Ciechanover, A., 405
Cirulis, J. T., 330
Cizeau, J., 239
Clark, B. F., 442
Clark, S. L., Jr., 262
Clausen, T. H., 182, 185, 193, 426
Cleary, P. P., 367
Cleveland, J. L., 70, 182, 229, 231, 233, 239
Clogston, J., 290
Cluzeaud, F., 49
Codogno, P., 47, 49, 50, 80, 89, 120, 127, 132, 200, 203, 305, 437, 442, 445, 446, 448
Coffino, P., 406
Cognet, L., 286
Cohen, K. S., 279
Cohen, N. A., 169
Colell, A., 286
Colins, N. L., 424, 431
Collazo, C. M., 337
Colombo, M. I., 86, 87, 88, 91, 93, 346, 347, 352, 364, 365, 384
Colvin, V., 278
Conley, S., 289
Connell, S., 70, 384, 400
Contento, A. L., 120
Cook, L. J., 58
Cooke, R. H., 249
Cooper, J., 152
Coppens, I., 347
Coram, M., 406
Corbett, E. L., 346
Cordenier, A., 58
Cossart, P., 339, 364, 365, 376
Cossins, A. R., 286
Coto-Montes, A., 457, 458, 459, 460, 462, 465, 470
Courty, S., 286
Cregg, J. M., 14, 145
Crespo, J. L., 166
Criollo, A., 126, 182, 240
Cristea, I. M., 182
Crocquet-Valdes, P. A., 384
Crotzer, V. L., 419
Cuddon, P., 178
Cue, D., 366
Cuervo, A. M., 48, 58, 65, 67, 74, 79, 120, 167, 200, 297, 298, 299, 300, 301, 302, 303, 308, 309, 311, 312, 314, 315, 320, 346, 404, 406, 442, 472
Cui, B., 286, 287
Cuie, Y., 287
Cummings, F. R., 333, 334
Cywes, C., 367

D

Dahan, M., 286
Dai, F., 236
Daigaku, Y., 206, 208, 209, 210, 211
Dalen, H., 229
Dalton, E., 329
D'Amelio, M., 182
Danev, R., 289
Dang, Y., 236
Danial, N. N., 432
Danielsen, M., 67
Danon, D., 241
Dantuma, N. P., 406
Dasgupta, S. K., 229, 239
Datta, L. W., 333
Davies, J. E., 178
Davis, A. S., 334, 346, 347, 352, 384, 400
De, A., 290
DeBerardinis, R. J., 48
Debnath, J., 423, 424, 425, 428, 429, 431, 433, 434, 435, 436, 437
de Chastellier, C., 384
de Duve, C., 43, 218, 278
Deffieu, M., 120
Degani, I., 133
Degenhardt, K., 437
De Haro, S. A., 347
Deiner, E. M., 230, 237
De La Vega, F. M., 333
Delgado, M. A., 345, 346, 347, 352, 384, 400
Del Piccolo, P., 57, 167
Del Roso, A., 442, 445, 446, 448
Deng, W., 290
Dengjel, J., 404, 406
Denk, H., 182
Dennis, A. M., 285
Dennis, P. B., 66, 78, 79, 97
Deo, R., 279, 287
de Pablo, R., 182
De Paoli, V., 288
Deretic, V., 334, 345, 346, 347, 352, 364, 365, 384, 400
deRoos, P., 405
De Rosa, S. C., 290
Dessen, P., 89, 437
De Tata, V., 442
Deter, R. L., 278
Devalapally, H., 286
Devenish, R. J., 347
de Wit, W., 466
Dexter, T. M., 229
Di, A., 327, 328, 329, 331
Dice, J. F., 120, 298, 299, 301, 302, 308, 311, 312, 316, 320
Dierich, A., 405
Di Lisi, R., 57, 167

Dillon, C. P., 70, 384, 400
Ding, W. X., 26
Ding, Y., 182
Dingle, J. T., 71
Dionne, C. A., 239
di Rago, J. P., 229
Djavaheri-Mergny, M., 120, 127, 182
Djonder, N., 67
Dodd, S. L., 442
Doherty, P. C., 404
Dolznig, H., 230, 237
Domann, E., 376
Dombrowski, F., 52, 56
Dominko, T., 229
Donald, S. P., 287
Donati, A., 441, 442, 445, 446, 448, 453
Dongre, A. R., 405
Dorfel, D., 407
Dorn, B. R., 364, 365
Dorsey, F., 70, 384, 400
Dorsey, F. C., 182, 229, 231, 233, 239
Dove, S. K., 249
Drezek, R., 278
Driessen, C., 404, 406
Dröge, W., 442
Drubin, D. G., 470
Drummond, H., 333, 334
Dubbelhuis, P. F., 50
Dubik, D., 239
Dubisson, J.-F., 383
Ducati, R. G., 346
Duchesne, L., 286
Duda, D. G., 279
Dunn, W. A., Jr., 14, 65, 74, 79, 144, 145, 202, 216, 248, 249, 364, 365, 442
Duss, I., 334
Dutta, S., 65, 120, 126, 474
Duvoisin, R. M., 229, 230
Dye, C., 346

E

Eaton, J. W., 229
Ecker, N., 229
Edens, H. A., 120
Edens, W. A., 120
Edgar, B. A., 167
Efe, J. A., 249, 258
Egner, R., 65
Eguchi, S., 166, 169, 170
Eguchi, Y., 239
Ehrbar, K., 334
Ehrensperger, M. V., 286
Eissa, N. T., 384, 400
Elangovan, B., 239
Elazar, Z., 58, 119, 120, 121, 122, 123, 125, 127, 128, 131, 133, 138
Elmaoued, R. A., 345, 346, 347, 352, 384, 400

Elmore, S. P., 229, 234, 241
Elofsson, U. M., 286
Elorza, A., 228, 229
Elsässer, H. P., 86, 87, 88, 90
Emmerich, N. P., 405, 406
Emmrich, F., 279, 281, 285, 287, 290
Emr, S. D., 14, 145, 249, 258
Eng, J., 405
Engelhard, V. H., 419
Engelhardt, H., 229, 230
Engqvist-Goldstein, A. E., 470
Epardaud, M., 405
Epple, U. D., 248, 249
Erdjument-Bromage, H., 166, 170, 173, 174, 175
Erichsen, E. S., 65, 406
Eriksson, S., 329
Ernst, S. A., 205
Eskelinen, E.-L., 8, 15, 16, 88, 132, 144, 152, 153, 154, 156, 159, 161, 162, 211, 262, 300, 301, 315, 347, 367, 462, 469
Ezaki, J., 7, 20, 182, 207, 215, 218, 238, 240, 289

F

Facchinetti, V., 167
Fadden, P., 169
Fader, C. M., 88, 93
Fahimi, H. D., 147, 156, 216
Fahlen, T., 326
Falkow, S., 328, 329
Fan, Y., 437
Farinas, J., 134
Farquhar, M. G., 193
Farr, A., 204
Farrar, C., 289
Fass, E., 58, 120, 123, 133, 138
Fazio, J. A., 346
Felgner, P. L., 67
Feng, H. M., 384
Fengsrud, M., 65, 66, 74, 77, 406
Fennell, D. F., 70
Feramisco, J., 318
Ferguson, D. J., 347
Fernandez, J. M., 287
Fernig, D. G., 286
Ferrer, I., 120, 127, 182
Ferrer-Miralles, N., 289
Ffrench, M., 442
Ficht, T. A., 326, 329
Fierabracci, V., 442, 445, 446, 448
Filburn, C. R., 123
Filimonenko, M., 182
Fimia, G. M., 26
Finch, C. A., 230
Fingar, D. C., 174
Finkel, T., 182

Finkel, Y., 333
Finlay, B. B., 326, 328, 329, 330, 332, 334, 338
Finn, P. F., 120, 308
Fischer, E. R., 384
Fischer, R., 404, 406
Fisher, E. M., 182
Fisher, S. A., 333, 334, 385
Fitzgerald, K. J., 167
Fitzgerald, P., 286
Flaherty, D. B., 120
Fleming, A., 178
Fleming, M., 232
Fletcher, G. C., 50, 120
Floto, R. A., 58, 178
Flynn, J., 346
Flynn, S. P., 54
Folkes, J. P., 337
Fonseca, B. D., 169, 170, 174
Fonteneau, J. F., 416
Forster, S., 69
Fortin, G. R., 287
Fraldi, A., 182
Franchi, L., 384
Franchini, D.-M., 384
Franke, A., 333
Franklin, J. L., 120
Freije, J. M., 182
Freundt, E., 65, 120, 126, 474
Frickey, T., 248, 249, 250, 251, 254, 255, 258
Fried, J., 442
Friedrich, C. L., Jr., 182
Frisch, S. M., 434
Fromholc, C., 473
Fuchsbichler, A., 182
Fuertes, G., 152, 306
Fueyo, A., 16
Fujiki, Y., 216
Fujimura, S., 53, 66, 77, 202, 206, 208, 209, 210, 211
Fujioka, K., 286
Fujita, E., 437
Fujita, H., 367
Fujita, N., 1, 27, 30, 58, 98, 133, 134, 183, 200, 236, 251, 388, 392, 471
Fujiwara, T., 239
Fujiwara, Y., 232
Fukumura, D., 279
Funao, J., 332, 347, 365, 366, 369
Fung, C., 424, 425, 434, 435, 436, 437
Furlan, M., 88
Furuta, S., 29
Furuya, N., 16, 53, 66, 77, 98, 103, 202, 211, 471
Futai, M., 106
Futaki, S., 284, 286, 289

G

Gadek, T. R., 67
Gal, J., 182

Galan, J. E., 337, 338
Gale, E. F., 337
Galluzzi, L., 126, 182
Galyov, E. E., 328, 329
Gambhir, S. S., 290
Gannagé, M., 403
Ganta, S., 286
Ganzoni, A., 230
Gao, S., 424, 425, 434, 435, 436, 437
Gao, X., 167, 288
Garami, A., 167
Garcia-del Portillo, F. Z. M., 328, 330
Garuti, R., 240
Garvis, S., 328, 329, 332
Gasko, O., 241
Gassull, M., 333
Gatenby, R. A., 287
Gaugel, A., 248, 249, 251, 254
Gaullier, J. M., 258
Gautier, F., 240
Geetha, T., 182
Gelinas, C., 437
Geneste, O., 240
Geng, J., 239
George, M. D., 236, 248, 249
Gerstel, B., 376
Getts, R. C., 289
Geuze, H. J., 191, 192, 273
Ghidoni, R., 49, 203
Gibson, S. B., 120, 126, 127, 229
Gidding, T. H., Jr., 91
Giepmans, B. N., 193
Gigengack, S., 192
Gil, J., 120, 127
Gil, L., 120, 123
Giles, D., 326
Gill, D. J., 182
Gillies, R. J., 287
Gillooly, D. J., 258
Gingras, A. C., 169, 170
Glauert, A. M., 146, 147, 148
Gleeson, C., 329
Gniadek, T. J., 5
Goemans, C. G., 91, 92, 133, 332
Goepfert, P., 290
Goldberg, A. L., 103, 167
Goldsmith, P., 178
Gomez-Santos, C., 120, 127
Gonzalez, F. J., 241
Gonzalez-Maeso, J., 288
Gonzalez-Polo, R. A., 89, 437, 469
Goosney, D. L., 326
Gordon, P. B., 49, 65, 66, 67, 69, 70, 71, 72, 73, 74, 75, 76, 77, 79, 445
Gorga, J. C., 405
Gori, Z., 442, 445, 446, 448
Gorvel, J. P., 328, 334, 338, 384
Gostick, E., 290

Gotch, F., 419
Goto, M., 54, 202, 204
Gotow, T., 465
Gottlieb, R. A., 21, 86, 88, 91, 92
Gouin, E., 339, 364, 365, 376
Gould, R., 258
Gounon, P., 328
Govaerts, C., 406
Goyette, P., 333, 385
Granfors, K., 326
Grasso, P., 278
Gray, P. W., 393
Green, D. R., 70
Green, T., 333
Greenberg, A., 239
Griendling, K. K., 120
Griffin, T. J., 167
Griffiths, A. M., 333
Grinde, B., 55, 65
Grinstein, S., 326, 328
Grissom, S. F., 229, 234, 241
Grobe, L., 376
Groc, L., 286
Groen, A. K., 52, 56
Groisman, E. A., 329
Gronowicz, G., 230
Gross, M. L., 406
Grunebach, F., 407
Guan, J., 248, 249
Guan, J. L., 3, 14, 182, 240
Guan, K.-L., 167, 170, 175
Guarente, L., 228
Gubbels, M. J., 384
Gudjonsson, T., 428
Guertin, D. A., 167, 170, 174, 175
Guio-Carrion, A., 286
Gunther, S., 333
Guo, D., 216
Guo, J., 285, 286
Guo, Z., 236
Gurer, C., 405
Gustafson, L. A., 53
Gustafsson, B., 229, 442
Gutierrez, M. G., 88, 346, 347, 352, 364, 365, 384
Gygi, S. P., 169, 170

H

Habibzadegah-Tari, P., 248, 249
Hafen, E., 167
Hafer, K., 126
Hagen, L. K., 76, 78, 79
Haigh, S. E., 228, 229
Haigh, T. A., 406
Hailey, D. W., 25
Hain, T., 363
Hakansson, S., 328

Hall, D. H., 15
Hall, M. N., 166
Hamada, H., 289
Hamada, S., 332, 347, 365, 366, 367, 368, 369
Hamazaki, J., 182
Hammarback, J. A., 437
Hammerschmidt, W., 406
Hampe, J., 333
Han, A. P., 232
Han, J., 120
Han, J. W., 239
Hanada, K., 3, 6, 15
Hanaoka, H., 15
Handa, Y., 371
Hanes, J., 289
Hangen, E., 126
Hara, K., 166, 169, 170, 173, 174
Hara, T., 3, 14, 20, 182, 240, 366
Hara, Y., 7
Haraga, A., 326, 328, 329
Harashima, H., 286, 289
Hardeland, R., 458, 459, 474
Hardt, W.-D., 328, 334
Harhaji, L., 290
Harper, F., 182
Harper, T. F., 290
Harris, J., 347
Harris, M. H., 437
Harris, T. E., 170
Hartmann, D., 152
Hashimoto, D., 16
Hashimoto, T., 217
Hatano, M., 7, 14, 16, 48, 240, 339, 386
Hatcher, J. F., 120
Häussinger, D., 52, 56
Hayat, M. A., 146, 147, 234
Hayes, S. F., 384
Hayman, M. J., 338
Haystead, T. A., 169
He, H., 236
He, J., 285, 286
Heath, J. K., 405
Hebner, C., 424
Hedges, S., 329
Heeren, J., 152
Heessen, S., 406
Heid, H., 182
Heine, A., 407
Heine, M., 286
Heinrich, J. M., 279, 281, 285, 287, 290
Heintz, N., 182
Heller, K. N., 405
Helmick, L., 279
Hemelaar, J., 236
Hendy, R., 278
Henke, S., 249
Hensel, M., 326, 329
Henseleit, K., 56

Heston, L., 410
Heuser, J. E., 346
Heynen, M. J., 230
Hibshoosh, H., 16
Hickman, J. A., 240
Hidayat, S., 166, 173, 174
Higgins, D. E., 339, 365, 376, 384
Higuchi, Y., 182
Hikoso, S., 182
Hillman, R. S., 230
Himeno, M., 367
Hino, S., 437
Hinton, J. C., 329
Hirota, M., 16
Hirschberg, K., 133
Hirsimaki, P., 147, 156
Hislop, A., 405
Ho, Y. S., 127
Hochella, M. F., Jr., 278
Hoekstra, D., 289
Hoekstra, M. F., 169
Hoffman, R. A., 458
Hofmann, F., 67
Holden, D. W., 326, 328, 329, 332
Holden, L., 78
Holm, M., 67
Holmes, A. B., 249
Holstein, G. R., 182
Homble, F., 328
Hong, S., 165
Hori, M., 182
Horie, Y., 50
Hoshino, A., 286
Hosokawa, N., 7, 108, 110, 114
Howard, C. V., 150, 151, 154, 155
Howland, K., 419
Høyer-Hansen, M., 43
Hoyt, M. A., 406
Høyvik, H., 66, 67, 72, 73, 74
Hsu, P. P., 166, 174
Hu, C.-A., 287
Hu, X. W., 14
Huang, C., 21, 86, 88, 91, 92
Huang, J., 132
Huang, Q., 167
Huang, Z., 178
Hueck, C. J., 326, 329
Huett, A., 333, 385
Hughes, D. C., 249
Hugot, J. P., 333, 385
Human, A., 254
Humme, S., 406
Hunt, D. F., 419
Hurley, J. P., 286
Huse, K., 333
Hutson, N. J., 445
Huynh, K., 322

I

Ichimura, H., 123
Ichimura, Y., 3, 26, 205, 236
Ida, S., 16
Idone, V. J., 327, 328, 329, 331
Iemura, S., 3, 14, 182, 240
Ikenoue, T., 165, 170
Illert, A. L., 152, 162, 469
Imarisio, S., 58
Imehara, H., 126
Imreh, S., 406
Ingram, D. K., 442
Inohara, N., 385
Inoki, K., 165, 167, 170, 175
Isaacs, A., 182
Isahara, K., 465
Isakovic, A., 290
Ishibashi, T., 54, 202, 204
Ishidoh, K., 77, 78, 79
Ishihara, N., 3, 205, 236
Ishii, T., 236
Ishiura, S., 3, 6, 15
Isoai, A., 437
Israels, S. J., 120, 126, 127, 229
Issekutz, A. C., 123
Iwai, K., 126
Iwai, N., 290
Iwai-Kanai, E., 21, 86, 88, 91, 92
Iwamaru, A., 290
Iwamoto, K. S., 126
Iwata, J. I, 7, 20, 182, 216, 217, 220, 238, 240, 289

J

Jäättelä, M., 43
Jacinto, E., 166, 167
Jackson, W. T., 91, 365, 384, 392
Jaffee, E. M., 407
Jagannath, C., 384, 400
Jäger, S., 88, 145, 152, 162, 367
Jahreiss, L., 14, 40, 178, 182, 347
Jain, R. K., 279
Jalil, A., 469
James, D. E., 192
Janue, A., 182
Jayachandran, G., 290
Jeffee, E. M., 65
Jefferies, W. A., 330
Jeffries, T. R., 249
Jenkins, J. R., 248
Jenoe, P., 166
Ji, L., 290
Jia, K., 442
Jiang, F., 288
Jiang, W., 347

Author Index

Jiang, X., 328, 329, 331
Jiang, Z., 288
Jin, S., 424, 437
Joaquin, M., 167
Johansen, T., 181, 182, 185, 187, 192, 193, 426
Joiner, K. A., 364
Jokitalo, E., 8, 132, 143, 211, 262, 315, 462
Jones, A. T., 367
Jones, B. D., 326
Jones, M. A., 329
Jones, M. D., 329
Joseph, J. A., 442
Jotereau, F., 416
Jou, M. J., 123
Juhasz, G., 15
Juin, P., 240
Julian, D., 442
Jung, J. U., 347
Jung, S. Y., 167

K

Kabeya, Y., 2, 3, 6, 7, 10, 14, 28, 29, 98, 115, 133, 145, 205, 207, 210, 236, 239, 330, 366, 367, 386, 425
Kadowaki, M., 53, 54, 66, 77, 78, 79, 132, 199, 200, 201, 202, 204, 206, 208, 209, 210, 211, 445, 453
Kakudo, T., 286, 289
Kalaany, N. Y., 167
Kalbacher, H., 404, 406
Kale, A. A., 286
Kaliappan, A., 66, 78, 79
Kamada, Y., 14, 239
Kamimoto, T., 332, 347, 365, 366, 369
Kamiya, H., 286, 290
Kanamori, S., 465
Kanazawa, T., 53, 66, 77, 202, 206, 208, 209, 210, 211
Kanemoto, S., 437
Kang, S. A., 167, 174
Kaniuk, N. A., 384
Kao, W. W., 127
Karantza-Wadsworth, V., 424
Karger, B. L., 448
Karim, M. R., 199, 200, 206, 208, 209, 210, 211
Kastan, M. B., 182
Kato, T., 15
Katz, S., 228, 229
Kaufman, R. J., 347, 437
Kaufmann, S. H., 346
Kaul, G., 289
Kauppinen, T., 182
Kaushik, S., 297, 299, 472
Kawai, A., 200
Kawakami, T., 289
Kazgan, N., 15
Keane, J., 347
Kenner, L., 182

Kerman, K., 286
Kern, H. F., 86, 87, 90
Kessin, R. H., 15
Kessler, B. M., 236
Khairallah, E. A., 202
Kido, H., 16
Kiel, J. A. K. W., 347
Kiffin, R., 120, 299
Kikkawa, U., 166, 169, 170
Kilty, R., 182
Kim, D. H., 166, 167, 170, 173, 174, 175
Kim, E., 175
Kim, H. S., 286
Kim, I., 241
Kim, J., 202
Kim, L., 21, 86, 88, 91, 92
Kim, S. O., 120
Kim, Y. P., 286
Kimchi, A., 120
Kimura, S., 1, 3, 8, 14, 27, 30, 58, 88, 98, 133, 134, 183, 200, 236, 251, 388, 392, 426, 471
King, J. E., 166, 173, 174, 175
King, M. C., 249
Kingsley, R. A., 326, 329
Kinkade, J. M., 120
Kirisako, T., 2, 3, 6, 14, 133, 145, 205, 210, 236, 330, 366, 367, 386, 425
Kirkegaard, K., 26, 90, 364, 384, 392
Kirkland, R. A., 120
Kisen, G. Ø., 65, 66, 70, 71, 74, 75, 76, 77, 79
Kishi, C., 3, 14, 16, 182, 240
Kissova, I., 120, 229
Kitano, T., 126
Kitchin, P. A., 473
Kjeken, R., 66
Klein, G., 406
Klein, L., 405
Kleinert, R., 182
Klionsky, D. J., 14, 26, 27, 42, 48, 51, 53, 57, 58, 65, 67, 74, 79, 80, 92, 98, 108, 110, 114, 132, 133, 145, 162, 167, 194, 200, 202, 211, 229, 236, 239, 240, 248, 249, 263, 278, 279, 298, 306, 333, 346, 347, 365, 386, 390, 393, 404, 425, 431, 460, 465, 470
Klotz, L. O., 123
Knabe, D. A., 54
Knecht, E., 16, 120, 152, 406
Kneen, M., 134
Knodler, L. A., 329, 338
Knoll, J. H., 289
Kobayashi, K. S., 384
Kobayashi, S., 262
Kobayashi, T., 67
Kobayashi, Y., 7, 14, 386
Köchl, R., 14
Kogure, K., 290
Koh, A. L., 290

Kohtz, D. S., 182
Koike, M., 20, 182
Kolodziejska, K. E., 384, 400
Komatsu, M., 7, 20, 48, 50, 70, 182, 207, 215, 218, 224, 238, 240, 289
Kominami, E., 2, 3, 6, 7, 14, 15, 20, 21, 23, 48, 50, 58, 65, 66, 77, 78, 79, 88, 133, 145, 152, 162, 182, 200, 205, 207, 208, 210, 211, 215, 236, 238, 240, 289, 330, 366, 367, 386, 425, 437, 465
Kondo, S., 290, 437
Kondo, T., 126
Konecki, D. S., 300
Kong, J., 120, 126, 127, 229
Kong, K. Y., 290
Kong, M., 437
Kopito, R. R., 91
Kopitz, J., 65, 66, 70, 71, 74, 75, 76, 77, 79, 98, 115
Kopsombut, P., 230
Korsmeyer, S. J., 432
Kotler, M. L., 458, 459
Koup, R. A., 290
Kouroku, Y., 437
Koury, M. J., 229, 230, 236
Koury, S. T., 230
Kovács, A. L., 53, 65, 66, 161
Kovács, J., 161, 445, 446
Kovats, S., 405
Kozma, S. C., 167
Kraft, C., 110
Kraus, M., 404, 406
Krause, K. H., 120
Krause, U., 50, 132
Kravchenko, L. V., 287
Kravchuk, O., 424
Krebs, H. A., 56
Kreymborg, K., 404, 406
Krick, R., 249
Krishnakantha, T. P., 216
Kristiansen, K., 182, 192
Kroemer, G., 26, 126, 240, 262, 469
Krystev, L. P., 287
Kuballa, P., 333
Kube, D., 406
Kubori, T., 337
Kuhle, V., 326
Kuma, A., 6, 14, 15, 16, 20, 26, 29, 48, 133, 134, 229, 240, 251, 339, 425
Kumagai, H., 437
Kundu, M., 227, 229, 231, 233, 239, 243, 262
Kurilla, M. G., 406
Kuroiwa, A., 236
Kuronita, T., 367
Kuroyanagi, H., 236
Kuusisto, E., 182
Kwok, W. W., 416
Kyei, G., 346, 347, 352, 384, 400

L

Laemmli, U. K., 204, 319
Lai, S. K., 289
Lake, B. G., 217
Lamark, T., 181, 182, 185, 187, 192, 193, 426
Lambeth, J. D., 120
Land, H., 426
Landthaler, M., 406
Lane, W. S., 405
Larochette, N., 89, 437, 469
Larquet, E., 328
Larsson, M., 416
Las, G., 228, 229
Lasne, D., 286
Lassegue, B., 120
Latek, R. R., 166, 170, 173, 174, 175
Lavieu, G., 49, 203
Lavine, B., 471
Lawrence, J. C., Jr., 169, 170
Lazarow, P. B., 216
Le, T. G., 240
Leboulch, P., 232
Leduc, P. R., 279
Lee, D. H., 103
Lee, I. H., 182
Lee, J., 286
Lee, J. W., 405
Lee, R. E., 346
Lee, S. I., 167
Lee, S. P., 406
Lee, S. S., 216
Lee, T., 120
Lee, V. H., 169, 170, 174
Leenstra, F. R., 466
Lees, C. W., 333, 334
Leeuwenburgh, C., 442
Legakis, J. E., 239
Leib, D. A., 347
Lelyveld, V. S., 236
Lemasters, J. J., 112, 229, 234, 241
Lemmon, M. A., 249
Lemons, R., 69
Lenardo, M., 65, 120, 126, 474
Leonchiks, A., 405
Lesage, S., 333, 385
Leung, A., 120
Leung, K. Y., 328
Levade, T., 49, 203
Levchenko, T. S., 279, 286
Leverve, X. M., 56
Levine, B., 15, 16, 26, 48, 65, 74, 79, 167, 240, 262, 298, 346, 347, 365, 384, 404, 442
Levitskaya, J., 405, 406
Levitsky, V., 406
Lewinski, N., 278
Lewis, P. R., 146, 147, 148
Li, C., 437

Li, C. W., 286
Li, D. N., 108, 114
Li, P., 405
Li, R., 288
Li, S., 288
Li, W., 229, 231, 233, 239
Li, W. P., 286, 287
Li, Y., 120, 134, 167
Li, Z., 121
Liang, C., 347
Liang, X. H., 240
Lienhard, G. E., 192
Lin, L., 288
Lin, Y., 167
Lindmo, K., 15
Lindner, J., 167
Lindner, W., 448
Lindquist, R. A., 167, 174
Lindsay, M., 258
Lindsten, T., 243, 437
Ling, W., 236
Ling, Y. M., 347
Lipinski, M. M., 329, 365, 376, 384
Lipman, M., 410
Lippincott-Schwartz, J., 25, 28, 29, 31, 37
Liston, D. R., 327, 328, 329, 331
Liu, D., 167
Liu, J., 182
Liu, J. O., 236
Liu, X. D., 384, 400
Liu, Y., 287
Liu, Z., 65, 120, 126, 474
Llopis, J., 193
Lock, R., 424, 425, 434, 435, 436, 437
Locke, M., 262
Loewith, R., 166
Lof, C., 52
Lombard, D. B., 182
Long, X., 167, 173, 174
Lopez, A., 338
Lopez, J. M., 458
Lopez-Otin, C., 16, 182
Lorberg, A., 166
Lorelli, W., 366
Lott, J. M., 405
Lounis, B., 286
Lovrié, J., 287
Lower, S. K., 278
Lowry, O., 204
Loyd, M. R., 229, 231, 233, 239
Lu, D. R., 285, 286
Lu, L., 232
Lu, Z. J., 285, 286
Luccardini, C., 286
Lucchini, S., 329
Lucocq, J. M., 152
Luiken, J. J., 53
Lukyanov, K. A., 28

Lüllmann-Rauch, R., 152
Lum, J. J., 26, 48, 437
Lupas, A., 248, 249, 251, 254
Lyamzaev, K. G., 241
Lynch, D. K., 436

M

Machesky, L. M., 375
MacKintosh, C., 169, 170, 174
Maclean, K. H., 182
Madden, E., 309, 310, 311, 316
Magnuson, M. A., 167
Maher, D., 346
Mailleux, A. A., 431
Maiuri, M. C., 126, 182, 240
Majeski, A., 299
Mak, T. W., 50
Maki-Ikola, O., 326
Malerod, L., 182
Malstrom, S., 239
Maltese, W. A., 467
Mammucari, C., 57, 167
Mangel, W. F., 67
Manning, B. D., 167
Manon, S., 120, 229
Marcus, S. L., 329
Mari, B., 286
Marino, G., 16, 182
Marjomaki, V. S., 50
Markard, A. L., 166, 174
Markovic, I., 290
Markovic, Z., 290
Marshansky, V., 106
Martinez-Fraga, J., 458, 459
Martinez-Vicente, M., 301
Martinou, J. C., 229
Marto, J. A., 419
Maruki, Y., 173, 174
Maruyama, K., 289
Marzella, L., 309
Masaki, R., 386
Masiello, P., 442, 445, 446, 448
Masiero, E., 57, 167
Masini, M., 442, 445, 446, 448
Masoro, E. J., 442
Massey, A., 299, 300, 301, 308
Massey, D. C., 333
Master, S. S., 346, 347, 352, 364, 365, 384
Masucci, M. G., 405, 406
Masuho, Y., 236
Mathew, R., 286, 424, 437
Mathieu, J., 120, 127
Mathis, D., 405
Mathur, N., 326
Matsubae, M., 14
Matsuda, Y., 218, 236
Matsui, A., 12

Matsui, M., 14, 15, 16, 17, 20, 48, 133, 205, 229, 240, 251, 339, 366, 377, 390, 425
Matsumura, Y., 182, 289
Matsushima, M., 239
Maurer, U., 286
Maurice, P. A., 278
Mautner, J., 65, 407
Mayo, J. C., 458, 459
Mayr, G., 333
Maysinger, D., 287
McCaffery, J. M., 193
McCormack, A. L., 405
McDonald, K. L., 470
McEwen, R. K., 249
McIntyre, T. M., 70
McKenney, J. B., 230
McLean, I. W., 398
McLeland, C. B., 290
McMichael, A., 419
McMillan-Ward, E., 120, 126, 127
McNeil, S. E., 278, 290
Medarova, Z., 289
Medina, D., 182
Medina, E., 367
Meibohm, B., 346
Meijer, A. J., 47, 50, 52, 53, 56, 73, 80, 106, 132, 200, 203, 305, 442, 445
Meiling-Wesse, K., 248, 249
Mel, H. C., 230
Melendez, A., 15
Meley, D., 89, 437
Menendez, G., 458, 459
Menendez-Palaez, A., 458, 459
Menzies, F. M., 14, 347
Mercer, C. A., 66, 78, 79, 97, 98, 99, 100, 101, 103, 106, 107, 108, 112, 114, 115
Meresse, S., 328, 329, 332, 334, 338
Métivier, D., 88, 437
Michaelsen, E., 182, 192
Michell, R. H., 249
Mikulits, W., 230, 237
Milan, G., 57, 167
Milasta, S., 286
Millan-Ward, E., 229
Miller, G., 410
Miller, J. L., 229, 231, 233, 239
Miller, S. I., 326, 328, 329
Mills, K. R., 424, 431, 436
Mills, S. D., 328
Milosevic, S., 65, 407
Minaguchi, T., 239
Mineki, R., 77, 78, 79
Minematsu-Ikeguchi, N., 3, 21, 23, 58, 65, 133, 200, 207, 208, 210, 330, 386
Minna, J. D., 290
Miotto, G., 200, 201, 206, 208, 209, 210, 211, 445, 453
Miron, M., 170

Mishima, K., 20, 366
Mitani, S., 236
Miyamoto, T., 166, 169, 170
Miyawaki, A., 193
Miyazaki, J., 15
Mizote, I., 182
Mizushima, A., 133, 134, 332, 334
Mizushima, N., 2, 3, 6, 7, 10, 13, 14, 15, 16, 17, 20, 28, 48, 58, 91, 98, 106, 115, 133, 145, 167, 182, 200, 205, 207, 210, 229, 236, 238, 240, 250, 251, 262, 274, 289, 330, 332, 333, 339, 346, 347, 365, 366, 367, 369, 371, 375, 377, 386, 388, 390, 393, 404, 425
Mobley, W. C., 286, 287
Modjtahedi, N., 126
Moffat, J., 167
Mohamed, H., 228, 229
Mohandas, N., 229, 230
Moldeus, P., 121
Molina, A. J., 228, 229
Møller, M. T. N., 78
Molofsky, A. B., 384, 385
Momoi, T., 437, 465
Monastyrska, I., 470
Mondon, C. E., 446
Moore, A., 289
Moore, M. R., 459
Mora, J. R., 289
Morel, M., 286
Moreno, D., 182
Moreno, E., 384
Moreno, R. D., 229
Morgenstern, J. P., 426
Moriguchi, R., 290
Morikawa, K., 437
Moriyama, Y., 386
Morrow, I. C., 258
Morselli, E., 182
Mortimore, G. E., 56, 132, 200, 201, 202, 207, 445, 446, 453
Moshiach, S., 70, 384, 400
Mostoslavsky, R., 182
Mousa, S. A., 367
Mousavi, S. A., 66
Mukherjee, C., 437
Mulinyawe, S., 91
Müller, A. J., 334
Müller, M., 404, 406
Müller, M. R., 407
Müller, W. J., 128
Mullner, E. W., 230, 237
Munafó, D. B., 86, 87, 88, 91, 92, 364, 365
Münz, C., 347, 403, 404, 405, 406, 407, 408, 412, 413, 415, 416
Murakami, M., 16
Murakami, T., 437
Muramatsu, M., 236

Murata, S., 7, 20, 182, 238, 240, 289
Murayama, K., 77, 78, 79
Muthuswamy, S. K., 424, 428, 429, 431, 433, 436
Mwandumba, H. C., 346

N

Naash, M., 289
Nagayama, K., 289
Nair, U., 239
Nakagawa, I., 332, 347, 363, 365, 366, 367, 368, 369
Nakagawa, T., 405
Nakahara, Y., 20, 366
Nakai, A., 182
Nakamura, K., 20, 366
Nakamura, N., 200
Nakamura, S., 286
Nakamura, Y., 239, 289
Nakane, P. K., 398
Nakase, I., 289
Nakashima, A., 166, 169, 170
Nakata, M., 332, 347, 365, 366, 367, 368, 369
Nakaya, H., 14, 16, 48, 240, 339
Namai, E., 289
Nannmark, U., 182
Nara, A., 332, 347, 365, 366, 369
Nasser Eddine, A., 346
Natsume, T., 3, 14, 182, 240
Neefjes, J. J., 191, 404
Neermann, J., 106
Negishi, Y., 289
Nelson, D. A., 437
Nelson, D. J., 327, 328, 329, 331
Nepryakhina, O. K., 241
Neu, S., 152
Neufeld, T. P., 15, 178
Neustadt, J., 228
Neuzil, J., 229
Ney, P. A., 227, 229, 231, 233, 239, 243
Nguyen, P. Q., 120
Nicolae, D. L., 385
Nie, S., 279, 287, 288, 290
Niebuhr, K., 328
Niedzwiecka, A., 170
Niemann, A., 88, 91
Niioka, S., 54, 202, 204
Nijhof, W., 229
Nikolic, N., 290
Nimmerjahn, F., 65, 407
Nimmo, E. R., 333, 334
Niranja, D., 91, 92
Niranjan, D., 133, 332
Nishida, K., 182
Nishito, Y., 182
Niwa, H., 15
Nixon, R. A., 132
Noble, N. A., 241

Nobukuni, T., 167
Noda, T., 1, 2, 3, 6, 14, 15, 27, 30, 58, 88, 98, 115, 133, 134, 145, 183, 200, 205, 210, 236, 251, 330, 366, 367, 386, 388, 392, 425, 426, 471
Nordheim, A., 248, 249, 250, 251, 254, 255, 258
Northrop, J. P., 67
Novelli, M., 442
Nowicki, M., 279, 281, 285, 287, 290
Nyberg, E., 71

O

Ochman, H., 329
Oda, Y., 289
Ogata, M., 112, 437
Ogawa, M., 333, 339, 347, 363, 365, 371, 373, 375, 384
Ogawa, S., 437
Ogier-Denis, E., 50
Ogura, Y., 385
Oh, Y. H., 286
Ohba, H., 286
Ohkuma, S., 200
Ohlson, M. B., 326, 328, 329
Ohmuraya, M., 16
Ohsawa, Y., 465
Ohshima, S., 50
Ohshima, Y., 236
Ohsumi, M., 2, 3, 6, 10, 205, 236
Ohsumi, Y., 3, 7, 12, 14, 15, 16, 17, 20, 48, 80, 106, 133, 145, 205, 207, 210, 229, 236, 238, 239, 240, 289, 311, 330, 332, 334, 339, 366, 367, 377, 390, 406, 425
Ohtani, S., 290
Okada, C. Y., 67
Okada, H., 3
O'Kane, C. J., 178
Okano, H., 20, 366
Okayama, H., 100
Okazaki, N., 236
Okazaki, T., 126
Oku, M., 347
Oku, T., 286
Okumura, Y., 16
Olive, M., 182
Omiya, S., 182
O'Morain, C. A., 333
Oorschot, V., 191
Opferman, J. T., 229, 231, 233, 239
Oppliger, W., 166
O'Regan, R. M., 279, 281, 285, 287, 288, 290
Oren, M., 120
Orkin, S. H., 232
Ortiz-Vega, S., 167
Oshiro, N., 166, 169, 170, 173, 174

Oshitani-Okamoto, S., 10, 14, 133, 205, 207, 236
Otto, G. P., 15
Outzen, H., 182, 185, 187, 192, 193, 426
Øverbye, A., 50, 58, 63, 66, 76, 77, 78, 79, 98, 132
Overholtzer, M., 431
Overmeyer, J. H., 467
Øvervatn, A., 181, 182, 185, 187, 192, 193, 426

P

Pace, J., 338
Packer, M., 240
Paharkova-Vatchkova, V., 405
Palfia, Z., 161
Palis, J., 229
Palmer, A. E., 134, 193
Paludan, C., 406
Pan, D., 167
Pandhare, J., 287
Pankiv, S., 181, 182, 185, 193, 426
Pardoll, D. M., 65, 407
Park, J. H., 239
Parker, P. J., 249
Parkes, M., 333, 334, 385
Parsot, C., 328
Parthasarathi, K., 123
Parton, R. G., 258
Pasco, M., 58
Pask, D., 178
Patel, S., 424
Patentini, I., 448
Patterson, G. H., 28, 29, 30, 31, 37
Pattingre, S., 240, 471
Pauleau, A. L., 469
Paulus, J. K., 436
Payne, A. M., 442
Payne, A. P., 459
Pei, Y., 236
Peng, T. I., 123
Penheiter, K. L., 326
Penn, R. L., 278
Perander, M., 182, 187, 192
Perfettini, J. L., 89, 437
Perfetto, S. P., 290
Perisic, O., 182
Perrin, A. J., 326, 328, 329, 331
Perry-Garza, C. N., 21, 86, 88, 91, 92
Persson, C., 328
Peters, P. J., 191
Petersen, O. W., 428
Peterson, T. R., 167, 174
Pethe, K., 347
Petiot, A., 50
Petkova, V., 289
Petnicki-Ocwieja, T., 384
Petros, J. A., 290
Pfeifer, C. G., 329

Pfisterer, S. G., 247
Pham, W., 289
Phang, J. M., 277, 287
Pieczenik, S. R., 228
Pierron, G., 89, 120, 127, 240, 437, 469
Pilwat, G., 67
Piper, R. C., 249
Pistor, S., 375
Pizarro-Cerda, J., 384
Pletjushkina, O. Y., 241
Ploegh, H. L., 191, 236, 404
Plomp, P. J., 56, 73
Pokrovski, A. A., 287
Polakiewicz, R. D., 169, 170
Pollera, M., 445, 446, 448
Ponpuak, M., 345
Popov, V. L., 384
Pösö, A. R., 132, 445
Prabhakar, J., 367
Prabhu, S., 290
Piagei, R., 326
Prchal, J. T., 229, 239
Prenant, M., 229, 230
Prescott, A. R., 152
Prescott, N. J., 333, 334, 385
Prescott, S. M., 70
Preston, T. J., 128
Priault, M., 229
Price, D. A., 290
Prinz, M., 182
Progulske-Fox, A., 364, 365
Proikas-Cezanne, T., 247, 248, 249, 250, 251, 254, 255, 258
Prost, L. R., 326, 328
Proud, C. G., 169, 170, 174
Pucciarelli, M. G., 330
Pujol, C., 327, 328, 329, 331
Punnonen, E.-L., 50, 147, 156
Purdy, G. E., 347
Py, B. F., 329, 365, 376, 384
Pypaert, M., 347, 405, 407, 408, 412, 413, 415

Q

Qian, T., 229, 234, 241
Qian, W., 286
Qin, J., 167
Qu, X., 16, 240
Quadri, S., 123
Quinn, M. T., 182
Quiros, P. M., 182

R

Rabinovitch, M., 91, 364, 365, 384
Rabsch, W., 326
Radolf, J. D., 205
Raiborg, C., 65, 182, 406
Raicevic, N., 290

Rain, J. C., 240
Ramalho-Santos, J., 229
Ramirez, A., 286, 287
Rammensee, H., 405, 406
Ramos, I. C., 52
Ramos, R., 385
Randall, M. S., 229, 231, 233, 239
Randall, R., 204
Rank, F., 428
Rao, J., 285, 286
Rathman, M., 328
Rattan, S. I., 442
Raught, B., 169, 170
Raviglione, M. C., 346
Ravikumar, B., 347
Razi, M., 261
Recchia, G., 448, 453
Rechsteiner, M., 67
Reddy, J. K., 216
Reed, M. G., 150, 151, 154, 155
Reef, S., 120
Reggiori, F., 239, 240, 248, 249, 470
Reginato, M. J., 424, 431, 436
Reich, M., 404, 406
Reith, A., 65, 66
Reitzer, L., 106
Reunanen, H., 50, 147, 156
Rhen, M., 329
Rhoades, E. E., 346
Rhoades, E. R., 346
Ricci, J. E., 286
Rich, K. A., 376, 384
Rickinson, A. B., 406, 419
Riemann, F., 67
Rinderknecht, E., 393
Ringold, G. M., 67
Rioux, J. D., 333, 385
Roberts, E. A., 347
Roberts, R. G., 333, 334
Robertson, R. B., 287
Robinson, J., 410
Roccio, M., 167
Rocha, N., 404
Roche, E., 406
Rodriguez, C., 458, 459
Rodriguez-Colunga, M. J., 458, 459, 460, 462, 465, 470
Rodriguez-Enriquez, S., 241
Roederer, M., 290
Rogan, P. K., 289
Rohde, M., 367
Rohrs, H. W., 406
Roman, R., 67
Romanello, V., 57, 167
Romano, P. S., 384
Ronnov-Jessen, L., 428
Rosebrough, N., 204
Rosen, J., 16

Rosenstiel, P., 333
Rosenzweig, N., 288
Rosenzweig, Z., 288
Rosqvist, R., 328
Roth, G. S., 442
Roth, J. A., 290
Rothbard, J., 419
Rout, M. P., 182
Roux, P. P., 167
Roy, D., 327, 328, 329, 331
Ru, B., 167
Rubinsztein, D. C., 14, 48, 58, 178, 182, 347
Ruckerbauer, S., 254
Rudensky, A. Y., 405
Rudolf, R., 57, 167
Ruf, F., 288
Ruggino-Netto, A., 346
Ruíz-Albert, J., 328, 329, 332
Russell, D. G., 67, 346, 347
Rusten, T. E., 15
Ryder, T. A., 328, 329, 332

S

Sabanay, H., 120
Sabatini, D. M., 166, 167, 169, 170, 173, 174, 175
Sa-Eipper, C., 239
Saetre, F., 50, 58, 63, 76, 78, 79, 98, 132
Saftig, P., 88, 144, 145, 152, 161, 162, 367, 469
Sagara, H., 333, 339, 347, 365, 371, 375
Sahai, N., 278
Saiki, S., 178, 347
Sainz, R. M., 458, 459
Saito, A., 437
Saito, I., 20, 366
Sakai, Y., 14, 145, 216, 347
Sakurai, K., 128
Sala, G., 49, 203
Salas, E., 424, 425, 434, 435, 436, 437
Salcedo, S. P., 328, 332
Salin, B., 229
Salmaso, S., 286
Salminen, A., 182
Salmon, P., 426
Salvador, N., 152, 319
Salvador-Montoliu, N., 16, 182
Samara, C., 182
Samari, H. R., 78
Samokhvalov, V., 120
Sancak, Y., 167, 174
Sanchez, D., 88
Sanders, C., 416
Sandoval, H., 229, 239
Sandoval, I. V., 14, 145
Sandri, C., 57, 167
Sanjuan, M. A., 70, 115, 384, 400
Sanowar, S., 326, 328
Sansonetti, P., 328, 364

Santidrian, A. F., 120, 127
Santos, D. S., 346
Saprunova, V. B., 241
Sarbassov, D. D., 166, 167, 170, 173, 174, 175
Sarkar, A., 287
Sarkar, S., 58, 178, 347
Sasakawa, C., 333, 339, 347, 363, 365, 371, 373, 375
Sasaki, K., 289
Sasaki, M., 239
Sass, M., 15
Sato, N., 465
Sato, S., 15
Sato, W., 50
Satomi, Y., 3, 205, 236
Saucedo, L. J., 167
Sawamura, K., 289
Sawant, R. M., 286
Sawyer, S. T., 229, 236
Sayen, M. R., 21, 86, 88, 91, 92
Scadden, D. T., 279
Scarlatti, F., 49, 203
Schaeffer, J., 229
Schaeper, U., 239
Schaible, U. E., 67
Schalm, S. S., 173, 174
Schatten, G., 229
Schellens, J. P., 50, 106, 132
Schepers, A., 406
Scherz-Shouval, R., 119, 120, 121, 122, 123, 125, 127, 128
Schesser, K., 328
Scheuner, D., 347
Schiaffino, S., 167
Schiestl, R. H., 126
Schlesinger, P. H., 67
Schliess, F., 52, 56
Schlumberger, M. C., 328, 334
Schmelzle, T., 166, 431
Schmid, D., 347, 404, 405, 406, 407, 408, 412, 413, 415
Schmidt, C. K., 152
Schmidt, D. E., 448
Schneider, M. D., 240
Schnoelzer, M., 182
Schoor, O., 404, 406
Schreiber, R. D., 393
Schreurs, F. J., 466
Schuck, S., 347
Schuldiner, O., 15
Schuller, H. J., 65
Schumacher, A., 229, 239
Schumacker, P. T., 121
Schwarzmann, G., 469
Schweers, R. L., 229, 231, 233, 239
Schwiers, E., 121
Schwörer, C. M., 200, 207, 445
Scidmore, M. A., 330

Scott, R. C., 15
Sealfon, S. C., 288
Seaman, M., 15
Sechi, A. S., 375
Seglen, P. O., 15, 49, 50, 52, 53, 55, 58, 63, 65, 66, 67, 68, 69, 70, 71, 72, 73, 74, 75, 76, 77, 78, 79, 80, 92, 98, 132, 200, 201, 262, 346, 406, 445, 449
Seibenhener, M. L., 182
Selak, M. A., 243
Seleverstov, O., 277, 279, 281, 285, 287, 290
Semenza, G. L., 241
Sengupta, S., 166, 174
Settembre, C., 182
Settlage, R. E., 419
Sgroi, D. C., 436
Shabanowitz, J., 419
Shahiwala, A., 286
Shaner, N. C., 193
Shao, Z., 121
Shapira, R., 120
Sharafkhaneh, A., 384, 400
Sharipo, A., 405
Sharling, L., 120
Shaul, Y. D., 167
Shaw, M. H., 347
She, X., 236
Shea, J. E., 329
Sheen, J. H., 166, 174
Shelton, K. D., 167
Shen, C., 290
Shertzer, H. G., 121
Shi, J., 120
Shi, L., 239, 288
Shi, Y., 437
Shibata, M., 465
Shiffer, K. A., 200, 207, 445
Shifman, O., 120
Shimoda, L. A., 241
Shimonishi, Y., 3, 205, 236
Shinohara, Y., 286
Shintani, T., 108, 110, 114, 132, 298, 470
Shiosaka, S., 437
Shiota, C., 167
Shirasawa, T., 236
Shorer, H., 120, 123
Shugart, Y. Y., 333
Shvets, E., 58, 120, 123, 131, 133, 138
Sibirny, A., 14, 145
Siddiqi, F. H., 178
Sierra, V., 459, 460, 462, 465, 470
Siliprandi, N., 201, 445, 453
Silverberg, M. S., 333, 385
Simeon, A., 65
Simerly, C., 229
Simonetti, I., 442
Simons, J. W., 290
Simonsen, A., 182

Author Index

Simpson, J. C., 367
Sinai, A. P., 364
Sinclair, R., 290
Singh, G., 128
Singh, S. B., 334, 346, 347, 352, 364, 365, 384
Sips, H. J., 52, 56
Sjaastad, M. D., 328
Skulachev, V. P., 241
Slot, J. W., 192, 273
Smallheer, J. M., 367
Smith, A. C., 328, 329, 330, 331, 332, 333, 337, 338, 339, 384
Smith, B. R., 289
Smith, E. M., 169, 170, 174
Snapp, E. L., 39
Snyder, L. M., 230
Snyder, S. H., 169
So, M. K., 285, 286
So, N. S., 328, 329, 331
Soars, D., 333, 334
Sohal, B. H., 229
Sohal, R. S., 229
Solheim, A. E., 55, 72
Sollott, S. J., 123
Solomon, F., 329
Somersan, S., 416
Soncini, F. C., 329
Sonenberg, N., 169, 170
Sonesson, A. W., 286
Song, D., 286
Sooparb, S., 300
Sorescu, D., 120
Soto, N., 167
Sou, Y. S., 182, 207
Souquere, S., 89, 120, 127, 437, 469
Southern, P. J., 367
Southern, S. O., 367
Spampanato, C., 182
Sparks, D. L., 278
Spike, R. C., 459
Spooner, E., 167, 174
Spronk, C., 53
Stacey, D. W., 67
Stangl, K., 230, 237
Stass, S. A., 288
Stecher, B., 334
Steck, T. L., 230
Steele-Mortimer, O., 326, 328, 329, 334, 338
Steigerwald-Mullen, P. M., 406
Steinbach, P. A., 134, 193
Steinberg, B. E., 384
Steinberg, T. H., 67
Steinman, R. M., 405
Stenmark, H., 15, 182, 187, 192, 258, 367
Stern, S. T., 278, 290
Stevanovic, S., 405, 406
Stevens, D. M., 167
Stierhof, Y. D., 254

Stiles, L., 228, 229
Stocker, H., 167
Stockinger, B., 406
Storrie, B., 309, 310, 311, 316
Strasser, A., 431
Straub, M., 65
Striepen, B., 384
Stroh, M., 279
Strom, A. L., 182
Strømhaug, P. E., 65, 66, 73, 74, 92, 240, 248, 249, 406
Strominger, J. L., 405
Stuffers, S., 182
Stumptner, C., 182
Su, B., 167
Subauste, C. S., 384
Subramani, S., 14, 145, 216
Subramanian, T., 239
Suga, M., 286
Suh, J., 289
Suh, Y. A., 120
Suk, J. S., 289
Sukhan, A., 337
Sullards, C., 120
Sun, A. Y., 128
Sun, G. Y., 128
Sun, J., 405
Surazynski, A., 287
Suri, A., 406
Surmacz, C. A., 445
Sutovsky, P., 229
Suyama, K., 16
Suzuki, K., 7, 14, 15, 80, 286, 386
Suzuki, M., 371
Suzuki, R., 289
Suzuki, T., 333, 339, 347, 364, 371, 373, 375, 384
Suzuki, Y., 236
Suzuki-Migishima, R., 20, 366
Suzumori, K., 239
Swanson, M. S., 364, 383, 384, 385
Swift, H., 230
Sykes, A. K., 262
Szotyori, Z., 473

T

Tabata, S., 15
Tacchetti, C., 182
Taddei, M., 454
Tadsemir, E., 240
Tagawa, Y., 386
Tager, J. M., 52, 56, 72
Taguchi, Y., 126
Tait, S. W., 70, 286, 384, 400
Takahashi, A., 126
Takahashi, E., 16, 239
Takahashi, R., 166, 169, 170
Takahashi, T., 466
Takamura, A., 3, 14, 182, 240

Takano, S., 200
Takao, T., 3, 14, 205, 236
Takatsuki, A., 91
Takeda, T., 182
Takehana, K., 166, 169, 170
Takeuchi, J., 406
Takikawa, H., 239
Takizawa, T., 289
Tal, R., 229
Tallóczy, Z., 15, 347
Tamiya, E., 286
Tanaka, K., 7, 20, 48, 70, 182, 238, 240, 289
Tanaka, Y., 152, 162, 236, 301, 367, 469
Taneike, I., 53, 202
Tang, B. C., 290
Tang, J., 466
Tang, P., 328, 332
Tang, X., 120, 152
Tanida, I., 3, 6, 7, 15, 20, 21, 23, 26, 42, 50, 58, 65, 78, 79, 88, 133, 145, 152, 162, 200, 205, 207, 208, 210, 218, 236, 238, 240, 289, 330, 367, 386, 437
Taniguchi, M., 437
Tanii, I., 437
Taniike, M., 182
Tanimura, K., 166, 169, 170
Tapuhi, Y., 448
Tarbutt, R. G., 230
Tasdemir, E., 27, 36, 42, 44, 126, 182
Tashev, T. A., 287
Tashiro, Y., 386
Tassa, A., 240
Tatsuno, I., 371, 373
Tavernarakis, N., 240
Taylor, G. A., 334, 346, 347, 352, 364, 365, 384
Taylor, G. S., 406
Taylor, K. D., 333, 385
Taylor, M. P., 91, 365, 384, 392
Tee, A. R., 167
Tempst, P., 166, 170, 173, 174, 175
Terlecky, S., 302, 316, 320
Terman, A., 229, 442
Teuber, M., 333
The Wellcome Trust Case Control Consortium 333, 334
Thiagarajan, P., 229, 239
Thoene, J., 69
Thomas, G., 167
Thomas, W. A., 406
Thompson, A., 329
Thompson, C. B., 48, 243, 262, 437
Thoreen, C. C., 167, 174
Thumm, M., 14, 65, 145, 248, 249
Thuring, J., 249
Till, A., 333
Tischler, M. E., 72

Todd, N. W., 288
Todorovic-Markovic, B., 290
Toh, B. H., 328, 334, 338
Tokuhisa, T., 7, 14, 16, 48, 240, 339, 386
Tokumitsu, H., 236
Tokunaga, C., 166, 173, 174
Tolcheva, E., 286
Tolivia, D., 458, 459, 460, 462, 465, 470
Tolivia, J., 458
Tolivia-Cadrecha, D., 458, 459
Tolkovsky, A. M., 50, 91, 92, 120, 133, 332
Tolleshaug, H., 67, 69, 72
Tolstrup, J., 249
Toma, C., 384
Tomas-Zapico, C., 458, 459, 460, 462, 465, 470
Tomemori, T., 236
Tomoda, T., 236
Tooze, S. A., 14, 261
Torchilin, V. P., 279, 286
Torrejon-Escribano, B., 182
Toscano, K., 442
Tour, O., 134
Towbin, H., 218
Townsend, A., 419
Townsend, S. M., 326, 329
Trajkovic, V., 290
Tremelling, M., 333, 334, 385
Tricot, G., 230
Triller, A., 286
Trono, D., 426
Troulinaki, K., 240
Troxel, A., 16
Troy, E. B., 339, 365, 376
Trump, B. F., 144
Tschape, H., 326
Tsien, R. Y., 134, 193
Tsokos, M., 182
Tsolis, R. M., 326, 329
Tsuda, K., 332, 347, 365, 366, 369
Tsujimoto, Y., 239
Tsukamoto, S., 16
Tsukamoto, T., 218
Ttofi, E. K., 178
Tucker, K. A., 240
Tunnacliffe, A., 178
Tuo, W., 54
Turley, S. J., 405
Tuschl, T., 406
Tutel'ian, V. A., 287
Twig, G., 228, 229
Twining, B. S., 278
Tysk, C., 333

U

Uchiyama, T., 126
Uchiyama, Y., 7, 20, 48, 238, 240, 289, 465

Ueda, K., 290
Ueno, N., 3, 205, 210
Ueno, T., 2, 3, 6, 7, 14, 15, 20, 21, 23, 48, 50, 58, 65, 66, 78, 79, 88, 98, 99, 103, 133, 145, 152, 162, 182, 200, 205, 207, 208, 210, 211, 215, 236, 238, 240, 289, 330, 366, 367, 386, 406, 425, 437
Ugalde, A. P., 182
Ullrich, H. J., 346
Unanue, E. R., 406
Unsworth, K. E., 328, 329, 332
Urase, K., 465
Urban, R. G., 405
Uria, H., 458, 459
Usuda, N., 218
Utoguchi, N., 289

V

Valdivia, R. H., 329
Valeri, C. R., 230
Valeva, A., 67
Vallette, F. M., 229
Vande, V. C., 239
Vanden Hoek, T. L., 121
Van Der, M. R., 72
Vander Haar, E., 167
van der Heide, D., 466
van der Klei, I. J., 65, 74, 79, 347
van der Meer, R., 56
van der Pluijm, I., 182
Vandewalle, A., 49
van Leeuwen, F. W., 182
van Leyen, K., 229, 230
van Sluijters, D. A., 50
Van Woerkom, G. M., 53
Varela, I., 182
Vazquez, C. L., 85, 364, 365
Vázquez, E., 289
Veenhuis, M., 14, 145
Vega-Naredo, I., 457, 459, 460, 462, 465, 470
Vellai, T., 161
Velours, G., 120
Venanzi, E. S., 405
Venerando, R., 201, 445, 453
Venkatesh, L. K., 239
Ventruti, A., 49, 446, 448
Venturi, C., 182
Vergne, I., 346
Verhoeven, A. J., 52
Verkman, A. S., 134
Vervoorn, R. C., 52, 56
Verwilghen, R. L., 230
Vignali, D. A., 405
Vihinen, H., 8, 132, 143, 211, 262, 315, 374, 462
Villadsen, R., 428

Villaverde, A., 289
Virgin, H. W. IV, 347
Vitale, I., 126, 182
Vittorini, S., 446, 448
Vockerodt, M., 406
vom Dahl, S., 52, 56
von Boehmer, H., 405
von Figura, K., 152, 162, 469
Vranjes-Djuric, S., 290
Vreeling-Sindelarova, H., 50, 132

W

Waddell, S., 248, 249, 251, 254
Wagner, R., 106
Waguri, S., 7, 20, 48, 182, 238, 240, 289, 465
Walev, I., 67
Walker, D. H., 384
Walker, N., 346
Walker, S. J., 433
Wallace, D. C., 228
Wallis, T. S., 329
Walter, P., 347
Walters, J. J., 406
Walzer, G., 228, 229
Wan, F., 65, 120, 126, 474
Wang, C. C., 285, 286
Wang, C.-W., 239, 248, 249
Wang, G. D., 279, 287
Wang, J., 229, 239
Wang, L., 152, 170
Wang, M. D., 287
Wang, P. N., 285, 286
Wang, Q. J., 182
Wang, Y. J., 127
Wang, Z., 286
Warburg, O., 287
Ward, W. F., 56
Warren, G., 5
Wasungu, L., 289
Watanabe, M., 126
Watanabe, S., 50
Watanabe, T., 465
Waterhouse, N. J., 286
Waterman, S. R., 328, 329, 332
Watkins, S. C., 279
Watson, P. R., 329
Watt, C. J., 346
Wattiaux, R., 43, 309
Waugh, R. E., 230
Weaver, V. M., 424
Webster, P., 365, 376, 384
Weck, M. M., 407
Wehland, J., 376
Wehrly, T. D., 384
Wei, S., 167
Weiss, S., 406

Welch, M. D., 364
Welch, W., 318
Welsh, C. J., 384
Welsh, S., 65, 120, 126, 474
Wenz, M., 67
Wert, J. J., Jr., 132
Wesley, J. B., 241
Wessels, M. R., 366
Wessendarp, M., 384
Whatley, R. E., 70
White, E., 286, 424
White, F. M., 419
WHO, 346
Wiedmann, M., 229, 230
Wierenga, P. K., 229
Wikstrom, J. D., 228, 229
Willenborg, M., 152
Williams, A., 178, 347
Williams, B. G., 346
Williams, J. A., 205
Williams, R. L., 182
Williamson, J. R., 72
Wilmanski, J. M., 384
Wilson, G. S., 289
Wilson, J., 337
Wilson, M. I., 182
Wing, S., 299, 301
Winnen, B., 334
Winnik, F. M., 287
Winter, G., 229
Withoff, S., 70
Wohlgemuth, S. E., 442
Wolf, D. H., 65
Wolf-Watz, H., 328
Woo, J. T., 167
Wood, M. W., 329
Wooten, M. W., 182
Wu, C., 236, 279, 286, 287
Wu, G., 54, 290
Wu, J., 236, 243
Wu, M., 228, 229
Wu, M. Y., 15
Wullschleger, S., 166
Wyslouch-Cieszynska, A., 170

X

Xavier, R. J., 333, 385
Xiao, G., 65
Xiao, Y., 288
Xie, Z., 27, 144, 248, 333
Xing, Y., 288, 290
Xiong, Y., 120
Xu, J., 120, 128
Xu, T., 167
Xu, X., 120
Xu, Y., 384, 400
Xue, L., 50, 120

Y

Yabu, T., 126
Yamada, Y., 286
Yamaguchi, H., 332, 347, 365, 366, 369
Yamaguchi, O., 182
Yamaguchi, Y., 286
Yamaji, T., 3, 6, 15
Yamamoto, A., 2, 3, 6, 7, 10, 14, 15, 16, 17, 20, 42, 48, 133, 145, 182, 205, 207, 210, 229, 236, 240, 330, 332, 334, 339, 347, 365, 366, 367, 369, 377, 386, 390, 425
Yamamoto, K., 286
Yamamura, K., 15
Yamamura, S., 286
Yamawaki, H., 290
Yan, J., 236
Yan, Y. S., 241
Yang, C., 243
Yang, H., 288
Yang, J. M., 288
Yang, Q., 170, 175
Yang, W. L., 285, 286
Yap, G. S., 347
Yasuda, M., 239
Yasuhara, M., 286
Yates, J. L., 406
Yates, J. R. III, 405
Yezhelyev, M. V., 279, 281, 285, 287, 288, 290
Yin, X. M., 26
Ylä-Anttila, P., 8, 132, 143, 211, 262, 315, 374, 462
Yokota, S., 215, 216, 217
Yokoyama, M., 20, 366
Yonekura, A. R., 405
Yonezawa, K., 166, 167, 169, 170, 173, 174
Yorimitsu, T., 80, 112, 200, 239
Yoshida, M., 50
Yoshikawa, Y., 363, 384
Yoshimori, M., 12
Yoshimori, T., 1, 2, 3, 6, 7, 8, 10, 14, 15, 16, 17, 27, 30, 48, 58, 88, 98, 115, 133, 134, 145, 182, 183, 200, 205, 207, 210, 236, 240, 250, 251, 328, 329, 330, 331, 332, 333, 334, 337, 338, 339, 347, 365, 366, 367, 369, 371, 374, 375, 377, 384, 386, 388, 390, 392, 393, 425, 426, 437, 471
Yoshimoto, K., 15
Yoshinaga, K., 437
Yoshino, K., 166, 169, 170, 173, 174
Yoshizawa, F., 53, 202
Yu, C., 232
Yu, J., 16, 290, 471
Yu, J. W., 249
Yu, L., 65, 120, 126, 236, 474
Yu, S., 128, 216

Yuan, H., 21, 86, 88, 91, 92
Yuan, J., 329, 365, 376, 384
Yue, Z., 182
Yuen, T., 288

Z

Zabirnyk, O., 277, 279, 281, 285, 287, 290
Zacharias, D. A., 134
Zaharik, M. L., 328, 332
Zajac, A., 286
Zalckvar, E., 120
Zatloukal, K., 182
Zeng, X., 467
Zhang, C., 301
Zhang, C. X., 470
Zhang, F., 182
Zhang, H., 241
Zhang, J., 227, 229, 231, 233, 239, 243
Zhang, K., 288
Zhang, L., 37
Zhang, M., 406
Zhang, Y., 167, 279, 285, 286
Zhao, F., 243
Zhao, J., 57, 167
Zhao, S., 236

Zhau, H. E., 290
Zhelev, Z., 286
Zheng, J., 290
Zhong, Y., 182
Zhou, D., 405
Zhou, X., 170
Zhu, C., 182
Zhu, H., 121, 182
Zhu, W., 110, 112
Zhukov, T., 286
Zich, J., 406
Zimmer, J. P., 279
Zimmer, M., 28
Zimmerman, G. A., 70
Zimmermann, U., 67
Zinkernagel, R. M., 404
Zolnik, B. S., 290
Zong, W. X., 110, 112
Zoppino, F. C., 364, 365
Zorov, D. B., 123
Zouali, H., 333, 385
Zscharnack, M., 279, 281, 285, 287, 290
Zuurendonk, P. F., 72
Zwartkruis, F. J., 167
Zweier, J. L., 123

Subject Index

A

N-Acetyl-L-cysteine, autophagy inhibition, 126
Actin, Syrian hamster Harderian gland autophagy analysis, 470
Aging, *see* Caloric restriction; Liver autophagy
Atg proteins
    Atg8
        green fluorescent protein fusion protein, *see* Autophagosome
        processing in autophagosome formation, 26
    Atg16L1 role in *Salmonella* Typhimurium autophagy and Crohn's disease, 333–334, 385
    Atg18, *see* WIPI-1
    functional overview in mammals, 14
    gene expression analysis in mitophagy, 236
Autophagic sequestration assays
    betaine:homocysteine methyl transferase
        fragment generation in autophagy, 77–78
        Western blot of fragment generation, 78–79
    cytosolic proteins as probes
        advantages, 65
        lactate dehydrogenase
            intralysosomal accumulation, 75
            sequestration assay, 76, 77
            prelysosomal accumulation, 77
    damage-induced cell permeabilization, 69–70
    electrodisruption, 70–71
    marker selection criteria, 65–67
    membrane-impermeable probes
        electroporation, 68–69
        hepatocyte preparation and incubation, 67–68
    sugars as probes
        advantages, 71–72
        lactose, 73–74
        raffinose, 72
        sucrose, 72–73
Autophagosome
    dye-quenched-bovine serum albumin and autolysosome formation assay
        confocal microscopy, 94
        lysosome labeling, 92–93
        overview, 86
    electron microscopy, *see* Electron microscopy
    lifetime monitoring with photoactivatable green fluorescent protein–Atg8 fusion protein
        applications, 43–44
        autophagy induction, 36–38
        biological system, 28–29
        chamber setup, 33–34
        controls, 42–43
        half-life determination, 40–42
        photoactivatable protein selection and characteristics, 29–31
        photoactivation
            laser targeting calibration, 31–32
            optimization, 32, 34–35
        photobleaching minimization and evaluation, 35–36
        principles, 27–28
        pulse-labeling induced autophagosomes, 38–40
    macrophage autophagosomes, *see* Macrophage
    microtubule-associated protein 1 light chain 3 interactions, 2–3, 57–58
    pathogen sequestration, *see Listeria* autophagy; *Mycobacterium tuberculosis* autophagy; *Salmonella* Typhimurium autophagy; *Shigella* autophagy; *Streptococcus* autophagy
Autophagy
    cancer cell line protein turnover analysis, 48–50
    chaperone mediation, *see* Chaperone-mediated autophagy
    detachment-induced autophagy, *see* Detachment-induced autophagy
    electron microscopy assays, *see* Correlative light and electron microscopy; Electron microscopy
    flow cytometry assays, *see* Flow cytometry
    functional overview, 48
    Harderian gland, *see* Syrian hamster Harderian gland
    hepatocytes, *see* Hepatocyte autophagy
    induction assays, *see* Monodansylcadaverine
    liver, *see* Hepatocyte; Liver autophagy
    major histocompatibility molecule class II antigen presentation after macroautophagy, *see* Major histocompatibility molecule class II
    mammalian cells, *see* Microtubule-associated protein 1 light chain 3
    mammalian target of rapamycin activity, *see* Mammalian target of rapamycin

Autophagy (cont.)
  nanoparticle activation, see Nanoparticle-mediated autophagy
  organellophagy, see Mitophagy; Pexophagy; Reticulophagy
  p62 degradation, see Sequestosome 1
  pathogen sequestration, see Listeria autophagy; Mycobacterium tuberculosis autophagy; Salmonella Typhimurium autophagy; Shigella autophagy; Streptococcus autophagy
  sequestration assays, see Autophagic sequestration assays
  starvation-induced oxidative stress, see Reactive oxygen species

B

Beclin1, protein complex immunoprecipitation for Syrian hamster Harderian gland autophagy analysis, 471–472
Betaine:homocysteine methyl transferase
  autophagic sequestration assays
    fragment generation in autophagy, 77–78
    Western blot of fragment generation, 78–79
  autophagy assays with glutathione S-transferase fusion protein
    advantages, 114–115
    amino acid depletion studies, 103–106
    cell culture, transfection, and extraction, 100–101
    expression levels and fragment accumulation, 106
    linker-specific cleavage site assay
      advantages, 115–116
      organellophagy assays, 110–114
      overview, 107–109
    overview, 98–100
    protease inhibitors, 102–103
    purification and Western blot of fragments, 101–102
    time-course analysis, 107
  cell distribution, 98
BF, see Bifunctional protein
BHMT, see Betaine:homocysteine methyl transferase
Bifunctional protein, Western blot analysis of autophagy, 218–220

C

Caloric restriction
  antiaging effects, 442
  liver autophagy assays, see Liver autophagy models
    daily dietary restriction paradigm, 443
    rats, 443
    timed meal feeding paradigm, 443

Cargo sequestration assays, see Autophagic sequestration assays
Caspase-3, cleavage assay to detect lumenal apoptosis in detachment-induced autophagy, 432–433
Catalase, autophagy inhibition, 128
Cathepsin, Syrian hamster Harderian gland autophagy analysis
  cathepsin D processing analysis, 467–469
  cell viability
    cathepsin B assay, 466
    cathepsin C assay, 466–467
  tissue homogenization, 465
Chaperone-mediated autophagy
  assays
    levels of components
      lysosome isolation, 309–311
      Western blot, 311–312
    protein degradation
      calculations, 307–308
      inhibition of other autophagic pathways, 306–307
      pulse–chase, 303–306
    subcellular localization of active lysosomes
      electron microscopy, 314–315
      immunofluorescence microscopy, 312–314
    substrate translocation assays
      overview, 315–316
      protease protection assay, 318–320
      protein degradation with isolated lysosomes, 317–318
      radiolabeling of substrates, 316–317
  lysosome-associated membrane protein–2A
    knockout mouse, 301
    levels and autophagic activity, 299
    receptor function, 298
    species distribution, 300
    transgenic mouse, 301
  model systems, 300–301
  prospects for study, 321
  regulation, 299–300
  stress induction, 299
  substrates
    consensus sequence, 298
    properties, 301–303
    types, 298–299
  Syrian hamster Harderian gland assays, 472–473
CLEM, see Correlative light and electron microscopy
CMA, see Chaperone-mediated autophagy
Confocal microscopy
  autolysosome formation assay with dye-quenched-bovine serum albumin, 93
  Mycobacterium tuberculosis autophagy studies
    lysosomal protease delivery, 358
    phagosome acidification assay, 356

WIPI-1 puncta formation assays
  autophagy inducers and inhibitors, 252–253
  cell culture, 251–252
  imaging, 254–255
  quantitative analysis, 255
Correlative light and electron microscopy autophagy analysis
  cell culture, 264–265
  green fluorescent protein–LC3 studies
    electron microscopy of autophagic vacuoles, 268–271
    interpretation, 271–272
    light microscopy, 266–267
    overview, 262–264, 274
  cryoimmunogold electron microscopy prospects, 273–274
  electron tomography, 273
Crohn's disease, Atg16L1 role, 333–334, 385
Cytokeratin, Syrian hamster Harderian gland autophagy analysis, 470

D

DCF, *see* 2',7'-Dichlorofluorescein
DEHP, *see* Di-(2-ethylhexyl) phthalate
Dendritic cell, recombinant lentiviral infection with green fluorescent protein–microtubule-associated protein 1 light chain 3, 409–410
Detachment-induced autophagy
  epithelial cell manipulation
    retroviral vectors encoding green fluorescent protein–microtubule-associated protein 1 light chain 3, 425
    tandem fluorescent protein microtubule-associated protein 1 light chain 3 chimeras, 425–426
    transduction with retroviral and lentiviral vectors, 426–428
  integrin-blocking antibody studies, 437
  measurement in traditional anoikis assays
    green fluorescent protein–microtubule-associated protein 1 light chain 3 visualization in detached cells, 436–437
    substrate detachment assay, 434–435
    Western blot of microtubule-associated protein 1 light chain 3 lipidation, 435–436
  overview, 424
  three-dimensional epithelial culture autophagy during lumen formation
    ethidium bromide staining, 433–434
    green fluorescent protein–microtubule-associated protein 1 light chain 3 analysis, 430–431
    immunofluorescence microscopy of cleaved caspase-3 to detect lumenal apoptosis, 432–433
    lumenal cell death upon autophagy inhibition measurement, 431–432
    overlay culture, 428–430
DHE, *see* Dihydroethidium
2',7'-Dichlorofluorescein, reactive oxygen species assay in autophagy, 121–125
Di-(2-ethylhexyl) phthalate, pexophagy induction in mice, 216–217, 223–224
Dihydroethidium, reactive oxygen species assay in autophagy, 121–122
DQ-BSA, *see* Dye-quenched-bovine serum albumin
Dye-quenched-bovine serum albumin, autolysosome formation assay
  confocal microscopy, 94
  lysosome labeling, 92–93
  overview, 86

E

Electrodisruption, autophagic sequestration assays, 70–71
Electron microscopy
  autophagic compartment identification
    autophagy quantification, 161–162
    counting of autophagic compartments
      abundant autophagic compartments, 154
      cell pellets, 152
      reference trap, 155
      tissue samples, 154–155
    electron tomography, 150, 159–162
    fine structure of autophagosomes and autophagic compartments, 155–161
    overview, 144–145
    pathway, 144–145
    resin embedding of aldehyde-fixed animal cells
      reduced osmium tetroxide, 148–149
      unbuffered osmium tetroxide, 146–147
    sampling
      cell pellets, 152
      organs, 151–152
  chaperone-mediated autophagy lysosomes, 314–315
  Listeria autophagy analysis, 376–379
  major histocompatibility molecule class II antigen presentation after macroautophagy analysis with immunoelectron microscopy, 415–416
  mitophagy analysis in reticulocytes, 234–236
  pexophagy analysis in mammalian cells, 220–223
  sequestosome 1 degradation analysis with immunoelectron microscopy, 191–192

Electron microscopy (cont.)
  Shigella autophagosome transmission electron microscopy, 374–375
  Syrian hamster Harderian gland autophagy analysis
    microtubule-associated protein 1 light chain 3 immunocytochemistry, 464–465
    tissue dissection, fixation, and sectioning, 463
Electroporation, autophagic sequestration assay with membrane-impermeable probes, 68–69
ELISA, see Enzyme-liked immunosorbent assay
Enzyme-liked immunosorbent assay, interferon-γ assay of T cell reactivity, 417–418
Epithelial cell, see Detachment-induced autophagy

F

Flow cytometry
  autophagy assays of green fluorescent protein-microtubule-associated protein 1 light chain 3 fusion protein
    advantages, 139–140
    amino acid deprivation analysis, 134–137
    decline quantification, 137
    fluorescence-activated cell sorting of response to specific treatments, 138–139
    mitophagy analysis in reticulocytes, 230–232
    nanoparticle-based technology for autophagy detection, 290

G

GFP, see Green fluorescent protein
Green fluorescent protein
  autophagosome lifetime monitoring with photoactivatable green fluorescent protein–Atg8 fusion protein
    applications, 43–44
    autophagy induction, 36–38
    biological system, 28–29
    chamber setup, 33–34
    controls, 42–43
    half-life determination, 40–42
    photoactivatable protein selection and characteristics, 29–31
    photoactivation
      laser targeting calibration, 31–32
      optimization, 32, 34–35
      photobleaching minimization and evaluation, 35–36
    principles, 27–28
    pulse-labeling induced autophagosomes, 38–40
  detachment-induced autophagy studies, see Detachment-induced autophagy

major histocompatibility molecule class II antigen presentation after macroautophagy monitoring with green fluorescent protein–microtubule-associated protein 1 light chain 3
  electroporation, 410–411
  recombinant lentiviral infection of dendritic cells, 409–410
  transfection, 407–409
microtubule-associated protein 1 light chain 3 fusion protein autophagy assays
  green fluorescent protein and red fluorescent protein tandemly tagged microtubule-associated protein 1 light chain 3 assay, 7–8
  puncta formation assay, 6–7
  transgenic mouse
    cryosectioning, 19
    fluorescence microscopy of puncta, 19–20
    genetic features, 15–16
    genotyping, 17–18
    mouse maintenance, 16–17
    precautions, 20–21
    tissue fixation, 18–19
sequestosome 1 degradation imaging with pH-sensitive mCherry–green fluorescent protein double tag, 193–194
Streptococcus autophagy subcellular localization
  green fluorescent protein fusion proteins, 367
  immunofluorescence microscopy, 368
  morphometric analysis, 368–369
  transfection, 367
WIPI–1 puncta formation assays with fusion proteins
  live cell imaging, 257–258
  quantitative analysis, 256
  stable transfection, 256
  transient transfection, 255–256
Group A Streptococcus, see Streptococcus autophagy

H

Harderian gland, see Syrian hamster Harderian gland
Hepatocyte, see also Liver autophagy
  autophagic sequestration assay with membrane-impermeable probes
    cell preparation and incubation, 67–68
    electroporation, 68–69
  microtubule-associated protein 1 light chain 3 cytosolic fraction ratio as quantitative index of macroautophagy
    calculation of ratio, 207
    cytosolic LC3-II characterization, 205, 207
    macroautophagic flux evaluation as quantitative index, 208–211

Subject Index

overview, 200
subcellular fractionation, 203–204
Western blot, 204–205
protein degradation analysis of autophagy
  cell preparation, 52–53
  incubation conditions, 53–57
  materials, 51–52
  overview, 50–51
  perfusion, 56
  pitfalls, 58–59
  proteolysis assay of valine release, 201–203
High-performance liquid chromatography, liver autophagy proteolysis assays of valine release
  hepatocytes, 201–203
  parenchymal cells, 449–450
HPLC, see High-performance liquid chromatography

I

Immunofluorescence microscopy
  caspase–3 cleavage to detect lumenal apoptosis in detachment-induced autophagy, 432–433
  chaperone-mediated autophagy lysosomes, 312–314
  macrophage autophagosome dynamics, 396–397
  major histocompatibility molecule class II antigen presentation after macroautophagy analysis, 411–415
  microtubule-associated protein 1 light chain 3 and puncta formation assay, 4–5
  mitophagy analysis in reticulocytes, 233–234
  pexophagy, 220
  Salmonella Typhimurium autophagy analysis of microtubule-associated protein 1 light chain 3, 335–337
  sequestosome 1 degradation, 190–191
  WIPI–1 puncta formation assay, 254
ISH, see *In Situ* hybridization

L

Lactate dehydrogenase, autophagic sequestration assays
  intralysosomal accumulation, 75
  prelysosomal accumulation, 77
  sequestration assay, 76, 77
Lactose, autophagic sequestration assays, 73–74
LAMP–2A, see Lysosome-associated membrane protein–2A
LC3, see Microtubule-associated protein 1 light chain 3
LDH, see Lactate dehydrogenase
Legionella, macrophage infection effects on autophagy, 388–390
Linker-specific cleavage site, see Betaine: homocysteine methyl transferase

Listeria autophagy
  bacterial multiplication assay in cells, 379
  electron microscopy, 378
  infection
    epithelial cells, 376
    macrophages, 377–379
  overview, 365
  pathology, 376
Liver autophagy, see also Hepatocyte
  aging effects, 442
  assays
    antipolytic drug maximization of autophagic proteolysis, 445–446
    minimally invasive procedure for rapid evaluation, 448–449
    overview, 444–445
    parenchymal cell assay, 449–450
    protein degradation assay
      liver perfusion, 446
      perfusion medium, 446–447
      protein precipitation and scintillation counting, 447–448
  caloric restriction
    antiaging effects, 442
    models
      daily dietary restriction paradigm, 443
      rats, 443
      timed meal feeding paradigm, 443
Lysosome-associated membrane protein–2A
  chaperone-mediated autophagy
    assay of levels
      lysosome isolation, 309–311
      Western blot, 311–312
    role, 298
  knockout mouse, 301
  levels and autophagic activity, 299
  species distribution, 300
  Syrian hamster Harderian gland autophagy analysis, 469
  transgenic mouse, 301

M

Macrophage
  autophagosome formation and maturation, 385–386
  autophagy pathway flux monitoring green fluorescent protein–microtubule-associated protein 1 light chain 3
    fluorescence microscopy, 386–388
    Western blot, 386–388
  Legionella infection effects on autophagy, 388–390
  Listeria infection for autophagy studies, 376–377
  microtubule-associated protein 1 light chain 3 localization in primary mouse macrophages

Macrophage (cont.)
    autophagosome dynamics quantification with immunofluorescence microscopy, 396–397
    bone marrow–derived macrophage isolation
        bone marrow cell collection, 395
        femur and tibia dissection, 394
        media preparation, 393–394
        replating, 395–396
    cell density, 396
    controls, 397–398
    Hank's buffer fixative preparation, 398
    overview, 390–393
    Western blot of lipidated microtubule-associated protein 1 light chain 3, 399–400
    Mycobacterium tuberculosis
        autophagy studies, see *Mycobacterium tuberculosis* autophagy
        infection, 346
    pathogen autophagy induction, 384–385
Major histocompatibility molecule class II
    antigen presentation after macroautophagy
        monitoring with green fluorescent protein–LC3 and antigen–LC3
            electroporation, 410–411
            recombinant lentiviral infection of dendritic cells, 409–410
            transfection, 407–409
        overview, 404–405, 419
        substrates, 405–407
        targeted antigen colocalization with class II-containing compartments
            immunoelectron microscopy, 415–416
            immunofluorescence microscopy, 411–415
        restriction, 404
        T cell assays to monitor macroautophagy substrate presentation
            interferon-γ enzyme-linked immunosorbent assay of T cell reactivity, 417–418
            T cell coculture with macroautophagy substrate-expressing target cells, 416–417
Mammalian target of rapamycin
    activation signaling, 166–167
    activity assay
        cell culture, 168
        kinase assays *in vitro*
            immunohistochemistry of tissues, 176–177
            incubation conditions, 175–176
            mammalian target of rapamycin complex purification, 173–175
            substrate preparation, 172–173
        lysate preparation, 168–169
        substrates of complexes, 169–170
        tissue sample preparation, 169
        Western blot of substrate phosphorylation, 170–172
    autophagy role, 167–168, 177–178
    complexes, 166
MDC, see Monodansylcadaverine
Microtubule-associated protein 1 light chain 3
    autophagosome interactions, 2–3, 57–58
    autophagy marker, 2, 133
    correlative light and electron microscopy of green fluorescent protein–microtubule-associated protein 1 light chain 3
        cell culture, 264–265
        electron microscopy of autophagic vacuoles, 268–271
        interpretation, 271–272
        light microscopy, 266–267
    cytosolic fraction ratio as quantitative index of macroautophagy
        calculation of ratio, 207
        cytosolic LC3-II characterization, 205, 207
        macroautophagic flux evaluation as quantitative index, 208–211
        overview, 200
        subcellular fractionation, 203–204
        Western blot, 204–205
    detachment-induced autophagy studies, see Detachment-induced autophagy
    flow cytometry autophagy assays of green fluorescent protein-microtubule-associated protein 1 light chain 3 fusion protein
        advantages, 139–140
        amino acid deprivation analysis, 134–137
        decline quantification, 137
        fluorescence-activated cell sorting of response to specific treatments, 138–139
    green fluorescent protein and red fluorescent protein tandemly tagged microtubule-associated protein 1 light chain 3 assay, 7–8
    major histocompatibility molecule class II antigen presentation after macroautophagy monitoring with green fluorescent protein–LC3 and antigen–LC3
        electroporation, 410–411
        recombinant lentiviral infection of dendritic cells, 409–410
        transfection, 407–409
    puncta formation assays
        green fluorescent protein fusion protein, 6–7
        immunofluorescence microscopy, 4–5
        qualitative analysis in cells, 133–134
    Salmonella Typhimurium autophagy analysis by immunofluorescence microscopy, 335–337

Subject Index 505

sequestosome 1 interactions, 182
Syrian hamster Harderian gland autophagy
   analysis
   electron microscopy, 464–465
   immunohistochemistry, 462–463
   Western blot, 471
   transgenic mouse expressing green fluorescent
      protein fusion protein
      cryosectioning, 19
      fluorescence microscopy of puncta, 19–20
      genetic features, 15–16
      genotyping, 17–18
      mouse maintenance, 16–17
      precautions, 20–21
      tissue fixation, 18–19
   Western blot analysis of autophagy, 3, 9–11
Mitophagy
   betaine:homocysteine methyl transferase
      linker-specific cleavage site assay, 110,
      112–114
   knockout mouse studies
      Nix, 239–241
      Ulk1, 239–240
   overview, 228–229
   reticulocyte model
      Atg gene expression analysis, 236
      depolarization monitoring with
         tetramethylrhodamine methylester,
         234
      electron microscopy, 234–236
      flow cytometry, 230–232
      immunofluorescence microscopy, 233–234
      reticulocyte production and maturation,
         229–230
      Western blot analysis, 236–239
Monodansylcadaverine
   autophagy induction assay
      autophagic vacuole labeling in cultured
         cells, 86–90
      fluorometry, 90–92
      overview, 86
      structure, 86
mTOR, *see* Mammalian target of rapamycin
Mycobacterium tuberculosis autophagy
   analysis
      autophagy induction, 350
      inhibitors
         preparation, 348–349
         treatment, 350
      lysosomal protease delivery monitoring to
         autophagosome
         confocal microscopy, 358
         infection and autophagy induction,
            357–358
         materials, 356–357
      macrophage
         culture, 348–349
         infection, 349–350

   materials, 348–349
   mycobacteria
      preparation, 348
      recovery and data analysis, 350–352
   phagosome acidification assay
      Alexa 488 labeling of mycobacteria, 355
      confocal microscopy of LysoTracker
         Red, 356
      infection and autophagy induction, 356
      macrophage preparation, 354–355
      materials, 354–355
   RNA interference studies
      materials, 351
      transfection of small interfering RNA,
         351, 353
      Western blot, 353–354
   epidemiology of disease, 346
   macrophage infection, 346
   overview, 346–347

N

Nanoparticle-mediated autophagy
   autophagic window, 279–282
   classification of cells, 280–281
   nanoparticle-based technology for autophagy
      detection
      bioenergetics, 288
      flow cytometry, 290
      fluorescence microscopy, 290
      gene and drug delivery, 289–290
      metabolism studies, 287–288
      products and prospects, 291–292
      protein microarrays, 286
      single molecule tracking, 286–287
      single organelle targeting, 286
      in situ hybridization, 288–289
      Western blot, 285–286
   nanoparticle characteristics, 278–279
   quantum dots
      guidelines for autophagy monitoring,
         284–285
      labeling quality assay, 283–284
      size and autophagy induction, 279
      transfection, 282–283
Nix, knockout mouse studies of mitophagy,
   239–241
NLRs, *see* Nucleotide oligomerization
   domainlike receptors
Nucleotide oligomerization domainlike receptors,
   pathogen autophagy regulation, 384–385

O

Organellophagy, *see* Mitophagy; Pexophagy;
   Reticulophagy
Oxidative stress, *see* Reactive oxygen species;
   Syrian hamster Harderian gland

P

p62, see Sequestosome 1
Pathogen autophagic sequestration, see Listeria
  autophagy; Mycobacterium tuberculosis
  autophagy; Salmonella Typhimurium
  autophagy; Shigella autophagy; Streptococcus
  autophagy
Peroxisomal thiolase, Western blot analysis of
  autophagy, 218–220
Pexophagy
  mammalian assays
    electron microscopy, 220–223
    immunofluorescence microscopy, 220
    induction of peroxisome degradation,
      217–218
    model systems
      induction and accumulation of
        peroxisomes, 216–217
      mouse treatments, 217
    Western blot of marker proteins, 218–220
  overview, 216
Photoactivatable green fluorescent protein, see
  Green fluorescent protein
Protein microarray, nanoparticle-based
  technology for autophagy detection, 286
PT, see Peroxisomal thiolase
Puncta formation assays, see Microtubule-
  associated protein 1 light chain 3; WIPI-1

Q

Quantum dot, see Nanoparticle-mediated
  autophagy

R

Raffinose, autophagic sequestration assays, 72
Reactive oxygen species
  autophagy induction, 120
  fluorescent detection
    antioxidant inhibition of autophagy
      N-acetyl-L-cysteine, 126
      antioxidant types, 127
      catalase, 128
    2',7'-dichlorofluorescein, 121–125
    dihydroethidium, 121–122
    quantitative and comparative
      determination, 123–126
    single-cell microscopy, 121–123
Reticulocyte mitochondria clearance, see
  Mitophagy
Reticulophagy, betaine:homocysteine methyl
  transferase linker-specific cleavage site assay,
  110–112
RNA interference, Mycobacterium tuberculosis
  autophagy studies
  materials, 351
  transfection of small interfering RNA, 351, 353
Western blot, 353–354
ROS, see Reactive oxygen species

S

Salmonella Typhimurium autophagy
  advantages as model system, 333
  analysis
    bacteria growth and infection, 334–335
    controls
      autophagy inhibitors, 338–339
      chloramphenicol inhibition of protein
        synthesis, 337
      SPI-1 type III secretion system-deficient
        bacteria, 337–338
    immunofluorescence microscopy of
      microtubule-associated protein 1 light
      chain 3, 335–337
  Atg16L1 role, 333–334
  canonical autophagy comparison, 332–333
  intracellular bacterial populations, 329–330
  kinetics, 332
  markers, 330–331
  pathology, 326–327
  Salmonella-containing vacuole, 326–328
  type III secretion systems, 328–329
Semiconductor nanocrystals, see Nanoparticle-
  mediated autophagy
Sequestosome 1
  autophagy and degradation, 182
  degradation assays
    antibodies, 183–185
    immunoelectron microscopy, 191–192
    immunofluorescence microscopy, 190–191
    live cell imaging
      ectopic overexpression of p62, 192
      pH-sensitive mCherry–green fluorescent
        protein double tag imaging,
        193–194
    prospects, 194–195
    pulse–chase measurement of half-life,
      187–190
    Western blot, 185–187
  pathology, 182
  protein–protein interactions, 182
Sequestration assays, see Autophagic sequestration
  assays
Shigella autophagy
  autophagosome
    immunofluorescent microscopy of green
      fluorescent protein fusion proteins,
      374–375
    transmission electron microscopy, 378
  bacterial multiplication assay in cells, 374
  evasion mechanisms, 372
  infection
    baby hamster kidney cells, 373
    Madin-Darby canine kidney cells, 372–373

mouse embryonic fibroblasts, 372–373
plaque formation assay, 373
overview, 364
pathology, 370–371
In Situ hybridization, nanoparticle-based technology for autophagy detection, 288–289
SQSTM1, *see* Sequestosome 1
Starvation-induced oxidative stress, *see* Reactive oxygen species
Streptococcus autophagy
  bacterial intracellular invasion and degradation rate analysis
    degradation assay, 369–370
    replication assay, 369–370
    viability assay, 369
  infection
    bacterial strains, 366
    incubation conditions, 366–367
    mammalian cell lines, 366
  overview, 365
  pathology of group A *Streptococcus*, 365
  subcellular localization
    green fluorescent protein fusion proteins, 367
    immunofluorescence microscopy, 368
    morphometric analysis, 368–369
    transfection, 367
Sucrose, autophagic sequestration assays, 72–73
Syrian hamster Harderian gland
  anatomy, 458
  autophagy assays
    Beclin1 protein complex immunoprecipitation, 471–472
    cathepsin analysis of cell viability
      cathepsin B assay, 466
      cathepsin C assay, 466–467
      cathepsin D processing analysis, 467–469
    chaperone-mediated autophagy detection, 472–473
    cytoskeletal protein expression analysis, 471
    dissection for optical microscopy, 460–462
    electron microscopy
      microtubule-associated protein 1 light chain 3 immunocytochemistry, 464–465
      tissue dissection, fixation, and sectioning, 463
    lysosome-associated membrane protein–2A expression analysis, 469
    microtubule-associated protein 1 light chain 3
      immunohistochemistry, 462–463
      Western blot, 471
    morphological characteristics, 459–460
    tissue homogenization for biochemical studies, 465
  oxidative stress, 459, 474

prospects for study, 473–474
sex differences, 458

T

T cell, *see* , *see* Major histocompatibility molecule class II
Tetramethylrhodamine methylester, mitochondrial depolarization monitoring in mitophagy, 234
Three-dimensional epithelial culture, *see* Detachment-induced autophagy
TMRM, *see* Tetramethylrhodamine methylester
Transgenic mouse
  lysosome-associated membrane protein–2A, 301
  green fluorescent protein-microtubule-associated protein 1 light chain 3 expression for autophagy analysis (reverse order of GFP and MAP1LC3)
    cryosectioning, 19
    fluorescence microscopy of puncta, 19–20
    genetic features, 15–16
    genotyping, 17–18
    mouse maintenance, 16–17
    precautions, 20–21
    tissue fixation, 18–19

U

Ulk1, knockout mouse studies of mitophagy, 239–240

W

Western blot
  betaine:homocysteine methyl transferase fragment generation analysis in autophagy, 78–79, 101–102
  lipidated microtubule-associated protein 1 light chain 3 studies
    detachment-induced autophagy, 435–436
    macrophage autophagy, 399–400
  lysosome-associated membrane protein–2A, 311–312
  mammalian target of rapamycin substrate phosphorylation, 170–172
  microtubule-associated protein 1 light chain 3 analysis of autophagy, 3, 9–11, 204–205
  mitophagy analysis in reticulocytes, 236–239
  Mycobacterium tuberculosis autophagy proteins, 353–354
  nanoparticle-based technology for autophagy detection, 285–286
  pexophagy marker proteins, 218–220
  sequestosome 1 degradation, 185–187
  Syrian hamster Harderian gland autophagy analysis
    Beclin1 protein complexes, 472
    cathepsin D processing, 467, 469

Western blot (cont.)
　　cytoskeletal proteins, 470
　　microtubule-associated protein 1 light chain 3, 471
WIPI-1
　　correlative light and electron microscopy of green fluorescent protein–microtubule-associated protein 1 light chain 3
　　　　cell culture, 264–265
　　　　electron microscopy of autophagic vacuoles, 268–271
　　　　interpretation, 271–272
　　　　light microscopy, 266–267
　　human WIPI proteins and yeast orthologs, 248–249
　　puncta formation assays
　　　　confocal microscopy
　　　　　　autophagy inducers and inhibitors, 252–253
　　　　　　cell culture, 251–252
　　　　　　imaging, 254–255
　　　　　　quantitative analysis, 255
　　　　green fluorescent protein fusion protein assay
　　　　　　live cell imaging, 257–258
　　　　　　quantitative analysis, 256
　　　　　　stable transfection, 256
　　　　　　transient transfection, 255–256
　　　　immunostaining of endogenous protein, 254
　　　　phospholipid–protein overlay assay, 258
　　　　principles, 250–251

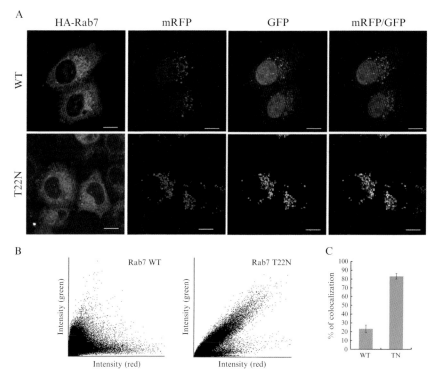

Shunsuke Kimura et al., Figure 1.2 The tfLC3 assay. (A) HeLa cells were co-transfected with plasmids expressing mRFP-GFP-LC3 and either wild-type or the T22N mutant of HA-Rab7. Twenty-four h after transfection, the cells were subjected to starvation for 2 h, fixed, and subjected to immunocytochemistry using anti-HA antibody. Antimouse antibody conjugated with Alexa405 was used as the secondary antibody. Bar indicates 10 μm. (B) Each correlation plot is derived from a field of view shown in panel A. (C) Colocalization efficiency of mRFP with GFP signals of tfLC3 puncta was measured using ImageJ software, and shown as the percentage of the total number of mRFP puncta. The value indicates average and standard deviation from at least 5 images.

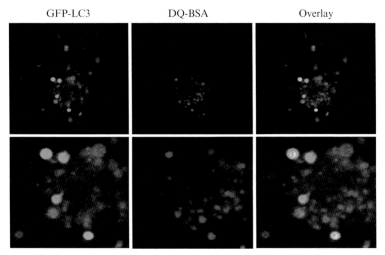

Cristina Lourdes Vázquez and María Isabel Colombo, Figure 6.4 Autolysosomes labeled with DQ-BSA. Stably transfected K562 cells overexpressing pEGFP-LC3 were incubated with DQ-BSA (10 μg/ml in RPMI + 10% fetal bovine serum) for 12 h at 37 °C to label the lysosomal compartment. Cells were then washed twice with PBS and incubated for 2 h at 37 °C in starvation media. Cells were mounted on coverslips and immediately analyzed by confocal microscopy. *Lower panels*: higher magnification of the upper panels (images from Fader and Colombo).

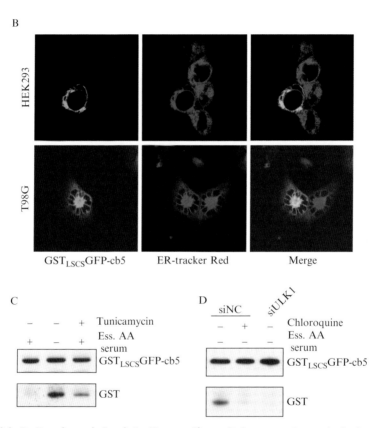

Patrick B. Dennis and Carol A. Mercer, Figure 7.6 Measuring reticulophagy by $GST_{LSCS}GFP$-cb5 cleavage. (A) Schematic showing the $GST_{LSCS}GFP$ reporter with the carboxy-terminal cb5 sequence and LSCS (*hatched region*). The primary structure of the cb5 sequence is shown. (B) HEK293 and T98G cells were transfected with the $GST_{LSCS}GFP$-cb5 construct (*green*) and cultured in full medium. ER-Tracker Red in $HBSS^{Ca2+Mg2+}$ was added to the transfected cells, to a final concentration of 0.5 μM, followed by incubation for 25 min at 37 °C before the live cells were analyzed by confocal microscopy. (C) The $GST_{LSCS}GFP$-cb5 construct was transfected into HEK293 cells and cultured for 13 h in either full medium (+) or starvation medium (−) in the presence of leupeptin (11 μM). One set of cells was cultured in full medium with 5 $\mu g/ml$ tunicamycin for the 13-h treatment time. Full-length $GST_{LSCS}GFP$-cb5 and released GST were enriched on glutathione agarose and expression was analyzed by Western blotting, using the B-14 antibody to detect GST. (D) $GST_{LSCS}GFP$-cb5 was cotransfected with nonsilencing siRNA (siNC) or siRNA against ULK1 and treated as in C, except the starvation time was reduced to 6 h. One set of cells was treated with 100 μM chloroquine during the 6-h starvation period as indicated. Levels of full-length $GST_{LSCS}GFP$-cb5 and released GST were analyzed as in C.

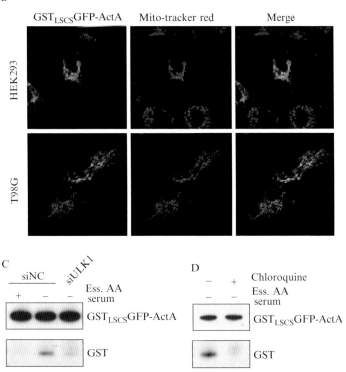

Patrick B. Dennis and Carol A. Mercer, Figure 7.7 Measuring mitophagy by GST$_{LSCS}$GFP-ActA cleavage. (A) Schematic showing the GST$_{LSCS}$GFP reporter with the carboxy-terminal ActA sequence and LSCS (*hatched region*). The primary structure of the ActA sequence is shown. (B) HEK293 and T98G cells were transfected with the GST$_{LSCS}$GFP-ActA construct (*green*) and cultured in full medium. MitoTracker Red was added to the transfected cells, to a final concentration of 50 nM, followed by incubation for 15–30 min at 37 °C before the live cells were analyzed by confocal microscopy. (C) The GST$_{LSCS}$GFP-ActA construct was cotransfected with nonsilencing siRNA (siNC) or siRNA against ULK1 in HEK293 cells that were cultured for 13 h in either full medium (+) or starvation medium (−) in the presence of leupeptin (11 μM). Light membrane fractions were prepared from the treated cells and extracted. Full-length GST$_{LSCS}$GFP-ActA and released GST were analyzed as in Fig. 7.6. (D) HEK293 cells were transfected with GST$_{LSCS}$GFP-cb5 and cultured for 13 h in starvation media. One set of cells was treated with 100 μM chloroquine during the starvation period as indicated.

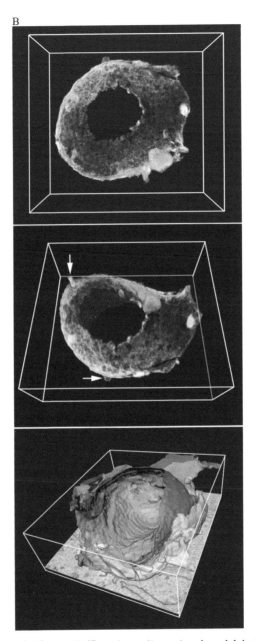

Päivi Ylä-Anttila *et al.*, Figure 10.6B Three-dimensional models built from the image stack after tracing the phagophore and ER membranes by hand. The upper panel shows the inner surface of the phagophore membrane. The bottom of the phagophore cup is missing, as it was not present in the semithick sections used to collect the tilt series. The middle panel shows the outside of the phagophore membrane, which is now tilted 150 degrees from the upper panel. Note the two extensions bulging from the phagophore (arrows). The lower panel shows the phagophore membrane in purple, the ER cistern outside the phagophore in yellow, and the ER cistern inside the phagophore in red. The orientation of the organelle is such that the reader is looking inside the phagophore from the open end.

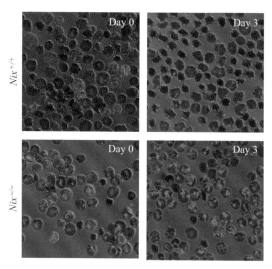

Ji Zhang et al., Figure 15.2 Immunofluorescence microscopy of MTR-stained red blood cells in culture. Images show overlay of the red channel (561-nm laser) and transmitted light in the DIC mode. $Nix^{+/+}$ reticulocytes clear their mitochondria in culture over 3 days (upper panels). In contrast, $Nix^{-/-}$ reticulocytes are unable to clear their mitochondria over the same period of time *in vitro* (Schweers et al., 2007).

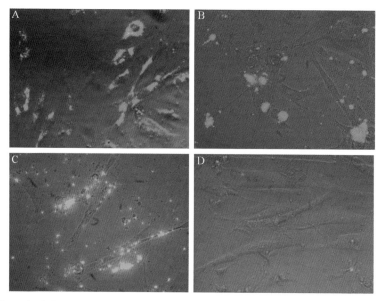

Oleksandr Seleverstov et al., Figure 18.2 Fluorescence signal and morphology of QD-labeled human mesenchymal stem cells. A and C, 24 h after labeling, B and D, 72 h after labeing. 2.5 nM labeling concentration of each (525 [green emitting, small-sized, autophagy activating NP] and 605 [red emitting, large-sized, cytoplasmic flux reference]) QD. Both green and red particles has a diffuse cellular distribution at the beginning, which rapidly changes to a granular one. Green fluorescence disappears from morphologically normal human mesenchymal stem cells at 72 h completely; dead cells may retain well-detectable green fluorescence (D). The red fluorescence is visible at 72 h (B) and up to 52 days (Seleverstov et al., 2006). The AW is calculated as: 52 days (1248 h) – 3 days (72 h) = 49 days (1176 h). Magnification 400x for all.

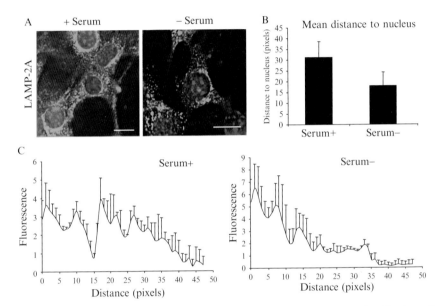

Susmita Kaushik and Ana Maria Cuervo, Figure 19.2 Intracellular redistribution of CMA-active lysosomes. (A) Indirect immunofluorescence for LAMP-2A in cultured mouse fibroblasts maintained in the presence/absence of serum. Bar: 10 μm. (B) Mean distance of the fluorescent puncta (lysosomes) to the nucleus. (C) Graph representing the intracellular distribution of fluorescent puncta with respect to the nucleus in the two indicated conditions. Values are mean + standard error of 4 different cells in each condition.

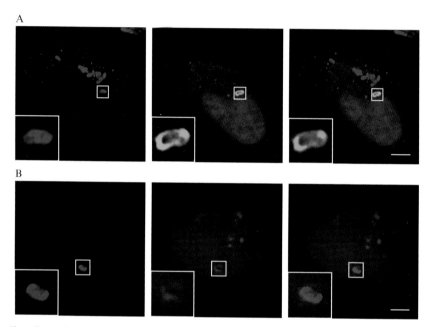

Cheryl L. Birmingham and John H. Brumell, Figure 20.2 Immunofluorescence of S. Typhimurium targeted by autophagy. (A) HeLa cells were transiently transfected with GFP-LC3 (green) and infected with S. Typhimurium. At 1 h p. i., cells were fixed and stained for bacteria (blue). Shown is a confocal Z-slice of a tight-fitting LC3$^+$ autophagosome around a bacterium. Magnified images of the boxed areas are shown in the bottom left of the corresponding panels. Size bars = 5 μm. (B) HeLa cells were transiently transfected with RFP-LC3 (red), infected, fixed and stained as in A. Shown is an epifluorescence image of a cup-shaped LC3$^+$ structure around one pole of a bacterium (image taken by S. Shahnazari, Hospital for Sick Children, Toronto). This structure presumably represents a forming autophagosome elongating to form a complete structure around the targeted bacterium.

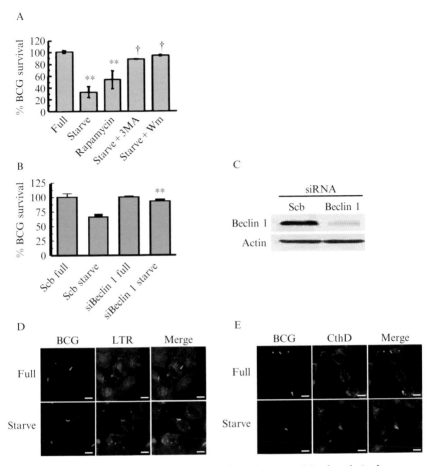

Marisa Ponpuak et al., Figure 21.1 Autophagy increases *M. tuberculosis* phagosome maturation and eliminates mycobacteria in macrophages. (A) Autophagy reduces mycobacterial survival. RAW264.7 macrophages were infected with BCG for 1 h and incubated with or without starvation medium in the presence or absence of rapamycin, 3-methyladenine (3MA), or wortmannin (Wm) for 2 h. Cells were lysed to determine mycobacterial viability. (B) and (C) Beclin 1 is important for autophagic killing of mycobacteria. RAW264.7 cells were transfected with siRNAs against Beclin 1 or scramble control. Protein knockdown was allowed to proceed for 48 h. Mycobacterial viability was determined as in A. Immunoblot analysis was performed to validate Beclin 1 knockdown level using Actin as a loading control. (D) and (E) Autophagic induction enhances acidification and acquisition of a lysosomal protease by the mycobacterial phagosome. RAW264.7 cells were transfected with siRNAs against proteins of interest. Cells were infected with BCG and stained with LysoTracker Red or antibodies against cathepsin D. Quantitative analysis of percent colocalization was performed. Confocal images of cells transfected with siRNAs scramble control subjected to starvation treatment are shown as examples. Data, means ± SEM from three different experiments, $\star\star p \leq 0.01$, $^{\dagger}p \geq 0.05$. Fig. 21.1A is modified from Gutierrez *et al.* (2004). Figs. 21.1B and 21.1C are modified from Delgado *et al.* (2008).

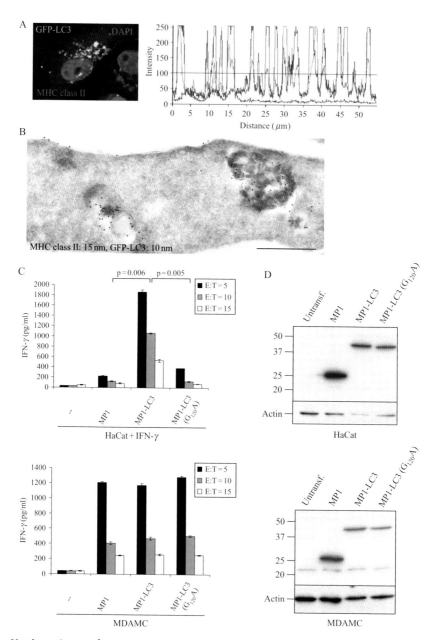

Monique Gannagé and Christian Münz, Figure 24.1 LC3 fusion proteins localize to MHC class II–containing compartments and are more efficiently presented on MHC class II molecules to CD4$^+$ T cells. (A) Example of a colocalization analysis of GFP-LC3 (green) and MHC class II molecules (red) in MDAMC breast carcinoma cells. Analysis with the LSM510 profile tool reveals double-positive vesicles along the path outlined in

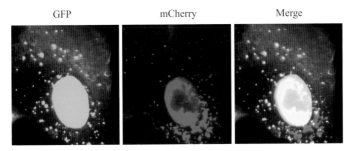

Jayanta Debnath, Figure 25.1 Stable expression of tandem mCherry-GFP-LC3 in mammary epithelial cells. MCF10A mammary cells were infected with a retrovirus encoding mCherry-GFP-LC3 to generate stable pools. Upon 2 h of starvation with Hanks's saline buffer, both double mCherry, GFP-positive LC3 puncta, which correspond to early autophagosomes, as well as single mCherry-positive LC3 puncta, which correspond to mature autolysosomes, are observed within cells.

red in the picture on the left. The histogram on the right shows that 12 of 18 MHC class II-positive vesicles contain significant amounts of GFP-LC3 (fluorescence intensity >100). (B) Immunoelectron microscopy, labeling MHC class II molecules with 15-nm gold particles and GFP of GFP-LC3 with 10-nm gold particles, reveals colocalization of GFP-LC3 and MHC class II molecules on intravesicular membranes of late multivesicular endosomes. In addition, some MHC class II staining is found on the cell surface. (C) The influenza matrix protein 1 (MP1)-specific $CD4^+$ T cell clone 11.46 or the MP1 specific $CD8^+$ T cell clone 9.2 (top and bottom, respectively) were stimulated at the indicated effector to target (E:T) ratios with HLA matched epithelial target cells expressing either wild-type MP1, MP1 fused to LC3 (MP1-LC3) or the fusion protein mutated at the C-terminal amino acid of LC3 (MP1-LC3 ($G_{120}A$)). T cell reactivity was measured the next day by IFN-γ secretion into the supernatant, as assessed by cytokine-specific ELISA. Error bars indicate standard deviations and p values are given for paired, one-tailed student T test statistics. (D) Expression levels of the different MP1 constructs were analyzed for the used target cells by Western blot analysis with anti-MP1 antiserum. Actin blots demonstrate the levels of protein loading for the analyzed SDS Page gels. Fig. 24.1B and parts of Fig. 24.1C have been published previously (Schmid *et al.*, 2007).

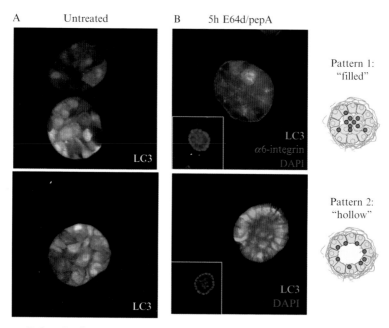

Jayanta Debnath, Figure 25.2 Analysis of LC3 within acini grown using 3D epithelial cell culture. (A) In day 8 MCF10A acini stably expressing GFP-LC3, puncta are observed throughout the structures. (B) MCF10A acini were treated with the lysosomal cathepsin inhibitors E64d and pepstatin A (pepA; 10 μg/ml each) to delineate sites of LC3 turnover in the lysosome. In these structures, two patterns emerge. First, in acini that have not cleared their lumen (top panel), punctate LC3 (green) is predominantly observed within central cells that lack contact with the underlying basement membrane ECM, which is delineated by stabilized α6 integrin receptor subunit located at sites of direct cell-matrix contact (red). Second, in acini in which the lumen has cleared (bottom panel), apically oriented clusters of GFP-LC3 are observed in the cells that contact the underlying basement membrane.